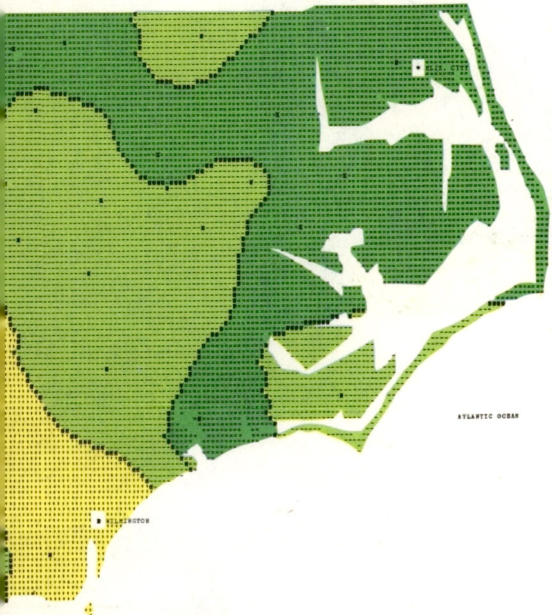

Arrival of spring 1970 across North Carolina, as indicated by the opening of the flowers on dogwood *(Cornus florida)* and redbud *(Cercis canadensis)*. The map represents the possibilities of combining computer printouts with colored overlays to enhance pattern visibility. The map was presented as output from the North Carolina Phenology Program in *Wildlife in North Carolina,* March 1971. For further information on computer mapping in phenology, see Chapter 2.2 in this volume by Caprio, Hopp, and Williams.

REDBUD

LITHO BY THE GRAPHIC PRESS, INC., RALEIGH, N. C.

Ecological Studies

Analysis and Synthesis

Edited by

J. Jacobs, München · O. L. Lange, Würzburg

J. S. Olson, Oak Ridge · W. Wieser, Innsbruck

Volume 8

Phenology and Seasonality Modeling

Edited by

Helmut Lieth

With 120 Figures

Springer-Verlag New York · Heidelberg · Berlin 1974

HELMUT LIETH, Ph.D.
Professor of Botany and Ecology
University of North Carolina at Chapel Hill
Chapel Hill, North Carolina 27514, USA

Contribution No. 85, Eastern Deciduous Forest Biome US-IBP.

Library of Congress Cataloging in Publication Data

Main entry under title:

Phenology and seasonality modeling.

(Ecological studies, v. 8)
Chiefly papers presented at a symposium held during the 25th annual AIBS meeting in Minneapolis, Aug. 1972.
1. Phenology—Congresses. I. Lieth, Helmut, ed.
II. American Institute of Biological Sciences.
III. Series.
QH544.P62 574.5′4 73-23022

Printed in the United States of America.

ISBN 0-387-06524-5 Springer-Verlag New York · Heidelberg · Berlin
ISBN 3-540-06524-5 Springer-Verlag Berlin · Heidelberg · New York
ISBN 0-412-13290-7 Chapman & Hall Limited London

Foreword

The pulse of life with the seasons is a classic theme of biology, equally capturing every man's curiosity about early and late milestones of every year's cycle and the critical physiologist's inquiry into life's subtle signals and responses. Natural historians of ancient and renaissance time as well as today have charted the commonsense facts behind inspired traditions of poetry and practical rules for growing food and fiber. This volume brings together several ways of organizing the basic principles of phenology. These find order in the otherwise overwhelming mass of detail that captures our fleeting attention, like the daily newspaper, and then tends to fade into the overstuffed archives of history.

Is this order so obvious and understandable that there is no longer any scientific challenge to "phenology" as a tradition? Or does apparent simplicity mask a complex and ultimately baffling obstacle to the understanding of seasonality in even those few indicator plants and animals we know best, not to mention the less known species or races making up the rest of each major landscape unit or ecosystem? Denying both these hasty opinions, we think that this volume well illustrates a range of questions and answers—from soundly established (but not trivial) doctrine to exciting inquiry about how ecosystems are organized.

Indeed, phenology has traditionally emphasized the "wrapping up" of seasonal history after it has occurred or just as it was occurring. In doing so, much has been learned about biological variables, which could be predicted (within limits) if key environmental factors were known, and preferably understood in terms of physical and physiological mechanisms. Such understanding and some mathematical expressions of mechanism, that are even better than mechanical, blind exercise of statistical technique now do seem to be improving our capacity to predict the future. Of course we are speaking of *conditional* probabilities: forecasting most probable outcomes and some upper and lower bounds (e.g., for time to 50% leafing of plants or migration of animals, *if* certain weather input is given (average year; early warmth; late frost). If the weather itself remains almost as unpredictable in the future as it has been in the past, this alone would limit the biological forecasts from the very best phenological models.

One reason that phenology was highlighted in 1966 as a distinct thrust in the United States National Committee for the International Biological Program (IBP) was a natural partnership with other models for biological productivity. The latter models might probe the reasons for differences in an ecosystem's intake of energy and cycling of nutrient from one place to another over the Earth or at one place, when a pioneer community is replaced by a mature one. Phenologic models are left to answer how much (and why) growth and biological activity varies around a typical year or between years with very atypical seasons.

Besides providing a crucial part of the few intensive ecosystem studies which IBP undertook, a second prospect that phenology offered was one basis for interpolating and extrapolating a few parameters on growth (or seasonal limits on growth) over wide areas in which no intensive analysis of total ecosystems could be expected. The emphasis on mapping and on ties to crops and other economic plants still offers such promise as part of "biome-wide" and regional studies.

A third reason for special, and at times almost separated emphasis on phenology (as distinct from highly instrumented local experimental study of ecological process), was the simple reality noted at the beginning of this foreword. A broad appeal is made for coordinated interest among teachers and many nonspecialists who might do extensive work well, but never have opportunity for the intensive work; also among students who might thereby become motivated toward more advanced studies that might never have come to their attention. We hope that this book will further interest just such a wide range of less specialized readers to explore the current "state of the art" of phenology, while helping specialists place their own new research in a broad perspective.

The Editors

Preface

In the first volume of the *Ecological Studies* series (D. Reichle, 1970) the editor of the present volume was charged with the task of explaining why phenological concepts were useful and necessary in ecosystems studies. At that time we stated that: (1) complete success in ecosystems modeling could never be achieved without the incorporation of phenology and seasonality considerations, and (2) a modern summary treatment of the subject matter in English did not exist and would be of top priority if the field were to receive adequate attention worldwide. This book is the result of the latter statement.

The concepts of this first monographic treatment of phenology, seasonality, and the ecosystem arose from discussions of the editor with Dr. Forest Stearns and the US/IBP Phenology Committee. It was decided that the editor and Dr. Stearns should organize a symposium to be held during the 25th Annual AIBS meeting in Minneapolis, Minnesota, in August 1972. Most of the chapters in this volume were presented as papers at the symposium.

The cooperation of the contributors has been remarkable. Some completely revised their manuscripts in light of other papers and discussions at the symposium. All were interested in complying with the proposed format for this book. Some contributions could not be discussed during the symposium but were subsequently treated in a similar way by mail. We mention here especially the contribution of Drs. Schnelle and Volkert.

Dr. Schnelle's interest has been gratifying. We intended originally to include a world list of phenologists in this book. Dr. Schnelle was very helpful in compiling such a list. However, after comparing this list with the context of the entire book it appeared that the list was close to complete only in the area of plant ecology whereas the book holds the total ecosystem in scope. After some discussion with the series editor we decided therefore to distribute this list as a biome memo report to obtain an improved list for publication at some later date. The report, entitled *International Catalogue of Agencies and Research Groups Working in Phenology* by Fritz Schnelle, Diana Watley, and Helmut Lieth *US/IBP-EDF BIOME MEMO REPORT* #73-3, is available through the Biome Office in Oak Ridge, or through the editor of this book.

This book is organized in the familiar pattern of a research paper: introduction, methods, research conducted (presented from various viewpoints like different trophic levels and different biomes), research synthesis in the form of models, and finally, applications of the research.

The different scientific areas are unequally represented for the following reason. Phenology is best developed in the autotrophic plant sciences, where intensive correlation studies on environmental parameters are available from many areas of the world and for most important conditions. Phenology is least developed for microbial decomposers. Considerable effort was made to gain participants from that field. In an attempt to open fruitful discussion among microbiologists and other specialists, the decomposer seasonality was presented in a panel discussion. This showed how much decomposer analysis is still very preoccupied with methodological problems, and how much more attention this field deserves. Most of this discussion is incorporated in Chapter 3.6.

Two other topics were presented in panel discussions, modeling of seasonality and future goals for phenological research and application. Both resulted in vigorous debate. The contributions to the modeling discussion are presented in Part 5. The various viewpoints presented in the future goals discussion are incorporated throughout this volume, but particularly in the first and last chapters. Overall consensus was evident in one respect: We all agreed that phenology is essential for both theoretical and practical purposes, and that an interdisciplinary program on the national/international scale is needed.

As is usual in an undertaking of this size, the editor must acknowledge his debt to the work of many people. Invaluable assistance was given by two of my students, Richard Reader and Barry Katz, both during the symposium and in the editing of the manuscripts. Miss Marion Sailor redrew many of the original figures. The editing of the papers was, in the beginning, jointly done with Diana Watley. The style in Part I and several other chapters in the book are her merit. We also acknowledge her assistance before and during the symposium.

During the final stage of proofreading, the editor stayed at the Nuclear Research Center (KFA) Jülich, German Federal Republic. We gratefully acknowledge the support granted by this institution and especially by the director of the Institut für Physikalische Chemie, Prof. Dr. K. Wagener. During this period, we had Ms. Cynthia H. Grossman as secretarial assistant, who together with Barry Katz helped compile the indices. The subject index was made from keywords and nouns suggested by the authors. It is far from perfect. As various international crises and strikes have delayed the book already, a further perfection of the index must be left for a future revision.

We certainly acknowledge the endorsement and financial support of this project by the Deciduous Forest Biome of the International Biological Program, Dr. Stanley Auerbach, Director, and Dr. Robert Burgess, Executive Director. We thank these gentlemen for their suggestions and administrative help.

The publisher, Dr. Konrad F. Springer, demonstrated his interest in this project by attending part of the symposium. We acknowledge with thanks his advice for the preparation of this volume. The expertise of his staff in the New York office has been essential in polishing this volume to the usual quality of Springer-Verlag.

The cooperation of Dr. Forest Stearns deserves special recognition. He was always at hand to smooth the waves created by editorial actions. Thanks go also to the series editor and my personal friend, Dr. Jerry Olson. His editorial comments were extremely helpful.

Jülich, April 1974 Helmut Lieth

List of Contributors

ACKERMAN, THOMAS L., Laboratory of Nuclear Medicine and Radiation Biology, University of California Los Angeles, Mercury, Nevada, USA

ANDERSON, ROGER C., Arboretum and Wildlife Refuge, 1207 Seminole Highway, Madison, Wisconsin, USA

BAKER, HERBERT G., Department of Botany, University of California, Berkeley, California, USA

BAMBERG, SAM A., Laboratory of Nuclear Medicine and Radiation Biology, University of California Los Angeles, Mercury, Nevada, USA

BLAIR, BYRON O., Department of Agronomy, Purdue University, Lafayette, Indiana, USA

BRADSHAW, WILLIAM E., Department of Biology, University of Oregon, Eugene, Oregon, USA

CAPRIO, JOSEPH M., Montana Agricultural Experiment Station, Plant and Soil Science Department, Montana State University, Bozeman, Montana, USA

CLAUSZ, JOHN C., Department of Biology, St. Andrews Presbyterian College, Laurinburg, North Carolina, USA

EDMISTEN, JOE A., Department of Biology, University of West Florida, Pensacola, Florida, USA

FENWICK, JACK R., Department of Agronomy, Purdue University, Lafayette, Indiana, USA

FLINT, HARRISON L., Department of Horticulture, Purdue University, Lafayette, Indiana, USA

FRANKIE, GORDON W., Department of Entomology, Texas A&M University, College Station, Texas, USA

FRENCH, NORMAN, Natural Resource Ecology Laboratory, Colorado State University, Ft. Collins, Colorado, USA

GARNER, J. H. B., Office of Criteria Development, Environmental Protection Agency, Research Triangle Park, North Carolina, USA

GREEN, FRANCES, Department of Biology, University of West Florida, Pensacola, Florida, USA

HOPP, RICHARD J., Department of Plant and Soil Science, University of Vermont, Burlington, Vermont, USA

KATZ, BARRY A., Department of Botany, University of North Carolina, Chapel Hill, North Carolina, USA

KRAMER, CHARLES L., Division of Biology, Kansas State University, Manhattan, Kansas, USA

KUMMEROW, JOCHEN, Laboratório de Botánica, Universidad Católica de Chile, Santiago, Chile

LETTAU, HEINZ, Department of Meteorology, University of Wisconsin, Madison, Wisconsin, USA

LETTAU, KATHARINA, Center for Climatic Research, University of Wisconsin, Madison, Wisconsin, USA

LIETH, HELMUT, Department of Botany, University of North Carolina, Chapel Hill, North Carolina, USA

MABEE, HAROLD F., Department of Biology, East Tennessee State University, Johnson City, Tennessee, USA

MALAISSE, FRANÇOIS P., Laboratoire de Botanique et de Climatologie, Université Nationale du Zaire, Lubumbashi, Zaire

MOONEY, HAROLD A., Department of Biological Sciences, Stanford University, Palo Alto, California, USA

MORAIN, STANLEY A., University of Kansas Center for Research, Inc., Remote Sensing Laboratory and Department of Geography, Lawrence, Kansas, USA

NEWMAN, JAMES E., Department of Agronomy, Purdue University, Lafayette, Indiana, USA

NIENSTAEDT, HANS, North Central Forest Experiment Station, Rhinelander, Wisconsin, USA

OPLER, PAUL A., Organization for Tropical Studies, Inc., P.O. Box 1499, Miami, Florida, USA

PARLANGE, JEAN-YVES, Connecticut Agricultural Experiment Station, New Haven, Connecticut, USA

PARSONS, DAVID J., Department of Biological Sciences, Stanford University, Palo Alto, California, USA

RADFORD, JOHN S., IBP Desert Biome, Department of Botany, Utah State University, Logan, Utah, USA

READER, RICHARD, Ecology Curriculum, University of North Carolina, Chapel Hill, North Carolina, USA

SAUER, RONALD H., Natural Resource Ecology Laboratory, Colorado State University, Ft. Collins, Colorado, USA

SCHLICHTING, HAROLD E., JR., Department of Botany, North Carolina State University, Raleigh, North Carolina, USA

SCHNELLE, FRITZ E., Deutscher Wetterdienst, Zentralamt (retired), Offenbach/Main, W. Germany

SMITH, MICHAEL H., Savannah River Ecology Laboratory, Aiken, South Carolina, USA

STEARNS, FOREST W., Department of Botany, University of Wisconsin, Milwaukee, Wisconsin, USA

TAYLOR, FRED G., JR., Environmental Sciences Division, Oak Ridge National Laboratory, Oak Ridge, Tennessee, USA

VOLKERT, ERIK, Forstliche Fakultät, Universität Göttingen, Hannover Münden, W. Germany

WAGGONER, PAUL E., Connecticut Agricultural Experiment Station, New Haven, Connecticut, USA

WEISE, CHARLES M., Department of Zoology, University of Wisconsin, Milwaukee, Wisconsin, USA

WIELGOLASKI, FRANS-EMIL, Botanical Laboratory, University of Oslo, Blindern-Oslo, Norway

WILLIAMS, JACK S., Montana Agricultural Experiment Station, Plant and Soil Science Department, Montana State University, Bozeman, Montana, USA

Table of Contents

Part 4 Representative Biome Studies

Part 5 Modeling Phenology and Seasonality

Part 6 Applications of Phenology

PART 1

Introduction to Phenology and the Modeling of Seasonality

1.1 Purposes of a Phenology Book[1]

Helmut Lieth

Department of Botany
University of North Carolina
Chapel Hill, North Carolina, USA 27514

1 History of this Volume

This volume's authors make a case for phenology—not so much for phenology as it is commonly understood (a field technique of agricultural meteorology), but as an aspect of analysis and management of ecological systems or ecosystems. We have indicated its importance earlier (e.g., Lieth, 1970, 1971; Lieth and Radford, 1971) but our experience is mostly confined to plant-environment relations. Ecosystems require more comprehensive study.

To obtain necessary phenological information for modeling of ecosystems, the United States contribution to the International Biological Program (US/IBP) established an IBP Phenology Committee. Various phenological projects carried out in connection with US/IBP work have been tabulated in their report (US/IBP Phenology Committee, 1972). Clearly, that compilation covers only a small section of the phenological research going on, even in North America, and is certainly not representative of all the pertinent work done by ornithologists, entomologists, agronomists, foresters, and wildlife managers.

The US/IBP Phenology Committee noted the need for study of problems of phenology and seasonality, worldwide in scope, oriented toward ecosystem analysis, suited to stimulate future national and international cooperation, and inclusive of as many pertinent disciplines as possible. Toward these ends we planned a working symposium that was held at the 25th Annual Meeting of the American Institute of Biological Sciences (AIBS) in Minneapolis, Minnesota, August 1972. As nearly as possible we hoped to achieve the following results:

1. To summarize the current research results.
2. To bring together competent people from different disciplines to discuss and present complementary viewpoints.

[1] Contribution No. 86 of the Eastern Deciduous Forest Biome, US/IBP.

KEYWORDS: Phenology; seasonality; definitions; flowchart; worldwide seasonality patterns; phenodynamic; North Carolina; summergreen deciduous forest.

3. To compile information in important subdisciplines of various fields to guide new research in each.
4. To create new focal points for synthesis involving phenology.
5. To critically review certain scientific approaches, and define terms where needed.

I would like to start briefly on the last point, presenting updated definitions. A brief outline of the regional and worldwide significance of phenology in ecosystems studies introduces a flowchart with a discussion of the logic and significance of phenological modeling.

2 Definitions

The title of this book contains three nouns: phenology, seasonality, and modeling. These nouns are used in slightly different ways in the various contributing disciplines. Therefore we give the following definitions and elaborations.

Phenology

The classic definition of phenology (Linnaeus, 1751; Schnelle, 1955) is given by Hopp in Chapter 2.1. A similar definition is also available (Lieth, 1970): "Phenology is generally described as the art of observing life cycle phases or activities of plants and animals in their temporal occurrence throughout the year."

Also stated is the prime purpose of many old phenological studies: "[to] permit a phenological calendar to be constructed and superimposed on the astronomic or civil calendar, such that the seasons of the year are not marked by calendar dates but rather by dated groups of phenological events." This book goes far beyond that stated purpose. The US/IBP Phenology Committee (1972) note, "Phenology is the study of the timing of biological events and their relationship to seasonal climatic changes," emphasizing correlation of temporal occurrence with climatic changes, triggered extensive discussion, both within the Phenology Committee and among the symposium attendees. Several zoologists stated that some important organisms in their fields had little direct dependence on the climate but depended rather on the developmental cycle of their hosts or food source organisms. G. W. Frankie (personal communications, 1972) suggested a definition of phenology as "the study of predictably occurring biological events and their relationship to biotic and abiotic environmental changes."

The US/IBP Phenology Committee, and a subsequent mail discussion, distilled the following definition:

> Phenology is the study of the timing of recurring biological events, the causes of their timing with regard to biotic and abiotic forces, and the interrelation among phases of the same or different species.

To put this definition into the frame of space and time, it was further suggested that:

> The unit of study may vary from a single species (or variety, clone, etc.) to a complete ecosystem. The area involved may be small (for intensive studies on all phenophases of entire ecosystems) or very large (for interregional comparison of significant phenophases). The unit of time is usually the solar year with which the events to be studied are in phase. The events themselves may cover variable time spans, often much shorter than the solar year.

This definition encompasses autecology (of individuals) and synecology (of communities), but focuses on study of organismal and community reactions to seasonally varying, natural environmental conditions.

Seasonality

The Random House Dictionary of the English Language (1969) defines seasonal to mean events, objectives, state variables, etc., that are "pertaining to, dependent on, or accompanying the seasons of the year or some particular season", a special case of "periodical" events.

If we apply this definition to phenology and ecology we need to know how seasonality varies in different parts of the world and what factors may identify a given season. The general temperate zone notations of spring, summer, fall, and winter do not apply in some subtropical regions, where temperature fluctuations are interrupted instead by alternating rain and drought seasons. As a definition for the term "seasonality" we suggest the following:

> Seasonality is the occurrence of certain obvious biotic and abiotic events or groups of events within a definite limited period or periods of the astronomic (solar, calendar) year.

This definition implies that the probability of certain events occurring is distributed about some mean date; occasional occurrence of phenophases at unexpected or unseasonal yeardays is possible but has a low probability. Events for which a recurrence within the same time periods of the calendar year cannot be distinguished are called nonseasonal.

Modeling

The number of definitions for the term "modeling" has gotten out of hand. With the increasing emphasis on model building in ecology, every activity may be pressed into this term, ranging from the "word model", which in most cases is a descriptive approach for a complicated situation, to the "computer model", which may be, mathematically, either an enormous amount of accounting or a sophisticated system of interlinked equations. A definition for our purposes may be stated as follows:

> Modeling is the attempt to quantify relations between biological processes (including phenological events) and environmental parameters or between different biological processes in order to more accurately describe, understand,

predict, simulate the correlations between, or the causes of, the biological processes and their driving or controlling forces.

Modeling here includes graphical, numerical or codified programmed analyses, but not verbal, descriptive, or pictorial presentation.

3 The Significance of Phenological and Seasonal Concepts in Ecosystems Studies

The observable interactions among the functional groups of organisms in the ecosystem, e.g., primary producers, herbivores, carnivores, decomposers, indicate that the species composition in each group and their structural and functional attributes must be analyzed over time as well as in space. In an environment of definite but gradual seasonal oscillation, the organism may have either a life cycle short enough to be completed during a certain combination of environmental conditions, or it may have a metabolic system that adapts to gross environmental changes; i.e., its developmental sequence may fit into seasonal environmental oscillation. To interpret the temporal organization the ecological phenologist must understand both the timing mechanisms of individual organisms and the role of organism interactions in the rate and pattern of development in the wider system. He is primarily concerned with mechanisms regulating the recurrent seasonal events rather than with those regulating long-term development such as succession.

In this volume we address two questions: (1) how interdependent, or at least synchronous, are developments of different trophic levels at a given location? (2) Are seasonal patterns a worldwide phenomenon? Table 1 shows the seasonal patterns of plant flowering in various biomes correlated with the temperature fluctuation and daylength, yet modified by the rainfall pattern in the Mediterranean climate. Assuming no other competition for this resource, we would expect to find insect volume and diversity limited by the size of the resource, i.e., the number of flowering taxa times the average number of flowers per taxon (not shown in Table 1). However, if certain insects are specialized for those flowers blooming only in August and September, the level of flowering during the other months is of no importance.

Such seasonality of flowering as shown in Table 2 and, for the Mediterranean climate, in Chapter 4.5 is suitable for correlating flowering with climatic variables, but are not precise enough to explain the interaction between particular flowers and particular insects in a seasonally oscillating environment.

Phenodynamic Profile of Selected Organisms in a Particular Region

The developmental sequence of different organisms throughout any year is, to a large extent, predictable in a climate with strong seasonal changes. This has been shown for plant development in southern Germany (Pfau, 1964). Table 2 shows a selection of phenophases used for forecasting by the West German Weather Service, for which Pfau calculated the degree of correlation between the occurrence of two

Table 1. Seasonal pattern of flowering in different biomes (adapted from Lee, 1971)

Locality	Biome	No. of Taxa	Percentage of Taxa in Bloom during Jan.	Feb.	Mar.	Apr.	May	June	July	Aug.	Sept.	Oct.	Nov.	Dec.
Northeast Greenland	Tundra	153	0	0	0	0	6.5	69.9	89.5	33.9	1.3	0	0	0
Iceland	Tundra	363	0	0	0	0.2	9.1	55.0	94.4	45.7	2.2	0	0	0
Denmark	Summergreen deciduous forest	1341	0	0.2	1.6	8.8	28.6	53.6	70.0	49.6	17.2	2.1	0.2	0
Korea	Warm temperate mixed forest	2856	.03	0.9	3.8	13.7	33.0	37.0	43.6	41.0	14.1	3.0	0.7	.03
Florida		1027	17.5	19.1	29.6	46.2	54.8	67.2	73.7	70.9	60.2	40.9	26.1	19.3
Portugal	Predominantly Mediterranean Sclerophyllous forest	2275	3.1	9.3	29.1	57.6	76.8	84.2	69.4	46.0	29.0	11.0	3.4	2.0
Cape Peninsula, South Africa		1541	67.1	50.8	36.3	32.5	31.6	30.5	47.3	66.9	77.2	84.8	81.9	71.9
Java	Predominantly Equatorial rain forest	4008	68.4	70.8	74.4	77.2	77.5	79.7	78.7	78.4	77.3	76.5	75.3	73.0

Table 2. Correlation coefficients between 21 consecutive phenophase dates reported from lower Bavaria, averaged for 1936–1959 (adapted from Pfau, 1964)

Species Phenophases	Average yearday	*Galanthus nivalis* (snowdrop) first flower	*Tussilago farfara* (colt's foot) first flower	*Avena sativa* (oats) start of sowing	*Solanum tuberosum* (potatoes, early var.) start of planting	*Avena sativa* (oats, early var.) tilling	*Taraxacum officinale* (dandelion) first flower	*Prunus cerasus* (sweet cherry) first flower	*Prunus cerasus* (sweet cherry) full bloom	*Pyrus malus* (apple) first flower	*Prunus cerasus* (sweet cherry) last flower	*Syringa vulgaris* (lilac) first flower	*Pyrus malus* (apple) full bloom	*Chrysanthemum leucanthemum* (daisy) first flower	*Pyrus malus* (apple) last flower	*Secale cereale* (winter rye) first flower	*Prunus cerasus* (sweet cherry) first mature fruit	*Secale cereale* (winter rye) start of harvest	*Avena cereale* (oats, early var.) start of harvest	*Tilia cordata* (winter linden) leaf coloration	*Tilia platyphyllos* (summer linden) leaf dropping	*Tilia cordata* (winter linden) leaf dropping
Average yearday	—	70	81	89	101	103	111	113	119	125	127	129	132	140	140	153	180	203	215	279	290	294
Snowdrop—first flower	70		0.88	0.74	0.83	0.75	0.85	—	—	—	—	—	—	—	—	—	—	—	—	—	—	—
Colt's foot—first flower	81	—		0.80	0.81	0.89	0.93	—	—	—	—	—	—	—	—	—	—	—	—	—	—	—
Oats—start of sowing	89	—	—		0.81	0.97	0.78	—	—	—	—	—	—	—	—	—	—	—	—	—	—	—
Potatoes (early var.)—start of planting	101	—	—	—		0.83	0.86	—	—	—	—	—	—	—	—	—	—	—	—	—	—	—
Oats (early var.)—tilling	103	—	—	—	—		0.87	—	—	—	—	—	—	—	—	—	—	—	—	—	—	—
Dandelion—first flower	111	—	—	—	—	—		0.91	0.88	0.75	0.83	0.70	—	—	—	—	—	—	—	0.82	0.88	—
Sweet cherry—first flower	113	—	—	—	—	—	—		0.99	0.86	0.96	0.70	0.83	0.71	0.76	—	—	—	—	0.84	0.85	—
Sweet cherry—full bloom	119	—	—	—	—	—	—	—		0.92	0.99	0.79	0.89	0.78	0.84	—	—	—	—	0.86	0.84	—
Apple—first flower	125	—	—	—	—	—	—	—	—		0.95	0.93	0.99	0.89	0.95	0.78	0.79	—	—	0.89	0.82	—
Sweet cherry—last flower	127	—	—	—	—	—	—	—	—	—		0.84	0.93	0.82	0.88	0.70	0.73	—	—	0.87	0.83	—
Lilac—first flower	129	—	—	—	—	—	—	—	—	—	—		0.92	0.81	0.89	0.77	0.78	—	—	0.84	—	—
Apple—full bloom	132	—	—	—	—	—	—	—	—	—	—	—		0.87	0.98	0.83	0.76	—	—	0.82	—	—
Daisy—first flower	140	—	—	—	—	—	—	—	—	—	—	—	—		0.94	0.95	0.91	—	—	0.83	—	—
Apple—last flower	140	—	—	—	—	—	—	—	—	—	—	—	—	—		0.87	0.81	—	—	0.89	—	—
Winter rye—first flower	153	—	—	—	—	—	—	—	—	—	—	—	—	—	—		0.85	0.81	—	—	—	—
Sweet cherry—first mature fruit	180	—	—	—	—	—	—	—	—	—	—	—	—	—	—	—		0.89	0.76	—	—	—
Winter rye—start of harvest	203	—	—	—	—	—	—	—	—	—	—	—	—	—	—	—	—		0.77	0.88	—	—
Oats (early var.)—start of harvest	215	—	—	—	—	—	—	—	—	—	—	—	—	—	—	—	—	—		0.89	—	—
Winter linden—leaf coloration	279	—	—	—	—	—	—	—	—	—	—	—	—	—	—	—	—	—	—		0.88	0.82
Summer linden—leaf dropping	290	—	—	—	—	—	—	—	—	—	—	—	—	—	—	—	—	—	—	—		0.92
Winter linden—leaf dropping	294	—	—	—	—	—	—	—	—	—	—	—	—	—	—	—	—	—	—	—	—	

particular phenophases. The timing of developmental stages in apparently unrelated elements of a given regional flora may be predicted. This predictability also exists across trophic levels, such that certain plant phenophases may serve as monitors of insect or fungus developmental rates (Chapters 2.1 and 6.4).

Nowhere is there a coverage of the phenology in all compartments of one ecosystem. Malaisse in Chapter 4.6 describes the phenology of the Miombo (dry woodland, raingreen forest) in Africa. Monographic treatments, once available, will no doubt fill books this size. To demonstrate the value of comparative analysis, we have compiled examples for the major ecosystem compartments from North Carolina (summergreen deciduous biome) (Fig. 1) and plotted their development and function as time-dependent occurrences, in order to demonstrate synchronism and sequential patterns. Further details are given in the legend to Fig. 1.

Most of the examples in Fig. 1 are from the North Carolina Piedmont, a good example of summergreen (or cold-deciduous) forest. Therefore, we show in Fig. 2 the topography of our main area, and a vegetation period there and in adjacent areas. The base map in Fig. 2 was created with the computer mapping routine SYMVU (1972) using our usual SYMAP output for the State of North Carolina (Reader, *et al.*, 1972). The other entries were made by hand using the pertinent SYMAP outputs as guidance.

Worldwide Seasonal Patterns

The importance of seasonality for interdependent ecosystem compartments in the temperate zones is well documented (see summaries in part of this volume). To what extent is seasonality evident in extra-tropical regions? What portion of the world's land areas are affected by seasonal climates? Wyatt and Sharp (unpublished class project) used the UNC Biosphere Model (Lieth, 1972) to demonstrate the length of a meteorologically defined vegetation period in number of hours (see map inside back cover) and in number of months (Fig. 3).

Since daylength is an important factor in the balance between photsynthesis and respiration, the number of daylight hours (taken from astronomic tables) were calculated for the months of the vegetation period. In this study the estimate was done by using a graph showing the latitude-dependent annual shift of daylength (Golley and Lieth, 1972). The map obtained in this way is shown in Fig. 3. For the criteria used, the lowest level and the highest level have the weakest seasonality inasmuch as they have either a negligible vegetation period or an almost continuous one. The area in between, however, covers almost 75% of the globe.

Strong seasonalities in some parts of the earth trigger seasonal bird migration, spore dissemination, and differential outputs of gaseous material—all of which display a possibility of long-range dispersal or diffusion. Virtually no place on earth may be assumed to be without a certain degree of seasonality.

4 Seasonal and Phenological Modeling

An ecosystem model incorporating phenological concepts can logically combine (1) the fluctuation of abiotic environmental parameters, (2) the demands of

SEASONALITY IN THE SUMMERGREEN FOREST OF THE SOUTHEASTERN U.S.

Environment

Solar Radiation
Wells, Whigham, & Lieth, 1972

Temperature
U.S. Dept. Commerce, 1971

Precipitation
Whigham, 1971

pH of Through Fall
Whigham, 1971

Soil Nitrogen
Whigham, 1971

Potassium
Whigham, 1971

Soil Phosphorus
Whigham, 1971

Soil Calcium
Whigham, 1971

Soil Magnesium
Whigham, 1971

Soil Iron
Whigham, 1971

Soil Copper
Whigham, 1971

Soil Manganese
Whigham, 1971

Soil Aluminum
Whigham, 1971

Soil Zinc
Whigham, 1971

Primary Producers

Liriodendron T.
Taylor, 1974

Cornus F.
Lieth, Radford, 1971

Cercis C.
Lieth, Radford, 1971

Pinus Taeda
Strain, Higginbotham, p.c.

Consumers

Birds
Duever, Fatora, 1970

Birds
Duever, Fatora, 1968

Insects
Rabb, Lawson, 1956

Mammals
Golley, 1961

Decomposers

Nematodes
Coleman, 1971

Soil Microarthropods
Celecia, Smythe, Metz, p.c.

Soil Microarthropods
Celecia, Smythe, Metz, p.c.

Myxomycetes (All) [2]
Frederick, Pendergrass, p.c.

Myxomycetes (Seasonal) [2]
Frederick, Pendergrass, p.c.

Nos. of Fungi Isolated from
Forest Litter, Katz & Lieth, 1974

Fungi on Litter
Maybee & Garner, 1974

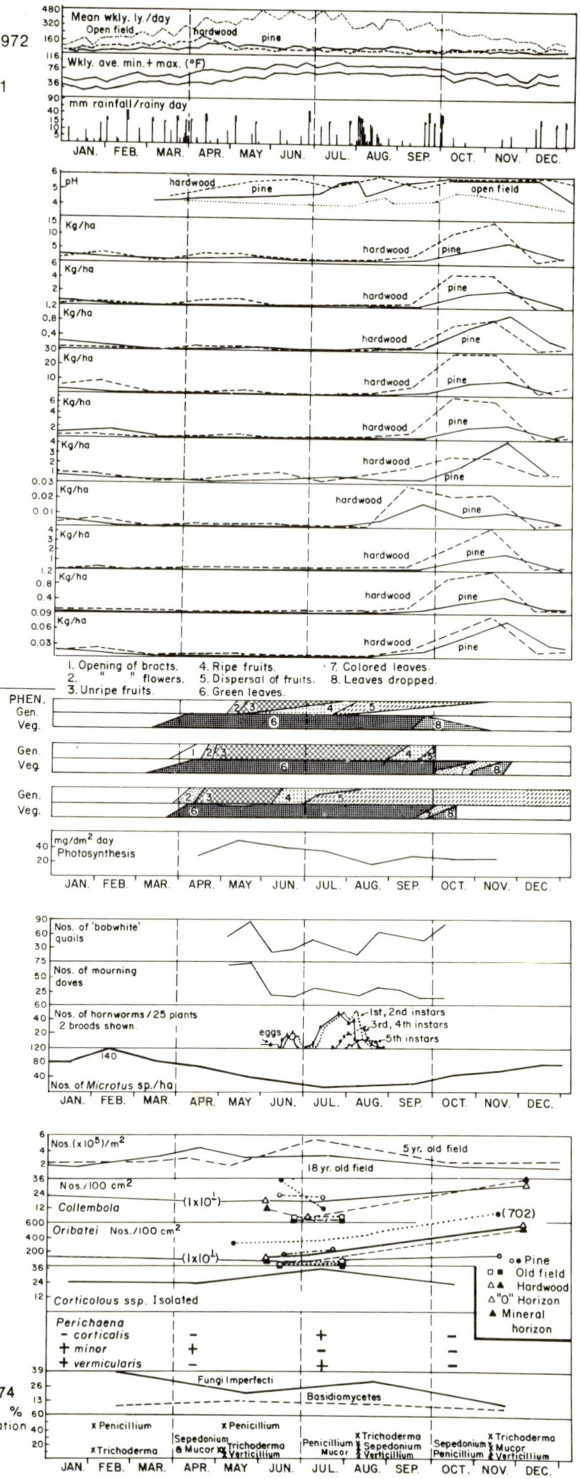

individual species for certain levels of abiotic environmental parameters, (3) the rhythmic properties of the whole environment, and (4) the developmental responses of individual species. To provide the potential for interaction among independent functions of different species, our model should logically handle growth and development of the species, alone and in connection with each other.

This book contains several partial models of phenological interactions. It therefore seems necessary to synthesize some concepts into one preliminary model. The flowchart in Fig. 4 outlines information we require to operate such a model, how it may be possible to connect the individual pieces, what may be the most practical way to handle the routines, and what output we might expect from such a model.

While this chart is basically constructed to demonstrate the problems and logic of phenological modeling, it may serve as one starting point for the development of an ecosystem model driven by environmental forces and additionally controlled by phenological requirements.

Our phenological model has three elements: a primary producer green plant population that metabolizes and develops within a given set of environmental circumstances; a consumer population that shares the same abiotic environment and is supported by the producer i.e., the producer may become checked by the growth and development of the consumer population. A predator provides an arbitrary manipulation of the consumer compartment size. The litter compartment is the recipient of all dead or decaying biomass. The levels of growth and development achieved for each compartment are written into a file called biotic environment on a daily basis, and the values in this file provide the feedback from producer loop to consumer loop and vice versa. This file also can be used to produce a graph of producer and consumer growth and development over time.

The growth and development of each of the producer-consumer partners is calculated by a series of loops which bring the individual from one phenophase to the next as defined in the usual form in the producer and consumer files, respectively. The intial phenophase loop starts with a value called "seed" or "egg", and checks whether this entity is ready to grow. If the environment supports growth, per the abiotic environment file, such growth is calculated and a biomass figure is entered in the biotic environment file. This biomass entry, together with the current abiotic environment conditions, is used to check whether a consumer (or predator) may possibly be supported according to the requirements given in the consumer (or predator) file. If no consumer can be supported at this time, a check is made for the readiness of the producer developmental sensor, for the next phenophase, and then for the suitability of the abiotic environment. At this point, it is possible for the organism to go on to the next phenophase. If a consumer had been possible, the

---

Fig. 1. Comparison of the phenodynamic of selected components of the main trophic levels in North Carolina ecosystems. Most data are from investigations in the Piedmont (*T* in Fig. 2 gives the approximate location). Data from Dillon (1971) are for the estuaries in North Carolina (near *M* in Fig. 2).

[2] This material was presented by Frederick at the symposium, see Pendergrass and Frederick (1973) and Ku and Frederick (1973) for full details.

VEGETATION PERIOD IN NORTH CAROLINA ENTERED IN A THREE DIMENSIONAL COMPUTER DRAWING

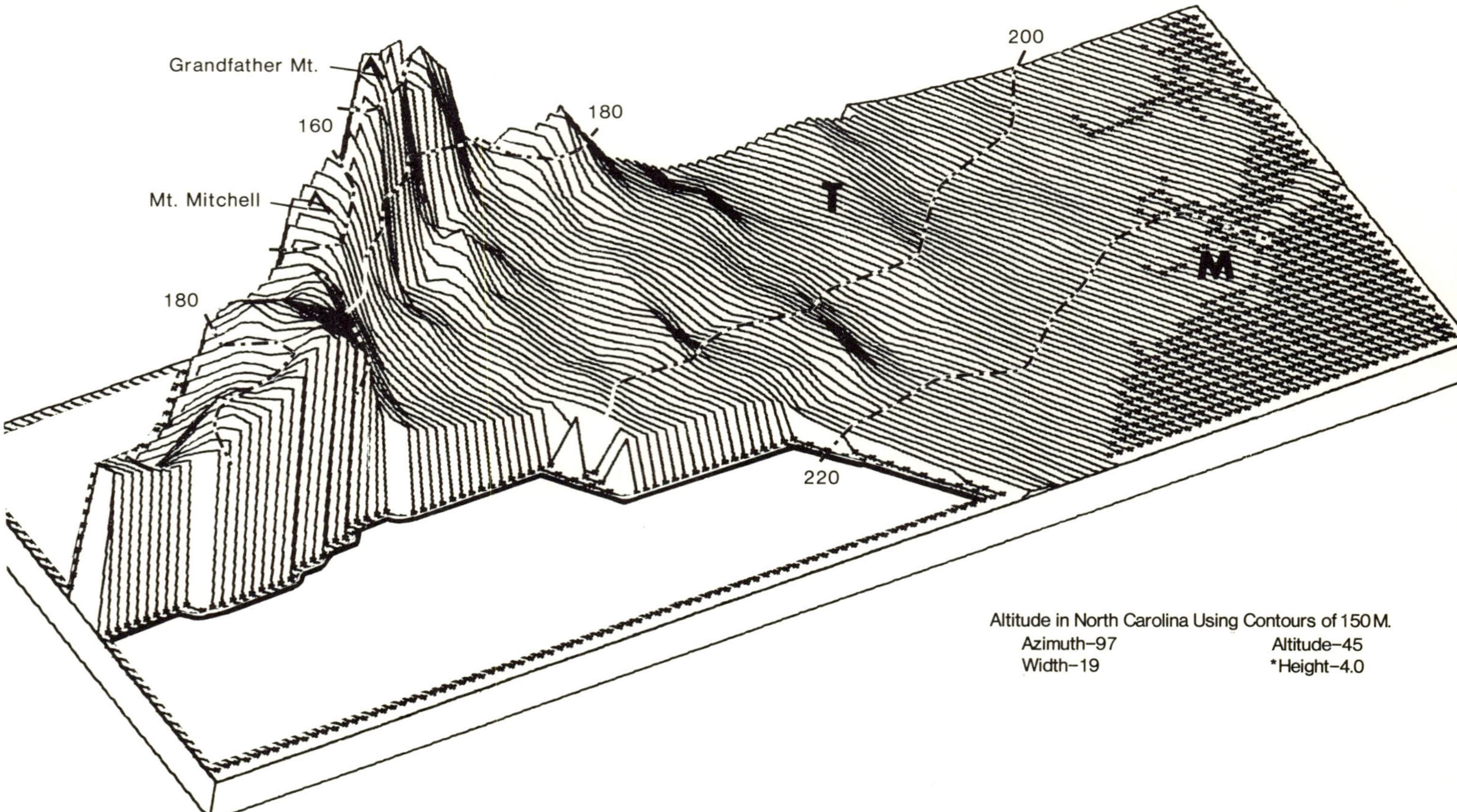

calculations would have shifted to the consumer loop to follow the development of the consumer up to that same point in time. The data calculated at the positions indicated by black points are entered daily into the "biotic environment" file, so that for each day separate producer and consumer biomass values are available along with one developmental value for each.

Each time the conditions are not yet complete for proceeding to the next step (in growth or in development), the calculator returns to the initial condition, at which point it adds a day to the time file and the biotic environment file. Both of these files serve as switches between the producer loop and consumer loop. In other words, the abiotic environment must be right for both producer and consumer, the producer must be at an edible stage, and the consumer must be ready to eat—all at the same time; otherwise, the loops keep running until the producer is partially consumed, the producer is completely consumed, or the consumer dies because there was not enough to eat, or because it was not ready to eat at the necessary time. As long as producer and consumer are alive, they can each pass through their life cycles.

The combined causality of a biologic sensor (called here "developmental sensor") and abiotic signal (called here "growth sensor") is usually circumvented in ecosystems modeling; the biologic sensor is assumed "on" and capable of attaining a steady level. The plant response to the abiotic signals is in the form of a differential or difference equation (O'Neill, 1971; Patten, 1971). An example of the use of growth equations for a seasonal partial ecosystem model of the tundra was presented by Timmin *et al.* (1972).

In Fig. 4 growth is calculated independently from, and prior to, the phenophase reached. It is generally known that cold periods in spring may depress the photosynthesis-respiration enzyme system in plants, or that diseases may inhibit the growth system of animals. The organism will then require several days to build up before continuing growth and development. Therefore, we must check whether the developmental stages of organisms can be advanced at any particular time and whether photosynthesis may be channelled into different organism compartments like root, leaf mass, seed, egg, offspring, or reserve fat.

The procedure described might simulate reasonably well the correspondence between exogenous control signals—external timing signal or Zeitgeber (Bünning, 1967)—and an endogenous sensor system. The two steps, "Call P1" and "Growth sensor ready?", check the levels of the five environmental aspects against the producer's (or, at equivalent point in the consumer loop the consumer's) requirements. If the conditions are tolerable for producer (or consumer) growth, the achievable biomass is calculated. "Call P2" and "Developmental sensor ready?" check the life cycle properties (the phenodynamics, as shown in Fig. 1, and the biomass

---

Fig. 2. Three-dimensional computergraph of simplified altitudinal patterns of North Carolina with length of vegetation period super-imposed. The graph was produced with *SYMVU* (1972). The length of the vegetation period in days was entered by hand according to regular *SYMAP* (Reader *et al.*, 1972) outputs. Approximate locations for data collection points listed in Fig. 1. are shown. T = Research Triangle Universities IBP Site; M = Morehead City.

NUMBER OF DAYLIGHT HOURS DURING THE GROWING SEASON

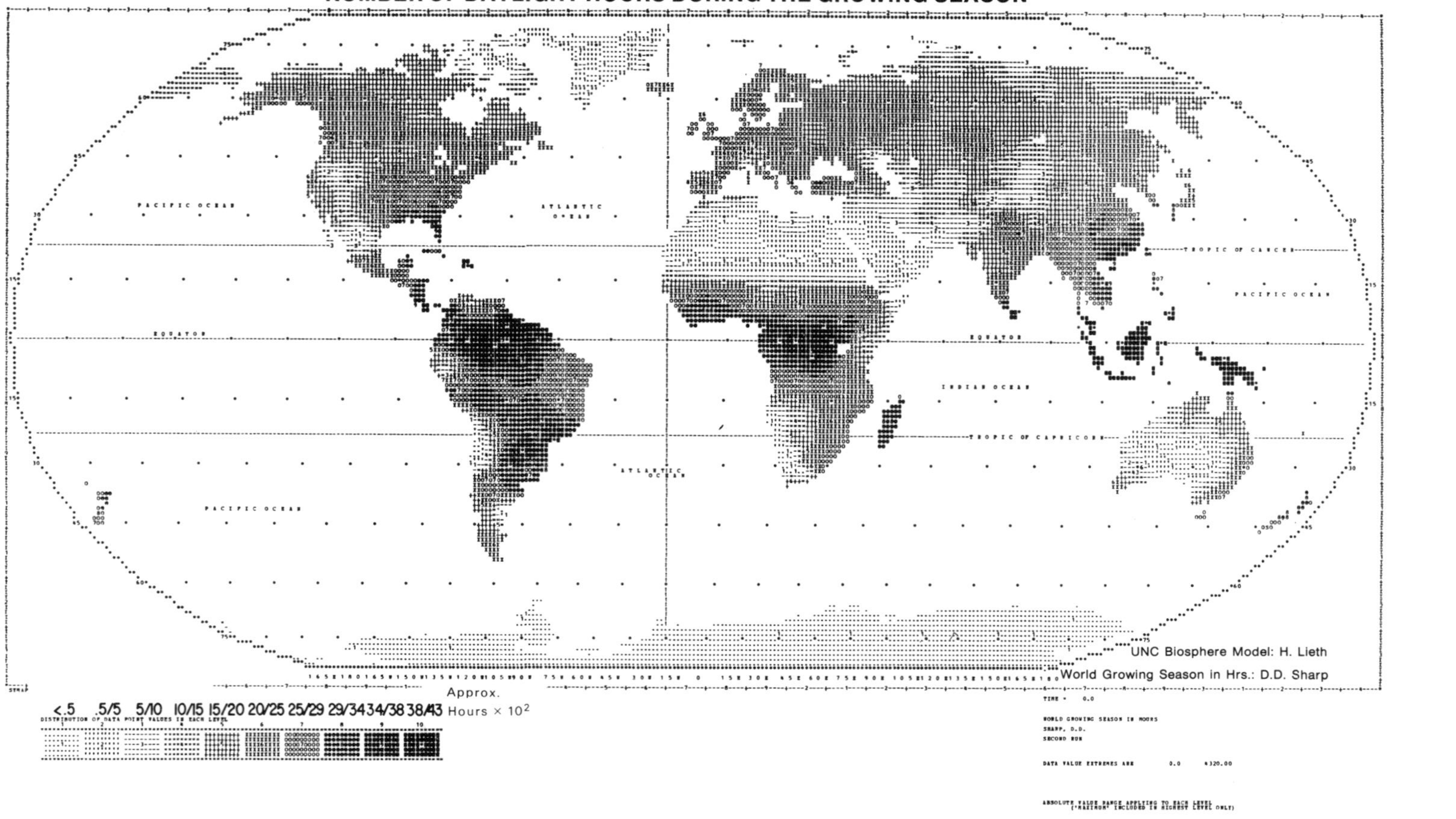

necessary before passing into the next phenophase) against the developmental signals to date. The significance of this approach and the importance of oscillating environmental forces, such as daylength, temperature changes, passage through threshold values, and others, have been substantiated in many papers in the past and are also reflected in several chapters of this book.

To complete the discussion on physiologically and environmentally based phenological modeling, we need to comment briefly on the role of the genome in the results of the models. It often happens that hypotheses formulated from observations made on one set of organisms must be modified to account for responses observed in other areas. Using a sample of the moth *Acromycta rumicis* collected at 43°N, Danilevskii (1957), see Danilevskii (1965) reported that the moths emerged only when the photoperiod was longer than 14.5 hr. When he repeated the experiment, using moths collected at 60°N, he found the critical photoperiod changed regularly by about 1 hr for every 5° change of latitude. Geographically distinct populations had to adjust their response to photoperiod to avoid encountering unfavourable temperatures and moisture conditions which otherwise would have prevented the occurrence of the event. Such adaptation to local conditions has often been accomplished by genetic adjustment, as is frequently reported in the literature for both plants and animals (see also Chapter 2.5).

It is doubtful that genomic effects can be incorporated into most ecosystem models. Most current ecosystem problems can be solved while assuming that genomic changes are negligibly slow. However, there are cases in bacterial, fungal, and other evolutions that warn us of the reality of adaptive evolution of species in response to management tools like pesticides, fungicides, or insecticides.

The authors express hope that this book will stimulate the modeler of ecosystems to incorporate phenological facts and principles into his models.

References

Bünning, E. 1967. The Physiological Clock. Rev. 2nd ed. 167 pp. New York: Springer-Verlag.

Clausz, J. C. 1969. Distribution of propagules of species of the Saprolegniaceae in the North Carolina Botanical Garden lake and some possible controlling factors. Ph. D. Thesis. Chapel Hill: Botany Department, University of North Carolina. 103 pp.

Fig. 3. Length of photosynthetic period in hours daylight for the vegetation period. Vegetation period in months is shown on the inside back cover. The criteria employed to define the growing periods beginning and end were (1) temperature thresholds starting from 0° north of 60° latitude, increasing linearly from there to 10°C to 30°C latitude south, for which no temperature-generated seasonality was assumed; (2) seasonal precipitation when amount in millimeters for any given month dropped below double the monthly average temperature. The number of months in which significant amounts of rain falls or the number of months exceeding 2 mm of rain per 1°C average monthly temperature were entered. About 300 stations were selected from the Climate Diagram World Atlas (Walter and Lieth; 1960–1967), evenly spread over the outline of the UNC Biosphere Model. The same criteria were applied to each station to calculate the number of months vegetation period. The resulting map on the inside back cover shows 6 different seasonality levels.

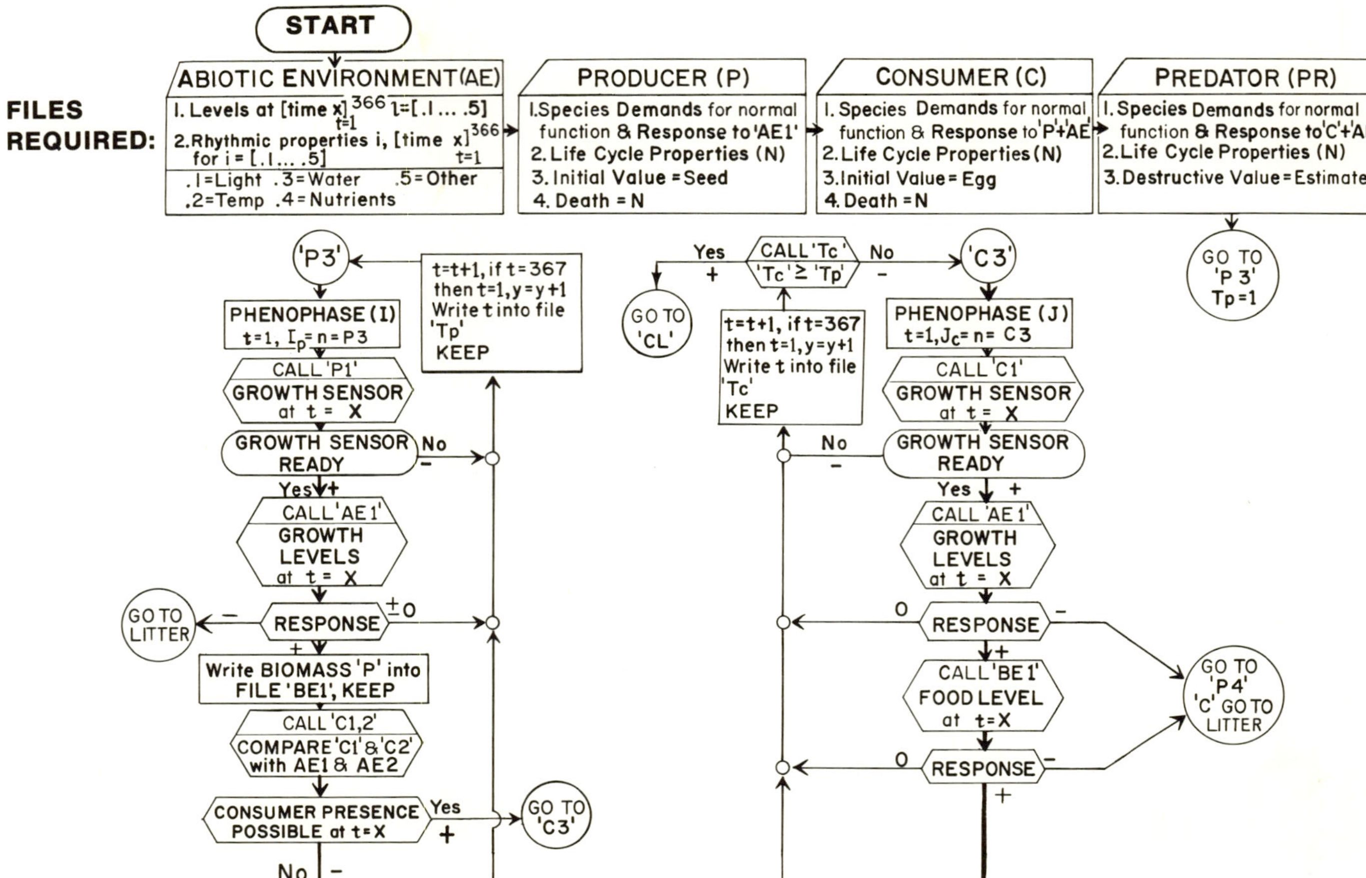
FILES REQUIRED:
START
ABIOTIC ENVIRONMENT (AE)
1. Levels at [time x] t=1 366 i=[.1 ... .5]
2. Rhythmic properties i, [time x] t=1 366 for i = [.1 ... .5]
.1=Light .3=Water .5=Other
.2=Temp .4=Nutrients
PRODUCER (P)
1. Species Demands for normal function & Response to 'AE1'
2. Life Cycle Properties (N)
3. Initial Value = Seed
4. Death = N
CONSUMER (C)
1. Species Demands for normal function & Response to 'P'+'AE'
2. Life Cycle Properties (N)
3. Initial Value = Egg
4. Death = N
PREDATOR (PR)
1. Species Demands for normal function & Response to 'C'+'AE'
2. Life Cycle Properties (N)
3. Destructive Value = Estimate
GO TO 'P 3' Tp=1
'P3'
PHENOPHASE (I)
t=1, Ip=n=P3
CALL 'P1'
GROWTH SENSOR at t = X
GROWTH SENSOR READY
No
Yes
CALL 'AE1'
GROWTH LEVELS at t = X
RESPONSE
GO TO LITTER
Write BIOMASS 'P' into FILE 'BE1', KEEP
CALL 'C1,2'
COMPARE 'C1' & 'C2' with AE1 & AE2
CONSUMER PRESENCE POSSIBLE at t=X
Yes
No
GO TO 'C3'
t=t+1, if t=367 then t=1, y=y+1 Write t into file 'Tp' KEEP
CALL 'Tc'
'Tc' ≥ 'Tp'
Yes
No
GO TO 'CL'
'C3'
t=t+1, if t=367 then t=1, y=y+1 Write t into file 'Tc' KEEP
PHENOPHASE (J)
t=1, Jc=n= C3
CALL 'C1'
GROWTH SENSOR at t = X
GROWTH SENSOR READY
No
Yes
CALL 'AE1'
GROWTH LEVELS at t = X
RESPONSE
CALL 'BE1'
FOOD LEVEL at t=X
RESPONSE
GO TO 'P4' 'C' GO TO LITTER

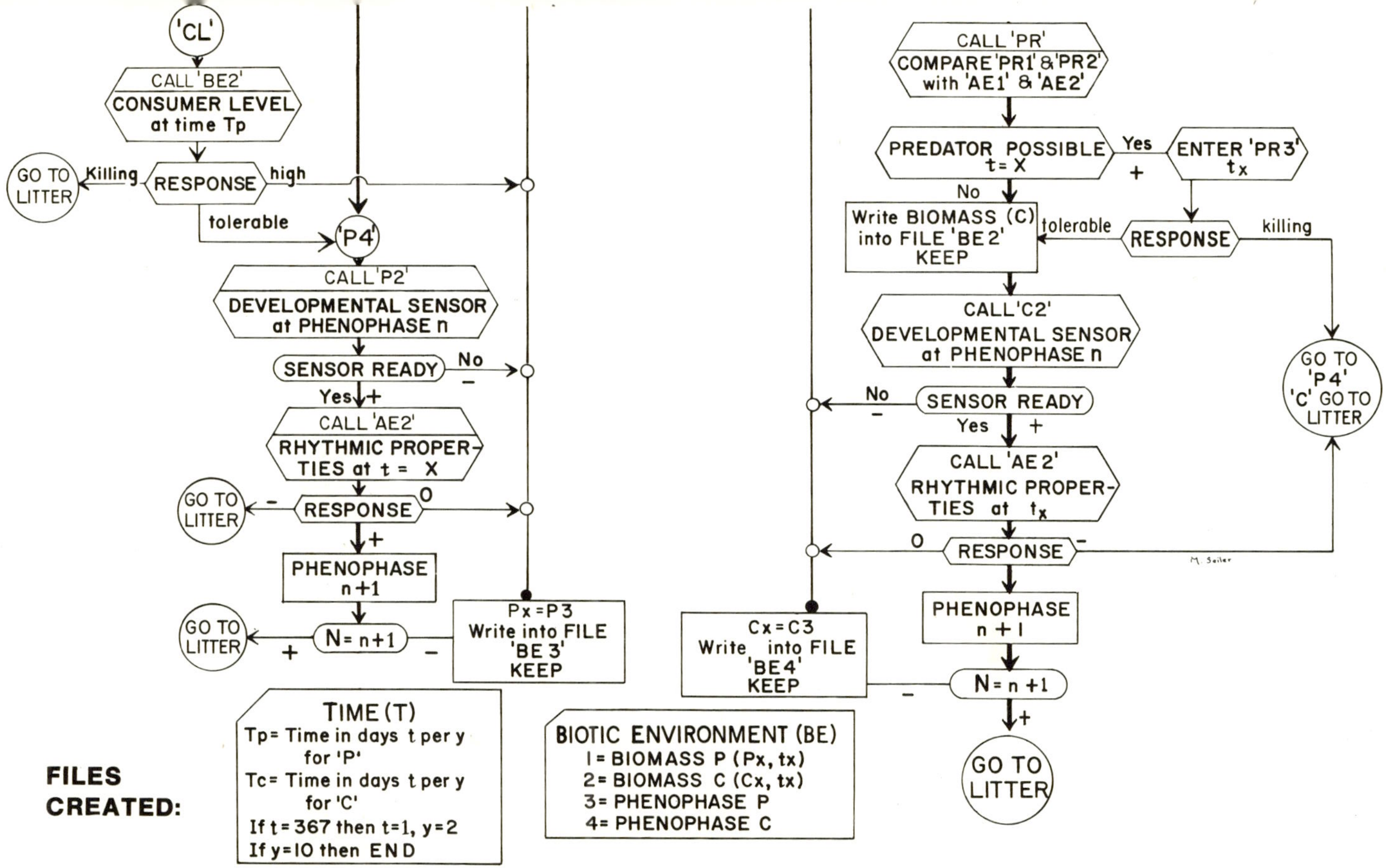

Fig. 4. Phenological flowchart of a producer-consumer system. For the detailed explanation of the elements in the chart, see text.

Coleman, D. C. 1971. Numbers and biomass of soil nematodes of two South Carolina old fields. Amer. Midland Naturalist. 85, 262–265.

Danilevskii, A. S. 1965. Photoperiodism and seasonal development of insects. 283 pp. Edinburgh and London: Oliver and Boyd.

Dillon, C. R. 1971. A comparative study of the primary productivity of estuarine phytoplankton and macrobenthic plants. Ph.D. Thesis. Chapel Hill: Botany Department, University of North Carolina. 112 pp.

Duever, M. J., and Fatora, J. R. 1968. Daily and seasonal activity patterns of mourning doves on the AEC Savannah River Plant. *Proc. 22nd Ann. Conf. Southeastern Assoc. Game and Fish Commissioners.* 99 pp,.

——— 1970. Daily and seasonal activity patterns of bob white quails on the AEC Savannah River Plant. *Proc. 22nd Ann. Conf. Southeastern Assoc. Game Fish Commissioners.*

Flint, H. L. 1974. Phenology and genecology in woody plants, p. 83. In: *Phenology and Seasonality Modeling*, H. Lieth, ed. New York: Springer-Verlag.

Golley, F. B. 1961. Interaction of natality, mortality, and movement during one annual cycle in a *Microtus* population. *Am. Midland Naturalist* 66, 152–159.

———, and H. Lieth. 1972. Bases of organic production in the tropics. In *Tropical Ecology with an Emphasis on Organic Production*, P. Golley and F. Golley, eds. pp: 1–225. Athens: University of Georgia Press.

Hopp, R. J. 1974. Plant phenology observation networks. p. 25. In: *Phenology and Seasonality Modeling*, H. Lieth, ed. New York: Springer-Verlag.

Katz, B. A., and H. Lieth. 1974. Decomposer seasonality, p. 163. In: *Phenology and Seasonality Modeling*, H. Lieth, ed. New York: Springer-Verlag.

Ku, C. L., and L. Frederick. 1973. Studies on Myxomycetes occurring on bark of living trees in Atlanta, Georgia. *Bull. Georgia Acad. Sci.* in press.

Lee, Y. N. 1971. Patterns of flowering periods in selected floras of the world. *J. Korean Res. Inst. Better Living*. 6, 41–51.

Lieth, H. 1970. Phenology in productivity studies. p. 29. In *Analysis of Temperate Forest Ecosystems*, D. Reichle, ed. Heidelberg: Springer-Verlag.

——— 1971. The phenological viewpoint in productivity studies. p. 71. In *Productivity of Forest Ecosystems.* pp. 71–84. Proc. Brussels Symposium, P. Duvigneaud, ed. Paris: UNESCO.

——— 1972. Modeling the primary productivity of the world. *Deciduous Forest Biome Memo Rep., 72–9.* 10 pp., offset.

——— and J. W. Radford, 1971. Phenology, resource management and synagraphic mapping. *BioScience* 21, 62–70.

Linnaeus, C. 1751. *Philosophia Botanica.* Stockholm: Kiesewetter.

Mabee, H. F., and J. H. B. Garner. 1974. Seasonal variations of soil fungi isolated from the rhizosphere of *Liriodendron tulipifera* L. In *Phenology and Seasonality Modeling*, H. Lieth, ed. New York: Springer-Verlag.

Malaisse, F. P. 1974. Phenology of the Zambezian woodland area with an emphasis on the miombo ecosystem. p. 269. In: *Phenology and Seasonality Modeling*, H. Lieth, ed. New York: Springer-Verlag.

Mooney, H. A., D. J. Parsons, and J. Kummerow. 1974. Plant development in Mediterranean climates. p. 255. In: *Phenology and Seasonality Modeling*, H. Lieth, ed. New York: Springer-Verlag.

O'Neill, R. V. 1971. Systems approaches to the study of forest floor arthropods. In *Systems Analysis and Simulation in Ecology.* B. C. Patten, ed. Vol. 1, pp. 441–477. New York: Academic Press.

Patten, B. C., ed. 1971. *Systems Analysis and Simulation in Ecology.* Vol. 1, 607 pp. New York: Academic Press.

Pendergrass, L., and L. Frederick. 1973. Myxomycetes from bark of living trees in Panola Mountain State Park, Georgia *Castanea*. in press.

Pfau, R. 1964. Varianz- und Korrelationsanalytische Untersuchungen an Phänologischen Phasen im Naturraum 06 (unterbayerisches Hügelland). *Meteor. Rundschau* 17 (4), 113–122.

Rabb, R. L., and F. R. Lawson. 1956. Some factors influencing the predation of *Polistes* wasps on the tobacco horn worm. *J. Econ. Entomol.* 50, 778–784.

Reader, R., and UNC Applied Programming Group. 1972. SYMAP (version 5.16A) *Instruction Manual.* Document No. LSR–139–0. 65 pp. offset. N.C.: Triangle Universities Computation Center, Research Triangle Park.

Schnelle, R. 1955. *Pflanzen-Phänologie.* (Probleme der Bioklimatologie Bd. III). Leipzig: Akad. Verlagsges.

SYMVU Manual (Version 1.0). 1971. 52 pp. mimeo. Cambridge, Mass.: Laboratory for Computer Graphics and Spatial Analysis, Harvard University.

Timmin, M. E., B. D. Collier, J. Zich, and D. Walters. 1972. A computer simulation of the arctic tundra ecosystem near Barrow, Alaska. 125 pp mimeo. San Diego, Calif.: Biology Department, San Diego State University.

U.S. Department of Commerce. 1971. *Climatological Data, North Carolina.* Vol. 76, 204 pp. Washington, D.C.: NOAA Environmental Data Service.

US/IBP Phenology Committee. 1972. *Report, July 1972.* 54 pp. offset. Austin, Tex.: US/IBP Environmental Coordinating Office.

Waggoner, P. E. 1974. Modeling seasonality. p. 301. In *Phenology and Seasonality Modeling*, H. Lieth, ed. New York: Springer-Verlag.

Walter, H., and H. Lieth. 1960–1967. *Klimadiagramm Weltatlas.* Jena: Fischer-Verlag.

Wells, C., D. Whigham, and H. Lieth. 1972. Investigation of mineral nutrient cycling in an upland piedmont forest. *J. Elisha Mitchell Sci. Soc.* 88, 66–78.

Whigham, D. F. 1971. An ecological study of *Uvularia perfoliata* in the Blackwood Division of the Duke Forest, North Carolina. Ph.D. Thesis. 183 pp. Chapel Hill: Botany Department, University of North Carolina.

PART 2

Methods for Phenological Studies

Introduction

Forest Stearns

Department of Botany
University of Wisconsin, Milwaukee,
Wisconsin, 53201, USA

and

Helmut Lieth

Department of Botany
University of North Carolina
Chapel Hill, North Carolina, USA 27514

Phenological work requires a multitude of methods and approaches. Each group of organisms needs special procedures for observation and experimentation. It is impossible to cover all these methods in this volume. The basic logic of scientific work—data acquisition, data handling, evaluation and interpretation, and experimentation to test hypotheses—is similar for all organisms. Phenological methods vary as to whether they aim to assess the properties of a single species over a wide spatial range or of a multitude of species in one community or biocoenosis.

The analysis of behavior of single species over large distances is best developed for higher plants and birds. Today the work includes the most modern techniques of data acquisition, handling, and interpretation. Moreover, the single observation of one phenophase at one location is also the basis for the complete phenodynamic analysis of an entire community (biocoenosis). We have therefore restricted this discussion of phenological methods to one group of organisms, the higher plants, and to basically one aim, phenological mapping.

The number of phenological events is legion, but not all are easy to observe or useful for mapping; some are relatively long, others short, and in many the beginning or the ending is not clearly delineated. The U.S. International Biological Program (IBP) has had, as one priority, the selection of adequately defined phenophases in species appropriate for widespread use as indicators. Preliminary steps have been taken to standardize phenological studies in the United States. The

US/IBP Phenology Committee (1972) has published a tabular report on current phenological studies in the United States that includes sites and locations of phenological observations, a brief description of phenophases recorded, and species observed at each site. Phenological data should be pooled for quick access and comparison by any interested worker. Sayrs and Taylor (1972) have suggested a data format for a computerized data pool. Other examples of computerized data handling for plant phenology are reported by Hopp (see Chapter 2.1). One of the best organized, ecologically oriented, computerized data files is that of the U.S. Bird Banding Laboratory (see *North American Bird Banding Manual*, Vol. I, 1972).

Methods suitable for the mapping of primary-producer phenology and descriptions of the goals and methods of observer networks and phenological gardens are presented in Chapters 2.1 and 2.2. Morain (see Chapter 2.3) treats remote sensing of vegetation, the latest development in high-speed data acquisition. Computer mapping methods are particularly valuable to phenology. In Chapter 2.4 Caprio, Hopp, and Williams compare computer and handmapping of phenological observations.

The symposium paper of James and Shugart (1972) about climatic influences on nesting of the robin (*Turdus migratorius*) is not included in this volume. It contained a computer map produced with the IBM stampede package. This package program can accept, interpolate, and map data for three coordinates. The resulting map looks very much like a handdrawn isophane map.

Techniques important for study of other important ecosystem units are partially covered in Part 3. Modeling of phenological data and computerized data evaluation are covered in Part 5, and in some other papers, e.g., in Chapter 4.3. Part 4, which presents intensive studies from several different biomes, includes methodological hints for phenodynamic studies, as well as remarks on the handling of special organism groups, e.g., bacteria and fungi.

Phenological responses to the environment are genetically coded. To analyze responses to different codes, experiments are conducted requiring transplants and controlled environmental conditions. In Chapter 2.5 Flint, presents a literature review of studies in genecology as related to response to environment as an example for phenological experimentation.

References

James, F. C. and H. H. Shugart, 1972. Seasonality modeling of the nesting season of the robin, (*Turdos migratorius*). submitted for publication to *Condor*.

North American Bird Banding Manual. 1972. Vol. I. U.S. Department Interior, Bureau of Sport Fisheries and Wildlife and Dep. Environment, Canadian Wildlife Service. Variable page numbers. Bird Banding Laboratory, Laurel, Md.

Sayrs, R. L., and F. G. Taylor. 1972. Preliminary phenology data format. *Eastern Deciduous Forest Biome Memo Rep.* 72–48. 13 pp. mimeo.

US/IBP Phenology Committee. 1972. *Report, July 1972.* N. R. French, Chairman and ed. Austin, Tex.: US/IBP Environmental Coordinating Office. 54 pp. offset.

2.1 Plant Phenology Observation Networks[1]

Richard J. Hopp

Department of Plant and Soil Science
University of Vermont
Burlington, Vermont, USA 05401

1 Historical Background

Phenological observations and calendars were used in agriculture thousands of years ago in both China and Rome. The term "phenology" was first proposed by the Belgian botanist Charles Morren in 1853, but the father of modern plant phenology and phenological observation networks is the Swedish botanist Carolus Linnaeus (or Carl von Linné). In *Philosophia Botanica* Linnaeus (1751) outlined methods for compiling annual plant calendars of leaf opening, flowering, fruiting, and leaf fall, together with climatological observations "so as to show how areas differ." Linnaeus' simple statement of the purpose of phenology is still valid today, although the potential uses of phenology have expanded considerably.

A partial history of phenological networks and observations is summarized in Table 1. This table is far from complete, as it only gives examples of the highlights of phenological activities and developments primarily in Europe and the Americas. More detailed information on the history of phenology and phenological observations is given by Ihne (1884), Abbe (1905), Schnelle (1955), Caprio (1966), and Wang (1967). Today phenological networks exist in many countries which are not included in the listing, such as Czechoslovakia, Finland, France, Italy, Netherlands, Russia, Turkey, Australia, Japan, India, South Africa, and others.

During the six decades from 1883 to 1941 Ihne was probably the most important proponent of phenology in Europe and was well known throughout the world. He was the first to use a single species (*Syringa vulgaris*) for phenological mapping (Ihne, 1885), and for 59 years he made his phenological observations available in a series of annual publications (Ihne, 1883–1941). His gravestone in Darmstadt, Germany, carries the inscription *Seine Lebensarbeit war Phänologie* (his lifework was phenology) (Schnelle, 1960).

[1] Vermont Agricultural Experiment Station Journal Article No. 300.

KEYWORDS: Phenology; plants; observation networks; history; bioclimatic law; phenophases; meteorological network; isophanes; volunteer observers; Germany; Poland; Norway; Spain; Argentina; United States; Wisconsin.

Table 1. A partial history of 200 years of plant phenology observation networks.

Period	Originator/Agency	Country/Region of Observations	Comments
1750	Linnaeus (1751)	Today's Sweden	Network of 18 sites
	Stellingfleet (Ihne, 1884)	Stratton, Norfolk, England	Published *Floral Calendar* (1755)
	Spruengli (Pfister, 1972)	Switzerland	Carried out over 4000 plant and animal observations, and precise observations of snow and frost phenomena (1759–1802)
	Scapoli (Ihne, 1884)	Today's Yugoslavia	Published *Calendarium Florae Carniolicae* (1762)
	Soc. Meteor. Palatina (1781–1792) in Mannheim, Germany	Europe	First international meteorological and phenological network established (1780)
	Haenke (Ihne, 1884)	Czechoslovakia	Published *Beobachtung der Bluetezeiten in Prag* (1786)
1800	New York Agri. Soc. (Abbe, 1905)	United States	Call for phenological observations (1807)
	Comm. of General Land Office (Abbe, 1905)	United States	Call for phenological observations (1817)
	Regents of Univ. of New York (Abbe, 1905)	United States	Call for phenological observations (1820)
	Quetelet (1849)	Europe	80 stations established in Belgium, Holland, Germany, Italy, France, England, and Switzerland (1839); data available for 1841–1872
1850	Fritsch (Schnelle, 1955)	Austria	Austrian Central Inst. for Meteorology & Magnetism instructed meteorological observers in making vegetation observations (1853)
	Hough (1864)	United States	Network organized by Smithsonian Institution. Observations on plants, birds, and insects (1851–1859) by 320 observers in 33 states; 86 plant species, including a few cultivated fruits
	Mikesell (Smith, 1915)	Wauseon, Ohio, USA	Records on native and cultivated plants (1873–1912)
	Hoffmann (1881)	Central Europe	First phenology maps
	Hoffman & Ihne (1882)	Central Europe	Appeal for first open flower date on 22 species, plus first ripe fruit date on six of them
	Ihne (1883–1941)	Central Europe	Published annual observations of 100-site network for 59 years
	Ihne (1884)	Germany	History of plant phenological observations for Europe in the 18th and 19th centuries
	Ihne (1885)	Central and Northern Europe	First time a single species used (*Syringa vulgaris*) at 500 sites
	Forestry personnel (Witterstein, 1971)	Germany	Forest-phenological network of 240 stations (1885–1894)
1900	Abbe (1905)	Europe	Account of 19th century phenologists, including Quetelet, Fritsch, Linsser, and Hoffmann
	Ihne (1905)	Central Europe	Published "Phenological map of beglnning of spring in Central Europe."

	U.S. Weather Bureau, USDA (1905, 1909)	Indiana, USA	20 cooperative weather observers and crop correspondents gathered information on forest trees; orchard trees, vines, shrubs; cereals, grasses, and vegetables (1904–1908)
	Hopkins (1918, 1938)	United States	Hopkins' law for spring progression (1918), refined in 1938
	Deam (Lindsey & Newman, 1956)	Bluffton, Indiana, USA	Records of his private arboretum and garden (1920–1952)
	Biological Inst. for Agr. & Forestry, Berlin (Witterstein, 1971)	Germany	Local and regional phenological activities organized into national phenological network in 1922
	Gates (Hulbert, 1963)	Manhattan, Kansas, USA	Records on 132 plant species (1926–1955)
	Printz (Wielgolaski, personal communication 1972)	Norway	Organized private network in 1928, extending from southern Norway to North Cape
	Leopold and Jones (1947)	Sauk and Dane counties, Wisconsin, USA	Phenological records for 1935–1945
	German Weather Bureau (Witterstein, 1971)	Germany	Took over natlonal phenological service in 1936; enlarged to 10,000 observers, i.e., a density of one observer per 50 km²
1940	National Meteorological Service (Bertinotti, personal communication 1972)	Buenos Aires, Argentina	Observations on 70 plant species since 1940; results to be published in 1973
	Hodson (1971)	Minnesota, USA	Leaf bud development and flowering dates of trees and shrubs (1941–1971)
	National Meteorological Serv. (Barasoain, personal communication 1972)	Spain	Organized network in 1943; currently 465 observers
	National Inst. for Hydrology & Meteorology (Sokolowska 1971)	Poland	Began rebuilding network after WW II (1945) with 300 observers; now about 700 observers; phenological yearbooks available for 1946–1960
1950	Agrometeorological Dept., German Weather Service (Witterstein, 1971)	Germany	Reorganized network in West Germany in 1952. Now 2700 observers, a density of one observer per 90 km²
	Schnelle (1955)	World	Survey of development of phenology and phenological networks
	Lindsey and Newman (1956)	Bluffton, Indiana, USA	Analyzed observations of Deam on 51 species (1920–1952) in relation to temperature records
	Schnelle and Volkert (1957, 1964)	Europe	Network of 70 phenological gardens in 19 European countries
1960	National Meteorological Serv. (Garabatos, 1969)	Argentina	Fruit phenological network in upper valley of Rio Negro and Neuquén River (1960); observations on 18 fruit cultivars
	Caprio (1966)	Europe, USA	Survey of phenological developments since Linnaeus
	Wang (1967)	World	Survey of phenological developments since Linnaeus
1970	Pfister (1972)	Switzerland	History of phenological observations (1759–1802)

A number of localized phenological observation series for various periods of time are on record; other data no doubt rest in the dark drawers of many nature lovers. Mikesell (Smith, 1915) kept phenological records on both native and cultivated plants in Wauseon, Ohio, between 1873 and 1912. In Bluffton, Indiana, Deam made "meticulous observations" in his private arboretum and gardens from 1920 to 1952. Gates left phenological records taken on 132 plant species from 1926 to 1955 at Manhattan, Kansas. "Because he always walked to and from work the plants were seen one or more times daily year after year" (Hulbert, 1963). A phenological record is available for two counties in Wisconsin from 1935 to 1945 (Leopold and Jones, 1947). In Minnesota, records were kept over a 30-year period (from 1941) on leaf bud development and flowering of a group of trees and shrubs (Hodson, 1971).

The efforts by dedicated individuals in various countries led to the recognition of a need for combining and coordinating phenological and meteorological observations, particularly for agriculture, horticulture, and forestry. Thus, since the turn of the century, phenology and phenological observation networks in many countries have become identified with government agencies, and the plant phenological networks usually with national meteorological services. Such arrangements insure uniformity and continuity of observations and their evaluations, which is not possible when efforts are likely to terminate with the retirement or death of an individual observer.

In 1904 the U.S. Weather Bureau adopted a limited plan for securing phenological observations when report forms were distributed in Indiana to 20 cooperative weather observers or crop correspondents. Observations were made on forest trees; orchard trees, vines, and shrubs; and cereals, grasses, and vegetables. The phenophases observed on forest trees included first leaves, woods green, leaves coloring, leaves falling, and woods bare; fruit trees and small fruits were observed for the dates of first leaves, first bloom, full bloom, and beginning and ending of harvest. The observations in the third group dealt with beginning and ending of soil preparation, planting, and harvest. These plant phenological data were published (U.S. Department of Agriculture, Weather Bureau, 1905, 1909) together with observers' miscellaneous notes, which included dates of bird migration; first appearances of robins, bluejays, blackbirds, buzzards, and other birds; bees and earthworms; and the croaking of frogs. After 1908 this project faded away.

Since De Witt advanced the idea of mapping dates of phenological events, as early as 1792, many workers attempted to relate such events to geographical coordinates and elevation (Caprio, 1966). Based on observations available from the various phenological networks in the northern hemisphere, Hopkins (1918, 1938) proposed and later refined and expanded a bioclimatic law that provides a framework within which to evaluate phenological observations. One section of this law states that, other conditions being equal, the south to north progression of spring phenophases in the temperate part of the North American continent is delayed by four days for each degree of latitude northward, for each 5° of longitude eastward, and for each 400-foot increase in elevation. This model is, of course, very general. It has certain geographical limitations and it is difficult to apply to individual plant species in any one season. Lieth and Radford (1971) compared data from the southeastern United States against Hopkins' model, and concluded that with improvement it might be useful in resource management models. Other models and mathematical treatments of phenological data are described in this volume.

2 Relationship of Phenological and Meteorological Networks

A phenological network consists of cooperators observing and recording the date of the occurrence of defined developmental stages or phenophases of indicator plants. Phenological observations may be carried out intensively at a limited number of sites or extensively at a great number of locations over a wide area. Considering the close relationship between climatological parameters and plant development, phenological networks are not only a parallel to, but often are an important supplement to, the network of weather observation stations in a country. An extensive phenological network corresponds to the system of volunteer weather observers for the various national weather services. Intensive phenological study sites such as phenological gardens can be compared in the United States to the reference climatological stations of the National Oceanic and Atmospheric Administration (NOAA).

Most cooperative weather observers record daily maximum and minimum temperatures and precipitation, from which maps are constructed showing zones of equal average monthly temperature, last occurrence of frost in spring, and so on. Plants are excellent indicators of climatic differences because the time of occurrence of phenological events of many plants is, to a large degree, controlled by local climatological factors. Indicator plants are living instruments that show in their growth response the composite effect of temperature, rainfall, humidity, wind, sunshine, etc. Phenological observations on selected indicator plants may help forecast or estimate the occurrence of succeeding phenological events in the same or other species. If carried out over a period of years, phenological observations characterize the annual variations in the dates of certain phenological events at any one location. Maps of long-term average dates identify zones of similar climatic conditions, as do meteorological maps; such maps are useful tools for decision making in agriculture, horticulture, or forestry.

Phenology thus represents a merging of the meteorological and biological sciences. Phenology has one advantage over meteorology as a zonal differentiator: plants are cheaper than instruments. The desired accuracy determines the density of a meteorological or phenological network, especially where the climate changes markedly over a short distance, such as in mountainous areas. The U.S. National Weather Service has approximately 6000 cooperative observers, an average density over the contiguous United States of about one observer per 1500 km^2 (600 sq. mi).

While phenological networks may supplement networks of weather stations, certain fundamental differences exist between the two types of observations. Temperature, for example, is measured with standardized equipment in specially designed instrument shelters at a uniform height above the surface so as to exclude near-ground influences. These measurements characterize the macroclimate of the area where the shelters are located. The near-ground influences, however, are very important in plant development. In a nine-year study in Burlington, Vermont, the average date of the last occurrence of 0°C at 7.5 cm above ground was 16 days later than at the 150-cm standard height of the instrument shelter. In fall the first near-ground date for 0°C was 14 days earlier than the shelter date. The averages were similarly affected for −2° and −4°C (Hopp, *et al.*, 1964). Thus the freeze-free period was 30 days shorter near the ground than at the 150-cm height. These considerations are important when low growing plants are included in observation networks, and attempts are made to correlate phenological observations on such plants with climatological data recorded in standard instrument shelters.

A second important difference is that weather-recording instruments are designed to respond to particular factors, while plants respond to the full range of continuously changing environmental influences. Soil types, exposure, topography, etc., in addition to meteorological parameters, affect plant response. If native plants are used as indicators in phenological networks, genetic differences may affect the response to environment. Thus each report of a phenological observation carries with it a possible inherent uncertainty in contrast to the exacting temperature or precipitation report of the weather observer. Schnelle (1955) stated appropriately as a rule for all phenological evaluations "that each observation from any one station, if at all possible, should be considered in relation to the data from a neighboring station."

3 International Phenological Networks

Patterns in plant development are independent of political boundaries. If appropriate plants are used as indicators, and uniform observations made by cooperating observers over large areas, the green wave can be monitored, i.e., the spring progression of plant development from south to north in the northern hemisphere and from north to south in the southern hemisphere. The corresponding fall regression, the brown wave, flows in the opposite direction.

The need for international cooperation in phenology has long been recognized (see Table 1). The first international phenological network with observation sites in several European countries, together with a meteorological network, was established in 1780 by the Societas Meteorologica Palatina (1781–1792). In the last 15 years or so, a system of international phenological gardens has developed in Europe to obtain comparable phenological observations among different countries. Their purpose and development are discussed by Schnelle and Volkert (see Chapter 6.2).

4 Selected National Phenological Networks

Plant phenological observation networks in many countries are administered as an integral part of the national meteorological services. Four representative examples of such national networks are discussed here—Germany, Poland, Spain, and Argentina—as well as a private network in Norway. Phenological data collected by volunteer observers supplement climatological data throughout the vegetative season for the benefit of either agriculture and forestry as a whole or, as in the case of Argentina, of a specific segment. Obtaining volunteer observers apparently presents no problem once the nature and importance of accurate phenological observations has been explained. These volunteers appear to be very willing to cooperate and seem to derive a satisfaction from being part of, and contributing to, an important nationwide endeavor.

Germany

Perhaps the most comprehensive reporting network is that in West Germany. Prior to 1939 a network of about 10,000 cooperative observers functioned in Germany under the German Weather Bureau, about one observer per 50 km^2 (20 sq mi). In 1952 the weather services of the three western zones were combined into the German Weather Service and the phenological service was reestablished, administered by the Agrometeorological Department. The phenological network includes about 2700 stations fairly evenly distributed throughout West Germany—about one observer for each 90 km^2 (about 36 sq mi). The aim is to bring the density of the network closer to the prewar figure (Witterstein, 1971; personal communication, 1972).

The phenological observers of the German Weather Service are lay cooperators, people that either through their occupation or by choice are in close contact with nature. They include farmers, orchardists, gardeners, foresters, lumbermen, or employees of highway departments. Entire school classes participate in the observations under the guidance of their teachers, and phenology becomes part of their study of biology and ecology. Retired people going for walks in the country observe plants with considerable interest and knowledge. Among these observers are idealists and nature lovers who have carried out the observations for 10, 15, or even 20 years and "could not live without them" (Witterstein, 1960, 1971).

The observers receive detailed instructions together with colored pictures of the phenophases to be observed (Deutscher Wetterdienst, 1962; Schnelle, 1966; Witterstein, 1968). Two report forms are used, one for spring observations and one for summer and fall dates. Nineteen phenophases are allowed for on various plant species, native plants, agronomic crops, and fruit crops. The spring report form (Fig. 1) lists 123 possible observations on 66 species and the second report allows for 80 observations on 42 species. The forms are arranged so that the data can be transferred directly to computer cards. The two reports, due by 10 July and 30 November, respectively, are, of course, not adequate for a continuous monitoring of plant development during the vegetative season. This information is provided by a system of immediate reporting (Sofortmeldung). About 350 selected observers mail their reports on the occurrence of 60 important stages of plant development on the day when the particular phenophase is observed.

In addition, a phytopathological-phenological rapid reporting service (Schnellmeldedienst) was established. It supplies the plant protection offices with details about plant diseases and insect pests. Based on the previous occurrence of certain phenophases and subsequent climatic conditions, predictions of plant disease and

Deutscher Wetterdienst

I. Phänologischer Meldebogen für Frühling 1968

Bei Anschriftänderung bitte Berichtigung vornehmen!

Abs.: Deutscher Wetterdienst, Zentralamt, 6050 Offenbach 1

Beobachtungsort: ....

(Bitte nur ausfüllen, wenn mit Wohnort nicht übereinstimmend)

Sp. 1-4		5-6	7-15	16
71	01 02 03	68										
KA	KN	Jahr	Stationsnummer	JB

(Bitte nicht ausfüllen!)

Rücksendung bitte bis spätestens 10. Juli 1968

Erklärung der Abkürzungen für die Phasen:

- ab = Vollblüte, allgemeine Blüte
- Ä = Ährenschieben
- Au = Aufgang
- b = erste Blüten offen, Beginn der Blüte
- Bg = Bestand geschlossen, Bestandsschluß
- BO = erste Blattoberflächen sichtbar, Blattentfaltung
- Bst = Beginn der Bestellung, Beginn der Aussaat
- e = Ende der Blüte
- M = Maitrieb, erste neue Nadeln sichtbar
- Sch = Beginn des Schossens

Bitte in den auszufüllenden Pflanzen-Spalten nur Sortenname und Eintrittstag eintragen; sonst nichts einschreiben!

Alle Eintrittstage bitte nur als Jahrestagszahlen eintragen!

Wildwachsende Pflanzen

	Pflanzenart	Phase	Sp.	Eintrittstag
01	Haselnuß, Haselstrauch	b	17-19	
	Schneeglöckchen (einheimisches)	b	20-22	
	Huflattich	b	23-25	
	Salweide	b	26-28	
	Spitzahorn	b	29-31	
	Forsythie	b	32-34	
	Roßkastanie	BO	35-37	
	Roßkastanie	b	38-40	
	Weißbirke	BO	41-43	
	Sommerlinde, Großblättrige Linde	BO	44-46	
	Schlehdorn, Schwarzdorn, Schlehe	b	47-49	
	Rotbuche	BO	50-52	
	Stieleiche, Sommereiche	BO	53-55	
	Stieleiche, Sommereiche	b	56-58	
	Flieder, Nägelchen, Syringe	b	59-61	
	Weißdorn, Hagedorn, Mehlfäßchen	b	62-64	
	Fichte, Rottanne, Schwarztanne	M	65-67	
	Holunder (Schwarzer Holunder)	b	68-70	
	Wiesenfuchsschwanz	ab	71-73	
	Knaulgras	ab	74-76	

Landwirtschaftliche Kulturpflanzen

	Pflanzenart	Sorte	Phase	Sp.	Eintrittstag
	Hafer		Bst	77-79	
02	Hafer		Au	17-19	
	Hafer		Sch	20-22	
	Hafer		Ä	23-25	
	Sommergerste		Bst	26-28	
	Sommergerste		Au	29-31	
	Sommergerste		Sch	32-34	
	Sommergerste		Ä	35-37	
	Frühkartoffeln, nicht vorgekeimt		Bst	38-40	
	Frühkartoffeln, nicht vorgekeimt		Au	41-43	
	Frühkartoffeln, nicht vorgekeimt		Bg	44-46	
	Spätkartoffeln		Bst	47-49	

Landwirtschaftliche Kulturpflanzen (Fortsetzung)

Pflanzenart	Sorte	Phase	Sp.	Eintrittstag	
Spätkartoffeln		Au	50-52		
Spätkartoffeln		Bg	53-55		
Wintergerste		Sch	56-58		
Wintergerste		Ä	59-61		
Winterroggen		Sch	62-64		
Winterroggen		Ä	65-67		
Winterroggen		ab	68-70		
Winterweizen		Sch	71-73		
Winterweizen		Ä	74-76		
Zuckerrüben		Bst	77-79		
Zuckerrüben		Au	17-19		03
Zuckerrüben		Bg	20-22		
Futterrüben		Au	23-25		
Beginn der Feldarbeiten			26-28		
Beginn des Weidegangs			29-31		
Wiese, Beginn des Heuschnitts			32-34		

Obst

Pflanzenart	Sorte	Phase	Sp.	Eintrittstag
Aprikosen		b	35-37	
Pfirsiche		b	38-40	
Johannisbeeren		b	41-43	
Stachelbeeren		BO	44-46	
Stachelbeeren		b	47-49	
Süßkirschen		b	50-52	
Süßkirschen		e	53-55	
Sauerkirschen		b	56-58	
Sauerkirschen		e	59-61	
Pflaumen		b	62-64	
Pflaumen		e	65-67	
Birnen		b	68-70	
Birnen		e	71-73	
Äpfel		b	74-76	
Äpfel		e	77-79	

gelocht: geprüft:

Fig. 1. Phenological checksheet for spring observations of the German Weather Service, an example of computer-oriented data handling. (Reproduced with permission of the West

Bemerkungen nur über besondere Witterungserscheinungen und deren Einflüsse auf die Pflanzenentwicklung:

Wildwachsende Pflanzen

Pflanzenart	Phase	Eintrittstag
Schwarzerle	b	
Schwarzerle	BO	
Gelber Hartriegel, Kornelkirsche	b	
Buschwindröschen	b	
Löwenzahn	b	
Lärche (Europäische Lärche)	BO	
Lärche (Europäische Lärche)	b	
Esche, Edelesche	b	
Esche, Edelesche	BO	
Winterlinde, Kleinblättrige Linde	BO	
Robinie, Scheinakazie	BO	
Robinie, Scheinakazie	b	
Rotbuche	b	
Fichte, Rottanne, Schwarztanne	b	
Tanne, Edeltanne, Weißtanne	b	
Tanne, Edeltanne, Weißtanne	M	
Kiefer, Föhre	b	
Kiefer, Föhre	M	
Traubenkirsche, Ahlkirsche	b	
Eberesche, Vogelbeere	b	
Heckenrose, Hundsrose, Heiderose	b	
Goldregen, Bohnenstrauch	b	
Falscher Jasmin	b	
Schneebeere	b	

Landwirtschaftliche Kulturpflanzen

Pflanzenart	Sorte	Phase	Eintrittstag
Sommerweizen		Bst	
Sommerweizen		Au	
Sommerweizen		Sch	
Sommerweizen		Ä	
Silo-Mais		Bst	
Silo-Mais		Au	
Silo-Mais		A	
Silo-Mais		ab	

Landwirtschaftliche Kulturpflanzen (Fortsetzung)

Pflanzenart	Sorte	Phase	Eintrittstag
Winterraps		b	
Frühkartoffeln vorgekeimt		Bst	
Frühkartoffeln, vorgekeimt		Au	
Frühkartoffeln, vorgekeimt		Bg	
Frühkartoffeln, vorgekeimt		b	
Frühkartoffeln, nicht vorgekeimt		b	
Futterrüben		Bst	
Futterrüben		Bg	
Grünpflück-Erbsen		Bst	
Grünpflück-Erbsen		Au	
Grünpflück-Erbsen		b	
Grünpflück-Bohnen		Bst	
Grünpflück-Bohnen		Au	
Grünpflück-Bohnen		b	
Tomaten		Auspflanzen	
Tomaten		b	
Weißkohl		Auspflanzen	
Rotklee, Beginn des Heuschnitts		X	
Luzerne, Beginn des Heuschnitts		X	

Obst

Pflanzenart	Sorte	Phase	Eintrittstag
Süßkirschen		ab	
Sauerkirschen		ab	
Pflaumen		ab	
Birnen		ab	
Äpfel		ab	
Erdbeeren		b	
Himbeeren		b	
Brombeeren		b	
Weinreben	X	Austrieb	
Weinreben	X	b	
Weinreben	X	ab	
Weinreben	X	e	

German Weather Service, Department of Agricultural Phenology, Prof. Dr. Seeman, Director, 805 Offenbach, West Germany.)

insect pest outbreaks are possible to aid in the scheduling of control measures (Schnelle, 1967).

Results of the phenological observations are published in monthly agrometeorological reports of the German Weather Service. The meteorological year books contain a section on phenology with maps and substantial data from 10 to 15% of the phenological network stations. The maps in the 1970 volume show isophanes for begin of bloom of snowdrops (*Galanthus nivalis* L.) and apples, as well as begin of bloom and harvest of winter rye for 1967 (Deutscher Wetterdienst, 1970). The map showing beginning of apple bloom is reproduced in Fig. 2. Based on long-term data from the German network stations and those in other European countries, the German Weather Service published a series of maps showing the isophanes of certain phenophases from early spring until fall (Schnelle, 1953, 1965, 1970).

Poland

The importance Poland attaches to phenological observations is demonstrated by the fact that as early as 1945 the Polish National Institute for Hydrology and Meteorology started to rebuild the phenological network destroyed during World War II. The Institute maintains a special meteorology and phenology section for agriculture and forestry.

Sokolowska (1971) discussed the organization of the Polish network in which phenophases of many native plants, fruits, vegetables, and agronomic crops are observed throughout the vegetative season. Zoophenology is represented by a smaller number of observations. The observers receive a calendar with detailed instructions how to make the observations together with a booklet containing 12 report forms on which certain phenophases are illustrated (Panstwowy Instytut Hydrologiczno-Meteorologiczny, 1971a, 1971b).

Based on observations from the Polish phenological network, Kozlowska-Szczesna and Sokolowska (1970) presented 10-year averages together with maps showing isophanes for 19 phenophases of various plant species. These maps characterize the vegetative season throughout Poland from April through October at approximately 10-day intervals.

Spain

In 1943 the climatological section of the National Meteorological Service in Spain organized phenological studies following the example of meteorological services in other countries. The network started with about 300 volunteer observers, farmers, teachers, etc. Since 1968 all phenological observers are directly attached to their corresponding meteorological centers, an arrangement that allows for closer contact and improved organization and functioning of the phenological network. Currently there are 465 observers, about one per 1000 km^2, in the phenological network. This compares with about 1400 meteorological stations at which temperature is recorded. The number of phenological observers was considerably larger prior to the reorganization, but it is expected to increase to a density of one observer per 500 km^2 (Barasoáin, personal communication, 1972).

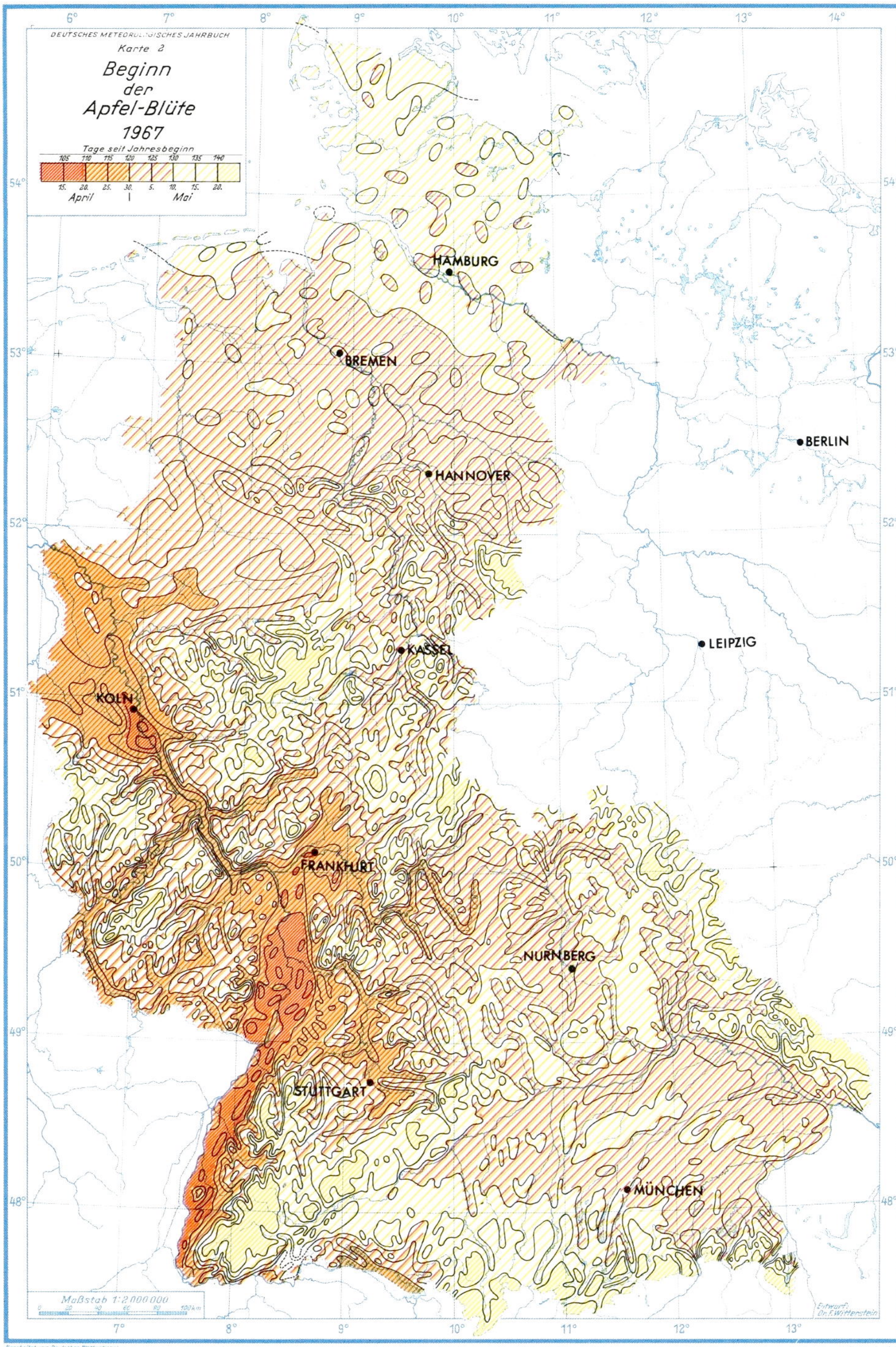

Fig. 2. Isophanes of beginning of bloom of apples in West Germany, 1967. (From Deutsches Meteorologisches Jahrbuch Bundesrepublik 1967; Offenbach a. M., Deutscher Wetterdienst, 1970.)

An annual meteorological-phenological calendar (Servicio Meteorológico Nacional, 1972) serves as a handbook for the cooperators, giving information and detailed instruction on observations and reporting of data. Thirty-two native species, 13 agronomic plants, and 11 fruit crops are listed on which several phenophases are to be observed with additional 27 species for certain regions. Requests for observations on several birds and insects are also included. Observers receive color illustrations of the most important plants on the list (Servicio Meteorológico Nacional, 1943).

The observers mail their reports at the end of each month. Dates of various phenophases together with meteorological data are listed in the monthly climatological bulletins published by the national meteorological service. From the accumulated data phenological maps are constructed which are a biological reflection of past weather conditions. These maps are considered of great importance for Spanish agriculture as they help to determine early and late zones and allow the division of the country into natural agricultural regions.

The 1972 meteorological-phenological calendar contains maps showing isophanes for the 1970–1971 agricultural year, September through August, of the flowering of almonds and apricots, of leaf fall of grapes, and the arrival of swallows. The map showing the bloom date of almonds is reproduced in Fig. 3. A detailed analysis of 6 phenophases for 1943–1954 was made by Ottmar (1960).

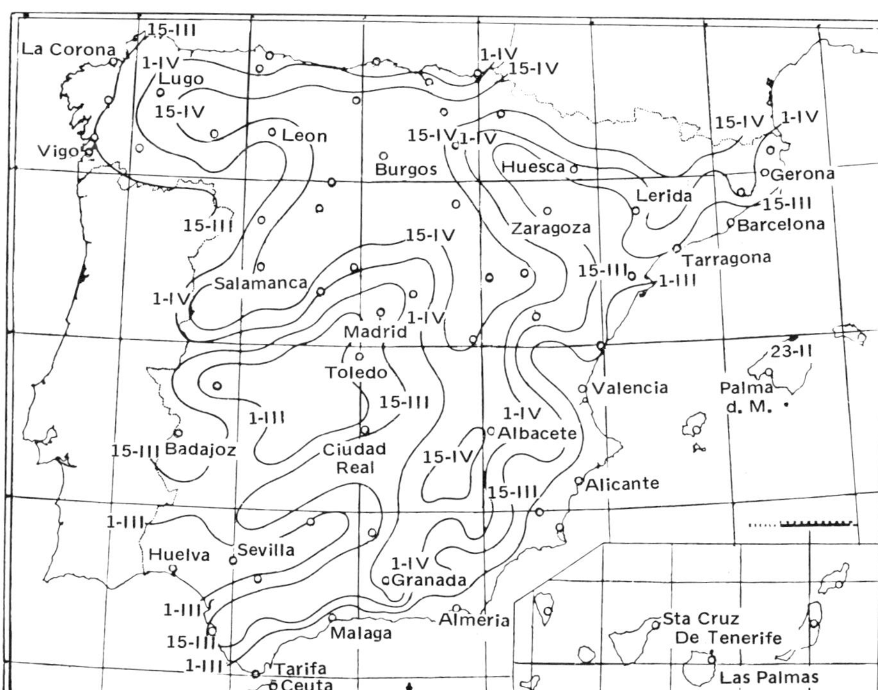

Fig. 3. Isophanes of apricot bloom in Spain, 1971. (From Calendario Meteoro-Fenológico, Madrid, Servicio Meteorológico Nacional, Sección de Climatología, 1972.)

Argentina

In 1960 a specialized fruit phenological service was organized by the National Meteorological Service in the fruit-growing region in the upper valley of the Rio Negro and the Neuquén river. This is an area located at approximately 39° S latitude and between 66°30′ and 68°30′ W longitude, about 120 km long and from 3 to 8 km wide. Sixteen observation sites were established in conjunction with meteorological observation sites. Ten cultivars each of *Vitis vinifera* and *Malus sylvestris*, and eight of *Pyrus communis*, are used for observation. The purpose of the study is to define the exact climatic requirements of these economically important grape, apple, and pear cultivars for the benefit of the fruit-growing industry of this region (Garabatos, 1969).

Norway

In Norway an extensive phenological network has existed since 1928 when it was organized by Henrik Printz. This network is not under the direction of the Meteorological Service, although they are interested in the work which supplements meteorological observations (Wielgolaski, personal communication, 1972). The locations in the network extend from southern Norway about 58° N latitude to near the North Cape at 71° N latitude. Interested local people serve as volunteer observers: gardeners, farmers, teachers, agronomists, foresters, engineers, students, and others.

The list of events to be observed includes phenophases for about 50 species of native and cultivated plants together with the arrival of migratory birds and other events that characterize the local climate such as the ice going out of rivers or lakes or the soil being frost free. The number of observers varied over the years. By 1952 data were available from nearly 600 locations throughout Norway (Lauscher *et al.*, 1955). In 1959 these workers published phenological means for 260 sites in the network which had at least five years of records. They also established 25-year means for date of occurrence of 22 phenophases during spring, summer, and fall on the basis of observations at 37 locations (Lauscher *et al.*, 1959) and showed the deviation of averages of various pentads from these long-term means. Positive and negative deviations usually cancelled each other out when means were combined into 10-year values. They concluded that 10 years of data constitute useful means for the Norwegian network.

5 Current Networks in the United States

Renewed interest in phenology in the United States during the 1950s resulted in the establishment of phenological networks as part of three regional projects of agricultural experiment stations: in the western region in 1957, in the north-central region in 1961, and in the northeastern region in 1965. Statewide networks were organized in Wisconsin in 1959 and in North Carolina in 1970. A network of phenological gardens was developed in Indiana by Purdue University, (see Chapter

2.2). A suggestion was made at the 1957 workshop on Agricultural Meteorology, sponsored by the American Meteorological Society and others, that phenological observations be included in the Weather Service's cooperative observer program, but this proposal has not been implemented (Hacia, 1969).

Western Region

The credit for the first effort in organizing a network for phenological observations over a large geographical area in the United States, since the project of the Smithsonian Institution in the 1850s, goes to the Montana Agricultural Experiment Station, Montana State University, where Caprio took the initiative in 1956. Initial surveys were made in Montana with the help from garden clubs and U.S. National Weather Service climatological observers using the common purple lilac as indicator plant (Caprio, 1957). Subsequently the study became part of a regional research project W-48 in which the agricultural experiment stations in the 11 states of the region cooperate (Arizona, California, Colorado, Idaho, Montana, Nevada, New Mexico, Oregon, Utah, Washington, and Wyoming). The regional climatologist of the Weather Service and county agricultural agents assisted in obtaining about 1000 volunteer cooperators (Caprio, 1966; personal communication, 1972). With the help of a grant from the Environmental Data Service of ESSA (Environmental Science Services Administration), now NOAA (National Oceanic and Atmospheric Administration), an additional network was started in 1967 for which two genetically uniform honeysuckle plants were selected: the Arnold Red honeysuckle (*Lonicera tatarica* cv. Arnold Red) and the Zabeli honeysuckle (*Lonicera korolkowii* var. *zabelii*). While lilacs are observed for leaf opening and several bloom phases, nine phenophases are observed on the honeysuckles from the time the first bud has leaved to the date when at least 95% of the mature berries are shriveled (Caprio *et al.*, 1970). Phenological observations thus extend over a considerable portion of the vegetative season. Four additional phases on fall foliage changes are described, but as of 1971 have not yet been extensively observed.The western network thus uses both established native plants and newly planted, genetically identical plants.

Plantings in Texas were added in 1969 to the W-48 lilac and honeysuckle networks, again with the cooperation of Weather Service climatological observers (Caprio, 1972; personal communication, 1972). The total number of phenological observers in 1972 for these 12 western states is about 2500. The distribution by states is shown in Fig. 4. The western network covers about one-half the area of the 48 contiguous states.

North-Central and Northeast Regions

The common purple lilac study in the western region provided the impetus for the establishment of a network in the north-central region. In 1961 the agricultural experiment stations of this region (Illinois, Indiana, Iowa, Kansas, Michigan, Minnesota, Missouri, Nebraska, North Dakota, Ohio, South Dakota, and Wisconsin) initiated regional research project NC-26. A phenology objective was

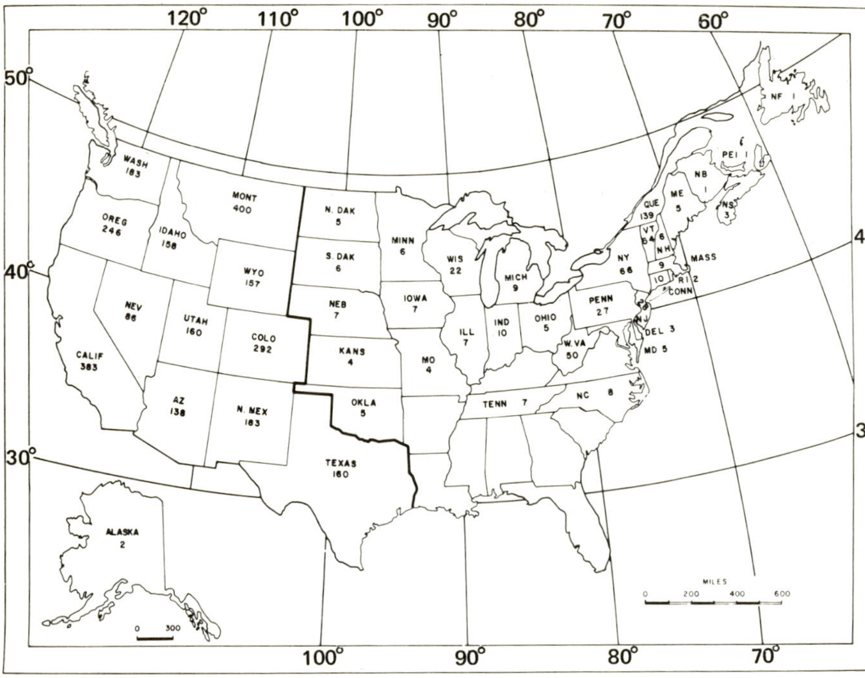

Fig. 4. Phenological networks in the western region of the United States (W-48) and in the north-central and northeastern regions (NE-69). Numbers indicate the number of observation sites in each state or province.

included in the project and observation sites were established in the 12 states of the region. Eventually the network was expanded to include sites in Oklahoma. The lilac cultivar Red Rothomagensis was chosen as indicator and genetically identical plants, vegetatively propagated, were distributed to the cooperators. At that time the plant was considered to be *Syringa persica* but this taxonomic designation has since been questioned. The important fact, however, is that genetically uniform plant material is used throughout the study.

In the 12 states of the northeastern region (Connecticut, Delaware, Maine, Maryland, Massachussetts, New Hampshire, New Jersey, New York, Pennsylvania, Rhode Island, Vermont, and West Virginia) a corresponding regional project of agricultural experiment stations, NE-35, added, in 1964, a phenology objective similar to the NC-26 study using the same Red Rothomagensis lilac cultivar as indicator plant. Tennessee cooperates informally in the study.

Standardized instructions were developed to insure uniformity in planting and maintenance of phenological sites and in recording the four phenophases that are being observed: first leaf, first flowers, full bloom, and end of bloom (Hopp *et al.*, 1969). Most observation sites in this network are located adjacent to stations of cooperative observers of the National Weather Service to facilitate comparisons between phenological data and meteorological observations. Many of the sites were chosen and established with the assistance of NOAA state climatologists.

When regional projects NC-26 and NE-35 terminated in 1970, the phenological studies of these two projects were combined as one of the objectives of a new regional project, NE-69. The program expanded, and Alaska, North Carolina, the Province of Quebec, and the Canadian Atlantic provinces were added to the original cooperating states. Honeysuckle cultivars as used in the W-48 project were added in some sections, especially in Alaska, North Carolina, and Canada. At present 28 agricultural experiment stations in the United States, Laval University, Quebec, and six experiment stations in the Atlantic provinces participate in the combined NE-69 network of nearly 700 sites. (Hopp *et al.*, 1972). The area covered by the NE-69 phenology sites in the United States is about 36% of the land surface of the contiguous United States. The distribution of the observation sites by states and provinces is shown in Fig. 4.

Wisconsin

Wisconsin provides an example of the possibilities for organizing statewide phenological observation networks. In 1959 a group of scientists recognized the need for systematically collecting phenological data for use in research in such fields as botany, entomology, microclimatology, horticulture, and agronomy. This led to the formation of the Wisconsin Phenological Society in which scientists and laymen cooperate. During the first 10 years of its existence the membership of the society fluctuated between 500 and 600.

A variety of phenological events is suggested for recordkeeping on a general data report form from which an observer may select those events that he can observe conveniently near his home and expects to be able to report year after year (Lettau, 1972). The spring 1972 report form lists about 80 events for observation on plants and animals. Several projects were carried out by the society on specific species such as common lilac (Dana *et al.*, 1963), crocus, scilla, and alfalfa. The society was instrumental in the establishment in 1971 of the first phenological garden in the United States, as part of the network of international phenological gardens organized in Europe, in the Arboretum of the University of Wisconsin at Madison. Due to strict import regulations the Madison garden is small in comparison to the European counterparts. It is limited at present to *Quercus robur*, *Fagus sylvatica*, *Sambucus nigra*, and the *Salix* species *glauca*, *smithania*, *aurita*, *viminalis*, and *acutifolia* (Lettau, 1972). In addition, the Wisconsin Phenological Society became affiliated with the NE-69 Red Rothomagensis lilac network.

North Carolina

In 1970 a phenology project was initiated by Helmut Lieth, Department of Botany, University of North Carolina, as part of the US/IBP Eastern Deciduous Forest Biome program. About 300 volunteer cooperators—college and high school students, garden club members, agricultural research stations, and other interested individuals—carry out spring and fall observations. Although this is an average density of about one observer per 400 km^2, the distribution is not very uniform.

The flowering of dogwood (*Cornus florida*) and redbud (*Cercis canadensis*) is used as indicator of the arrival of spring, while leaf coloration of tulip tree (*Liriodendron tulipifera*) and red maple (*Acer rubrum*) is observed to indicate the arrival of fall (Lieth *et al.*, 1970; Reader *et al.*, 1971, 1972).

From the data obtained from this network phenological maps were prepared by hand and also with the SYMAP computer mapping program (Lieth and Radford, 1971; Radford *et al.*, 1971). Computer mapping may become an important tool in the evaluation of data from high-density phenological observation networks (see Chapter 2.4).

6 The Future of Phenological Networks

Details of the existing phenological networks in the United States were tabulated in a report of the US/IBP Phenology Committee (1972). These networks could form the basis for a phenological organization within North America. The possibility of a nationwide coordination of phenological efforts and international cooperation was initially visualized as part of the U.S. contribution to the International Biological Program (U.S. National Committee for the International Biological Program, 1967, 1971). The proposal, however, was not activated on the originally intended broad basis. The funding and personnel policies of the agricultural experiment station projects carry with them the inherent possibility of termination of the phenology projects. Thus the need remains for a unification of the diverse phenological efforts in the United States under some agency adequately funded to assure continuation and expansion of phenological networks, as well as collection and evaluation of local data.

An expansion of the number of indicator plants and of the phenophases to be observed in a nationwide network could provide the basis for seasonal phenological maps as part of an overall phenological atlas of North America. This would also serve to encourage studies using phenological observations on indicator plants to forecast or estimate the occurrence of subsequent phenological events on plants, insects, or other organisms, for application in the agricultural management decision-making process.

Finally, phenological networks can be expected to play an important role in providing additional ground truth for remotely sensed phenological data from earth-orbiting satellites such as *ERTS*–1 (Earth Resources Technology Satellite) which was launched in July 1972 (see Chapter 2.3).

A long-range view envisions international coordination of observations from phenological networks. Long-term phenological norms could be established to parallel meteorological norms, and would serve as one tool for global and local environmental monitoring.

Acknowledgment

The discussions of the national phenological networks would not have been complete without additional communications received from J. A. Barasoaín (Spain), R. A. Bertinotti (Argentina), Emil Chroboczek (Poland), F. E. Wielgolaski (Norway), and F. Witterstein (West Germany). Their contributions are gratefully acknowledged.

References

Abbe, C. 1905. A first report on the relations between climates and crops. *U.S. Dept. Agr. Weather Bureau Bull. 36.* Washington, D.C.: U.S. Govt. Printing Office.

Blair, B. O., J. E. Newman, and J. R. Fenwick. 1974. Phenology gardens in Indiana. In *Phenology and Seasonality Modeling*, p. 45 (H. Lieth, ed.) New York: Springer-Verlag.

Caprio, J. M. 1957. You can help in plant-climate survey. *Montana Farmer-Stockman* 44 (15), 1, 24.

——— 1966. Pattern of plant development in the western United States. *Montana Agr. Exp. Sta. Bull. 607.*

——— 1972. A report of the phenological survey in the western region of the United States, 1971. *Weekly Weather Crop Bull.* 59 (4), 10–11.

———, R. J. Hopp, and J. S. Williams. 1974. Computer mapping in phenological analysis. In *Phenology and Seasonality Modeling.* p. 77. (H. Lieth, ed.) New York: Springer-Verlag.

———, M. D. Magnuson, and H. N. Metcalf. 1970. Instructions for phenological observations of purple common lilac and red berry honeysuckle. *Montana Agr. Exp. Sta. Circ. 250.*

Dana, M. N., J. Zimmerman, and K. Lettau. 1963. Lilac blossoming and Wisconsin weather. *Wisconsin Acad. Rev.* Winter, 33–36.

Deutscher Wetterdienst. 1962. *Pflanzentafeln für den Phänologischen Dienst.* Germany: Offenbach a.M.

——— 1970. *Deutsches Meteorologisches Jahrbuch der Bundesrepublik 1967.* Germany: Offenbach a.M.

Garabatos, M. 1969. Servicio fenológico frutal en el Alto Valle del Rio Negro y Neuquén. *Serv. Meteorol. Nacional Publ. D No. 6.*

Hacia, H. 1969. Phenology. *Weekly Weather Crop Bull.* 56 (31), 12–13.

Hodson, A. C. 1971. Minnesota springs—A thirty year record. *Minnesota Horticulturist* 99 (4), 56–58.

Hoffmann, H. 1881. Vergleichende phänologische Karte von Mitteleuropa. *Petermanns Geog. Mitt.* 27, 19–26.

———, and E. Ihne. 1882. Aufruf. *Bot. Centralbl. 9*, 287.

Hopkins, A. D. 1918. Periodic events and natural laws as guides to agricultural research and practice. *U.S. Dept. Agr. Monthly Weather Rev. Suppl.* 9.

——— 1938. Bioclimatics—A science of life and climate relations. *U.S. Dept. Agr. Misc. Publ. 280.*

Hopp, R. J., K. E. Varney, and R. E. Lautzenheiser. 1964. Late spring and early fall low temperatures in Vermont. *Vermont Agr. Exp. Sta. Bull. 639.*

———, M. T. Vittum, and N. L. Canfield. 1969. Instructions for phenological observations: Persian lilac. *Vermont Agr. Exp. Sta. Pamphlet 36.*

———, and B. E. Dethier. 1972. Regional phenological studies with Persian lilac (*Syringa persica*). *New York State Agr. Exp. Sta.* (*Geneva*) *Food Life Sci. Bull. 17.*

Hough, F. B. 1864. Observations upon periodical phenomena in plants and animals from 1851 to 1859, with tables of the dates of opening and closing of lakes, rivers, harbors, etc. In Results of Meteorological Observations, Made Under the Direction of the United States Patent Office and the Smithsonian Institution, from the year 1854 to 1859, Inclusive. *Rept. of the Commissioner of Patents Vol. 2, Part 1. Exec. Doc. 55. 36th Congress, 1st Session.* Washington, D.C.: U.S. Govt. Printing Office.

Hulbert, L. C. 1963. Gates' phenological records of 132 plants at Manhattan, Kansas, 1926–1955. *Trans. Kansas Acad. Sci. 66*, 82–106.

Ihne, E. 1883–1941. Phänologische Mitteilungen 1–59. Arb. Landwirtschaftskammer Hessen; Selbstverlag, Darmstadt, Germany; and others.

——— 1884. Geschichte der pflanzenphänologischen Beobachtungen in Europa nebst Verzeichnis der Schriften, in welchen sie niedergelegt sind. In *Beiträge zur Phänologie.* Giessen, Germany: J. Rickersche Buchh.

——— 1885. Karte der Aufblühzeit von Syringa vulgaris in Europa. *Bot. Centralbl.* 21, 85–88, 116–121, 150–155.

——— 1905. Phänologische Karte des Frühlingseinzuges in Mitteleuropa. *Petermanns Geog. Mitt.* 51, 96–108.

Kozlowska-Szczesna, T., and J. Sokolowska. 1970. Materialy do Klimatologii Polski: Fazy fenologiczne wybranych roslin w Polsce. *Inst. Geog. Polskiej Akad. Nauk Warsaw Dok. Geog.* 2/70, 31–91.

Lauscher, A., F. Lauscher, and H. Printz. 1955. Die Phänologie Norwegens. I. Allgemeine Übersicht. *Skrifter Norske Videnskap-Akad. Oslo. Mat.-Naturv. Klasse* No. 1.

——— 1959. Die Phänologie Norwegens. II. Phänologische Mittelwerte für 260 Orte. *Skrifter Norske Videnskaps-Akad. Oslo. Mat.-Nat. Klasse No. 1.*

Leopold, A., and S. E. Jones. 1947. A phenological record for Sauk and Dane counties, Wisconsin 1935–1945. *Ecol. Monogr.* 17, 81–122.

Lettau, K. 1972. The Wisconsin Phenological Society's first ten years. *Wisconsin Acad. Rev.* 19(2).

Lieth, H. 1971. The phenological viewpoint in productivity studies. In *Productivity of Forest Ecosystems.* Proc. Brussels Symp. 1969, 71–84. UNESCO.

———, and J. S. Radford. 1971. Phenology, resource management, and synagraphic computer mapping. *BioScience* 21, 62–70.

———, and D. Courtney. 1970. Plotting the seasons. *Wildlife North Carolina*, Sept. 14–15.

Lindsey, A. A., and J. E. Newman. 1956. Use of official weather data in spring time—Temperature analysis of an Indiana phenological record. *Ecology* 37, 812–823.

Linnaeus, C. 1751. *Philosophia Botanica.* Stockholm: Kiesewetter.

Ottmar, D. 1960. Beiträge zur Phänologie Spaniens auf Grund der phänologischen Beobachtungen des spanischen Wetterdienstes 1943–1954. *Ber. Deutsch. Wetterdienst.* 60.

Panstwowy Instytut Hydrologiczno-Meteorologiczny. 1971a. *Kalendarz observatora fenologicznego na rok 1972.* Warsaw.

——— 1971b. *Arkusze do notowan spostrzezen fenologicznych.* Warsaw.

Pfister, C. 1972. Phänologische Beobachtungen in der Schweiz der Aufklärung. *Informationen Beiträge Klimaforschung.* 8, 15–30.

Quetelet, A. 1849. Sur le climat de la Belgique. In *Phénomènes périodiques des Plantes.* Chap. IV. Brussels, Belgium. Annales de l'Observatoire.

Radford, J. S., S. M. Rogers, and H. Lieth. 1971. Plotting the seasons: Fall 1970/Spring 1971. *Wildlife North Carolina.* Feb. 10–11.

Reader, R., H. Lieth, and J. Radford. 1971. Looking for spring. *Wildlife North Carolina.* Sep. 16–18.

Reader, R., D. Rumpp, and H. Lieth. 1972. The arrival of fall, 1971. *Wildlife North Carolina.* Feb. 14–15.

Schnelle, F. 1953. Beiträge zur Phänologie Deutschlands. III. 6 Mittelwertskarten (1936 bis 1944) Vorfrühling bis Herbst. *Ber. Deutsch. Wetterdienst.* 1.

——— 1955. Pflanzen-Phänologie. *Probleme der Bioklimatologie.* Vol. 3. Geest & Portig, Leipzig, Germany: Akademische Verlagsgesellschaft.

——— 1960. Probleme der Phänologie im Lichte der Arbeiten Egon Ihnes. *Ber. Oberhess. Gesellsch. Natur- und Heilkunde Giessen. N. F.* 30, 53–66.

——— 1965. Beiträge zur Phänologie Europas. I. 5 Mittelwertskarten, Erstfrühling bis Herbst. *Ber. Deutsch. Wetterdienst.* 101.

——— 1966. Abriss einer Methodik der Phänologie. *Mitt. Deutsch. Wetterdienst.* 38.

——— 1967. Der agrarmeteorologische Dienst des Deutschen Wetterdienstes der Bundesrepublik Deutschland. *Agr. Meteorol.* 4, 1–10.

——— 1970. Beiträge zur Phänologie Europas. II. 4 Mittelwertskarten: Gesamtvegetationszeit und 3 Vegetationsabschnitte. *Ber. Deutsch. Wetterdienst.* 118.

———, and E. Volkert. 1957. Vorschläge zur Einrichtung "Internationaler Phänologischer Gärten" als Stationen eines Grundnetzes für internationale phänologische Beobachtungen. *Meteorol. Rundschau* 10, 130–133.

——— 1964. Internationale phänologische Gärten: Stationen eines Grundnetzes für internationale phänologische Beobachtungen. *Agr. Meteorol.* 1, 22–29.

——— 1974. International Phenological Gardens in Europe, basic network for international phenological gardens. In *Phenology and Seasonality Modeling* p. 383. (H. Lieth, ed.). New York: Springer-Verlag.

Servicio Meteorológico Nacional. 1943. Atlas de plantas para las observaciones fenológicas. *Publ. Ser. C No.* 9. Madrid.

——— 1972. *Calendario Meteoro-Fenológico.* Madrid: Sección de Climatología.

Smith, J. W. 1915. Phenological dates and meteorological data recorded by Thomas Mikesell between 1873 and 1912 at Wauseon, Ohio. *Monthly Weather Rev. Suppl. 2*, 23–93.

Societas Meteorológica Palatina. 1781–1792. *Ephemerides Societatis Meteorologicae Palatinae.* Mannheim, Germany.

Sokolowska, J. 1971. Stan obecny sieci fenologicznej i opracowan w zakresie fenologii w PIHM. *Zeszyty Problemowe Postepow Nauk Rolniczych* 120, 113–119. English summary: Present state of phenological network and of studies in phenology in National Institute for Hydrology and Meteorology in Poland.

U.S. Department of Agriculture, Weather Bureau. 1905. Phenological record, 1905. *Climate and Crops, Indiana Sect.* 6–8.

——— 1909. Phenological records for 1908. *Climatological Rep. Indiana Sect.* 4.

US/IBP Phenology Committee (N. R. French, Chairman). 1972. *Report, July 1972.* 54 pp. Austin, Tex.: US/IBP Environmental Coordinating Office.

U.S. National Committee for the International Biological Program. 1967. *Report No. 3 Part 1.* Research Studies Constituting the U.S. Contribution to the International Biological Program. Washington, D.C.: National Academy of Sciences.

——— 1971. *Report No. 4.* Research Programs Constituting U.S. Participation in the International Biological Program. Washington, D.C.: National Academy of Sciences.

Wang, J. Y. 1967. *Agricultural Meteorology.* San Jose, Calif.: Agricultural Weather Information Service.

Witterstein, F. 1960. Bedeutung und Aufgaben des phänologischen Dienstes. *Mitt. Deutsch. Wetterdienst.* 19.

——— 1968. *Anleitung für den phänologischen Beobachter.* Offenbach a.M., Germany: Deutscher Wetterdienst.

——— 1971. Die Phänologie. Informationen für den Fachdienst. Beil. *Mitt. Deutsch. Wetterdienst.* 2 (2), 1–16.

2.2 Phenology Gardens in Indiana[1]

Byron O. Blair, James E. Newman and Jack R. Fenwick

Agronomy Department
Purdue University
Lafayette, Indiana 47907, USA

1 Introduction

Plant phenology observations taken over a long history for various purposes share a common interest in evaluating seasonal influences of weather on different species. Because comprehensive review of literature has been cited by Caprio (1966) and Caprio *et al.* (1970), no such effort will be made here.

The present phenology garden system in Indiana is an outgrowth of a north-central regional research effort in climatology developed by the NC–26 Agricultural Experiment Station committee. The initial study, started in 1961 by W. L. Colville (1961), sought to correlate the phenological events of a plant species, from the same genetic source, with climatic changes from season to season over the 12-state area, and when and where possible to relate and predict from these events relationship to other biological phenomena of economic significance. Four phenology sites, of the 51 in the north-central region, were located in Indiana. The north-central region consists of 12 states from North Dakota south to Kansas and east to Ohio.

Persian lilacs (*Syringa chinensis—Syringa persica* × *vulgaris*) were transplanted in October 1960 and observations started in 1961 and 1962 at the four locations in Indiana. Caprio used this species in the western U.S. as it appeared to be satisfactory. This choice also had the advantage of enlarging the geographical and climatic area for obtaining information on the same species. The plant material was supplied by the Plumfield Nurseries, Freemont, Nebraska, and all cuttings were started from the same parent plant.

Phenology readings on first leaf emergence were started the season following transplanting and during the following season floral observations were also made.

[1] Journal Paper No. 4872, Purdue University, Agricultural Experiment Station. This research was supported by Purdue University AES Hatch Regional Research Funds under Project 1388, entitled "Phenological Response of Plants in Environmental Phenomena" and under Project 1685, entitled "The Climatic Resources of the North Central Region."

KEYWORDS: Phenological gardens; Indiana; flowering dates; planting; maintenance; agricultural meteorology.

Locations were at Wanatah, northwest; Lafayette, west-central; Farmland, east-central; and Dubois, south-central Indiana. All locations were also official weather observation stations.

2 Developing the Indiana Phenological Garden System

Prior to initiating the regional phenology study in 1960, J. E. Newman at Purdue had been responsible for the variety crop testing program on small grains. Test sites were located on various soil types at different elevations representing variations in climatic conditions over the state. Observations from the first four years of the regional study of lilac data confirmed these climatic variations and prompted interest in expanding into a program which would more accurately evaluate climatic differences over the state. This new program included the expansion of both a weather observation network and additional site locations, as well as adding other plant species native to the area which would provide phenological data for other seasons.

An agricultural forecasting network system in cooperation with the National Oceanics Atmospheric Administration (NOAA), National Weather Service (NWS) proposed in 1960, was founded in 1965. This program required the development of several well-instrumented weather stations in Indiana, offering an opportunity to establish additional phenological observation locations at these sites.

Developing and expanding a phenology garden network, which includes the addition of new species each from a common genetic source, pose several problems: longevity of the program, researchers' interests, locating useful sites, finding satisfactory observers, obtaining support and cooperation from other agencies, choosing suitable plant species, mechanically implementing the expansion, and maintenance after establishment. Expansion was undertaken with the intent that it could be planned as a sequence of five-year renewable programs which would continue for as long as 20 years. A program with such longevity immediately requires self-evaluation by the project leader, especially since there is a good possibility that his efforts will become the reward (and burden) of those who follow.

Locating sites requires consideration of native vegetation, elevation, soil types, economic crop distributions, and land use patterns.

To find observers willing to cooperate, assistance from NOAA, NWS state climatologist, and the agricultural meteorologist was of much help, as their records gave clues regarding interest and dependability of persons who were already volunteer weather observers. If an observer is near retirement age, chances for an equally competent replacement after the phenology project has started are often not good.

The cooperative support from other organizations can be of great help, but cooperation from the NWS is essential. The USDA-State University Cooperative Extension Service, with its ongoing interest in entomology and pathology, and other organizations having a direct or indirect interest in weather patterns, can provide valuable assistance. Other professional assistance can often be found at state or federal forestry stations.

The Indiana system was expanded with the intent of finding two perennial species which flower during each month of the growing season. The location of the original, and the added sites since 1964, are shown in Fig. 3. There are as many soil types as gardens. Differences in physical and chemical characteristics of the soils influence the root environment, which, in turn can affect phenological behavior of different species among locations. Complete physical and chemical analyses of three soil cores per garden from all locations were completed except from those established since 1969.

The following examples illustrate the description and variation between locations from north to south in Indiana.

Pinney Purdue Agricultural Center. Established in 1960, Wanatah, Porter County, 41.26° latitude, 86.56° longitude, 221 m elevation on a sandy loam profile of Oshtemo series; a member of the coarse loamy, mixed mesic family, Typic Hapludalf, Alfisols.

Johnson Horticulture Farm. In the southwest area of Indiana, Owensville, Gibson County, established in 1964, 38°16′ latitude, 87°45′ longitude, 135 m elevation, a member of the sandy mixed mesic family of Alfic Udipsamments, Entisol.

The following points are essential when choosing species suitable for establishing a garden planting near an official weather station:

1. Long-lived perennials are essential.
2. Species capable of providing cuttings, rootstocks, or clones must be easily and quickly propagated from the selected parent plant.
3. Species should easily be established over a broad spectrum of soil and climatic conditions.
4. Phenological observation indices should be produced soon after transplanting to facilitate useful data collection.
5. Floral or phenological indices must be meaningful.
6. Observations should be easily noted and readily identified by all observers.
7. Either a naturally low growing or a species which can be maintained at less than 1.5 m should be used.
8. Species should be attractive and of interest to the observer.
9. Species should not require any special care for establishment and survival.
10. Species should have natural resistance to insect and pathogen attack.

Many interesting and colorful species often desired by the naturalist or an amateur observer are ruled out by these specifications.

Individual plants which best meet all these criteria were selected for initiating nursery stock. A minimum propagation of 40 plants per species was started for transplant to the nursery. The species selected and their flowering dates are listed in Table 1.

Materials started from cuttings were usually taken from green wood tissue in late spring. Use of a root hormone usually aids in rooting of fresh cuttings. Shoots were started in flats of wet sand. After two to three weeks, when rooting was evident, plants were transplanted in potting soil for additional growth prior to retransplanting. Easily started cuttings can be transplanted from the sand flats

Table 1. Species selected for testing for phenology garden expansion and their flowering date at Lafayette, Indiana.

Genus	Species	Flowering date
Forsythia	× *intermedia* Zabel 'Spectabilis'	April 1–April 15
Syringa	*chinensis* (*S. persica* × *vulgaris*)	April 20–May 1
Spirea	sp.	April 25–May 10
Paeonia	*suffruticosa* Andr.	May 15–May 30
Dactylis	*glomerata* L.	May 25–June 10
Hemerocallis	*fulva* L.	June 10–July 10
Hibiscus	*syriacus* L.	June 5–August 15
Saponaria	*officinalis* L.	June 5–Sept. 15
Andropogon	*gerardii* Vitman	July 20–August 10
Sorghastrum	*nutans* (L.) Nash	Aug. 10–Aug. 30
Aster	*azureus* Lindl.	Sept. 1–Sept. 15
Solidago	*canadensis* L.	Aug. 25–Sept. 15
Lonicera	*tatarica* L. Arnold Red	May 1–Sept. 30
Lonicera	*korolkowii* Stapf. var. *zabelii Rehd.*	May 1–Sept. 30

directly to a nursery in spring where facilities for watering during the first two weeks after transplanting are available. Less vigorous species will require extra care and usually are more easily established when started in the fall. Some species such as *Spirea* will root equally well from dry wood shoots in the fall, but must be maintained over the winter in the greenhouse.

Rootstocks, such as peony, aster, and solidago, start successfully by separation of the parent plant. Direct planting of these species to the nursery is satisfactory in the fall or as soon as spring growth starts, provided the parent plant is sufficiently large to provide the needed plants. Species started from clones can be managed in a manner similar to rootstock materials.

Transplanting from the nursery to the garden was found most successful in late October or early November after plants were dormant. Spacing within the phenological gardens was most satisfactory on 3 × 3 m centers which allowed for maintenance of sod between plants with large mowing equipment. All plantings in Indiana were made in bluegrass sod, which is maintained by the official weather station at the nearby weather installations. Low growing species were located nearest the instrumentation to avoid micrometeorological influences. Each species is represented by three plants. Examples of Indiana Gardens are shown in Fig. 1 and 2.

Successful establishment of plants in the gardens requires one to two years depending upon the species. During this period the project leader should spend time working with the official observer in an educational program to familiarize him with all species, and the desired phenophase observations. Some plant loss can be expected during establishment. A nursery stock should be maintained for replacing plants not successfully established.

Fig. 1. The Purdue University Agronomy Farm, Weather Station, and Phenological Garden.

3 Phenological Observations

Most species are ready for observation after the second year. A visual aids manual for the observers is helpful for obtaining uniform data on all species (Hopp *et al.*, 1969). A manual with colored photo illustrations of all phenophases of each species to be observed was prepared for each observer in Indiana.

Fig. 2. The Davis Purdue Agricultural Center Weather Station and Phenological Garden.

Phenological events observed are recorded on a self-addressed postal card for each species. The observers are asked to mail the cards to the project leader as soon as observations are completed on a species. This permits posting and keeping data current as the season progresses. Two master mimeographed sheets listing all species and observations to be made are also provided. At the end of the season one copy of the master sheet is mailed to the project leader and the second retained by the observer for his record.

4 Maintenance

After establishment of the gardens a maintenance program was found essential. All gardens are visited twice each season and a third time if needed. The first visit is in late May to inspect for winter injury and insect or pathogen attack. Lilac and forsythia in some areas are subject to oyster shell insect scale (*Lepidosaphes ulmi* L.); if present, they are sprayed with recommended controls. Spring flowering species are pruned back to 1.5 m after observations are complete. Summer growth on these species is then sufficient for the next season's observations.

The second visit is made in late July or early August. Any plants with scale or pests are again sprayed. Weed competition is eliminated and an area around each plant 30 cm from the base is tilled to eliminate sod competition until plants are successfully established. If spring transplants were attempted and not successful, a third trip is made in late October for replacing these plants. Contact with the observer at each visit is very important to answer any questions which may have arisen and to reassure him that his volunteer work is appreciated and essential to the total program.

5 Meteorological Observations

Daily weather data are made available through the National Weather Service records. Daily minimum meteorological observations at all garden locations include the following: maximum-minimum temperature and humidity at screen height (150 cm), precipitation, and soil temperature at 10 cm under sod and bare soil. Most sites have had the meteorological measurements expanded since the gardens were established to include free water evapotation, water temperature, total run of wind over Class A pan, precipitation intensity and duration, and bare soil maximum and minimum temperatures at 2.5, 5, 10, 20, 40, and 100 cm. Wind velocity and direction at 10 m, incoming radiation, back radiation, total Langley units and dew amount and duration are also measured at the four major stations. This information is available through the Agricultural Climatology Office.

This information makes possible the development of a daily heat budget, determination of degree-days, and predicted soil moisture levels where soil moistures are not measured, as well as evaluation of the independent and multiple influences of these parameters on phenological events of each species. These facilities and phenology data facilitated the evaluation of 55 parameters correlated with the phenological pattern of spirea, orchard grass, big blue stem, and fall aster in 1968 and 1969 by Benacchio (see Benacchio and Blair, 1972).

6 Results and Discussion

To date, 14 species have been tried in the phenological garden program. Four have not been continued for various reasons. Peony was difficult to propagate in the quantity needed; and did not establish and flower consistently under the extremes of soil variation in Indiana. Hibiscus was subject to winter kill unless mulched, also, rabbits peeled the bark during heavy snow cover, killing this species at several locations. Aster produced abundant early growth in the spring at southern locations, sufficient to set flower buds and flower in late May and June as well as in early September. This confounded the results and confused the observers.

Solidago was added to replace aster, and in 1971 two species of *Lonicera* were introduced. These species, in time, may replace several of the species now in use. Not only early growth and flowering can be observed, but also fruit development, ripening, and maturity dates can be recorded, thus eliminating need for the numbers of species now used to measure mid- and late seasonal environmental differences.

Data collected on lilac from the four original stations established in 1961 and summarized by Hopp (1969) are shown in Fig. 3. Date of first bloom for this species varies 20 days from April 20 to May 10 from south to north over Indiana (Table 2). Leaf appearance may vary more than appearance of first flower, with less variation in the south than in central or northern Indiana. Central locations show the most uniformity of first flower appearance with only 10 days variation in May. The accumulation of heat units as represented by growing degree-days (GDD) varied from 1300 to 800 south to north in 1971 over a period of March 1 to June 19, as illustrated (solid lines in Fig. 3). The method used in computing the heat units is described by Barger (1969).

Data from five years of flowering observations on fall aster (*Aster azureus*) show that in some years the number of days to appearance of first leaf are similar to lilac yet in others very dissimilar. These variations may be explained by differences in growth habit. The first leaf of lilac arises from secondary tissue, whereas aster leaves develop each year from root crown buds near the soil surface. Soil and air temperatures from season to season vary considerably. Further study of these observations and existing soil physical data measurements are needed for a clearer understanding of their contribution to seasonal phenophase differences.

The appearance of first flower in the fall aster varies from 10 days to two weeks from south to north. This suggests that the longer day lengths of midsummer in the north and early autumn may play a role in heat unit accumulation sufficiently to narrow this range in flowering. Accumulation of heat in the soil may also accommodate more uniform flowering. To date, data are too meager for interpretation to be other than speculative.

At the Wanatah location, rather striking differences in the date of first flower of the fall aster was recorded between the upland and the muck stations. Differences in date of appearance of first flower are as great between these two stations as from south to north over the state. The date of appearance of first leaf appears to have little bearing on the appearance of first flower in the fall. The temperature differences of as much as 8° C between these stations occur in mid-summer, when the low drops to 10° C in mid-August at the muck station. Midwinter temperatures between these two stations on the same morning have been as great as 15° C. These

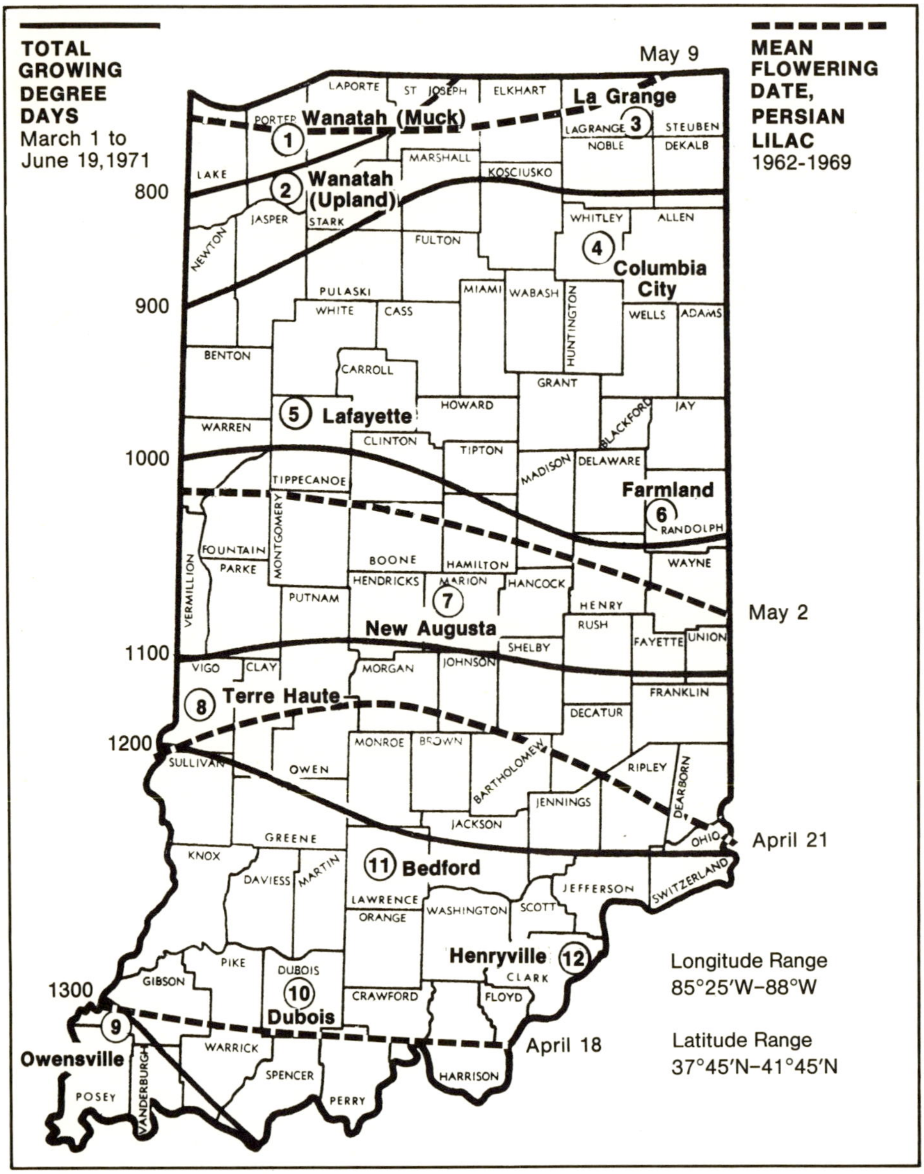

Fig. 3. Map of Indiana showing locations of phenological gardens. Also shown are the mean flowering dates of Persian lilac in Indiana 1962–1969 (*dashed line*), and total growing degree-days from March 1 to June 19, 1971 (*solid line*).

Table 2. Summary of Persian lilac phenology at four Indiana gardens, 1962–1972, day of appearance of first leaf and first flower

	Locations and Phenophase							
	Dubois		Lafayette		Farmland		Wanatah	
Year	First leaf	First flower	First leaf	First flower	First leaf	First flower	First leaf	First flower
1962	4/1	4/28	4/23	5/4	4/18	5/16	4/21	5/9
1963	4/2	4/16	4/1	4/27	4/1	5/6	4/2	5/6
1964	3/11	4/20	4/15	5/5	4/15	5/3	4/16	5/5
1965	3/25	4/24	4/2	5/5	4/7	5/7	4/22	5/9
1966	3/15	4/27	3/23	5/6	3/23	5/11	3/25	5/16
1967	3/13	4/10	4/10	4/27	3/28	5/1	4/3	5/17
1968	3/28	4/18	3/29	4/29	3/26	5/12	3/31	5/1
1969	3/21	4/23	4/14	5/3	4/15	5/6	4/9	5/6
1970	—	—	4/23	5/7	4/21	5/1	4/23	5/10
1971	3/15	4/24	4/19	5/14	4/19	5/15	4/26	5/18
1972	3/25	4/28	4/17	5/7	4/17	5/6	4/30	5/14
Mean	3/25	4/18	4/9	5/3	4/7	5/7	4/10	5/9
Earliest	3/11	4/10	3/23	4/27	3/23	5/1	3/25	5/1
Latest	4/2	4/28	4/23	5/7	4/21	5/11	4/23	5/17
Extreme days	22	18	31	10	29	10	29	16

Accumulative difference in mean days from first leaf and to first flower between locations

Leaf → 15 → −2 → 3

Flower → 15 → 4 → 2

Mean total days

First Leaf → 16 days

First flower → 21 days

temperature differences suppress warm season annuals, and encourage the generation of new growth of cool season perennials.

Responses to phenological plantings at this location have guided some cultural changes in crop production practices. Warm season annuals are being replaced by a mixture of both cool and warm season perennials (grass and legume) which extend the growing season by 40 days. Culture is shifting from cash cropping to an animal grazing program.

Southern Indiana can support a double cropping system by using a combination of cool season spring annuals followed by warm season annuals. But the colder areas in northern Indiana are not well adapted to such a cropping program for annuals because of the short growing season.

Further analysis of accumulating phenological observations and physical measurements may provide a better understanding of how species now under observation, and others, utilize their environment.

References

Barger, C. L. 1969. Total growing degree days. *Weekly Weather Crop Bul.* 56 (18), 10.

Benacchio, S. S., and B. O. Blair. 1972. A new approach to phenological research relationships between environmental factors and days to the appearance of the first leaf in four perennial species. *Agron. J.* 64, 297–302.

Caprio, J. M. 1966. Patterns of plant development in the western United States. *Montana Agr. Exp. Sta. Bull. 607.*

———, M. O. Magnuson, and H. N. Metcalf. 1970. Instructions for phenological observations of purple common lilac and red berry honeysuckle. *Montana Agr. Exp. Sta. Circ. 250.*

Colville, W. L. 1961. *USDA-AES Annual Report of Cooperative Regional Projects* (*NC-26*). Supported by Allotments of the Regional Research Fund Hatch Act, as Amended August 11, 1955.

Hopp, R. J. 1969. Persian lilac phenology. *Progr. Rep.*, USDA-AES Regional Project NE-35.

———, M. T. Vittum, and N. L. Canfield. 1969. Instructions for phenological observations: Persian lilac. *Vermont Agr. Exp. Sta. Pamphlet 36.*

2.3 Phenology and Remote Sensing

Stanley A. Morain

University of Kansas Center for Research, Inc.
Remote Sensing Laboratory, and Department of Geography
Lawrence, Kansas 66045, USA

1 Introduction

The more we extend the frontiers of remote sensing for earth resources, the more we begin to realize that, despite the wide array of sophisticated remote sensing devices, one of our major problems is still the unraveling of complex spectral relationships within the biological skin of the earth. Until we learn to detect, identify, and categorize vegetational patterns with remote sensors, we will not have fully achieved the promise of remote sensing.

Two concepts help satisfy this need. The *spectral concept* states that each entity on the landscape has a spectral "fingerprint" that can be uniquely isolated by partitioning and recombining portions of the electromagnetic spectrum. The *temporal concept*, on the other hand, states that even though some spectral fingerprints may be unique at one point in time, remote sensing can be improved if we know the object's spectral characteristics through time. This second concept commands our attention in studies of phenology.

Remote sensing offers a valuable stimulus for advancing our interest in temporal change. The data can be provided as frequently as needed over any particular study area. Given sequential data, we may now capitalize on three kinds of phenological variations: that associated with boundary mapping, as is essentially the case with the advance of the so-called green wave; point identifications relating primarily to taxonomic mapping; and area identifications, as, for example, growth stages in agricultural crops. These three variations are discussed further in the section on Applications in Phenology.

KEYWORDS: Multispectral sensing; remote sensing; temporal sensing; microdensitometer; photo interpretation; image interpretation; space technology; ERTS imagery; aerial reconnaissance; biogeography; agricultural sensing; natural resources; vegetation mapping; phytocenology; plant ecology; stress; crops.

2 Sensor Review

All natural phenomena absorb, reflect, and emit energy at distinctive wavelengths, both within and outside the visible region of the electromagnetic spectrum (Fig. 1). This is true for plant species and communities as well as for other natural bodies. Depending upon the wavelengths used, different kinds of matter and energy interactions are involved. For example, energy in the ultraviolet (UV) region is

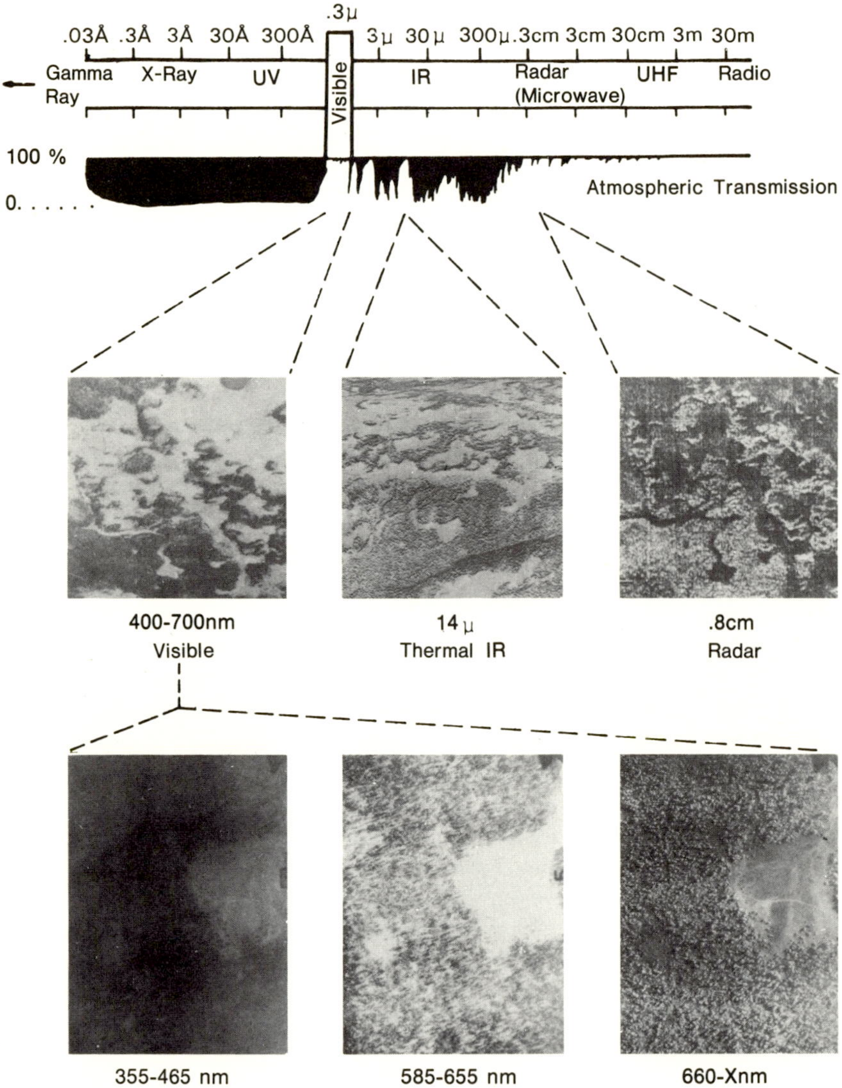

Fig. 1. Percent atmospheric transmission from gamma to radio wavelengths. Major remote sensing windows are located in the visible, near-infrared, parts of the thermal infrared, and radar regions. Comparative imagery from narrow bands within these windows is also shown (redrawn from Colwell, 1963).

related to electron shifts (e.g., fluorescence), whereas the radar wavelengths are sensitive to molecular rotations. A more complete discussion of these relationships is given in Colwell (1963) and Parker and Wolff (1966). Two basic questions arise from this premise: (1) Which portions of the spectrum are best suited to obtain given kinds of vegetational information (see Vinogradova, 1955); and (2) once obtained, can we be certain that the information is unique and specific?

The following discussion will be confined to a review of sensors operating between about 0.4 μm or 400 nm (blue light) and 3 cm (x band radar). Much below 400 nm, atmospheric attenuation is so severe that useful imagery is difficult to obtain. Wavelengths longer than 3 cm will not be considered because imagery is simply not available. The types of airborne sensors discussed will also be confined to those less widely known and used; namely, multiband cameras, multispectral scanners, thermal infrared imagers, and side-looking radar imagers. Panchromatic and color photography is amply and widely treated in Smith (1968), Heller (1969), Sorem (1967), and Colwell (1960). As will be evident, however, most of the remote sensing applications in phenology have so far utilized photographic sensors.

Multiband Photography

Multiband photography relies primarily on the signature concept, but is potentially far more useful when used in a time sequential framework. The technique employs film-filter combinations to achieve spectral partitioning, as shown at the bottom of Fig. 1. The number of film types and filters is constantly increasing, and since each gives a range of usable results under different exposure and development situations, this approach to mapping seasonal change has considerable versatility. Even unlikely film-filter combinations have been found, sometimes by accident, to be useful for very specific tasks. For example, in a recent aerial survey of the California coast by NASA, the camera crew combined Ektachrome (SO-397) film with a Wratten 25A filter and overexposed by one full stop. In reviewing the results, they discovered an exceptional ability to penetrate water and to discern shallow submarine details.

Multiband photography is usually obtained from low flying aircraft; hence the images are usually large scale (1:3000–1:20,000). For many routine forest and range mapping requirements, where the final objectives are for local timber typing, volume estimates, or estimates of animal carrying capacity, large-scale photographs that maximize tonal distinctions are best. The multiband approach should prove equally useful in special-purpose mapping for identifying phenological change or physiological stress. At present, however, our knowledge of how to interpret these more subtle conditions on imagery is limited. Images are obtained directly through a camera or lens cluster designed so that each lens simultaneously photographs exactly the same area. By using different filters on each lens, narrow slices of the visible spectrum can be recorded while retaining a metric mapping base. Moreover, camera systems have finer resolution than any of the other sensors, so that at large and medium scales image texture and scene context are preserved as aids for interpretation.

The greatest problem in the use of multiband cameras for phenological change

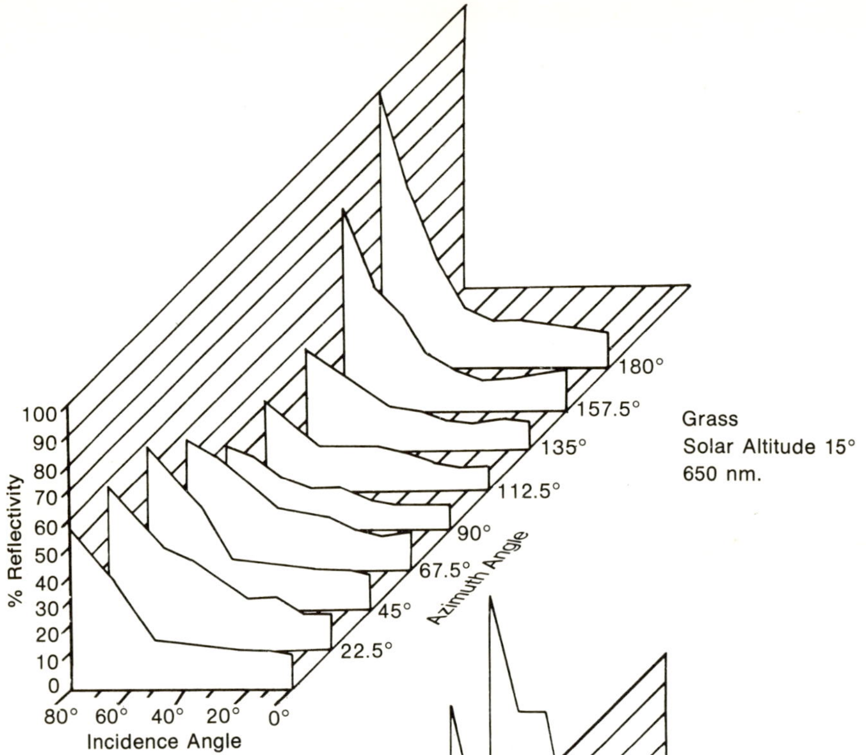

100
90
80
70
60
50
40
30
20
10
0
% Reflectivity
80°
60°
40°
20°
0°
Incidence Angle
22.5°
45°
67.5°
90°
112.5°
135°
157.5°
180°
Azimuth Angle
Grass
Solar Altitude 15°
650 nm.

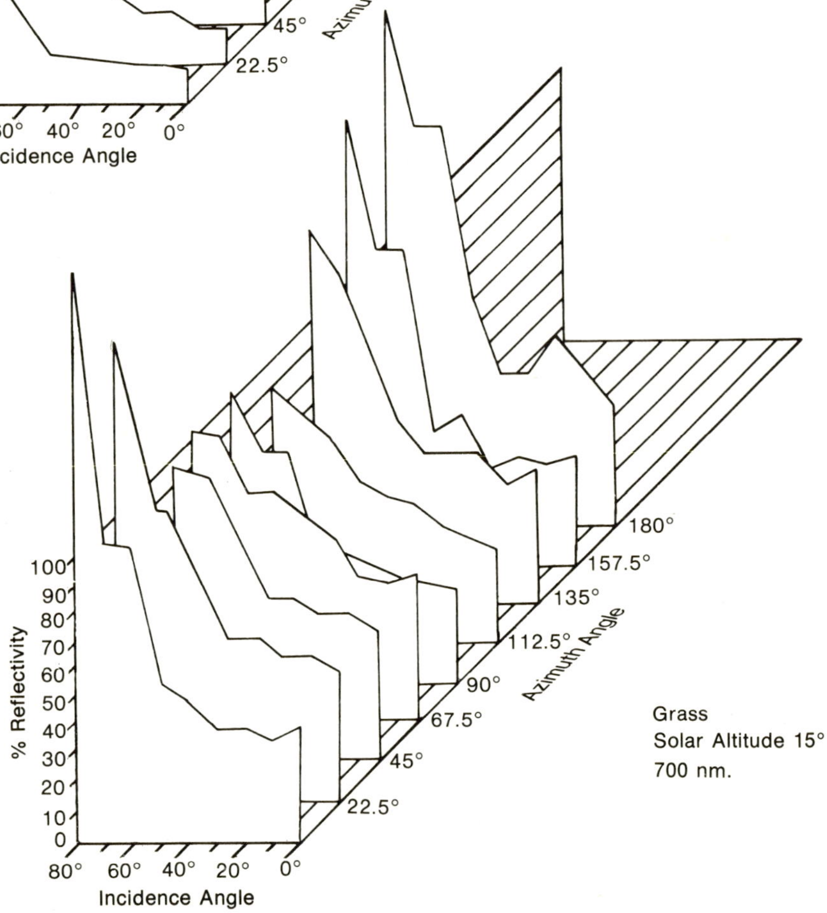

100
90
80
70
60
50
40
30
20
10
0
% Reflectivity
80°
60°
40°
20°
0°
Incidence Angle
22.5°
45°
67.5°
90°
112.5°
135°
157.5°
180°
Azimuth Angle
Grass
Solar Altitude 15°
700 nm.

is that of selecting film-filter combinations to maximize tonal differences. This can be done best after careful consideration of the reflectance properties of the vegetation in the scene. Reflectances for most species and communities have not been measured; and in those cases where they have (e.g., Gates *et al.*, 1965; Olson and Good, 1962), we are quite uncertain about the relationships linking ground or laboratory measurements with tonal renditions on aerial and space photographs. In short, the scope for qualitative and quantitative measurements of tone is relatively great, but the scope for meaningful interpretation without considerable supporting fieldwork remains low (see, for example, Colwell and Lent, 1967; Kondrat'yev and Buznikov, 1970).

Detection of subtle tonal variations and their conversion to meaningful information is extremely difficult when performed for large areas. Every species and community may be expected to have directional and temporal reflectivities that can confuse an interpreter (Rinker and Frost, 1968). Angular dependencies involving blue grass sod (*Poa pratensis*) and live sycamore leaves (*Platanus occidentalis*), may serve to illustrate the point. Mary (1964) found that blue grass with randomly oriented blades 5 cm. long had a narrow range of reflectivities, regardless of viewing angle, in the spectral band between 670 nm and 1.03 μm. Sycamore leaves, however, showed a reflectance peak of 0.545 v at 20° and a null of 0.2 at 84°. When these variations are evaluated in terms of solar altitude, as well as viewing angle, the picture becomes even more complicated.

Egbert and Ulaby (1972) have calculated the changes in grass reflectivity as a function of solar altitude, incidence angle (camera angle), and azimuth angle (e.g., pointing into or away from the sun) for various spectral bands. An example of their research is given in Fig. 2. Calibrating all these relationships for each species and community, for each angle scanned, and for each spectral channel employed will be difficult and time consuming. Yet, how can we produce large-scale maps of subtle tone differences between categories if those differences are confused with the normal spectral variation within categories? The answer is not easy. It requires the use of total hemispherical, as well as specular reflectance data and the use of sequential imaging (perhaps seasonal) to aid discrimination.

If we had spectral curves for each phenological state of every species, group of species, and ecological situation, we could predict which of them should be distinguishable within the bandwidth of the spectrum and choose film-filter combinations accordingly. In the absence of these curves, we must accept the fact that some desired categories will be lost (and some undesirable ones included) in the data obtained by any selection of films and filters.

A fundamental question arises regarding how many categories are really meaningful in a set of multiband images. Six bands flown over a heterogeneous

---

Fig. 2. Reflectance from grass as a function of incidence angle (viewing angle), azimuth angle (camera looking into or away from sun), solar altitude, and spectral wavelength. The amount of reflectivity from grass at 650 nm and 700 nm when the sun is 15° above the horizon varies considerably, depending upon camera and solar geometry. By knowing these relationships for each of the major growth form categories in a vegetated scene, one can more effectively design remote sensing experiments to capitalize on spectral differences. (From Egbert and Ulaby, 1972.)

landscape may be sufficient to distinguish growth forms and some habitats in the scene, but may be insufficient for making detailed taxonomic distinctions. Six additional bands may permit a few floristic identifications but introduce ambiguities with regard to their phenology or vigor. The same 12 bands, when employed over a more or less homogeneous landscape, e.g., a broad expanse of sagebrush (*Artemisia tridentata*), may provide no more discrimination than a single well chosen band. If these two environments are now compared for the same 12 bands, or for any combination thereof, interpretation and categorization become extremely complex.

Multispectral Scanners

The strategy behind multispectral scanning is that if data can be sufficiently finely partitioned, a large part of the spectral redundancy inherent in natural scenes can be eliminated. The technique aims to divide the spectrum into narrow parts and then to recombine selected bands in order to isolate the entities under study. Since data are not collected in image format, the number of data bits for any scene is very large, and their sorting and recombination can only be handled by a computer.

In general, three kinds of vegetational information can be obtained in the spectral region between 400 nm and 2.6 μm. This is the region most finely partitioned by present scanning systems (Table 1). Pigmentation changes are detectable in the visible bands (400 to 700 nm); structural differences in leaf mesophyll are indirectly observable in the near-infrared (800 to 1000 nm); and moisture stress is best detected between 1.6 and 2.6 μm (Rohde, 1971; Rohde and Olson, 1971).

In practice, tonal differences related to physiological stress are detectable on large-scale imagery. Rohde (1971), for example, observed that false color images produced from scanner data permitted easier detection of needle cast disease on scotch pine (*Pinus sylvestris*) and *Fomes annosus* damage on red pine (*P. resinosa*) than either conventional color or color infrared films. Weber and Polcyn (1971) reported similar success in detecting stress in forests.

The most obvious limitation in adopting the scanner technique to large areas is the numerous causes of color change in vegetation, many of which are directly related to the recent history of each habitat. While it may be possible to detect

---

Plate 1. This series of photographs traces the color (*left column*) and color infrared (*right column*) appearances of an oak-hickory forest throughout its growing season. All photos are of the same area as viewed from 500 ft. Slight differences in coverage are due to camera viewing angle. The top pair of photos was obtained on April 15, 1969; the second pair on April 22; the third pair on April 29; and the fourth on May 9. Beyond this point in the growing season (the green wave) both the green and near-infrared reflectances are so strong that little tone discrimination is possible with these films. Nor is it feasible to make consistent identifications during the yellow wave, as shown in the bottom two pairs (October 2 and October 16, respectively). At this time of year trees of the same species assume their fall coloration at different times, depending upon individual site characteristics. In constrast, since the green wave is highly correlated with photoperiodicity and degree-days, phenology of species is more easily traceable. Maximum color variation on infrared film seems to be late April for the series shown here, but this may change from year to year.

15/4/69
170 m (500′)
15 + cc30B
1/250 @ f5.6

22/4/69
170 m (500′)
15 + cc30B
1/250 @ f5.6

29/4/69
170 m (500′)
15 + cc30B
1/250 @ f5.6

9/5/69
170 m (500′)
15 + cc30B
1/250 @ f5.6

2/10/69
170 m (500′)
15

16/10/69
170 m (500′)
15 + 82 A

Plate 1

Table 1. Comparison of multispectral scanner bands and their significance in sensing vegetation.

System	Spectral Regions (μm)	Number of Channels per Region	Vegetational Importance
Bendix 24-channel scanner;[a] scan ∢ 80°	0.34– 0.5	3	Green reflectance; chlorophyll absorption; IR reflectance; forest fire mapping
	0.53– 1.05	7	
	1.18	1	
	1.73	1	
	2.1– 4.75	3	
	6.0–13	7	
	calibration	2	
University of Michigan 18-channel scanner[b]	0.32– 0.5	5	Green reflectance; chlorophyll absorption; IR reflectance; forest fire mapping
	0.5– 1.0	8	
	1.5– 1.8	1	
	2.0– 5.5	3	
	8.0–13.5	1	
NASA Earth Resources Satellite—1;[c] scan ∢ 3°	0.5–0.6	1	Center of green reflectance; center of chlorophyll absorption; vegetation stress; IR reflectance
	0.6–0.7	1	
	0.7–0.8	1	
	0.8–1.1	1	

[a] Stein (1971) and Zaitzeff *et al.* (1970). [b] Holter (1971). NASA (1971).

differences, therefore, the need for field observations, at least on a sampling basis, is still essential. It is reported that in stressed vegetation the visible reflectance increases but the direction and degree of infrared reflectance vary (Knipling, 1969). The largest differences within and between species are stated to occur at wavelengths between 1.4 and 2.6 μm (Rohde and Olson, 1971).

Pigmentation changes associated with phenological states may be detectable on small-scale scanner imagery if such changes are interpreted partly on the basis of ecology. As a rule, however, one might anticipate that at high or space altitudes and coarse resolution (both spectrally and spatially) only gross pigmentation changes would be detectable. Extremely subtle color changes, such as those shown in Plate 1 (top set), probably could not be seen from very high altitudes because of atmospheric attenuation. Equally difficult might be the problem of distinguishing between phenologically, physiologically, and ecologically induced changes. How does one categorize such differences as those shown, for example, in Plate 2, *A* and *B*?

For examples of current multispectral scanners, Table 1 lists the spectral regions and selected design parameters of three systems. Two of these operate from aircraft altitudes and the third from space altitudes aboard the *ERTS-1* satellite. It can be seen that the aircraft versions more or less concentrate on the visible and near-infrared regions of the spectrum. Channels beyond 3 μm are fewer in number and more specialized in application. The satellite scanner on *ERTS-1* employs only four spectral bands, especially selected because of their vegetational content (Park, 1969). The reported ground resolution of the satellite scanner, when first launched, was about 80 m; however, linear features as narrow as 10 m are detectable if they have high contrast ratios with their surroundings.

Processing and displaying data from scanners is complicated and costly and would become even more so if international mapping standards had to be met. The data are stored on magnetic tape and later converted to image format. Individual bands, or combinations of bands, can be displayed with color added to aid the delineation of categories as shown in the *ERTS-1* image in Plate 2. The most pressing difficulties still to be resolved are geometric distortion and data registration. Nevertheless, for rapid special-purpose surveys of small areas, where the resulting maps are not intended for long-term use, these problems are surmountable. The object in most of these practical applications is to gain as current a picture as possible of short-term vegetational phenomena. Usually, as in the case of disease detection or moisture stress, the potential savings warrant the expense of collecting and processing the data.

Thermal Infrared Scanners

Imagery may de produced directly from radiation in the 2- to 14-μm spectral band. Energy incident upon a detector is converted to an electrical impulse and displayed on a cathode-ray tube (CRT). Film is then passed over the tube at a rate proportional to the speed of the aircraft and inversely proportional to its height; each scan line is recorded in sequence after the fashion of a television picture. In airborne systems the scan lines are perpendicular to the aircraft track, as shown in Fig. 3. The resulting imagery contains a number of geometric distortions inherent in the system design; nevertheless, for reconnaissance purposes, such imagery is sometimes preferred over the format obtained by comparable bands of multispectral scanners because it can be "read" more easily. For both approaches the basic electromagnetic principles are the same.

Plate 2. *A*, *B;* Ektachrome and Ektachrome infrared photos of portion of the Flint Hills, Kansas (near Clyde on the Republican River west of Manhattan). Both taken with handheld 35-mm cameras from 30,000 ft on April 23, 1969, 10 am local sun time. Bright red fields on *B* are wheat and alfalfa. Natural grasslands appear bluish-red with some areas beginning to show blushes of pinkish-red. Riparian vegetation is just beginning to turn red. Dark blue and gray areas are fallow or bare. In addition to showing the effects of haze attenuation the Ektachrome photo does not permit consistent distinction between wheat fields and bare fields (*a*, *b* versus *a'*, *b'*).

C; Portion of an ERTS-1 multispectral scanner image taken from the Monterey Bay, California, frame (E-6556-0023-7). This is a color combination of three data channels (0·5 to 0·6 μm; 0·6 to 0·7 μm; and 0·8 to 1·1 μm) obtained on June 6, 1972 (original scale 1: 1,000,000). Several shades of red and pink can be distinguished in the Central Valley agricultural region. Grasslands of the coast ranges appear in bluish gray with blushes of pink in the higher elevations. The coastal pine and redwood forests north of Santa Cruz appear in deep red and purple tones. Sequential images would permit comparison of season aspects in this scene, but for detailed phenological studies calibrated images collected daily or weekly are required. (Image courtesy of NASA and USGS.)

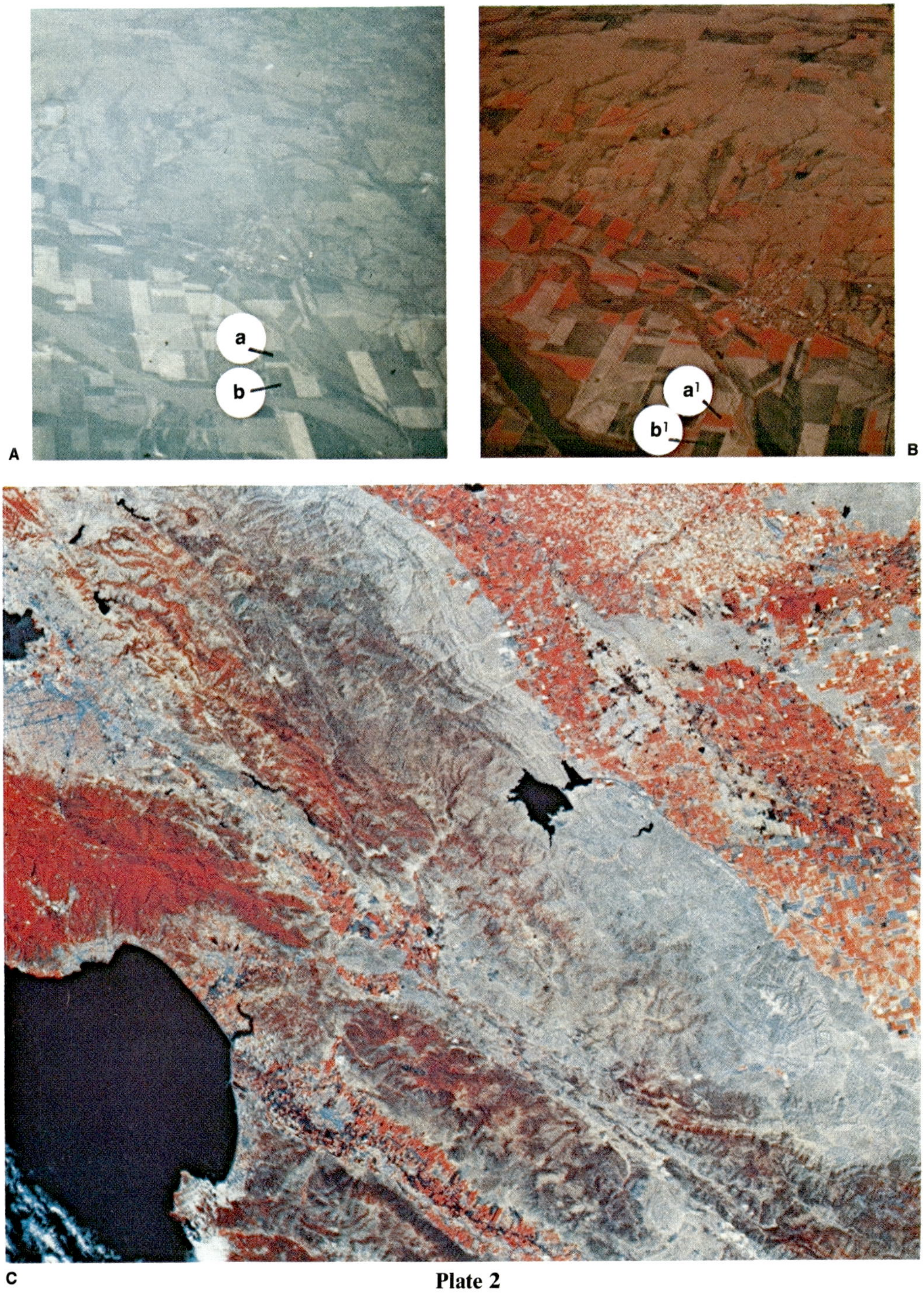

Plate 2

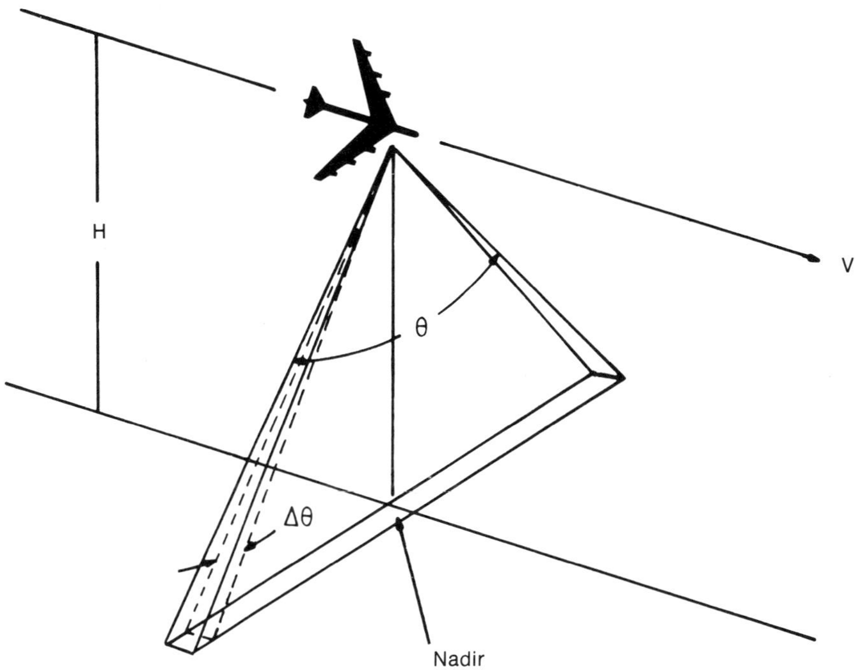

Fig. 3. Configuration for a typical infrared scanner. A scan mirror is rotated from side to side having a field of view of θ and resolution of $\Delta\theta$. The scan is roughly perpendicular to the flight direction. The detector is cooled below zero and is thus sensitive to very minor changes in terrestrial emissivities. Relative emissivities of the scene below the aircraft are converted first to electrical signals, then to an intensity modulated light source, and recorded on film. The higher the emissivity (warmer temperatures), the brighter will be the light source and hence the lighter the tonal rendition on positive imagery.

Instead of reflected energy, the thermal infrared region records that part of the absorbed energy that is reemitted. For simplicity, this is referred to as temperature sensing. It is made possible by the fact that an object's total radiation is temperature dependent, according to the Stefan-Boltzmann law.[1] Ostensibly, the information being recorded on film is the total radiation incident upon a highly sensitive detector. Warm objects in the landscape appear in light tones, cooler ones in darker tones. More sophisticated interpretation is constrained by interpreter experience and other variables, the most important of which relate to the atmosphere and the emmissive characteristics of specific objects.

Within the region from 2 to 14 μm the atmosphere has two major 'windows': one spans the 3- to 5-μm band and the other spans the 7- to 14-μm band (Fig. 4). The remainder of the infrared spectrum is largely opaque because of atmospheric effects. Inside the windows the atmosphere is not uniformly transparent. Rather, it acts as a selective attenuator of energy emitted from surface phenomena. More-

[1] The Stephan-Boltzmann law states that the total radiation from an object is equal to the product of the fourth power of its absolute temperature, emissivity, radiation area, and the Stephan-Boltzmann constant.

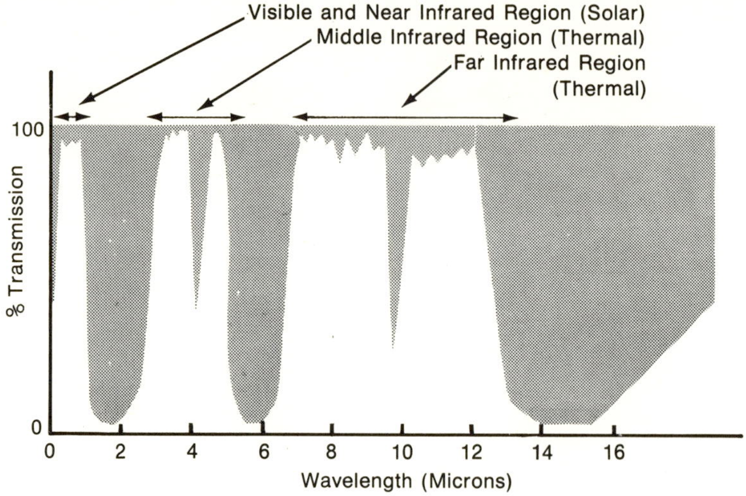

Fig. 4. Percent atmospheric transmission from visible to far infrared. Major "windows" are located in the visible region from approximately 400 to 1000 nm in the middle infrared from about 3 to 5 μm, and in the far infrared from about 7 to 13 μm. Compare the detail of this diagram with that shown in Fig. 1.

over, the atmosphere itself is a strong radiator in certain bands and contributes to part of the signal received by the sensor. The sum of these atmospheric effects is not easily calculated, so at present there is no way to convert the recorded grey levels of vegetation to absolute temperatures (Holter, 1971).

Another deterrent to calculating object temperature from imagery arises from the directional components of an object's radiation. If all objects were perfectly diffuse and radiated a uniform hemisphere about themselves, the amount of radiation would be the same no matter what the viewing angle. Similarly, if all objects were specular radiators, they would be invisible except at critical angles. In reality, natural surfaces are neither perfectly diffuse nor perfectly specular; thus the energy they radiate and reflect varies as a function of wavelength, viewing angle, surface roughness, diurnal cycle, and moisture content. One should treat with caution any conclusions drawn from specular data such as those shown in Fig. 5. While it seems possible to distinguish grass meadow from sedge meadow at three o'clock in the afternoon, we do not know that this is true for all viewing angles within the scanning area.

Tonal contrasts are essential in creating image pattern, but to regard these tones as unique signatures is naive. Emissivities are constantly changing functions of the diurnal cycle as shown in Fig. 5 and 6; hence, relative tonal differences between objects are of little inherent value unless sequential images are available. The obvious exceptions, of course, are extreme temperature anomalies like forest fires, and it is precisely here that thermal imaging has found a unique application in phytocenology.

In spite of these difficulties, we should not despair for future applications of infrared line scanners. Present-day scanners are sensitive to temperature differences finer than 1°K, very near the theoretical sensitivity of the detectors employed (Holter, 1971). With continued research into the emmissive properties of species and species groups, into ways of calibrating the imagery, and into image rectification techniques, we may someday be able to map and interpret extremely subtle ecological and phenological differences and apply these to energy budget studies of ecosystems. It is even conceivable that such measures as community productivity, as registered by CO_2 turnover, may be qualitatively estimated by comparing image intensities through an atmospheric window (perhaps at 3.5 to 4 μm) with those obtained in a CO_2 absorption band (say at 2.8 to 4.3 μm) (Cooper, 1964). While these ideas always seem to be accompanied by frustrating problems, they are not without foundation in theory.

Active Microwave (Radar)

Side-looking airborne radar (SLAR) imagers are the best sensors available for temporal sensing, and for this reason alone they deserve serious consideration in phenological studies. Several attributes lead to their unique status among the family of sensors described here. First, because they operate by transmitting a signal and receiving that portion reflected by the surface, they are described as active rather

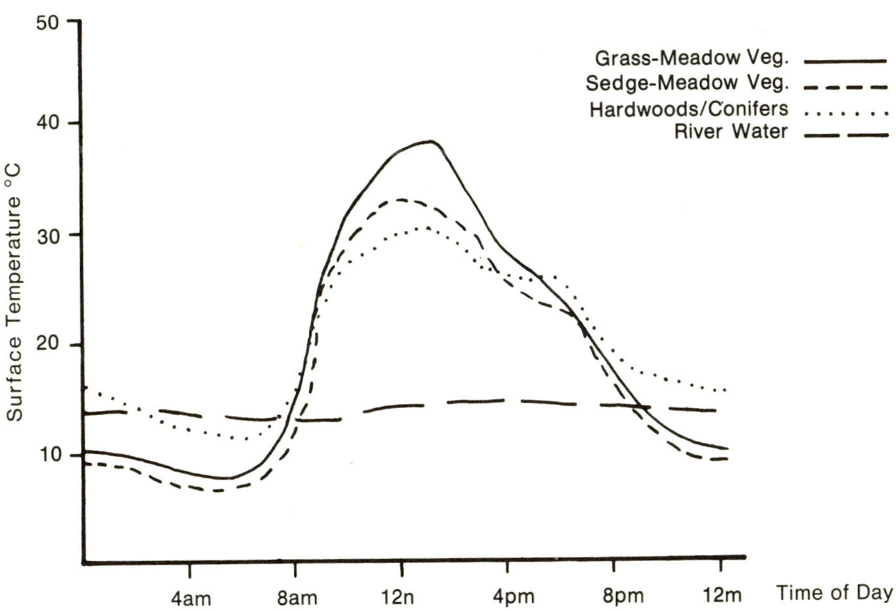

Fig. 5. Diurnal variations in vegetational temperatures as measured by a precision radiation thermometer (PRT-5). River water (long dashes) remains relatively temperature constant. Vegetated surfaces vary in relation to one another, with the greatest fluctuation in this case being grass meadow (solid line). Temperature crossover points, located at roughly 8 am, 3: 30 pm, and 6 pm, aid in discriminating the various types.

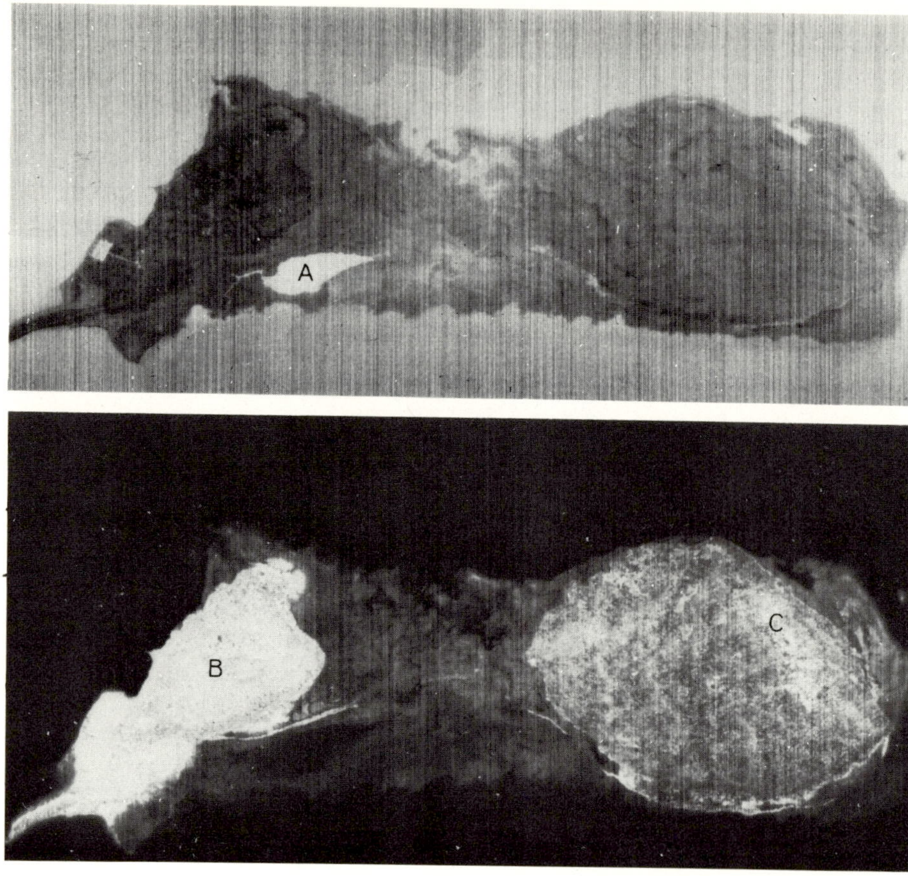

Fig. 6. Thermal infrared images of Big and Little Goat islands off Jamaica. *Above;* A nighttime image showing relatively warm marine and cool terrestrial surfaces. Note that the lagoon at *A* is warmer than the surrounding offshore water. *Below;* A daytime image showing a temperature reversal to warmer land and cooler water. The vague pattern between the warmer land masses at *B* and *C* is a shallow underwater reef. Note that *B* and *C* have different temperatures (emissivities) as a result of differences in soil-vegetation patterns. Images courtesy of NASA (Mission 172).

than passive sensors[2] (Fig. 7). Second, because they do not rely on solar illumination, they can acquire imagery when photographic sensors cannot. Such times include both diurnal and seasonal periods of low sun angle. Third, as a consequence of their relatively long wavelength, they can obtain imagery through thick cloud layers and, depending upon their precise wavelength, through many kinds of precipitating clouds. All these attributes suggest that geographic regions hitherto inaccessible, either because of remoteness or because of persistent cloud cover, could be imaged using radar sensors.

[2] Passive sensors, such as the aerial camera and more sophisticated devices like the multispectral and thermal infrared scanners, record reflected or emitted radiation from objects. Active sensors irradiate the scene under study with a transmitted signal of measurable strength and record the magnitude of return signal.

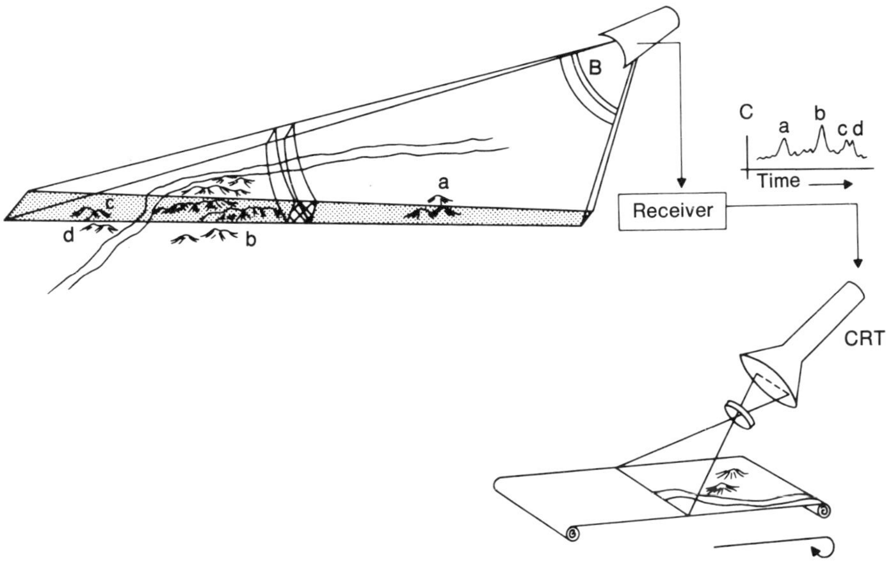

Fig. 7. Configuration for a typical side-looking radar system. A signal is transmitted, impinges on the terrain, and is reflected to the receiver. The variable reflectance (σ°) of terrain elements *a*, *b*, *c*, and *d* is used to intensity modulate a "flying-spot" on the cathode-ray tube. Film is passed in front of the CRT and exposed in proportion to the intensity modulated light. The stronger the return from a given terrain element, the brighter the light and the lighter will be its tonal rendition.

For image analysis the parameters most influencing gray tone are (1) surface roughness at the scale of the wavelength employed; (2) signal polarization; (3) viewing angle; (4) scene moisture content; and (5) homogeneity of the reflecting medium. The individual and combined effects of these variables for interpreting vegetation are poorly understood; as a result, many published accounts treat radar imagery as an exotic form of photography rather than as imagery portraying microwave properties. Contributing to the interpretation problem are other limitations inherent in the design of most systems. Foremost among these is their design based on a single frequency.[3] Others include their moderate to coarse resolution (12 to 15 m for present systems) and the fact that, at present, all SLARs are uncalibrated. These problems and their possible solutions are detailed by Moore (1970), but deserve brief comment here.

Use of single-frequency (monospectral) imaging systems is extremely limiting for object identification.[4] At present, a pseudo multispectral capability is achieved by altering the polarization of transmitted and received signals from horizontal to vertical. Thus a scene imaged at any given frequency may be displayed in four

[3] The statement by Olson (1971) that "most radars available today are broad band systems that transmit pulses including a range of frequencies and therefore a range of wavelengths . . ." (p. 16) is incorrect. Such systems are only now in the experimental stages.

[4] Current operational SLARs employ single wavelengths ranging from 0.8 cm (*K*-band) to about 3 cm (*X*-band).

polarizations: horizontal transmit and receive (HH); horizontal transmit, vertical receive (HV); (VV) or (VH). Unfortunately, there is scant theoretical information regarding the reflectance properties of species and natural communities as a function of signal polarization, and as a result most of the current ability to interpret imagery is derived from empirical observations supported by incomplete or (sometimes) inappropriate ground data. For similar reasons there would be difficulties surrounding the interpretation of multifrequency imagery. Nevertheless, there is some evidence to indicate that the broad band microwave response of vegetation is as variable as its visible response; and, since sensors operating in the visible region will always be time dependent, it would seem highly desirable to develop a true multispectral radar capability for use in phenology.

Aside from frequency limitations, problems of calibration are perhaps most serious. Even if we knew, quantitatively, the microwave reflectance characteristics of different vegetation types, we could not interpret them quantitatively from radar imagery. There are so many uncalibrated data transformations from point reflectance to processed imagery that only qualitative statements can be made about the contents of a single image. Often, even these are questionable unless the sensing aircraft is equipped with roll, pitch, and yaw recorders. In other words, part of the unique advantage of radar as an active sensor is lost because systems have not been designed to accurately measure differences in received signal strength.

In terms of recorded information the parameters that most control radar reflectivity are scene geometry (roughness) and moisture content (including surface films such as dew). This means that measurable changes in signal backscatter[5] can occur instantaneously with the passage of a single storm or gradually as growth cycles progress. For any given view of the terrain, therefore, one has to distinguish between phenological stages and spurious short-term attributes. This is true not only for radar but for all sensors. In the final analysis these distinctions may prove the most difficult for remote sensors applied to phenology. For radar none of these problems is insurmountable. Fully calibrated, multifrequency, fine resolution systems are technically feasible. The need for such devices has apparently not yet been shown to be worthwhile.

3 Applications in Phenology

A review of the remote sensing literature dealing with spectral and temporal aspects of vegetation reveals that most research has focused on determining spectral differences between species (see, for example, Krumpe, 1972; Sapp *et al.*, 1971). Far less attention has been given to either diurnal or seasonal spectral trends, although these are increasingly recognized as being highly useful in phytocenology. Among the better known and more recent contributions to an evaluation of the time dimension are articles by Brunnschweiler (1957), Boesch and Brunnschweiler (1960), Olson and Good (1962), Haralick *et al.* (1969), Steiner (1969), and Driscoll

[5] Backscatter is the term normally employed to distinguish that portion of the transmitted signal reflected back to the receiver from that portion reflected forward or to the sides (forward and sidescatter, respectively).

and Francis (1970). Only those treating agricultural topics fall into the category of applied remote sensing. Furthermore, it is significant that of those listed only the article by Haralick *et al.* addresses a sensor other than the airborne camera.

The question of ground-based sensors for monitoring phenological change has hardly been raised. The research by Johnson and Vogel (1968) on quantifying forest canopy closures using a vertically pointing hemispherical camera is one of the few recent contributions to this area. Much needed data from nonimaging sensors, such as spectrophotometers, radiometers, and radar scatterometers, are rare in the literature, and, in any case, have only been systematically gathered for a few selected species and environmental situations.

Point Identification (Taxonomic Mapping)

By earth-orbiting satellites, man is able for the first time to view and record any part of earth sequentially. For vegetation sciences this means that phenology will become an ever-increasing part of the routine analysis of phytocenoses; it also suggests that future vegetation maps might include these data as part of their classification schemes. Historically, investigations have been forced to treat phenology implicitly; hence mappers have previously not needed to develop sophisticated phenological mapping strategies. Almost all modern vegetation maps are based on physiognomic, floristic, or ecological criteria or combinations of these. It is little wonder that phenological maps derived from remote sensor data do not exist.

As an example of the potential for phenological mapping of point identifications, and at the same time to indicate the kinds of problems that arise, see Plate 1. These photographs represent the same area of an oak-hickory forest in eastern Kansas. There are six sets of photos, each containing one Ektachrome photograph obtained from a low flying aircraft and a simultaneously obtained color infrared photograph. The first set was obtained in mid-April 1969 and the last set was obtained on October 16 of that year. Thus there are color and color infrared records of this particular forest for six times throughout its growing season.

In the first set, one is just able to see a blush of pink in the infrared photograph, with slightly stronger red tones for some of the understory vegetation. The forest canopy is obviously quite open at this time of the year, and it seems that there is opportunity for identifying elm (*Ulmus*) trees by observing the red reflectance in their upper branches. Elm is one of the first species in this phytocenosis to initiate spring growth. The second set, one week later on April 22, shows most individuals in a state of vigorous growth; there is some ability to distinguish between species using relative green reflectance. In the color infrared photograph, comparison of color and color infrared tones might identify tree species fairly well at specific times of the year. This has been shown previously by Northrup and Johnson (1970) and Krumpe *et al.* (1971). In the third set of photographs (acquired April 29, 1969), the green and infrared reflectances are already "saturated." Several individuals are still lighter or darker than their surroundings, but the certainty of identifying them at either the specific or generic level has declined. By the early part of May (set 4), all possibility for identifying tree species on the basis of their green or infrared reflectances has vanished. This situation persists throughout the remainder of the growing season and into the fall. By October 2 (set 5) there is considerable spectral variation between the individual trees. However, identification is extremely unreliable, for the simple reason that the onset of autumnal colorations is entirely a result of local environment and individual genotypes. Individuals of the same species often change color at different rates and at different times, depending upon how wet or dry their local soil is, or depending upon the severity of insect attack, etc. The last pair of photographs, October 16, contains a great range of tones, but there is little scope for species identification.

Species identification is thus most reliable in temperate latitudes during the vernal sequence or when imagery is available for both the greening and yellowing phases. Determining the actual dates of onset for phenological stages requires an ability to identify species from imagery and to correlate the tones with the stages of growth. In other words, for phenological studies involving other than gross determinations of deciduousness, imagery is required daily.[6] From these data a subset of images must be selected for taxonomic purposes. By implication, a multispectral, all-weather sensor is required.

Boundary Mapping

The delineation of major phenological boundaries should be substantially aided by judicious consideration of remote sensing theory coupled with common sense. Boundaries, including zones too wide to perceive visually (or those having spectral differences too narrow to trace), should be mappable if the right sensor and platform are used. The advance of the so-called green wave (and its seasonal opposite, the yellow wave), for example, represents a seasonal boundary of considerable importance. Since it advances as a zone detectable as a spectral shift, it should be observable from space with sensors operating in the 750- to 1000-nm range. For these purposes, spatial resolution and image scale are less important than spectral sensitivity, as shown by Aldrich *et al.* (1971). Spectral sensitivity is essential to record very subtle blushes of pink as vegetation initiates its spring growth, or as it takes on fall coloration. Rouse and associates at Texas A&M are presently involved in mapping vernal and autumnal change along the Great Plains corridor using data from the multispectral scanner of *ERTS-1*. In addition to using the remote sensor data, they are making ground observations of biomass and weather conditions in hopes of providing timely information relevant to crop and rangeland decisions. The high altitude infrared photo of the Flint Hills region in Kansas (Plate 2, *A*) illustrates the basis for their rationale. In this photo the bright red fields are wheat and alfalfa. Most of the grassland appears bluish-red, but subtle increases in red chroma reveal areas that are slightly advanced (either because of exposure, fertilizer, or intentional burning).

Using the green wave strategy, one can eliminate much of the uncertainty described by Motovilov (1966) for the compilation of phenological maps. Motovilov's method for calculating the average date of onset of phenological phases (Eq. 1) could be made more precise by using remote sensor imagery, because the actual

[6] In tropical semi-evergreen forests even such coarse distinctions as periodic deciduousness would be useful.

date of onset of a given phase (b_n) could be monitored by a satellite whose orbital parameters provide a continuous view of the study area. Historical files of such data would also be useful in calculating average dates ($A_\alpha;a_n$).

$$B = A_\alpha + \frac{\Sigma(b_n - a_n)}{L} \tag{1}$$

where

B = average date of onset of phenological phase (greening, yellowing, etc.) expressed in map form as an isoline. A_α = average date of onset of phenological phase calculated by summing the dates for each observation point in a given year and dividing by the number of observation points. b_n = actual date by year for onset of a given phase. a_n = average date of onset at a given observation point. $(b_n - a_n)$ = difference between actual date and the average date. L = number of observations per sample point.

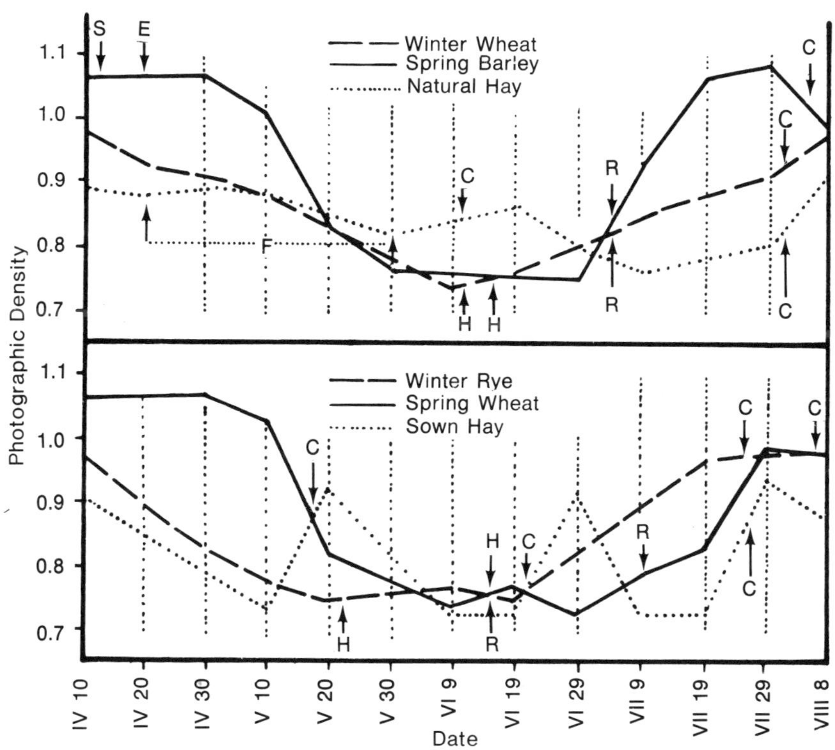

Fig. 8. Seasonal changes of average photographic densities of winter wheat, spring barley, natural hay, winter rye, spring wheat, and sown hay, measured on low-altitude panchromatic aerial photographs; valid for the Zürich area, Switzerland. The letters indicate the average dates or time periods for the following phenological aspects or farming activities: *C*, cutting (harvesting); *E*, emerging; *F*, flowering; *H*, heading; *P*, planting; *R*, beginning of ripening; *S*, sowing. (From Steiner, 1969, p. 190.)

Area Identifications

Growth stages in crops have been used in complex agricultural scenes to identify specific crop types and to calculate crop statistics. Farming attempts to insure uniform growth stages for a given crop in terms of dates of germination, maturation, and harvest. Depending upon the particular mix of crops and their calendars, phenology may thus be useful at several points throughout the growing season to assist agricultural determination (see Fig. 8).

Steiner (1969) has proposed a five-step model for conducting temporal crop surveys from space. First, the study area must be regionalized on the basis of climate and land use and then redivided into homogeneous subareas. Within these subareas, samples should be selected for mapping and ground observation. For phenological studies one would sample on the basis of lithology, exposure, drainage, and edaphic conditions. The third and fourth steps require repeated coverage by the appropriate sensors. Both the frequency of coverage and the specific sensor, or bands, employed should be determined empirically. For crops, the frequency of coverage might be on the order of one week, because planting time, germination, etc., are more or less controlled by management practices. For natural phenomena, however, weekly coverage would probably not be sufficient, as evidenced by the rapidity of change shown in Plate 1. When a complete temporal data set becomes available, reflectances and crop calendars could be compared, combined with weighting functions to account for the crop mix, and interpreted for crop identification and other information. Basically, the same approach could be applied to any seasonal vegetation, although the types and quantities of supporting data would be more critical. The major departure in adopting this model to phenological studies of natural vegetation would be in the final stage. Interpretation would perhaps be concerned more with the dynamics of vegetational development than with the identification of types.

4 Conclusions

The use of remote sensors for tracing phenological change calls for combining temporal and spectral concepts. Mere imaging on a sequential basis will be inadequate for most purposes, unless the interpreter knows *a priori* how his phenomena should appear at each point in time. Gross aspect changes may be an exception to this in certain regions, but even these will be troublesome if recorded on coarse resolution, small-scale imagery of poorly known areas.

Sequential spectral data, both specular and total hemispherical, collected from directly overhead, are crucial in deciphering data from more distant sensors. Furthermore, to be quantitative, these data will have to be calibrated against a standard for meaningful interpretation of subtle changes in reflectance. Finally, depending upon the research goals, it will be necessary to distinguish between physiological stress and normal phenological change. For all of these needs the problem still to be resolved is the actual date of onset versus the date of first spectral detectability; the latter varies according to sensor, resolution, and scale.

The situation with regard to sensors is similarly complex. Our best spectral sensors are the poorest temporal sensors because of atmospheric attenuation and, often, persistent cloud cover. If sensing is required daily, or even less frequently, cameras and scanners will prove inadequate. At the opposite extreme, our best temporal sensor (side-looking radar) is presently our poorest spectral sensor. There is no sensor ideally suited to collecting both spectral and temporal data. This perhaps limits remote sensing in phenology to broad-scale studies of vernal (autumnal) advance or gross aspect.

Applications that employ phenological change in their experimental design most often use it as a surrogate for nonphenological ends. We can expect in the future to see increased use of the time dimension in vegetation, particularly for monitoring forestry, rangeland, and agricultural phenomena. However, the direct application of air or spaceborne remote sensors to phenology will probably be overshadowed by detailed studies using ground-based sensors. Strangely, this approach has not been widely used.

References

Aldrich, S. A., F. T. Aldrich, and R. D. Rudd. 1971. An effort to identify the Canadian forest-tundra ecotone on weather satellite imagery. *Remote Sensing Environ.* 2, 9–20.

Boesch, H., and D. Brunnschweiler. 1960. Seasonal changes of the agricultural landscape interpreted from aerial photographs. *Geog. Helv.* 15(4), 257–261.

Brunnschweiler, D. H. 1957. Seasonal changes of the agricultural pattern: A study in comparative airphoto interpretation. *Photogrammetric Eng.* 23(1), 131–139.

Colwell, R. N., ed. 1963. Basic matter and energy relationships involved in remote reconnaissance. *Photogrammetric Eng.* 39(5), 761–799.

——— 1960. *Manual of Photographic Interpretation.* 868 pp.: American Society of Photogrammetry. Hanover, Pa: Everybody's Press.

———, and J. D. Lent. 1967. A test of the concept and practical application of multiband reconnaissance. *Ann. Rep.* Forestry Remote Sensing Laboratory, University of California, Berkeley, to the NASA Office of Space Sciences and Application. 62 pp.

Cooper, C. F. 1964. Potential applications of remote sensing to ecological research. In *Proc. 3rd Symposium on Remote Sensing of Environment.* pp. 601–606. Ann Arbor: University of Michigan.

Driscoll, R. S., and R. E. Francis. 1970. Multistage, multiseasonal and multiband imagery to identify and qualify non-forest vegetation resources. *Ann. Progr. Rep.* Forestry Remote Sensing Laboratory, University of California, Berkeley, to NASA Earth Resources Survey Program, 65 pp.

Egbert, D. D., and F. T. Ulaby. 1972. Effect of angles on reflectivity. *Photogrammetric Eng.* 38(6), 556–564.

Gates, D. M., H. J. Keegan, J. C. Schleter, and V. R. Weidner. 1965. Spectral properties of plants. *Appl. Optics* 4(1), 11–20.

Haralick, R. M., F. Caspall, and D. S. Simonett. 1969. Using radar imagery for crop discrimination: A statistical and conditional probability study. *Remote Sensing Environ.* 1, 131–142.

Heller, R. C. 1969. Imaging with photographic sensors. In *Remote Sensing with Special Reference to Agriculture and Forestry*, R. Shay, ed. Chap. 2. Washington, D.C.: National Academy of Sciences.

Holter, M. R. 1971. Infrared and multispectral remote sensing. *Proc. NATO AGARD*, June meeting, Colorado Springs. Colo.

Johnson, P. L., and T. C. Vogel. 1968. Evaluation of forest canopies by photography. *Res. Rept. 253.* U.S. Army Materiel Command, Cold Regions Research and Engineering Laboratory (CRREL), Hanover, N.H. 20 pp.

Knipling, E. B. 1969. Leaf reflectance and image formation. In *Remote Sensing in Ecology*, P. L. Johnson, ed. Athens: University of Georgia Press.

Kondrat'yev, K. Ya., and A. A. Buznikov. 1970. Some results of spectrophotometry of the earth from the Soyuz 7 spacecraft, *Dokl. Akad. Nauk SSSR* 195(5), 1084–1087. Also available as *U.S. Dept. Commerce JPRS Doc. 54047.*

Krumpe, P. F. 1972. Remote sensing of terrestrial vegetation: A comprehensive bibliography. Knoxville: University of Tennessee Graduate Program in Ecology.

———, H. R. DeSelm, and C. C. Amundsen. 1971. An ecological analysis of forest landscape parameters by multiband remote sensing. *Proc. 7th International Symposium on Remote Sensing of Environment.* 1, 715–730. Ann Arbor: University of Michigan.

Mary, D. J. 1964. Directional reflectance of certain materials in the near infrared. *Tech. Man. 64–29*, U.S. Army Materiel Command, 25 pp.

Moore, R. K. 1970. Ground echo. In *Radar Handbook*, M. I. Skolnik, ed. New York: McGraw-Hill.

Motovilov, G. P. 1966. Aerial photography and mapping of the Siberian forest. *Aero. Yemka Kartografrarovaniye Lesov Sibiri*, 3–171. For technical translation see *Tech. Transl. FSTC-HT-23-46-68.*

NASA Goddard Space Flight Center. 1971. *Data Users Handbook.* Doc. No. 715D249.

Northrup, K. G., and E. W. Johnson. 1970. Forest cover type identification. *Photogrammetric Eng.* 36(5), 483–490.

Olson, C. E., Jr. 1971. Collection and processing of multispectral imagery. In *Applications of Remote Sensors in Forestry.* International Union of Forest Research Organizations (IUFRO). Joint report by the working group on "Applications of Remote Sensors in Forestry," 25th IUFRO Congress, Druckhaus Romback Co.

———, and R. E. Good. 1962. Seasonal changes in light reflectance from forest vegetation. *Photogrammetric Eng.* 28(1), 107–114.

Park, A. B. 1969. What earth resource satellites can do for the agriculture community. *Paper No. 69–1083.* American Institute of Aeronautics and Astronautics, 6th Ann. Meeting and Tech. Display.

Parker, D. C., and M. Wolff. 1966. Remote sensing. *Intern. Sci. Technol.* 43, 20–31.

Rinker, J. N., and R. E. Frost. 1968. Environmental analysis and remote sensing. *Tech. Note.* U.S. Army Terrestrial Science Center, Hanover, N.H. 147 pp.

Rohde, W. G. 1971. Multispectral enhancement of disease in forest stands. In *Color Aerial Photography in the Plant Sciences.* 3rd Biennial Workshop. 288 pp. (see pp. 131–143). American Society of Photogrammetry. Hanover, Pa.: Everybody's Press.

Rohde, W. G., and C. E. Olson, Jr., 1971. Estimating foliar moisture content from infrared reflectance data. In *Color Aerial Photography in the Plant Sciences.* 3rd Biennial Workshop. 288 pp. (see pp. 144–164): American Society of Photogrammetry. Hanover, Pa.: Everybody's Press.

Sapp, C. D., *et al.* 1971. *Selected Bibliography of the Terrain Sciences.* Wayland, Mass.: Raytheon Company.

Smith, J. T., ed. 1968. *Manual of Color Aerial Photography.* 550 pp.: American Society of Photogrammetry. Hanover, Pa.: Everybody's Press.

Sorem, A. L. 1967. Principles of aerial color photography. *Photogrammetric Eng.* 33, 1008–1018.

Stein, K J. 1971. Multispectral scanner promising. *Aviation Week Space Technol.* May 24, 39–41.

Steiner, D. 1969. Time dimension for crop surveys from space. *Photogrammetric Eng.* 35(2), 187–194.

Vinogradova, A. I. 1955. Using aerial photography in different spectrum intervals to study vegetation and soils. *Geog. Sbornik* 7, 59–74. Also as *Tech. Transl. FSTC-HT-23-309-70.*

Weber, F. P., and F. C. Polcyn. 1971. Remote sensing with optical mechanical scanners to detect stress in forest. *Proc. 37th Ann. Meeting.* pp. 123–143.: American Society of Photogrammetry.

Zaitzeff, E. M., C. L. Wilson, and D. H. Ebert. 1970. MSDS: An experimental 24-channel multispectral scanner system. *Bendix Tech. J.* 3(2), 20–32.

2.4 Computer Mapping in Phenological Analysis

Joseph M. Caprio

Montana Agricultural Experiment Station
Plant and Soil Science Department
Montana State University, Bozeman, Montana 59715

Richard J. Hopp

Vermont Agricultural Experiment Station
University of Vermont, Burlington, Vermont 05401

and Jack S. Williams

Montana Agricultural Experiment Station
Plant and Soil Science Department
Montana State University, Bozeman, Montana 59715, USA

1 Introduction

The development of computer methods for mapping basic geographical information provides a much needed tool for phenological analysis. In the past, phenological maps have been drawn by hand which can be both time consuming and dependent on the subjective interpretation by the analyst. The speed and complete objectivity in future analyses of phenological data through computer mapping should make it possible to convert quickly great quantities of phenological data into map form which can then be used in many decision-making processes.

Applications of computer mapping in phenological analysis have been made by Lieth and Radford (1971) (see the inside front cover for the colored computer map). They used the SYMAP method to analyze phenological data on dogwood and redbud at sites throughout North Carolina. More recently, Caprio and Williams (US/IBP Phenology Committee, 1972) produced a SYMAP analysis of the onset of spring in 1971 across the conterminous United States based on the first bloom date of purple

KEYWORDS: phenology; computer mapping; SYMAP; lilac; dogwood; redbud; spring; bloom; flower; isophanes.

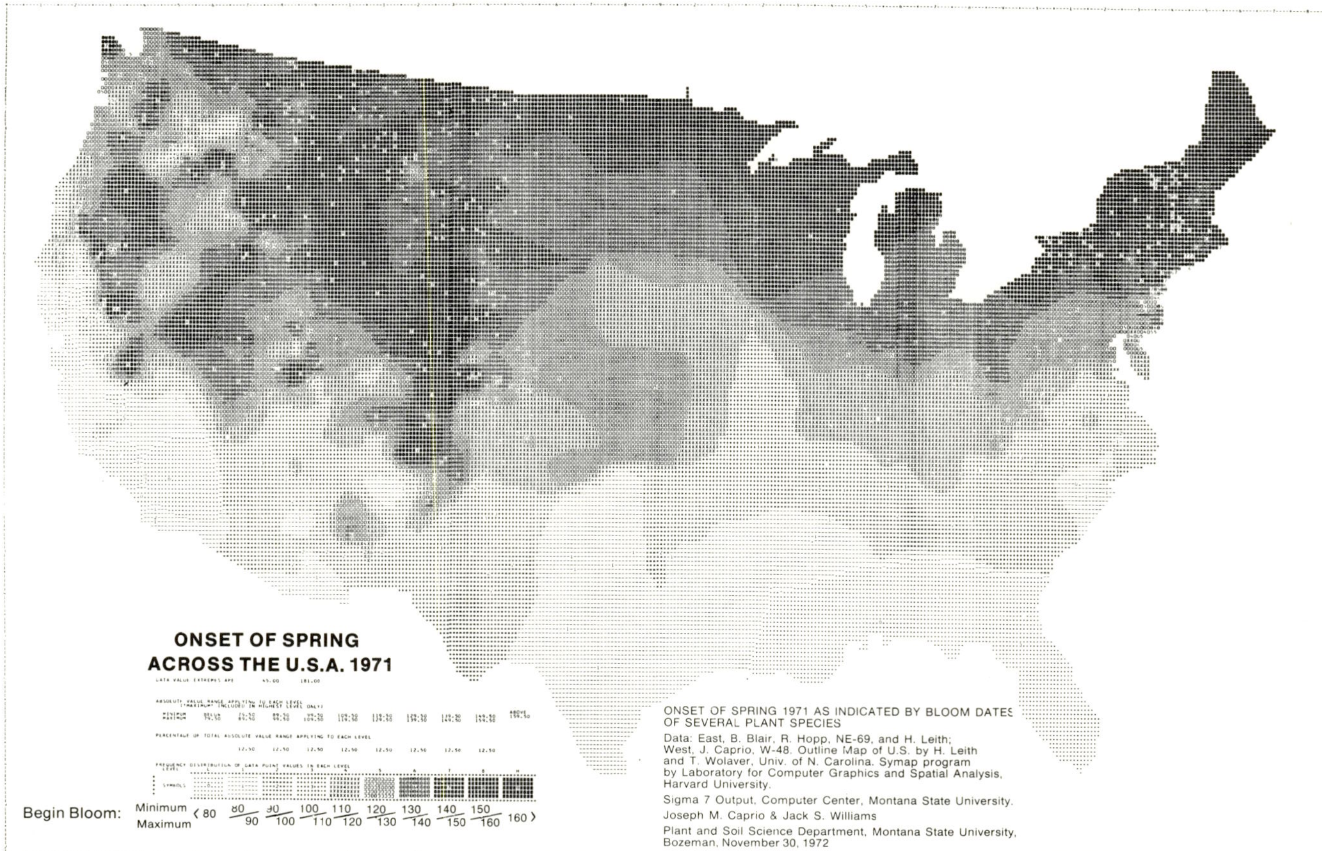

Fig. 1. Computer map of onset of spring across the United States.

common lilac in the western United States, "Red Rothomagensis" lilac in the eastern United States, and US/IBP data on the first bloom of dogwood and redbud in the southeastern part of the country. Data on purple common lilac and "Red Rothomagensis" lilac were contributed by Agricultural Experiment Stations Regional Projects W-48 and NE-69, respectively. The map had several data gaps, especially in the midwestern states. Additional information was therefore incorporated, particularly data from the north-central states, which were supplied by Byron Blair of Purdue University. The computer base map of the United States was provided by Wolaver and Lieth, who developed it in connection with the US/IBP Eastern Deciduous Forest Biome Studies (Wolaver and Lieth, 1972). The revised map of the onset of spring in 1971 is shown in Fig. 1.

The map shows how spring generally begins in southern areas and how it progresses northward: from February in Florida to northern New England in June. The pattern in the western United States is highly influenced by the mountainous terrain, as evidenced by the irregular pattern of isophanes in that area. While the general pattern of the onset of spring can be expected to be somewhat similar each year, changes in the isophanes will occur owing to annual weather variations.

2 Use of SYMAP

The SYMAP program, which produced the map, calculates all unknown value points by taking the values of nearby known value data points. The program assumes a relationship in which known value data points have an effect that is inversely proportional to the square of the distance between the known value data points and the unknown value points. The SYMAP program assigns a value to an unknown value point by taking a specified search radius and interpolating from the values of the data points within that radius. The program provides for an adjustable search radius so that the minimum and maximum number of data points used to calculate each unknown point between data points is 4 to 10, respectively. Details about the SYMAP program may be obtained from the SYMAP reference manual (1971) or from Reader *et al.* (1972).

3 Problems in Computer Mapping

An analysis of this kind has certain limitations. It does not account for local effects of mountains, valleys, or large bodies of water that might influence phenological events. It is recognized that the various plant species comprising the data base do not bloom simultaneously at any one location. Purple common lilac, for example, usually blooms several days earlier than "Red Rothomagensis" lilac. In addition, the density of the observation network is quite variable with some extensive areas devoid of any information. In view of the rather meager data base, this map can only be considered an approximation of the geographical pattern of the onset of spring throughout the United States in 1971.

Mapping of phenological events in mountainous areas presents many problems because the time of phenological occurrence usually changes very rapidly with small changes in elevation and because of the great influence of slope and aspect. On an average, phenological events at a given elevation come about one day later for every 20 miles of northward distance, while the same one-day difference also generally occurs with a 100-ft change in elevation. Also, it is not uncommon for a phenological event at a given elevation to differ by more than a week due solely to differences in slope and aspect.

Caprio (1966) developed a manual procedure for mapping phenological events in mountainous areas based on elevation stratification of information and final

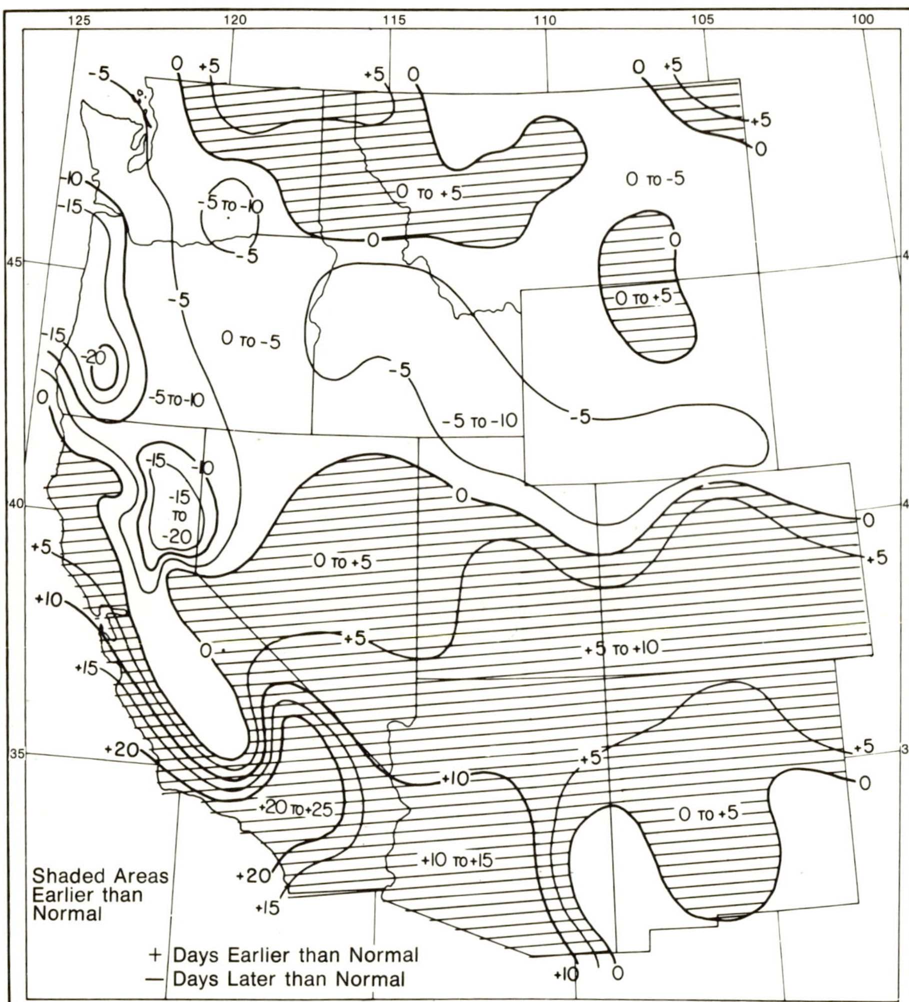

Fig. 2. Handdrawn map of departure of begin lilac bloom date from normal in the western United States, 1971 (in days).

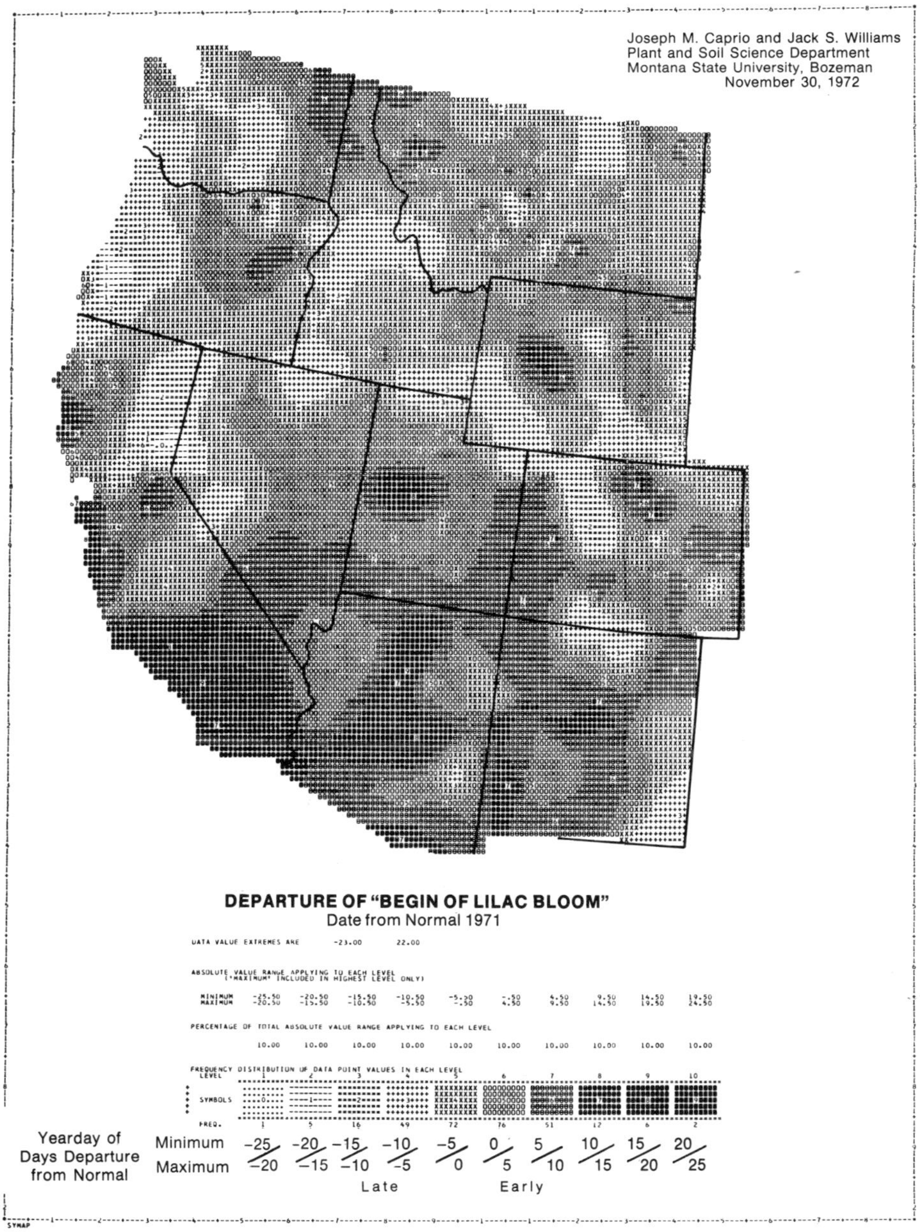

Fig. 3. Computer map of departure from normal of begin lilac bloom date in the western United States, 1971 (in days).

manual integration of equal-elevation maps. Similar types of three-dimensional analytical procedures could no doubt be developed for computer analysis of phenological information.

The computer mapping method was compared with the manual mapping procedure by using information from throughout the western United States on departure of 1971 begin lilac bloom from the 10-year (1957–1966) average. The manually drawn anomaly map of this information was published previously (Caprio, 1972). The anomaly data base consists of about 300 points. The manually drawn anomaly map is shown in Fig. 2 and the SYMAP analysis in Fig. 3.

It is evident that the two maps are quite similar, although some discrepancies appear. The manual and computer-generated maps would undoubtedly correspond even more closely if information from a denser network of stations were available.

References

Caprio, J. M. 1966. Pattern of plant development in the western United States. *Montana State Univ. Ag. Exp. Sta. Bull. 607.* 42 pp.

———. 1972. *A Report to Cooperators of the Phenological Survey in the Western Region of the United States*, 1971. *Weekly Weather and Crop Bulletin, 59*, 4, 10–11, January 24, 1972.

Lieth, H., and J. S. Radford. 1971. Phenology, resource management and synagraphic computer mapping. *BioScience* 21, 62–70.

Reader, R., UNC Applied Programming Group. 1972. SYMAP (version 5.16A) *Instruction Manual.* Doc. No. LSR–139–0. Research Triangle Park, N.C.: Triangle Universities Computation Center. 65 pp. offset.

SYMAP. 1971. *Reference Manual for Synagraphic Computer Mapping.* Version V. Cambridge, Mass.: Laboratory for Computer Graphics and Spatial Analysis, Harvard Graduate School of Design. 60 pp. mimeo.

US/IBP Phenology Committee. 1972. *Report, July 1972.* N. R. French, chairman and ed. Austin, Tex.: US/IBP Environmental Coordinating Office. 54 pp. offset.

Wolaver, T., and H. Lieth. 1972. *The Distribution of Natural and Anthropogenic Elements and Compounds in Precipitation across the U.S.: Theory and Quantitative Models.* Chapel Hill, N.C.: UNC Duplicating Shop. 75 pp. offset.

2.5 Phenology and Genecology of Woody Plants[1]

Harrison L. Flint

Department of Horticulture
Purdue University, Lafayette, Indiana 47901, USA

1 Introduction

Phenology and genecology, far from being separate disciplines, cover much ground in common. Until recently, phenology, the study of periodic phenomena in plants and animals, has been limited in practice largely to observation of visible phenomena. The relationship of plant physiological processes to phenological events seems obvious, but physiological bases for many phenological events have yet to be uncovered. To an increasing extent this is now being done.

Since interest in the species is often greater than that in the individual organism, the relationship between phenology and genecology also is of interest. Turesson (1923) first applied the term genecology to the ecology of races, i.e., to intraspecific variation in relation to their response to environment. Genecology is of interest to phenology not only because of coincidence of subject matter, but also because of the crucial need in phenology to separate genetic and environmentally induced phenological responses. Usefulness of phenological data obtained in the past sometimes has been limited by inattention to the fact of nonrandom genetic variation. Genecology also has much in common with biosystematics in methods; it differs in that its primary objective is to uncover and describe habitat-correlated genetic variation within species (Turesson, 1923) rather than to classify organisms. The history and current problems of genecology have been described in detail by Heslop-Harrison (1964) and Roche (1968).

Further review here of genecological and physiological aspects of phenology will be grouped under these somewhat arbitrary phenophase designations: (1) seed germination, (2) seedling growth, (3) shoot growth in mature trees, (4) photosynthesis, (5) induction of dormancy, (6) cold acclimation, (7) breaking of dormancy, and (8) flowering and fruiting.

[1] Journal Paper 4979, Purdue University Agricultural Experiment Station, West Lafayette, Indiana 47907.

KEYWORDS: woody plants; phenology; genecology; provenance; seeds; germination; growth; shoots; photosynthesis; dormancyinduction; dormancybreaking; acclimatization; flowering; fruiting; bud break.

2 Seed Germination

The phenomenon of seed dormancy in temperate zone woody plants and its adaptive significance are well known, and directly affect the phenology of seed germination. Research over the last 20 years has shown that dormancy in many seeds can be overcome not only by low temperature but also by long days accompanied by high temperature (Isikawa, 1954; Stearns and Olson, 1958; Johnson and Irgens-Moller, 1964).

When geographical variations in chilling requirements have been examined, the most common finding has been that seeds of northern origin require more chilling than seeds of the same species from southern sources. This is true with *Abies alba* (Messeri *et al.*, 1963), *Acer saccharum* (Kriebel, 1958), and *Liquidambar styraciflua* L. (Wilcox, 1968; Winstead, 1969). With *Pinus strobus* L., however, the situation is reversed. Two investigations (Mergen, 1963; Fowler and Dwight, 1964) have shown that chilling requirement for breaking dormancy in seeds of northern origin is less than for seeds of southern origin. Little is known about geographical variation in chilling requirement in many tree species. When more species have been studied, the case of *Pinus strobus* L. may prove not to be the anomaly that it appears to be at present.

Some woody species are known to require a combination of treatments, e.g., warm stratification followed by cold stratification (Giersbach and Crocker, 1929; Giersbach, 1937; Villiers and Wareing, 1965). The function of warm stratification apparently can be either (1) to hasten fungal decomposition of resistant seed coats or (2) to effect radicle growth, prerequisite in some species to breaking dormancy in the epicotyl.

In at least one case, *Fraxinus mandshurica* Rupr., incomplete chilling can be compensated for by the use of alternating temperatures during germination (Asakawa, 1957). The temperature range over which germination occurs in *Liquidambar styraciflua* L. is much greater for fully chilled than for partly chilled seeds, with abnormally high temperature (30° to 38°C) apparently substituting for part of the chilling requirement (Bonner and Farmer, 1966).

Germination of unchilled seed of *Tsuga canadensis* L. varies with temperature, photoperiod, and the interaction of these two factors (Stearns and Olson, 1958), i.e., optimum photoperiod varies from 17° to 22°C at an 8- or 12-hr day to 27°C at a 16-hr day. Full chilling (10 weeks) eliminates most light and temperature effects in this species. Germination of excised embryos is rapid and independent of both photoperiod and temperature within these ranges, suggesting the presence of an inhibitor in the endosperm or nucellus. Rate of germination of unchilled seed of *Pseudotsuga menziesii* (Mirb.) Franco increases significantly with a short night treatment (2-hr light break in the middle of the normal dark period), but only below 25°C (Johnson and Irgens-Moller, 1964). Like other photoperiodic phenomena, this is controlled by the reversible red – far red light reaction. No genetic variation with respect to this requirement has been found.

Several types of photoperiodic responses have been found in germination of seeds of herbs and grasses (Isikawa, 1954). One type is characterized by an optimum photoperiod, ranging from very short (3 hr in the case of *Silene armeria* L.) to very long (21 hr in the case of *Eragrostis ferruginea* Beauv.). A second type involves a critical photoperiod, below which essentially no germination occurs, and just above it an optimum photoperiod, beyond which germination is progressively less, as in *Rumex* sp. Certain seeds previously thought to be inhibited by light appear to constitute a special case within the latter type: Even though they are in fact inhibited by light over most of the photoperiodic range, they are stimulated beyond their performance in darkness by 1 to 10 min of light per cycle. Examples of this response include *Eschscholzia californica* Cham., *Nigella damascena* L., and *Veronica persica* Poir. It may be that similar phenomena occur in woody species as well, to be found in further research.

Differences in rate of germination with geographical origin have been noted in several tree species, but the effects of delay for lack of chilling usually have confounded observations of germination velocity. More attention to separating these effects seems warranted in future work. In the case of *Picea mariana* (Mill.) B.S.P., Morgenstern (1969) has found that the germination rate of fully chilled seeds increases with increasing latitude, logical in light of the very short growing season in the northern part of this species' range.

Optimum temperatures for seed germination have been investigated in many species (USDA, 1948). Relationship of optimum temperature to geographical source has been determined for only a few tree species, with variable results. Perner (1966) germinated seeds of 11 species of *Pinus* at temperatures from 8° to 32°C and found, with a few exceptions, that those having the most northerly distributions tended to have optimum germination temperatures near 16°C, while those from more southerly regions had optima closer to 24° or 32°C. In *Liquidambar styraciflua* L. (Winstead, 1969), germination of seeds of northern origin is more temperature-sensitive than that of southern (Mexican) sources.

3 Seedling Growth

Growth of many temperate zone tree species is controlled by photoperiod (Downs and Borthwick, 1956; Wareing, 1956; Vaartaja, 1957; Nitsch, 1957, 1968). Nitsch (1957, 1968) has reviewed the photoperiodic growth responses of many tree species and has grouped them according to type of response. More recently, photoperiodic responses have been documented in additional species, including *Pinus roxburghii* Sarg. (Bhatnagar *et al.*, 1970), *Pseudotsuga menziesii* (Mirb.) Franco (Lavender *et al.*, 1968), and *Sequoia sempervirens* (D. Don) Endl. (Hellmers and Pharis, 1968).

Tree seedlings of northern origin have been noted frequently to stop growth earlier than those of southern origin growing in the same plantations: *Acer saccharum* Marsh. (Kriebel, 1958), *Tsuga canadensis* (L.) Carr. (Nienstaedt and Olson, 1961), *Pinus strobus* L. (Genys, 1968), *Picea mariana* (Mill.) B.S.P. (Morgenstern, 1969), and *Picea abies* (L.) Karst. (Wright, 1944; Robak and Magnesen,

1970). A similar relationship exists with respect to elevational differences, i.e., seedlings from highest altitudes stop growing first (Genys, 1968; Robak and Magnesen, 1970).

The relationship of photoperiodic response to geographical origin has been worked out for several species through experiments in controlled environment chambers. The usual relationship has been that seedlings of northern origin respond more strongly to photoperiod than those of southern origin, as in *Liriodendron tulipifera* L. and *Quercus macrocarpa* Michx. (Vaartaja, 1961); *Populus deltoides* Bartr. and *P. tremuloides* Michx. (Vaartaja, 1960); *Pinus echinata* Mill., *P. palustris* Mill., and *P. taeda* L. (Allen and McGregor, 1962); and *P. sylvestris* L. (Jensen and Gatherum, 1965), or respond at a different critical photoperiod as with *Picea abies* (L.) Karst. (Dormling *et al.*, 1968)—confirming the supposed role of photoperiod in causing intraspecific differences in time of cessation of growth in summer. This relationship can be seen at different elevations as well as at different latitudes; e.g., *Pseudotsuga menziesii* (Mirb.) Franco, in which seedlings of continental origin in the Pacific Northwest respond more strikingly to photoperiod than seedlings of coastal origin (Irgens-Moller, 1958), a phenomenon also found in *Cornus stolonifera* Michx. (Smithberg and Weiser, 1968). Seedlings of different origin do not always show differences in response to photoperiod when they represent only modest differences in latitude, e.g., *Pinus densiflora* Sieb. & Zucc. and *P. luchuensis* Mayr from origins about 5° latitude apart (Ikemoto, 1960); but in other cases noticeable differences can be seen even at such small latitudinal intervals, e.g., *Taxodium distichum* (L.) Rich (Flint, 1972b).

Optimum temperatures for growth of seedlings of a number of tree species have been determined under controlled conditions, but in only a few cases has geographical variation been observed. Differences in ability to grow at low temperatures have appeared among provenances of *Pinus sylvestris* L. differing in latitude by 10° (Jensen and Gatherum, 1965). On the other hand, Arizona, California, and South Dakota provenances of *Pinus ponderosa* Dougl. & Laws. differ in growth rate but not in their response to temperature (M. Larson, 1967).

Seedlings of several conifers appear to have definite thermoperiodic requirements, but so few species have been investigated that generalization is difficult. *Abies magnifica* A. Murr., which occurs at elevations of 2000 to 3000 m in the Cascade and Sierra Nevada ranges, grows best with a day-night temperature differential of about 13°C, and it apparently makes little difference whether the day or the night temperature is higher (Hellmers, 1966a). *Acer rubrum* L. (Perry, 1962) and *Pinus taeda* L. (Kramer, 1957) also have been found to have high thermoperiodic requirements. A contrasting example, *Sequoia sempervirens* (D. Don) Endl., occurring at elevations mostly below 1000 m., shows very little response to day-night temperature differential (Hellmers 1966b). Perry (1962) has found that optimum day-night temperatures for *Pinus taeda* L. and *Acer rubrum* L. correspond closely to growing-season day and night temperatures at geographical origin.

Since individual species appear to have become adapted to the thermoperiod of their habitat, it seems likely that intraspecific populations also have done so, but there is little evidence to support or reject this idea.

4 Shoot Growth in Mature Trees

Nienstaedt (see Chapter 6.3), reviews in detail the phenology of shoot growth in provenance plantations of tree species; thus the review here will be cursory.

Photoperiodic growth responses in most cases do not change qualitatively after the juvenile period. Differential response to latitude or photoperiod, related to geographically based genetic variation, has been found in *Cornus stolonifera* Michx. (Smithberg and Weiser, 1968), *Liquidambar styraciflua* L. (Winstead, 1969), *Pinus contorta* Dougl. (Sweet and Wareing, 1968; Hagner 1970b), *Pinus elliottii* Engelm. (Bengston *et al.*, 1967), *Pinus sylvestris* L. (Hagner, 1970a), *Pinus strobus* L. (Santamour, 1960), and *Pinus taeda* L. (Perry *et al.*, 1966). Differential response related to elevation of origin has been found in *Pinus jeffreyi* Grev. & Balf. and *P. ponderosa* Dougl. & Laws. (Callaham and Metcalf, 1959) and in *Eucalyptus regnans* F. Muell. (Ashton, 1958).

Breaking of quiescence in buds in spring is controlled more by temperature than by photoperiod (Pauley and Perry, 1954; Irgens-Moller, 1957; Mándy and Kárpáti, 1958; Wang and Perry, 1958; Smithberg and Weiser, 1968; Perry, 1971). But there is evidence in some species that breaking of buds is inhibited by short days (Wareing, 1953; Nagata, 1968). Bud breaking and first growth in *Picea abies* (L.) Karst. are independent of photoperiod, but succeeding growth is highly photoperiod-dependent (Leikola, 1970).

Variation, where it exists in the pines, follows a close relationship with latitude and/or elevation of the origin, except in *P. elliottii* Engelm. (Bengston *et al.*, 1967), whose range is limited to southern latitudes. *Pinus resinosa* Ait. exhibits very little variation in this respect (Hough, 1967; P. Larson, 1967); it is wide-ranging with respect to longitude but not with respect to latitude or elevation.

Larix decidua Mill. (Leibundgut, 1959), *L. sibirica* Ledeb. (Dafis, 1962), *Picea abies* (L.) Karst. (Dietrichson, 1969), and *Tsuga canadensis* (L.) Carr. (Ferchau, 1959) also exhibit considerable variation in growth, but with little or no latitudinal or elevational pattern.

5 Photosynthesis

Photosynthesis functions not only by providing food materials for support of growth and development, but also in direct support of the photoperiodic reaction, since for accomplishment of its regulatory function the dark period must be preceded by a period of sufficiently high intensity illumination (Wareing, 1956).

The usual observation on seasonality of photosynthesis in deciduous trees is that from a winter minimum the process quickly increases to a maximum by late spring, persists at about that level for two to three months, and then decreases steadily until defoliation in autumn. Evergreen trees differ primarily in that a low rate of photosynthetic activity persists in winter during periods near or above 0°C (Saeki and Nomoto, 1958; Bourdeau, 1959; Parker, 1961a; McGregor and Kramer,

1963; Krueger and Ferrell, 1965; Bamberg, 1967). There is no evidence that the periodicity of the process varies within species in any geographically related manner.

Pinus strobus L. seedlings of southern origin are more efficient photosynthetically at low light intensities, but less efficient at low temperatures, than northern seedlings (Bourdeau, 1963). Optimum light intensity for photosynthesis in *Tsuga canadensis* (L.) Carr. is higher in southern (Tennessee) than in northern (New York) seedlings (Ferchau, 1959). In the Pacific Northwest seedlings of *Pseudotsuga menziesii* (Mirb.) Franco var. *glauca* from western Montana have a lower photosynthetic rate, and require more light to reach the compensation point at a given temperature, than seedlings of *P. menziesii* var. *menziesii* from Vancouver Island. Light for saturation is similar for both varieties (Krueger and Ferrell, 1965).

6 Induction of Dormancy

More or less concomitant with cessation of active growth in summer is entry into the dormant condition, characterized by cessation of cell division in apical and axial meristems and the presence of a winter chilling requirement which must be met before normal growth will resume. Vegis (1964) and Perry (1971) have reviewed in considerable detail the literature on the nature and control of dormancy, so the present review will be limited to aspects of interest to phenology and genecology.

Induction of dormancy appears to be controlled primarily by photoperiod, but the critical photoperiod varies among and within species. Within a species, trees of northern origin usually enter dormancy earlier (at longer photoperiods) than those of southern origin. This has been documented in *Liquidambar styraciflua* L. (Winstead, 1969); *Populus deltoides* Bartr., *P. tacamahacca* Mill., and *P. trichocarpa* Torr. & Gr. (Pauley and Perry, 1954); *Picea abies* (L.) Karst. (Robak and Magnesen 1970); *Pinus contorta* Dougl. (Sweet and Wareing, 1968; Hagner, 1970b); *P. strobus* L. (Genys, 1968); and *Pseudotsuga menziesii* (Mirb.) Franco (Irgens-Moller, 1958). Regarding *Picea abies* and the two *Pinus* species, trees from greater elevations have been found to enter dormancy earlier than those from lesser elevations. Photoperiod also controls induction of dormancy in *Larix kaempferi* (Lamb.) Sarg. (Kondo *et al.*, 1962).

The effect of long days in delaying dormancy may be modified, inhibited or nullified by low temperature (Downs and Borthwick, 1956), and dormancy can be inhibited by high temperature, as seen in species of *Populus* growing in Australia (Paton and Willing, 1968). More commonly, however, short days seem to operate independently of temperature within the range of about 15° to 30°C, but much more work is needed in this area.

7 Cold Acclimation

Cold resistance of woody plants has been reviewed by several investigators, most recently by Weiser (1970) and Alden and Hermann (1971). Hardening, a periodic phenomenon usually beginning with the advent of dormancy, but not directly related to dormancy (Irving and Lanphear, 1967), is now conceived as occurring in two steps in nature (van Huystee, 1965). The first phase, causing hardening to levels of $-15°$ to $-20°C$ (more or less) is induced in late summer and early autumn by short days at temperatures that would favor growth under long days. The second phase, causing hardening to levels of $-40°$ to $-80°C$, or sometimes lower, is brought on by exposure to low temperature (below $\sim 5°C$) for at least part of the daily cycle. The basic pattern of hardening and dehardening with hardiness maxima in mid- to late winter and minima in early to midsummer is qualitatively similar in temperate zone woody plants (Moschkov, 1935; Ulmer, 1937; Pisek, 1950; Parker, 1955, 1961b; Till, 1956; McGuire, *et al.*, 1962; Sakai, 1970).

Intraspecific differences in hardiness related to geographical origin were observed early in this century in *Juglans nigra* L. (Emerson, 1906) and *Pseudotsuga menziesii* (Mirb.) Franco (Frothingham, 1909), and have since been found in a variety of tree species: *Acer saccharum* Marsh. (Kriebel, 1957); *Fraxinus americana* L. (Wright, 1944); *Liquidambar styraciflua* L. (Winstead, 1969); *Liriodendron tulipifera* L. (Limstrom and Finn, 1956; Funk, 1958); *Pinus banksiana* Lamb. (Yeatman, 1967); *Pinus ponderosa* Laws. (Mirov *et al.*, 1952; Squillace and Silen, 1962); *Pinus echinata* Mill. and *P. taeda* L. (Minckler, 1951); *Pinus pinaster* Ait. (Bouvarel, 1960); *Pinus strobus* L. (Maronek and Flint, 1972); *Pinus sylvestris* L. (Hagner, 1970b); *Quercus rubra* L. (Flint, 1972a); *Tsuga canadensis* (L.) Carr. (Nienstaedt, 1958); and others. Differences in hardiness have been related to climate of the geographical origin in some species: *Liriodendron tulipifera* L. (Funk, 1958), *Pinus pinaster* Ait. (Bouvarel, 1960), *Pinus strobus* L. (Maronek and Flint, 1974), and *Quercus rubra* L. (Flint, 1972a).

Hardiness also has been related to latitude, assuming that higher latitudes have severer climate, for the following species: *Liquidambar styraciflua* L. (Winstead, 1969), *Pinus banksiana* Lamb. (Yeatman, 1967), and *Pinus contorta* Dougl. (Hagner, 1970b). Mergen and Thielges (1967) have related nuclear volume in *Picea glauca* (Moench) Voss, *P. sitchensis* (Bong.) Carr., *Pinus banksiana* Lamb., and *P. sylvestris* L. to latitude of origin, and suggested the possible relationship of this phenomenon to cold hardiness. On the other hand, three-year-old seedlings of *Abies concolor* (Gord. & Glend.) Lindl. of northern origin have been found to suffer more winter injury than those of more southern origin (Conkle *et al.*, 1967), apparently because of differential susceptibility to fluctuating temperature in winter.

Elevation of the origin has been shown to be related to differences in hardiness of different provenances in *Pseudotsuga menziesii* (Mirb.) Franco (Frothingham, 1909), *Eucalyptus* spp. (Ashton, 1958; Boden, 1958; Green, 1969), *Quercus rubra* L. (Flint, 1972a), and, at least by implication, to differences in time of induction of dormancy in *Picea abies* (L.) Karst. (Robak and Magnesen, 1970) and *Pinus strobus* L. (Genys, 1968). The unexpected latitudinal response of *Abies concolor* (Conkle *et al.*, 1967) was accompanied by a similar response to altitude, i.e., three-year-old seedlings from lower elevations sustained less winter injury than those from higher elevations. When altitudinal variants of *Pinus ponderosa* Dougl. & Laws were cross-transplanted, no differences in winter injury were seen despite substantial differences in growth (Mirov *et al.*, 1952).

8 Breaking of Dormancy

It has long been known that winter dormancy is ended in nature by accumulation of low temperatures. Exposure to 5°C for 6 to 10 weeks is sufficient, under experimental conditions, to fully break dormancy in *Liquidambar styraciflua* L. (Farmer, 1968), *Picea abies* (L.) Karst. (Worrall and Mergen, 1967), *Picea glauca* (Moench) Voss (Nienstaedt, 1966), *Pinus densiflora* Sieb. & Zucc. (Nagata, 1967), *Pinus strobus* L. (Berry, 1965), *Pinus sylvestris* L. (Hoffmann and Lyr, 1967), and *Tilia platyphyllos* Scop. (Lyr *et al.*, 1970); but *Tilia americana* L. requires a somewhat longer chilling period to break dormancy fully (Ashby, 1962). Partial chilling in some cases allowed resumption of growth but at less than the rate following full chilling (Hoffmann and Lyr, 1967; Worrall and Mergen, 1967).

The period of winter chilling at northern latitudes is far greater than the minimum required for full breaking of dormancy, suggesting that some other mechanism(s) must act to prevent premature growth during warm weather in late winter. Perry (1971) points out that some species require a minimum number of hours at about 25°C before growth will resume. Short days can inhibit growth of fully chilled plants of *Pseudotsuga menziesii* (Mirb.) Franco (Irgens-Moller, 1957) and *Pinus densiflora* Sieb. & Zucc. (Nagata, 1968), but under natural conditions flushing of *Picea abies* (L.) Karst. in early summer seems not to be greatly influenced by photoperiod during the first two or three weeks (Leikola, 1970).

Forest geneticists have attempted to take advantage of the relatively short minimum chilling period to accelerate the life cycle and thus shorten the juvenile stage of certain tree species. Lavender (1962) was able to accomplish two full growth cycles per year with *Pseudotsuga menziesii* (Mirb.) Franco and *Pinus ponderosa* Dougl. & Laws., and Dormling and co-workers (1968) were able to shorten the annual growth cycle of *Picea abies* (L.) Karst. to 15 weeks, thereby making three cycles per year possible in controlled environment. Under experimental conditions, long days can be substituted for the chilling requirement in *Liquidambar* (Farmer, 1968), *Pinus densiflora* Sieb. & Zucc. (Ikemoto, 1961; Nagata, 1967), *Pinus sylvestris* L. (Hoffmann and Lyr, 1967), and *Pinus coulteri* D. Don and *Pseudotsuga menziesii* (Mirb.) Franco (Hellmers, 1959). Critical photoperiod for this substitution was found to be between 12 and 14 hr for *Pinus densiflora* and around 13 hr for *P. sylvestris*. In other cases, long days can substitute for part but not all of the chilling requirement, as in *Tilia platyphyllos* Scop. (Lyr *et al.*, 1970) and *Ulmus americana* L. (Roberts and Main, 1965).

The dormant period has been extended in some tree species by applying gibberellic acid in autumn (Brian, *et al.*, 1959). Bud break was delayed one to three weeks the following spring in *Acer pseudoplatanus* L., *Betula pendula* Roth, *Fagus sylvatica* L., *Fraxinus excelsior* L., and *Sorbus aucuparia* L. In other species—*Acer rubrum* L., *Castanea sativa* Mill., *Parthenocissus tricuspidata* (Sieb. & Zucc.) Planch., *Rhododendron* x *kosterianum* Schneid., *Taxodium distichum* (L.) Rich., and *Ulmus procera* Salisb.—no such effect was seen.

Intraspecific differences in chilling requirement for breaking dormancy have been observed in several tree species. In most cases, e.g. *Betula papyrifera* Marsh.

and *B. alleghaniensis* Britt. (Wang and Perry, 1958), *Acer rubrum* L. (Perry and Wang, 1960), *Acer saccharum* Marsh. (Kriebel and Wang, 1962), several species of *Picea* (Nienstaedt, 1967), and *Pseudotsuga menziesii* (Mirb.) Franco (Wommack, 1960), chilling requirement has been correlated with severity of cold at the geographical origin.

9 Flowering and Fruiting

Phenology of flowering in trees is of obvious interest to the tree breeder. Flowering of a species probably is controlled by temperature in most tree species, e.g., *Pinus monticola* Dougl. (Bingham and Squillace, 1957); *Corylus avellana* L. (Sokolowska, 1962); *Betula pendula* Roth, *Picea abies* (L.) Karst., *Pinus sylvestris* L., and *Quercus robur* L. (Chalupa, 1964); and *Larix decidua* Mill., *Pinus sylvestris* L., and *Populus tremula* L. (Sarvas, 1969). Considerable information is available on phenology of flowering in tree species (Schnelle, 1955), but relatively little is known about intraspecific genetic variation in response of flowering to temperature or other environmental factors. Linsser (1867) has been given credit (Sarvas, 1969) for discovering the principle that the mean temperature sum (degree days) required for a given phenological event is a constant fraction of the average annual temperature sum for the locality in question. More recently, this relationship has been expressed mathematically in the study of anthesis in *Pinus sylvestris* L. and the geographical limits of the relationship have been established (Sarvas, 1969). Burk (1966) has made the interesting observation, although not on a tree species, that camphorweed (*Heterotheca subaxillaris*) flowers in a uniform garden at times correlated with periodicity of rainfall in native habitat.

Although botanical garden records doubtless contain considerable information on intraspecific variation in phenology of fruiting and seed ripening, very little of it has found its way into the scientific literature.

10 Conclusions

It is clear from the literature that phenology and genecology are complementary and in some cases inseparable. The phenologist who uses naturally occurring species as indicator plants runs the risk that intraspecific genetic variation may confound his phenological observations. Genecology can provide a means to lessen that risk by evaluating the genetic contribution to phenological differences observed. Other solutions include the use of genetically uniform populations, such as cultivars, as has often been done in phenological studies (Schnelle, 1955; Caprio, 1971; Hopp, 1971) or the use of cultivated exotic species, in which genetic variation is assumed to be more or less randomly distributed over the experimental area, as in the case of *Syringa vulgaris* L. in the western United States (Caprio, 1966).

Phenology and genecology also are important components of autecology, that interdisciplinary arena in which plant physiology, genetics, and morphology are combined to answer complex questions of adaptation, bioclimatology, plant geography, and reproductive biology. It seems inevitable, considering the increasingly complex and diverse problems that phenology is faced with, that in the future this field will become more dependent upon other disciplines.

References

Alden, J., and R. K. Hermann. 1971. Aspects of the cold hardiness mechanism in plants. *Bot. Rev.* 37, 37–142.

Allen, R. M., and W. H. D. McGregor. 1962. Seedling growth of three southern pine species under long and short days. *Silvae Genet.* 11, 43–45.

Asakawa, S. 1957. Studies in the delayed germination of *Fraxinus mandshurica* var. *japonica* seeds. 7. Thermoperiodicity in germination. *Bull. For. Exp. Sta. Meguro, Tokyo No. 103.* 25–44.

Ashby, W. C. 1962. Bud break and growth of basswood as influenced by daylength, chilling, and gibberellic acid. *Bot. Gaz.* 123, 162–170.

Ashton, D. H. 1958. The ecology of *Eucalyptus regnans* F. Muell.: the species and its frost resistance. *Australian J. Bot.* 6, 154–176.

Bamberg, S. 1967. Influence of daylength on the photosynthetic capacity of stone pine (*Pinus cembra* L.). *Ecology* 48, 264–269.

Bengston, G. W., W. H. D. McGregor, and A. E. Squillace. 1967. Phenology of terminal growth in slash pine: Some differences related to geographic seed source. *Forest Sci.* 13, 402–412.

Berry, C. R. 1965. Breaking dormancy in eastern white pine by cold and light. *Southeast Forestry Exp. Sta. Res. Note No. SE-43, U.S. Forest Serv.*, 3 pp.

Bhatnagar, H. P., B. B. Gupta, and B. S. Rauthan. 1970. Response of Chir pine (*Pinus roxburghii* Sarg.) seedlings to various photoperiods. *Indian Forester* 96, 600–606.

Bingham, R. T., and A. E. Squillace. 1957. Phenology and other features of the flowering of pines, with special reference to *Pinus monticola* Dougl. *Intermountain Forest and Range Exp. Sta., Res. Paper No. 53*, U.S. Forest Serv. 26 pp.

Boden, R. W. 1958. Differential frost resistance within one *Eucalyptus* species. *Australian J. Sci.* 2, 84–86.

Bonner, F. T., and R. E. Farmer, Jr. 1966. Germination of sweetgum in response to temperature, moisture stress, and length of stratification. *Forest Sci.* 12, 40–43.

Bourdeau, P. F. 1959. Seasonal variations of the photosynthetic efficiency of evergreen conifers. *Ecology* 40, 63–67.

——— 1963. Photosynthesis and respiration of *Pinus strobus* L. seedlings in relation to provenance and treatment. *Ecology* 44, 710–716.

Bouvarel, P. 1960. Note sur la résistance au froid de quelques provenances de pin maritime. *Rev. For. Franc.* 12, 495–508.

Brian, P. W., J. H. P. Petty, and P. T. Richmond. Extended dormancy of deciduous woody plants treated in autumn with gibberellic acid. *Nature* 184, 69.

Burk, C. J. 1966. Rainfall periodicity as a major factor in the flowering of races of camphorweed (*Heterotheca subaxillaris*). *Am. J. Bot.* 53, 933–936.

Callaham, R. Z., and W. Metcalf. 1959. Altitudinal races of *Pinus ponderosa* confirmed. *J. Forestry* 57, 500–502.

Caprio, J. M. 1966. Pattern of plant development in the western United States. *Western Reg. Res. Pub., Montana Agr. Exp. Sta. Bull. No. 607.* 42 pp.

——— 1971. Pretty shrubs and vines predict crop progress. *Weekly Weather Crop Bull.* Feb. 22, p. 11.

Chalupa, V. 1964. [The flowering of forest trees.] *Práce Výzkum Úst. Lesn. ĈSSR No. 28*, 139–173 (1964); abstr. in *For. Abstr.* 26, 537 (1965).

Conkle, M. T., W. J. Libby, and J. L. Hamrick, III. 1967. Winter injury among white fir seedlings: Unusual pattern in seed source study. *Pacific Southwest. Forest and Range Exp. Sta. Res. Note No. PSW-138*, U.S. Forest Serv., 7 pp.

Dafis, S. 1962. [The effect of additional light on height growth of different provenances of Siberian larch.] *Schweiz Z. Forstw.* 113, 333–337; Abstr. in *For. Abstr.* 24, 23 (1963).

Dietrichson, J. 1969. [The geographic variation of spring frost resistance and growth cessation in Norway spruce (*Picea abies* (L.) Karst.)] *Medd. Norske Skogforsøksv.* 27, (96), 91–106; Abstr. in *For. Abstr.* 31, 261 (1970).

Dormling, I., A. Gustafsson, and D. von Wettstein. 1968. The experimental control of the life cycle in *Picea abies* (L.) Karst. I. Some basic experiments on the vegetative cycle. *Silvae Genet.* 17, 44–64.

Downs, R. J., and H. A. Borthwick. 1956. Effect of photoperiod on growth of trees. *Bot. Gaz.* 117, 310–326.

Emerson, R. A. 1906. The relation of early maturity to hardiness in trees. *Nebr. Agr. Exp. Sta. Ann. Rept.* for 1905, 19, 101–110.

Farmer, R. E., Jr. 1968. Sweetgum dormancy release: effects of chilling, photoperiod, and genotype. *Physiol. Plant.* 21, 1241–1248.

Ferchau, H. A. 1959. Ecotypic variability of *Tsuga canadensis*: Photosynthesis and related processes. Abstr. in *Dissert. Abstr.* 19, 2440.

Flint, H. L. 1972. Cold hardiness of twigs of *Quercus rubra* L. as a function of geographic origin. *Ecology* 54, 1163–1174.

——— 1972b. Photoperiodic dependence of stem growth in two seed sources of *Taxodium distichum* (L.) Rich. Unpublished data.

Fowler, D. P., and T. W. Dwight. 1964. Provenance differences in the stratification requirements of white pine. *Can. J. Bot.* 42, 669–675.

Frothingham, E. H. 1909. Douglas-fir: A study of the Pacific Coast and Rocky Mountain forms. *U.S. Forest Serv. Circ. 150*. 38 pp.

Funk, D. T. 1958. Frost damage to yellow-poplar varies by seed source and site. *Sta. Note No. 115*. U.S. Forest Serv. Centr. States Forest Exp. Sta. 2 pp.

Genys, J. B. 1968. Geographic variation in eastern white pine. *Silvae Genet.* 17, 6–12.

Giersbach, J. 1937. Germination and seedling production of species of viburnum. *Contrib. Boyce Thompson Inst.* 9, 79–90.

———, and W. Crocker. 1929. Germination of *Ilex* seeds. Abstr. *Am. J. Bot.* 16, 854–855.

Green, J. W. 1969. Temperature responses in altitudinal populations of *Eucalyptus pauciflora* Sieb. ex Spreng. *New Phytol.* 68, 399–410.

Hagner, M. 1970a. A genecological investigation of the annual rhythm of *Pinus sylvestris* L. *Stud. For. Suec. Skogshøgsk. No. 80*. 40 pp.

——— 1970b. A genecological investigation of the annual rhythm of *Pinus contorta* Dougl. and a comparison with *Pinus sylvestris* L. *Stud. For. Suec. Skogshøgsk No. 81*. 26 pp.

Hellmers, H. 1959. Photoperiodic control of bud development in Coulter pine and bigcone Douglas-fir. *Forest Sci.* 5, 138–141.

——— 1966a. Temperature action and interaction of temperature regimes in the growth of red fir seedlings. *Forest Sci.* 12, 90–96.

——— 1966b. Growth responses of redwood seedlings to thermoperiodism. *Forest Sci.* 12, 276–283.

———, and R. P. Pharis. 1968. Influence of photoperiod and photoperiodic cycles on the growth of coastal redwood seedlings. *Bot. Gaz.* 129, 53–57.

Heslop-Harrison, J. 1964. Forty years of genecology. *Advan. Ecol. Res.* 2, 159–247.

Hoffmann, G., and H. Lyr. 1967. Über die Wirkung der winterlichen Thermoperiode auf das Wurzel- und Sprosswachstum von *Pinus sylvestris* L. *Flora* 158A, 373–383.

Hopp, R. J. 1971. Charting the progress of spring. *Weekly Weather Crop Bull.* Apr. 19, p. 12.

Hough, A. F. 1967. Twenty-five year results of a red pine provenance study. *Forest Sci.* 13, 156–166.

Huystee, R. B. van. 1965. Cold acclimation and accompanying metabolic changes in red-osier dogwood with emphasis on proteins. Abstr. *Dissert. Abstr.* 25, 4898–4899.

Ikemoto, A. 1960. Effect of daylength and temperature on the elongation of main axis in first-year seedlings of *Pinus densiflora* and *P. luchuensis.* (Engl. summary) *J. Jap. For. Soc.* 42, 172–175.

——— 1961. Effect of day-length upon breaking of dormancy in one-year-old seedlings of *Pinus densiflora. J. Jap. For. Soc.* 43, 162–165.

Irgens-Moller, H. 1957. Ecotypic response to temperature and photoperiod in Douglas-fir. *Forest Sci.* 3, 79–83.

——— 1958. Genotypic variation in the time of cessation of height growth in Douglas-fir. *Forest Sci.* 4, 325–330.

Irving, R. M., and F. O. Lanphear. 1967. Environmental control of cold hardiness in woody plants. *Plant Physiol.* 42, 1191–1196.

Isikawa, S. 1954. Light sensitivity against the germination. I. "Photoperiodism" of seeds. *Tokyo Univ. Bot. Mag.* 67, 51–56.

Jensen, K. F., and G. E. Gatherum. 1965. Effects of temperature, photoperiod, and provenance on growth and development of Scotch pine seedlings. *Forest Sci.* 11, 189–199.

Johnson, L. C., and H. Irgens-Moller. 1964. Effect of photoperiod and light quality on germination of Douglas-fir seed. *Forest Sci.* 10, 200–205.

Kondo, K., K. Muto, and S. Takikawa. 1962. [The effect of photoperiodic treatment on the growth of Japanese larch seedlings.] *Res. Bull. Exp. For. Hokkaido Univ.* 21, 283–300; Abstr. in *For. Abstr.* 24, 389 (1963).

Kramer, P. J. 1957. The effects of various combinations of day and night temperature and photoperiod on the height growth of loblolly pine seedlings. *Forest Sci.* 3, 45–55.

Kriebel, H. B. 1957. Patterns of genetic variation in sugar maple. *Ohio Agr. Exp. Sta. Res. Bull. No. 791.* 56 pp.

——— 1958. Geographic differentiation in seed dormancy and juvenile growth rate of Ontario sugar maple. *Proc. 6th Mtg. Comm. For. Tree Breeding Canada.* Montreal, Part 2.

———, and C.-W. Wang. 1962. The interaction between provenance and degree of chilling in bud-break of sugar maple. *Silvae Genet.* 11, 125–130.

Krueger, K. W., and W. K. Ferrell. 1965. Comparative photosynthetic and respiratory responses to temperature and light by *Pseudotsuga menziesii* var. *menziesii* and var. *glauca* seedlings. *Ecology* 46, 794–801.

Larson, M. M. 1967. Effect of temperature on initial development of ponderosa pine seedlings from three sources. *Forest Sci.* 13, 286–294.

Larson, P. R. 1967. Effects of temperature on the growth and wood formation of ten *Pinus resinosa* sources. *Silvae Genet.* 16, 58–65.

Lavender, D. P. 1962. The growth of seedlings of some coniferous species in a controlled environment. Abstr. in *Dissert. Abstr.* 23, 1892.

Lavender, D. P., K. K. Ching, and R. K. Hermann. 1968. Effect of environment on the development of dormancy and growth of Douglas-fir seedlings. *Bot. Gaz.* 129, 70–83.

Leibundgut, H. 1959. Untersuchungen uber europäische Lärchen verschiedener Herkunft.

2. Mitteilung. Der jährliche Verlauf des Höhenwachstums. *Schweiz. Z. Forstw.* 110, 309–314.

Leikola, M. 1970. The effect of artificially shortened photoperiod on the apical and radial growth of Norway spruce seedlings. *Ann. Bot. Fenn.* 7, 193–202.

Limstrom, G. A., and R. F. Finn. 1956. Seed source and nursery effects on yellow-poplar plantations. *J. Forestry* 54. 828–831.

Linsser, C. 1867. Die periodischen Erscheinungen des Pflanzenlebens in ihrem Verhältniss zu den Wärmeerscheinungen. *Mém. Akad. Imp. Sci. S. Pétersbourg* VII, sér. XI, No. 7.

Lyr, H., G. Hoffmann, and R. Richter. 1970. On the chilling requirements of dormant buds of *Tilia platyphyllos* Scop. *Biochem. Physiol. Pflanzen* 161, 133–141.

Mándy, G., and I. Kárpáti. 1958. [Determination of the temperatures required for flushing by various tree species.] *Idöjárás* 62, 261–266. Abstr. in *For. Abstr.* 21, 558 (1960).

Maronek, D., and H. L. Flint. 1974. Geographic variation in cold hardiness in *Pinus strobus* L. Unpublished data.

McGregor, W. H. D., and P. J. Kramer. 1963. Seasonal trends in rates of photosynthesis and respiration of loblolly pine and white pine seedlings. *Amer. J. Bot.* 50, 760–765.

McGuire, J. J., H. L. Flint, and E. P. Christopher. 1962. Cold storage of Christmas trees. *Rhode Island Agr. Exp. Sta. Bull. No. 362.* 12 pp.

Mergen, F. 1963. Ecotypic variation in *Pinus strobus* L. *Ecology* 44, 716–727.

———, and B. A. Thielges. 1967. Intraspecific variation in nuclear volume in four conifers. *Evolution* 21, 720–724.

Messeri, A., M. Innamorati, and Z. Jacopi. 1963. [Germination and growth of *Abies alba* in different temperatures and photoperiods.] *G. Bot. Ital.* 70, 287–302. abstr. in *For. Abstr.* 26, 369 (1965).

Minckler, L. S. 1951. Southern pines from different geographic sources show different responses to low temperature. *J. Forestry* 49, 915.

Mirov, N. T., J. W. Duffield, and A. R. Liddicoet. 1952. Altitudinal races of *Pinus ponderosa*: A 12-year progress report. *J. Forestry* 50, 825–831.

Morgenstern, E. K. 1969. Genetic variation in seedlings of *Picea mariana* (Mill.) B.S.P. I. Correlation with ecological factors. *Silvae Genet.* 18, 151–161.

Moschkov, B. S. 1935. Photoperiodismus and Frosthärte ausdauernder Gewächse. *Planta* 23, 774–803.

Nagata, H. 1967. Studies on the photoperiodism in the dormant bud of *Pinus densiflora*. II. Effects of temperature on the breaking of winter dormancy of first-year seedlings. *J. Jap. For. Soc.* 49. 415–420.

——— 1968. Studies on the photoperiodism in the dormant bud of *Pinus densiflora*. IV. Thermo-sensitivity of the terminal buds in the stage of predormancy. *J. Jap. For. Soc.* 50, 211–216.

Nienstaedt, H. 1958. Height growth is indicative of the relative frost resistance of hemlock seed sources. *Tech. Note No. 525.* U.S. Forest Serv., Lake States Forest Exp. Sta., 2 pp.

——— 1966. Dormancy and dormancy release in white spruce. *Forest Sci.* 12, 374–384.

——— 1967. Chilling requirements in seven *Picea* species. *Silvae Genet.* 16, 65–68.

——— 1974. Genetic variation in some phenological characteristics of forest trees. In *Phenology and Seasonality Modeling*, p. 389. H. Lieth, ed. New York: Springer-Verlag.

———, and J. S. Olson. 1961. Effects of photoperiod and source on seedling growth of eastern hemlock *Forest Sci.* 81–96.

Nitsch, J. P. 1957. Photoperiodism in woody plants. *Proc. Am. Soc. Hort. Sci.* 70, 526–544.

——— 1968. Photopériodisme et dormance chez les végétaux ligneux. *Mém. Soc. Bot. France* 1966, 55–66.

Parker, J. 1955. Annual trends in cold hardiness of ponderosa pine and grand fir. *Ecology* 36, 377–380.

——— 1961a. Seasonal trends in carbon dioxide absorption, cold resistance, and transpiration of some evergreens. *Ecology* 42, 372–380.

——— 1961b. Seasonal changes in cold resistance of some northeastern woody evergreens. *J. Forest.* 59. 108–111.

Paton, D. M., and R. R. Willing. 1968. Bud dormancy in *Populus. Australian J. Biol. Sci.* 21, 157–159.

Pauley, S. S., and T. O. Perry. 1954. Ecotypic variation of the photoperiodic response in *Populus. J. Arnold Arb.* 35, 167–188.

Perner, H. 1966. [Relationship between germination at various temperatures and distribution of *Pinus* species.] *Geobot. Mitt.* 41, 169–182; abstr. in *For. Abstr.* 28, 624–625. (1967).

Perry, T. O. 1962. Racial variation in the day and night temperature requirements of red maple and loblolly pine. *Forest Sci.* 8, 336–344.

——— 1971. Dormancy of trees in winter. *Science* 171, 29–36.

———, and C.-W. Wang. 1960. Genetic variation in the winter chilling requirements for date of dormancy break for *Acer rubrum. Ecology* 41, 790–794.

———, ———, and D. Schmitt. 1966. Height growth for loblolly pine provenances in relation to photoperiod and growing season. *Silvae Genet.* 15, 61–64.

Pisek, A. 1950. Frosthärte und Zusammensetzung des Zellsaftes bei *Rhododendron ferrugineum, Pinus cembra,* und *Picea excelsa. Protoplasma* 39, 129–146.

Robak, H., and S. Magnesen. 1970. Contribution to the knowledge of the ecology of growth termination in spruce seedlings of Norwegian and central European provenances. *Silvae Genet.* 19, 188–190.

Roberts, B. R., and H. V. Main. 1965. The effect of chilling and photoperiod on bud break in American elm. *J. Forestry.* 63, 180–181.

Roche, L. 1968. The value of short term studies in provenance research. *Commonw. Forestry Rev.* 47, 14–26.

Saeki, T., and N. Nomoto. 1958. On the seasonal change of photosynthetic activity of some deciduous and evergreen broadleaf trees. *Bot. Mag. Tokyo* 71, 235–241.

——— 1970. Freezing resistance in willows from different climates. *Ecology* 51, 485–491.

Santamour, F. S., Jr. 1960. Seasonal growth in white pine seedlings from different provenances. *Forest Res. Note No. 105.* U.S. Forest Serv., Northeast. Forest Exp. Sta., 4 pp.

Sarvas, R. 1969. Genetical adaptation of forest trees to the heat factor of the climate. *Proc. 2nd World Consult. on Forest Tree Breeding.* Washington, D.C.; F.A.O., Intern. Union Forest Res. Organizations.

Schnelle, F. 1955. Pflanzen-Phänologie. *Probleme der Bioklimatologie.* Bd. III. Leipzig: Akad. Verlagsges.

Smithberg, M. H., and C. J. Weiser. 1968. Patterns of variation among climatic races of red-osier dogwood. *Ecology* 49, 495–505.

Sokolowska, J. 1962. [The isophenes of flowering of *Corylus avellana* and temperature.] *Roczn. Dendrol. Polsk. Tow. Bot.* 26, 137–151; abstr. in *For. Abstr.* 24, 398 (1963).

Squillace, A. E., and R. R. Silen. 1962. Racial variation in ponderosa pine. *Forest Sci. Monogr.* 2. 27 pp.

Stearns, F., and J. S. Olson. 1958. Interactions of photoperiod and temperature affecting seed germination in *Tsuga canadensis. Am. J Bot.* 45, 53–58

Sweet, G B., and P. F. Wareing. 1968. A comparison of the rates of growth and photosynthesis in first-year seedlings of four provenances of *Pinus contorta* Dougl. *Ann. Bot.* 32, 735–751.

Till, O. 1956. Über die Frosthärte von Pflanzen sommergrüner Laubwälder. *Flora* 143, 499–542.

Turesson, G. 1923. The scope and impact of genecology. *Hereditas* 4, 171–176.

U.S. Department of Agriculture. 1948. *Woody Plant Seed Manual.* Forest Serv., Misc. Publ. No. 654. 416 pp.

Ulmer, W. 1937. Über den Jahresgang der Frosthärte einiger immergrüner Arten der alpinen Stufe sowie der Zirbe und Fichte. *Jahrb. Wiss. Bot.* 84, 553–592.

Vaartaja, O. 1957. Photoperiodic responses in seedlings of northern tree species. *Can. J. Bot.* 35, 133–138.

——— 1960. Ecotypic variation of photoperiodic response in trees, especially in two *Populus* species. *Forest Sci.* 6, 200–206.

——— 1961. Demonstration of photoperiodic ecotypes in *Liriodendron* and *Quercus. Can. J. Bot.* 39, 649–654.

Vegis, A. 1964. Dormancy in higher plants. *Ann. Rev. Plant Physiol.* 15, 185–224

Villiers, T. A., and P. F. Wareing. 1965. The possible role of low temperature in breaking the dormancy of *Fraxinus excelsior* L. *J. Exp. Bot.* 16, 519–531.

Wang, C.-W., and T. O. Perry. 1958. The ecotypic variation of dormancy, chilling requirements, and photoperiodic response in *Betula* species. Abstr. in *Proc. 10th Intern. Congr. Genet., Montreal* 2, 307.

Wareing, P. F. 1953. Growth studies in woody species. V. Photoperiodism in dormant buds of *Fagus sylvatica* L. *Physiol. Plant.* 6, 692–706.

——— 1956. Photoperiodism in woody plants. *Ann. Rev. Plant Physiol.* 7, 191–214.

Weiser, C. J. 1970. Cold resistance and injury in woody plants. *Science* 169, 1269–1278.

Wilcox, J. R. 1968. Sweetgum seed stratification requirements related to winter climate at seed source. *Forest Sci.* 14, 16–19.

Winstead, J. E. 1969. Ecotypic differentiation in *Liquidambar styraciflua* L. Abstr. in *Dissert. Abstr.* 29B, 3646.

Wommack, D. E. 1960. Effect of winter chilling and photoperiod on growth resumption in Douglas fir (*Pseudotsuga menziesii* (Mirb.) Franco). Abstr. in *Bull. Ecol. Soc. Am.* 41, 57–58.

Worrall, J., and F. Mergen. 1967. Environmental and genetic control of dormancy in *Picea abies. Physiol. Plant.* 20, 733–745.

Wright, J. W. 1944. Genotypic variation in white ash. *J. Forestry* 42, 489–495.

Yeatman, C. W. 1967. Geographic variation in Jack pine (*Pinus banksiana* Lamb.) seedlings. Abstr. in *Dissert. Abstr.* 28B, 11.

PART 3

Seasonality in Trophic Levels

Introduction

Helmut Lieth

Botany Department
University of North Carolina
Chapel Hill, North Carolina, USA 27514

The phenological concept in this volume is most strongly developed in plant sciences and ecology, and has some recognition in the field of meteorology. Some biological sciences ignore the term phenology, and studies of seasonal behavior appear under various, discrete headings such as migration, hibernation, or estivation. So far, no attempt has been made to compile information under the heading of phenology on the seasonal aspects of all the important trophic levels of the ecosystem.

The basic components (producer, consumer, decomposer) considered here are found in any ecosystem. It was necessary to subdivide these standard divisions in order to discuss logically the problems that are different in each subgroup. The primary producers are therefore represented by carbon dioxide fixers (Chapter 3.1) and nitrogen fixers (Chapter 3.2), realizing of course that nitrogen fixers may be decomposers as well. Chapter 3.1 also serves as an introduction to Part 4 of this book, which deals primarily with higher plants in various parts of the world. Chapter 3.2 is a research report rather than a review. Substantial treatments of this subject are available elsewhere (Fachverband Stickstoffindustrie E. V. Düsseldorf, 1961; Runge, 1964; Takahashi, 1970; Lemée, 1971; Copley, 1972); however, seasonality of the nitrogen fixation plays a secondary role in most of these treatments. It is difficult to separate nitrogen fixation from nitrogen release due to decomposing plant and animal material. Many studies examine the release by mineralization of nitrogen from top soil into the soil and soil water, without attempting to analyze the usually negligible input of the primary nitrogen fixers. The lack of a review paper may stimulate more research in this field.

The consumer group is represented by insects (Chapter 3.3), birds (Chapter 3.4), and mammals (Chapter 3.5). Two recent publications show the increasing interest in seasonal studies of birds and mammals (Fretwell, 1972; South *et al.*, 1972). The consumer papers provide an entry into the literature of consumers' seasonality. Bradshaw's paper (Chapter 3.3) includes, in addition, an emphasis on physiological modeling, and can also be considered as a contribution to Part 5 on phenological modeling. Since insects are so important as pests and disease carriers, entomologists' cooperation in regional and extraregional programs may increase beyond the notable examples, such as locust migration.

Amphibians and lizards have been omitted from the consumer discussions here. The marked seasonality of these animals is described by James MacMahon of Utah State University for publication elsewhere. During the symposium he presented new data from semidesert regions relating the seasonality of amphibians and lizards to the balance of desert ecosystems.

The decomposers are apparently the least investigated of the three main groups, and are represented in this book almost entirely by fungi (Chapters 3.6 and 3.7). The entire complex of soil microflora was discussed by a panel chaired by Dr. Edmisten. The discussion on terrestrial decomposers is part of Katz and Lieth's Chapter 3.6.

No person was available to discuss soil invertebrates; reference to pertinent papers on their seasonality is made in the general introduction.

Aquatic components occupy a special position in the landscape and in the ecosystem. While they are known for their strong seasonality, they have seldom been included within the concept of phenology. The seasonality of aquatic plankton, of course, is realized in seasonal blooms of certain species. In contrast to the seasonal adaptation of higher plants, the lifespan of planktonic organisms is much shorter than the annual seasonal cycle; a comparison of the seasonal occurrences among dominant aquatic species to the synchronous sequence of phenophases in a higher plant is more informative than is a comparative study of life cycles in the two species groups. The pattern of seasonal successions among plankton species is similar to those found in bird migration and should therefore be included in the concept of phenology. Since this book incorporates both seasonal and phenological considerations, we have included one paper on the seasonality of aquatic fungi (Chapter 3.8), for which there is little information. For aquatic systems current textbooks on limnology cover the seasonality of primary producers and consumers very adequately.

References

Copley, P. W. 1972. Evaluation of biological N_2 fixation in a grassland ecosystem. *US/IBP Grassland Biome, Tech. Rep. No. 152.* Ft. Collins: Colorado State University.

Fachverband Stickstoffindustrie E. V. Düsseldorf. 1961. *Der Stickstoff. Seine Bedeutung für die Landwirtschaft und die Ernährung der Welt.* 480 pp. Oldenburg: Stalling AG.

Fretwell, S. D. 1972. Populations in a seasonal environment. In *Monographs in Population Biology*—5. 217 pp. Princeton, N.J.: Princeton University Press.

Lemée, G., and N. Bichart. 1971. Recherches sur les écosystèmes des réserves biologiques de la Forêt de Fontainebleau. I. Production de litière et apport au sol d'eléments minéraux majèurs. *Oecol. Plant.* 6, 133–149.

Runge, M. 1964. Untersuchungen über die Mineralstickstoff-Nachlieferung an nordwestdeutschen Waldstandorten. 124 pp. Dissertation zur Erlangung des Doktorgrades der Mathematisch-Naturwissenschaftlichen Fakultät der Universität Hamburg.

South, F. E., J. P. Hannon, J. R. Willis, E. T. Pengelley, and N. R. Alpert, eds. 1972. *Hibernation and Hypothermia, Perspectives and Challenges.* 764 pp. New York: American Elsevier.

Takahashi, H., ed. 1970. *Proceedings of the 2nd Symposium on Nitrogen Fixation and Nitrogen Cycle.* Record of activities of JIBP/PP-Nitrogen Fixation and Nitrogen Cycle Group during 1968 and 1969. 130 pp. Sendai: Tohuku University.

3.1 Seasonality in Terrestrial Primary Producers

Roger C. Anderson[1]

Arboretum and Wildlife Refuge
1207 Seminole Highway
Madison, Wisconsin, 53711 USA

1 Introduction

Phenology concerns the visible response of an organism to a seasonally changing environment of adaptation resulting from natural selection operating on an array of genetically determined responses. Observational or experimental studies and models are being developed to predict phenophases of individual plant species. This chapter's intent is not to suggest all the variables that should be included in a model of seasonality; rather, a few previously published models of plant response to temperature are discussed from several complementary viewpoints.

2 Observations on Single Plants

Observational studies have provided a record of seasonal organismic response to one or several environmental factors, at a single location, between contrasting microenvironments within a region or across regional biomes. Phenophases examined have included vegetative growth, flowering, and dormancy. For a single organism these phases may occur at different times of the year and under the influence of differing environmental regimes. Figure 1 shows the vegetative life cycle of *Trientalis borealis*. Shoot development and flowering are spring events, rhizome growth occurs in the summer, and tubers are formed in the summer and fall (Anderson and Loucks, 1973).

Regardless of the sophistication attained by a modeling or experimental study, an observational phase provides a basis for comparing growth under transplant or

Contribution Number 87 from the Eastern Deciduous Forest Biome, US/IBP.
[1] Present address: Department of Biology, Central State University, Edmond, Oklahoma, 73034, U.S.A.

KEYWORDS: Phenophase; flowering; blooming; dormancy; heatsums; photoperiod; phenology; development; temperature; seasonal; plants.

experimental conditions or for developing or testing models. Distinct phenophases that are easily recognizable, microclimatic variations, and genetic variability that will alter plant response to environmental parameters are important in the observational phases (Leopold and Jones, 1947).

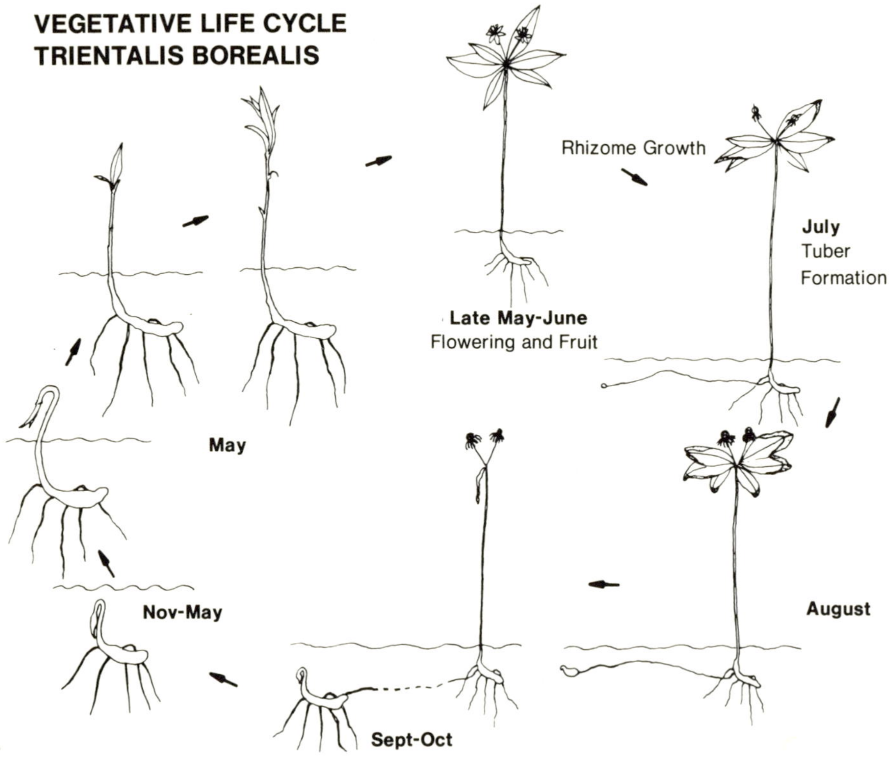

Fig. 1. Seasonal aspects of the vegetative life cycle of *Trientalis borealis* in Wisconsin.

3 Mapping of Single Phenophases

Synagraphic mapping techniques (Lieth and Radford, 1971) have provided an objective means for the synthesis of observational data. Figure 2 shows the flowering dates for lilac in Wisconsin during the spring of 1963, it was prepared with a computer program which interpolated blooming dates between observational points by using the adjacent observed values. The importance of any observed value in the calculation is inversely proportional to the square of its distance from the point for which a blooming date is being determined. Correlation of the timing of phenological events with environmental variables that have been subjected to similar mapping calculations is also possible, and may serve as a basis for model development.

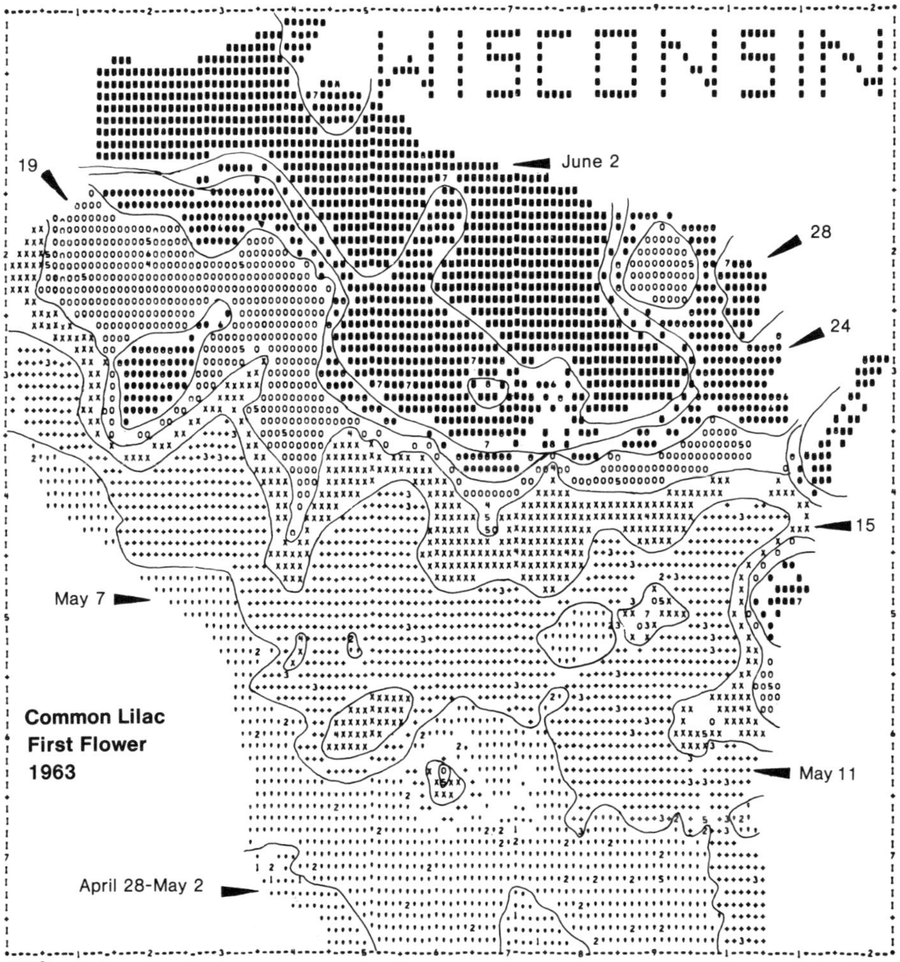

Fig. 2. The flowering dates of common lilac (*Syringa vulgaris*) in Wisconsin during the spring of 1963.

4 Converting Observations into Models

Heatsums, degree-days or degree-hours, and above-threshold temperatures have been used to predict the timing of specific plant phenophases. Lindsey and Newman (1956) present an objective procedure to determine threshold temperatures and the number of days prior to the phenophase to be included in the summation. The availability of maximum-minimum temperature data from U.S. Weather Bureau stations has resulted in their widespread use in the calculation of heatsums. Frequently, the daily mean is multiplied by 24 to obtain degree-hours. Lindsey and Newman (1956) compare this method, the remainder index, with a technique called the summation-duration method which estimates the duration of temperatures between the daily maximum and minimum. Further studies by Lindsey (1963) revealed that the duration-summation technique gives results within 1 to 3% of the true value, whereas the remainder index yields results within 8 to 15% of actual degree-hours as determined by hygrothermographs. A thorough discussion of the method for calculating degree-hours is provided by Arnold (1960).

Lettau (1965) used air and soil heatsums to predict the blooming of pink lady's slipper. Soil temperature frequently influences plant development before air temperature, and a method is presented whereby the heatsums accumulated in the soil above 0°C threshold are used in conjunction with air temperatures. Upper and lower thresholds as well as optimum temperatures, may change under different photoperiods or radiation or soil moisture levels, or during the life cycle of the plant (Wang, 1960; Anderson and Loucks, 1973). Moreover, the plant-temperature response during the day is likely to be different than it is at night. Figure 3 shows the change in thermal response for the garden pea during its life cycle (Wang, 1960).

The summation of degree-hours to predict plant phenophases assumes a linear relationship between temperature and the plant response which is characteristic of only a segment of the response curve. Furthermore, heatsums provide no indication of the temperature patterns during the period for which the summation was obtained. A potentially accurate but time-consuming technique would involve the

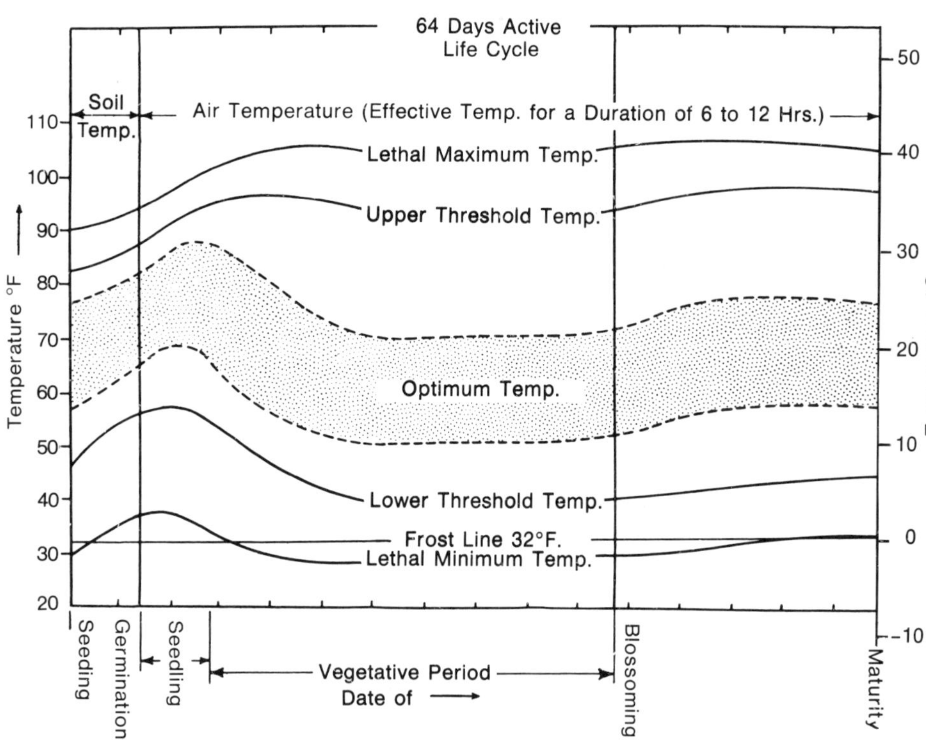

Fig. 3. The change in thermal response for garden peas during its life cycle. Abscissa: date after seeding in 4-day units; ordinate: temperature. (From Wang, 1960.)

determination of coefficients which express the growth rate at each temperature (Fig. 3). Models assuming linear response would be especially erroneous if the optimum temperature for the species growth was exceeded. However, these conditions are unlikely to occur for long periods of time in the temperate region during the spring (see Chapter 4.4).

A further refinement in predicting plant phenophase in the spring is suggested by Caprio (1971), who combined heatsums with solar radiation (between 0.3 and 4 μ) into a single term, solar thermal units (STU) According to Caprio, lilacs require 380,000 STU to bloom (see Chapter 5.4). In Norway, where springtime temperature is relatively high but radiation is low, more degree-days are required to begin lilac bloom than in portions of western United States where temperature is relatively low but radiation is high.

Frequently, photoperiod rather than temperature initiates a phenophase (see also Chapter 2.5). For example, temperature was found to control bud break, greening of bark, and flowering in red osier dogwood, while red bark color, leaf abscission, and hardening from about 22° to 0°C were determined by photoperiod. However, hardening from 0° to −120°C was triggered by the first frost (Smithberg and Weiser, 1968). Because of the consistency of photoperiod from one year to the next, phenophases under the control of photoperiod will occur at approximately the same time each year. Studies of native grass species have shown that for many of the wide-ranging species, photoperiodic induction of flowering is genetically fixed, and in transplant gardens or under experimental conditions they will require a specific photoperiod to flower or for proper vegetative growth (Olmsted, 1944; McMillan, 1959).

Simple models for predicting spring phenophases have been presented by Hopkins (1918), Caprio (1971), and Lieth and Radford (1971). The usefulness of these models seems to be in their application to spring events that are determined by temperature and in the use of environmental parameters that predict spacial variation in temperature, such as elevation, latitude, and longitude. The Lieth-Radford (1971) model is shown in the upper part of Fig. 4 and is compared with a model derived from Hopkins' law (Hopkins, 1918) in the lower part of the figure. The upper part of the figure shows the relationship between elevation and flowering time of dogwood in North Carolina. The Lieth-Radford model assumes the form $Y = 36.5\ X - 346$, where X is yearday of observed dogwood flowering and Y is altitude in meters. Both models provide an acceptable means for predicting flowering of dogwood, with the Lieth-Radford model being superior for the Piedmont and mountain regions of North Carolina.

Phenophases such as bud burst and spring blooming are frequently controlled by temperature, and many species are opportunistic in their response to temperature Moreover, the spring events deal with just a portion of the life cycle during which the plant's response to environmental parameters may be quite uniform. Temperatures are likely to occupy a range in which the plant species would show a linear response to heatsums. Predictive models of an annual phenological cycle may be more difficult to achieve because they necessitate the determination of the species' response to a greater number of variables and also a differing response of the organism to these variables.

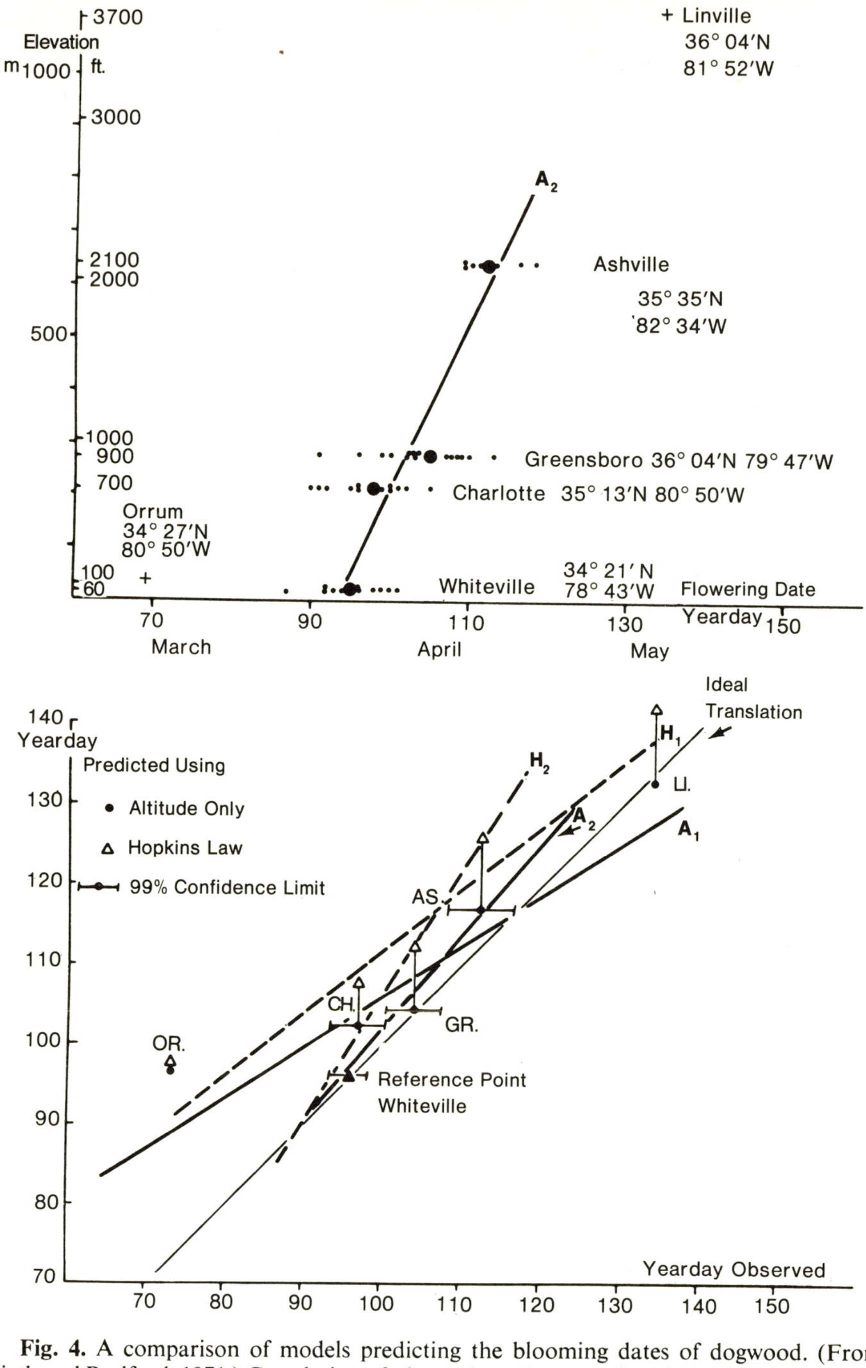

Fig. 4. A comparison of models predicting the blooming dates of dogwood. (From Lieth and Radford, 1971.) Correlation of phenophase "average first dogwood flower" with altitude and Hopkins' model, respectively. The upper part of the figure contains the portion of the data selected from the observer network in North Carolina and shows the relation between altitude and flowering time. *Abscissa:* yearday of dogwood flowers opening

5 Genetic Variability

Genetic differences within a species have resulted in strikingly different phenological patterns (see also Chapter 2.5). For example, in the mountains of Utah, certain clones of aspen are in full leaf two to three weeks before adjacent ones (Cottam, 1954). Similarly, McMillan and Pagel (1958) demonstrated that clones of *Symphoricarpos* that showed early bud burst in the field retained this quality under experimental conditions. Red maples from seed collected across a large segment of its geographic distribution from New York to Florida showed a pronounced variation in the number of cold-treatment days required to break dormancy. Plants from the southern portion of the range have no requirement for cold treatment, whereas northern populations did not break dormancy without chilling (Perry and Wang, 1960).

Differing phenological responses within a species that are controlled by genetic differences demonstrate the role of selection in producing populations adapted to local environmental regimes. Once established, these differences encourage genetic segregation and eventual ecotypic variation or speciation (Clausen, *et al.*, 1940; Vasek and Sauer, 1972). Adaptive response of the plant to changing environmental conditions must frequently precede the change in environmental conditions, i.e., the plant must be preadapted to survive. Therefore, in the case of *Trientalis borealis*, shoot development can be linked to temperature. However, tuber development (Fig. 1) would more logically be triggered by a shortening photoperiod, since it must begin before summer temperature conditions have lowered; the tuber is the overwintering structure and must be well developed before freezing temperatures occur (Anderson, 1970).

For many species later phases of plant development are controlled by photoperiod rather than by temperature, and wide-ranging species are often composed of populations that are adapted to local photoperiod. For example, clones of red osier dogwood, collected across a latitudinal range of 40° to 65°N, were found to respond similarly to temperature in the spring; however, in autumn climatic races responded to different photoperiods, depending upon their provenances (Smithberg and Weiser, 1968). Later phases of development would reflect a changing photoperiod, fluctuating soil moisture levels, as well as temprature and evapotranspiration patterns, that may alter the rate of development (Fritts, 1958).

---

recorded; *ordinate:* altitude in meters and feet. Cities with enough returns to enable a correlation calculation were selected: Whiteville, Charlotte, Greensboro, and Asheville. Within the range of these cities, the regression line (A_2) assumes the form Y(meters) = 36·5 m/day X(days) – 346(m). The lower part of the figure compares the regression lines calculated for altitude (A) and Hopkins' model (H), including (A_1, H_1) and excluding (A_2, H_2) the two extreme locations. The regression line A_2 is the same as A_2 in the upper part of the figure; however, the two ordinates are not drawn to the same scale. The R_2 values are for H_1 = 0·83, H_2 = 0·91, A_1 = 0·83, and A_2 = 0·98. This shows that the altitude gives, in this case, a better fit than Hopkins' complete law, but is satisfactory for the Piedmont and mountain regions only.

6 Community-Microclimate Feedback

Established plant communities generate a microclimate by altering radiation, temperature, moisture, and air current patterns; phenology, in turn, is influenced by the community-created microenvironment. For example, the spring ephemerals that occur in the deciduous forests have adapted to escape deep forest shade, by appearing early in the spring before the canopy closes and becoming dormant once the tree leaves have reduced radiation levels on the forest floor (see Chapter 4.4). Other deciduous forest species, such as the Adam and Eve orchid (*Aplectrum hymenales*), produce leaves in the fall and utilize the relatively high light in the fall and spring to carry on photosynthesis. It also loses its leaves when the forest becomes shady. Similarly, studies of the biennial woodland herb *Hydrophyllum appendiculatum* revealed that this species produces two kinds of leaves: pinnately shaped leaves that are adapted to high light levels and palmately shaped leaves that are adapted to low light levels and appear as the forest canopy closes (Morgan, 1971).

Thus seasonal models will predict the genetic response of plant species to a set of seasonally varying environmental parameters. The models must have built-in mechanisms to account for the changing response of the plant throughout its life cycle. Finally, to have ecological significance the phenophases of individual species should be interpreted in relation to the community in which they occur.

References

Anderson, R. C. 1970. The role of daylength and temperature in tuber formation and rhizome growth of *Trientalis borealis*. *Bot. Gaz.* 13, 122–128.

———, and O. L. Loucks. 1973. Some aspects of the biology of *Trientalis borealis*. *Ecology*, 54, 798–808.

Arnold, C. Y. 1960. Maximum-minimum temperatures as a basis for computing heat units. *Proc. Am. Soc. Hort. Sci.* 76, 682–692.

Caprio, J. M. 1970. The solar-thermal unit theory in relation to plant development and potential evapotranspiration. *Montana Agr. Exp. Sta. Circular 251*. Montana State Univ., Bozeman. 10 pp.

———. 1974. The solar-thermal unit concept in problems related to plant development and potential evapotranspiration. p. 353. In *Phenology and Seasonality Modeling*. H. Lieth ed. Springer-Verlag: New York.

Clausen, J., D. D. Keck, and W. M. Hiesey. 1940. Experimental studies on the nature of species. I. Effect of varied environments on western North American plants. *Carnegie Inst. Wash. Publ. 520*.

Cottam, W. P. 1954. Prevernal leafing of aspen in Utah mountains. *J. Arnold Arb.* 25, 239–247.

Fritts, H. C. 1958. An analysis of radial growth of beech in a central Ohio forest during 1954–1955. *Ecology*, 39, 705–719.

Hopkins, A. D. 1918. Periodical events and natural laws as guides to agricultural research and practice. *US Monthly Weather Rev. Suppl. No. 9* (*Weather Bull. No. 643*), 42 pp.

Leopold, A., and S. E. Jones. 1947. A phenological record for Sauk and Dane counties Wisconsin. *Ecol. Monograph 17*, 81–122, 1935–1945.

Lettau, K. 1965. A new method for predicting the blooming date of spring flowers. *Trans. Wisconsin Acad. Sci. Arts Letters* 54, 135–142.

Lieth, H., and J. S. Radford. 1971. Phenology, resource management, and synagraphic computer mapping. *BioScience* 21, 62–70.

Lindsey, A. A. 1963. Accuracy of duration temperature summing and its use for *Prunus serrulata*. *Ecology* 44, 149–151.

———, and J. E. Newman. 1956. Use of official weather data in spring time temperature analysis of an Indiana phenological record. *Ecology* 37, 812–823.

McMillan, C. 1959. The role of ecotypic variation in the distribution of the central grassland of North America. *Ecol. Monograph* 29, 285–308.

———, and B. F. Pagel. 1958. Phenological variations within a population of *Symphoricarpos occidentalis*. *Ecology* 39, 766–770.

Morgan, H. 1971. Life history and energy relationships of *Hydrophyllum appendiculatum*. *Ecol. Monograph* 41, 329–349.

Olmsted, C. E. 1944. Growth and development in range grasses. IV. Photoperiodic responses in twelve geographic strains of sideoats grama. *Bot. Gaz*. 106, 46–74.

Perry, T. O., and C.-W. Wang. 1960. Genetic variation in the winter chilling requirements for date of dormancy break for *Acer rubrum*. *Ecology* 41, 790–793.

Smithberg, M. H., and C. J. Weiser. 1968. Patterns of variation among climatic races of red osier dogwood. *Ecology* 49, 495–504.

Taylor, F. G., Jr. 1974. Phenodynamics of production in a mesic deciduous forest. p. 237. In *Phenology and Seasonality Modeling*. H. Lieth, ed. New York: Springer-Verlag.

Vasek, F. C., and R. H. Sauer. 1972. Seasonal progression of flowering in *Clarkia*. *Ecology* 52, 1038–1045.

Wang, J. Y. 1960. A critique of the heat unit approach to plant response studies. *Ecology* 41, 785–790.

3.2 Seasonality of Nitrogen Fixation in Gulf Coast Salt Marshes

Frances Green and Joe Edmisten

Biology Department
University of West Florida, Pensacola, Florida, 32500 USA

1 Introduction

One of the potential sites for nitrogen fixation within the salt marsh is the surface of living stems and leaves of marsh plants. These stems and leaves are capable of supporting a luxuriant growth of epiphytes. These bacteria and algae are replenished by floating cells which become attached during high tide inundations of the marsh. Many of these epiphytes have been identified as potential nitrogen fixers (Pomeroy, 1959), but have not been tested for actual fixation.

Another important potential site for nitrogen fixation in the marsh is on the dead and decaying stems and leaves of marsh plants. These materials have been observed to remain in the marsh for periods up to a year, with 50% of dead *Spartina* stems present after 300 days (Odum and de la Cruz, 1967). These stems are gradually broken down by microbial action, but the protein values of the materials actually increase during decomposition (Burkholder, 1956). This increase has been attributed to colonization by aerobic heterotrophic bacteria which are themselves relatively protein rich (Burkholder and Bornside, 1957). However, continued growth of such bacteria would seem to require a source of protein richer than the decaying grass which serves as the substrate. Teal (1962) reported that over a 16-week period the protein concentration in decomposing marsh grass was doubled, while carbohydrates declined by 75% and caloric content by 33%. He does not mention nitrogen fixation in relation to this increase, but notes that in addition to bacteria, blue-green algae were active in the breakdown of marsh grasses. This increase in protein is especially important because detritus serves as the major link between primary and secondary production (Teal, 1959).

The third available substrate for the growth of nitrogen-fixing organisms in the marsh is the mud surface. This mud is extremely rich in detrital particles, and would therefore be expected to have a nitrogen-fixing population similar to that of the

KEYWORDS: Seasonality; nitrogen fixation; salt marsh; Florida; ethylene production method; Spartina marsh; *Rhodospirillum*; *Rhodopseudomonas*; *Calothrix*; *Anabaena*; *Lyngbya*.

decaying stems and leaves. The mud surface also supports a rich, visible community of filamentous blue-green algae. Many of these cyanophytes have been identified as potential nitrogen fixers and are an important component of the marsh ecosystem (Moul, 1959).

Another possible site for nitrogen fixation is the water column. Nitrogen fixation in natural ecosystems has been reported in both freshwater (Brezonic and Harper, 1969) and marine (Stewart, 1965) environments. The waters that flood the marsh at high tide could carry a complement of estuarine nitrogen fixers, and could be further enriched by organisms from soils and stems. However, this water can have an important additional effect on nitrogen fixation within the salt marsh ecosystem.

One of the commonest types of estuarine pollution is nitrogen enrichment, often in the form of nitrate or ammonia. Both of these compounds inhibit nitrogen fixation when present in sufficient concentrations (Stewart, 1966). Therefore nitrogen fixation was measured in a marsh with high concentrations of nitrates, as well as in a relatively unpolluted marsh, so that comparisons of fixation rates under natural and artificial conditions could be made. Both quantity and seasonality of fixation were observed, and the results were used to determine the overall pattern of salt marsh nitrogen fixation.

2 Materials and Methods

Description of Sampling Sites

This study was done mainly at two marshes with vegetation patterns typical of the northern Gulf of Mexico, approximately 10 mi apart on separate bays in the Pensacola Bay system, Escambia Bay, and East Bay near Pensacola, Florida (Fig. 1, *A* and *B*). The outer margin of the Escambia Bay marsh has a solid stand of *Spartina alterniflora* 1 to 2 m high. These plants are usually partially inundated, with their bases completely exposed only during extreme low tides. Above the mean low tide line, *S. alterniflora* is replaced by *Juncus roemerianus*, which is the dominant species across most of the marsh. Within this large *J. roemerianus* zone, however, there are openings of from one to several square meters where no *Juncus* grows. These openings are, for most of the year, covered only by small vascular plants, the most typical of which is *Lilaeopsis chinensis*. In late summer, however, at least two composites bloom profusely in these openings: *Solidago mexicana* and *Pluchea camphorata*. Within the *Juncus* zone there may also occur occasional patches of *Kosteletzkya virginica*. Beyond the *Juncus* zone is a slightly higher and much narrower zone of mixed grasses. Species present include *Panicum virgatum*, *Distichlis spicata*, *Spartina patens*, and *Sporobolus virginicus*. The upper edge of this marsh is a causeway area where a highway crosses the marsh. Near the causeway the mixed grass zone is replaced by a stand of *Phragmites communis* mixed with shrubs such as *Baccharis halimifolia* and *Iva fructescens*.

The East Bay sampling site is located in a somewhat smaller marsh at the mouth of Fundy Bayou.

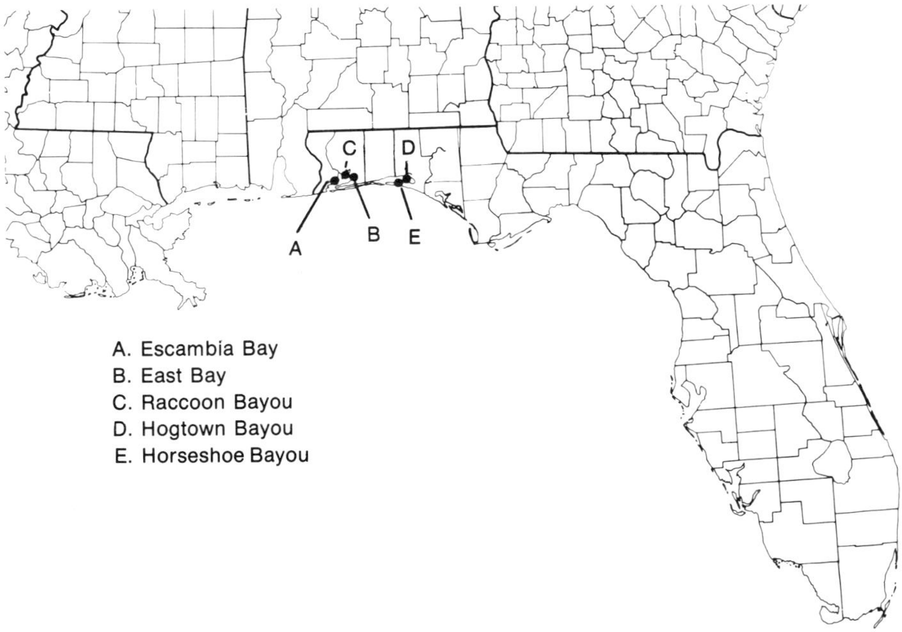

Fig. 1. The Pensacola Bay system. A = Escambia Bay, B = East Bay, C = Raccoon Bayou, D = Hogtown Bayou, E = Horseshoe Bayou.

The lower marsh areas A and B with which the sampling was concerned are basically similar. Tidal amplitudes, salinities, temperatures, and other environmental factors showed no significant variations between the two marshes during the period of this study. However, in one important respect the waters inundating the two marshes were significantly different. The waters of Escambia Bay are subject to an artificially high nutrient input from a number of sources, including several industries and sewage treatment facilities. In contrast, East Bay is a comparatively unspoiled estuary with little man-caused nitrification (USDI, 1970). Since the existence of high levels of nutrients, in general, and of nitrogenous compounds, in particular, has been shown to have a marked effect on levels of nitrogen fixation (Stewart, 1966), the two sampling sites were chosen for comparison of fixation levels under differing conditions of nitrification (Green, 1972).

Sample Collection

Sampling to determine the extent of nitrogen fixation was carried out from January 1971 through February 1972 (14 months). Sampling trips were made on a biweekly schedule, with both marshes always sampled on the same day and under as similar conditions as possible. In the *Spartina* and *Juncus* zones, materials regularly sampled were water (when present), surface mud, (top 1 cm), young stems, and old stems. The term "old stems" is used here to designate dead and decaying stems which were either still standing or recently fallen but not yet reduced to detritus. In the barren zone, the mud surface and any standing or tidal water present were regularly sampled. In addition, in the period from late summer to early winter, living or decaying *Pluchea* and *Solidago* stems were present, and were regularly tested for fixation.

In each plot two samples of each material to be tested were taken from the corners of the plot, so that four pairs of duplicate samples were made of each material sampled. One member of each pair was used as a control. Water samples were taken first to avoid mixing of bottom muds with the water column. Ten-ml water samples were placed in 25-ml Ehrlenmeyer flasks and sealed with serum stoppers. Samples of both old and young stems were prepared by cutting stems at the soil line and placing entire stems in 125-ml Ehrlenmeyer flasks. Mud samples were prepared by removing the top centimeter of mud from an area of several square centimeters and placing the mud in 50-ml Ehrlenmeyer flasks. The mud surface often bore a thick mat of primarily blue-green algae, which was always included in the mud sample. Additional-samples of various materials were taken from elsewhere within the plots as time permitted.

Testing

Analysis of ethylene production, or acetylene reduction, was carried out by use of gas chromatrography with a Beckman GC-5 chromatograph with flame detector. The column used for separation of the gases was packed with Porapak-R. Columns of varying lengths were used, but a column 6 ft in length and $\frac{1}{8}$-in. wide was found to be most satisfactory. Helium at 20 psi was used as a carrier gas. At a column temperature of 50°C, retention times were just over 2 min for ethylene and 3 min for actylene. Samples of 100 μl and an attenuation of 1×10^2 gave peaks which were on scale in most cases, although in cases of extremely active samples, attenuations as 2×10^5 were sometimes necessary. Figure 2 shows a typical peak obtained under such conditions. Control samples were run together with their equivalent samples.

The information obtained from these procedures was analyzed to yield quantitative information concerning levels of nitrogen fixation in the marsh. Ethylene peak heights were compared to peak heights yielded by known volumes of pure ethylene at known attenuations so that the actual volume of ethylene present in each 100 μl sample could be calculated. Since all peak heights represented ethylene present in 100 μl of sample gas, it was necessary to determine the exact volume of gas in each sample flask so that total ethylene produced could be calculated.

Following this procedure, mud and stem samples were oven dried so that ethylene production could be reported on a gram dry weight basis. Using the theoretical figure of 1.5 molecules of ethylene produced per molecule of nitrogen potentially fixed (Postgate, 1971), it was possible to calculate potential nitrogen fixation per gram dry weight per day for each substrate and zone. Fixation in water samples was calculated similarly but reported on a volume basis.

Finally, measurements were made in each zone of the marsh to determine the dry weight of stems and of surface muds per unit area, so that an estimate of total fixation per square meter could be made. This permitted a rough calculation of total input into the marsh by nitrogen fixation in the period of a year.

In order to identify some of the microorganisms responsible for nitrogen fixation in the marsh, both blue-green algae and bacteria were isolated from muds and stem samples found to fix nitrogen. Blue-green algae were always found growing abundantly on the exposed mud surfaces, and these algae were keyed from intact samples. In addition, many of the algae were transferred to nutrient broth medium and isolated in pure or near pure culture. Bacteria from both stems and mud were first plated onto nutrient or tripticase-soy enriched agar. Colonies thus isolated were transferred to a nitrogen-free medium. Some plates were incubated aerobically and others were kept under anaerobic conditions in a Brewer jar. Both bacteria and blue-green algae isolated from nitrogen-fixing samples were subsequently tested for nitrogen fixation using the test described above.

3 Results

The mean monthly values for actylene reduction in sample materials from our main sites (*A* and *B* in Fig. 1) are given in Tables 1 to 3. They are presented under two different categories: floral differences and seasonal patterns.

Floral Differences

In the *Spartina* zone, only one of the hundreds of water samples taken showed any detectable degree of fixation. This sample included a portion of a blue-green algal mat which had floated up from the mud surface, and was found to reduce acetylene activity at a rate of 3.0 μM of ethyelene per gram dry weight per day. For the most part, however, the waters of the *Spartina*, *Juncus*, and barren zones showed no detectable nitrogen-fixing ability in any season of the year. Since sampling times were so arranged that high, low, and intermediate tides would be encountered, it appears that estuarine waters as associated with the salt marsh have little, if any, role in marsh enrichment through biological nitrogen fixation.

The young *Spartina* stems in the two marshes showed no acetylene reduction during the first three months of sampling and only slight amounts during the fourth month. However, during the subsequent months ethylene production levels were relatively high, with a peak reached in August and a gradual decrease during the fall and winter months. The young stems of *Juncus* showed no acetylene reduction at all from January through May, and only slight fixation in summer and fall, with highest values occurring in October and November.

Old *Spartina* stems showed some acetylene reduction in all but the first months of sampling. A peak was reached in July with a mean of 294.9 μM ethylene per gram dry weight per day, which was the maximum value reported for any substrate or zone in the entire sampling period. Old stems of *Juncus* showed much lower levels of acetylene reduction, with a maximum in October. In both zones, however, for any given sampling period the old stems showed more acetylene-reducing ability than any other substrate. The single exception was one occasion in early December when young *Spartina* stems gave a value more than double that from old stems. In the barren zone, the composites *Solidago mexicana* and *Pluchea camphorata* begin blooming in midsummer, and as flower stalks die they fall onto the mud surface.

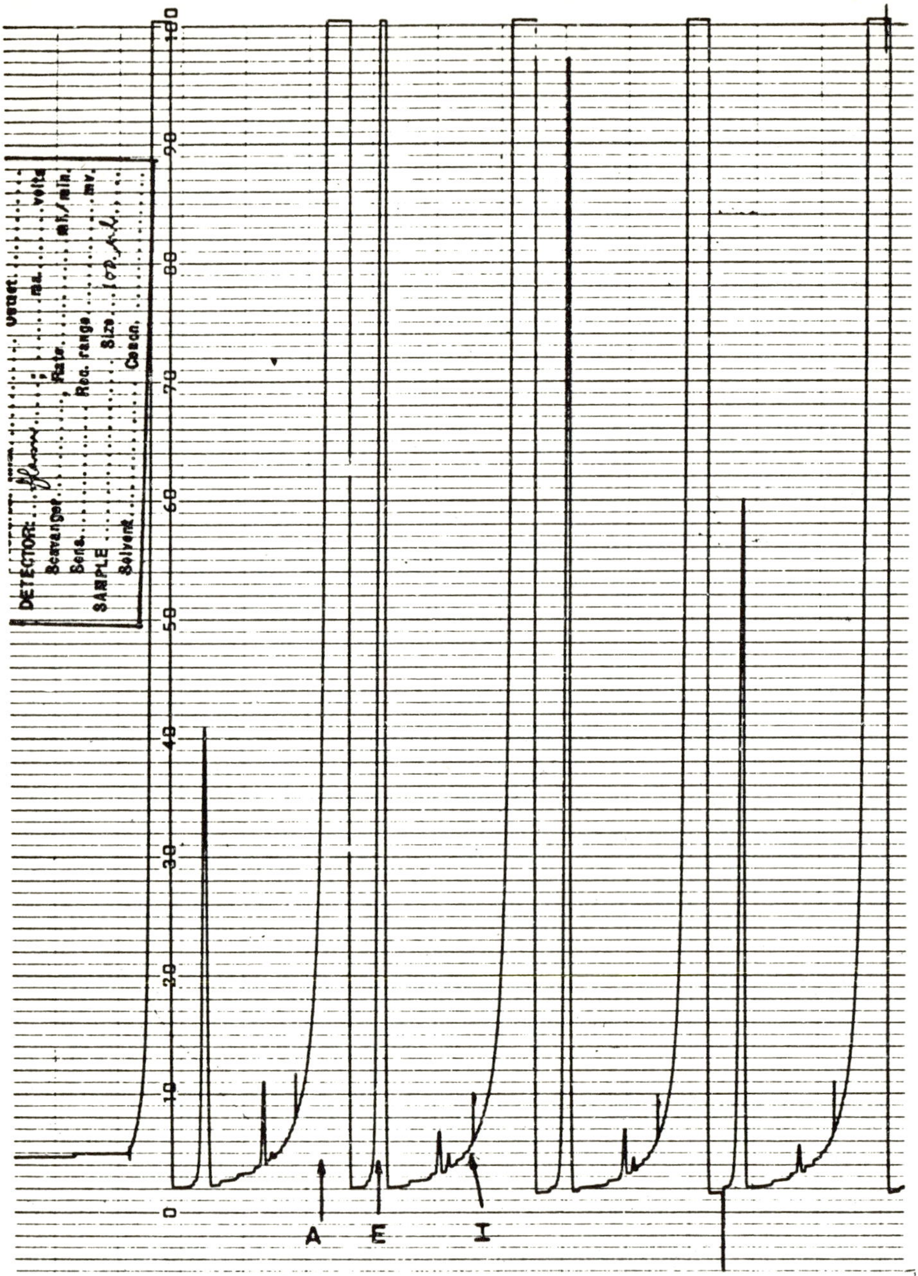

ACETYLENE REDUCTION TEST
Spartina (Old Stem)

Fig. 2. Results from gas chromatography of old *Spartina* stem sample. I = injection point, E = ethylene peak, A = acetylene peak.

Table 1. Mean monthly distribution of ethylene production (μM per gram dry weight per day) in the *Spartina* zone.

Area Sampled	Jan.	Feb.	Mar.	Apr.	May	June	July	Aug.	Sept.	Oct.	Nov.	Dec.	Jan.	Feb.
Seasons:	Winter		Spring		Summer					Fall		Winter		
Water	0	0	0	0	0	0	0	0.1[a]	0	0	0	0	0	0
No. of samples	(16)	(16)	(16)	(16)	(16)	(16)	(24)	(16)	(16)	(16)	(16)	(16)	(16)	(16)
Range	—	—	—	—	—	—	—	(1.6)	—	—	—	—	—	—
Mud surface	0.2	0.1	0.4	1.2	7.8	8.6	8.9	9.0	8.7	4.0	2.8	1.0	0.5	0.3
No. of samples	(16)	(16)	(16)	(16)	(16)	(16)	(24)	(16)	(16)	(16)	(16)	(16)	(16)	(16)
Range	(0.4)	(0.4)	(1.6)	(2.8)	(7.9)	(5.6)	(6.3)	(12.3)	(16.7)	(9.8)	(5.9)	(1.7)	(0.6)	(0.6)
Old stem	0	0.1	0.9	2.7	54.1	147.1	294.4	214.5	95.9	17.5	24.1	1.9	1.2	0.8
No. of samples	(16)	(16)	(16)	(16)	(16)	(32)	(48)	(32)	(32)	(32)	(32)	(32)	(16)	(16)
Range	—	(0.7)	(2.1)	(2.1)	(35.1)	(72.4)	(369.0)	(742.8)	(26.6)	(7.7)	(15.0)	(4.6)	(3.8)	(1.1)
Young stem	0	0	0	0.1	10.8	14.9	30.6	35.0	17.2	5.4	4.2	3.8	0.3	0.4
No. of samples	(16)	(16)	(16)	(16)	(16)	(16)	(48)	(32)	(32)	(32)	(24)	(16)	(16)	(16)
Range	—	—	—	0.2	(4.0)	(3.5)	(21.2)	(17.1)	(31.9)	(3.3)	(1.4)	(4.6)	(1.3)	(3.6)

[a] Based on the inclusion in one sample of a small floating mat of blue-green algae.

Table 2. Mean monthly distribution of ethylene production (μM per gram dry weight per day) in the *Juncus* zone.

Area Sampled	Jan.	Feb.	Mar.	Apr.	May	June	July	Aug.	Sept.	Oct.	Nov.	Dec.	Jan.	Feb.
Seasons:	Winter		Spring		Summer					Fall		Winter		
Water	0	0	0	0	0	0	0	0	0	0	0	0	0	0
No. of samples	(16)	(16)	(16)	(16)	(16)	(24)	(16)	(16)	(16)	(16)	(16)	(16)	(16)	(16)
Range	—	—	—	—	—	—	—	—	—	—	—	—	—	—
Mud surface	0	0	0.6	0.8	0.7	0.6	0.6	0.6	2.5	7.2	6.0	0.5	0.5	0.4
No. of samples	(16)	(16)	(16)	(16)	(16)	(24)	(16)	(16)	(16)	(16)	(16)	(16)	(16)	(16)
Range	—	—	(2.1)	(1.8)	(3.7)	(6.2)	(7.1)	(5.5)	(5.7)	(10.3)	(12.3)	(2.6)	(3.9)	(1.1)
Old stem	0	0	0.3	0.3	0.8	0.7	1.7	7.8	10.0	10.3	7.1	2.1	1.3	0.8
No. of samples	(16)	(16)	(16)	(16)	(16)	(24)	(16)	(16)	(16)	(16)	(16)	(16)	(16)	(16)
Range	—	—	(1.6)	(0.9)	(2.4)	(3.9)	(4.4)	(9.6)	(10.0)	(13.8)	(4.4)	(5.9)	(4.7)	(1.3)
Young stem	0	0	0	0	0	0.1	0.1	0.1	0.1	0.3	0.3	0.3	0.1	0.1
No. of samples	(16)	(16)	(16)	(16)	(16)	(24)	(16)	(16)	(16)	(16)	(16)	(16)	(16)	(16)
Range	—	—	—	—	—	(0.6)	(1.7)	(1.6)	(2.6)	(5.8)	(4.2)	(3.3)	(1.4)	(0.3)

Table 3. Mean monthly distribution of ethylene production (μM per gram dry weight per day) in the barren zone.

Area Sampled	Jan.	Feb.	Mar.	Apr.	May	June	July	Aug.	Sept.	Oct.	Nov.	Dec.	Jan.	Feb.
Seasons:	Winter		Spring		Summer					Fall		Winter		
Water	0	0	0	0	0	0	0	0	0	0	0	0	0	0
No. of samples	(16)	(16)	(16)	(16)	(16)	(24)	(16)	(16)	(16)	(16)	(16)	(16)	(16)	(16)
Range	—	—	—	—	—	—	—	—	—	—	—	—	—	—
Mud surface	2.1	2.5	3.3	3.1	47.3	58.4	74.1	68.4	68.3	57.2	25.6	5.6	2.2	2.4
No. of samples	(16)	(20)	(24)	(32)	(32)	(48)	(32)	(51)	(48)	(32)	(32)	(32)	(16)	(16)
Range	(4.6)	(3.9)	(7.2)	(5.4)	(62.1)	(71.6)	(44.0)	(27.2)	(33.3)	(69.0)	(41.0)	(2.2)	(2.4)	(0.8)
Old composite stems	—	—	—	—	—	—	—	—	14.1	71.2	21.6	21.0	—	—
No. of samples									(30)	(32)	(16)	(16)		
Range									(6.6)	(91.1)	(37.2)	(5.6)		

These old stalks were tested for nitrogen-fixing ability from September, when they were first observed, until the end of December, when they had either decayed beyond identification or had been transported from the marsh. Acetylene reduction from these stalks was lower than that from the mud surface except during October, when values from the former were considerably higher.

The mud surface in the *Spartina* zone showed acetylene reduction on each sampling occasion, but with low values less than 10 μM ethylene per gram dry weight per day even during the peak summer months. Mud surface values were even lower in the *Juncus* zone, and during the first two months of sampling no acetylene reduction was recorded there. The barren zone, like the *Spartina* zone, showed acetylene reduction in the surface mud during each sampling period, but values were considerably higher than those in the *Spartina* zone. In contrast to all other substrates and zones measured, no negative samples were ever found, and the lowest value ever recorded for any barren zone mud sample was 1.8 μM ethylene per gram dry weight per day.

Seasonal Patterns

In addition to these floral patterns, there were definite seasonal patterns in nitrogen-fixing ability in the salt marsh. Acetylene reduction in the *Spartina* zone was very little during the winter months, with a slight increase during the spring. Peaks in acetylene reduction on all substrates were reached during the summer months. During the fall the levels were lower than in summer but still much higher than those recorded during the spring. Winter fixation levels in the *Juncus* zone were even lower than those in the *Spartina* zone, and were still extremely low during the spring season. An increase was seen in values recorded for the summer, but in this zone the highest fixation on all substrates occurred during the fall. The barren zone showed a seasonal pattern similar to that of the *Spartina* zone, but with higher values in winter and spring than those found in either of the other zones. Therefore it would appear that other than strictly seasonal factors affect salt marsh nitrogen-fixing potential; these factors are such that the seasonal pattern of the *Juncus* zone is recognizably different from those of the other zones.

Observation of the seasonal pattern of acetylene reduction in all three zones also reveals a variation in results between the two winters included in the sampling period. During the first winter, no fixation was recorded in the *Juncus* zone, and only slight fixation, confined to old stems and mud surface, was found in the *Spartina* zone. Fixation was also low on the mud surface in the barren area. However, during the second winter of sampling fixation levels were measurably higher than those of the first winter. Comparison of seasonal results over the entire sampling period indicates that values for the second winter approximate more closely values obtained during the previous spring.

Samples from Other Florida Marshes

In order to compare the results from the main test sites over a wider region, we obtained samples from other marshes in Florida (*C*, *D*, and *E* in Fig. 1). The results are shown in Table 4. While these results do not include enough data to give seasonal fixation patterns at these sites, they compare favorably with results obtained during the same periods in the permanently established sampling sites.

During February and March 1972, tests were made to determine the effects of cold on the process of nitrogen fixation. Marsh materials known to show high levels of nitrogen fixation (barren zone mud and old *Spartina* stems) were collected and placed in a lighted refrigerator at 5°C. At two-day intervals, 10 samples were removed from the refrigerator and kept at room temperature. On alternate days, pairs of these samples were tested for nitrogen fixation using the technique described previously. This process was continued until samples had been received an exposure period ranging from 2 to 14 days of continuous cold. Results (Table 5) indicate that increasing periods of cold may inhibit nitrogen fixation for increasing lengths of time, and that even such cold as would be encountered in Florida marshes could temporarily bring the process of nitrogen fixation to an apparent halt. In more northern marshes, where ice may persist for days or weeks, this inhibition could cause a considerable delay in resumption of nitrogen fixation after thawing occurs.

Attempts to identify potential nitrogen-fixing organisms in the various zones of the salt marsh had limited success. The old stems of *Spartina alterniflora* were found to contain a dense population of bacteria of many species. Of these, the genera *Rhodospirillum* and *Rhodopseudomonas* were isolated and demonstrated to reduce acetylene in pure culture. The mud surface, particularly in the barren zone, yielded a wide variety of blue-green algae, many of which have been found by other researchers to fix nitrogen. However, only three genera which definitely showed acetylene reduction in pure culture were isolated: *Calothrix*, *Anabaena*, and *Lyngbya* —the latter was almost certainly *Lyngbya aestuarii* (J. T. Hardy, personal communication).

This latter species is of special interest for several reasons. First of all, it is the predominant component of the algal mat found on the surface of the barren zone. In addition, it is a nonheterocystous form, and until recently it was believed that only heterocyst-bearing blue-green algae could fix nitrogen. However, according to Stewart and Lex (1970), nonheterocystous forms can fix nitrogen under anaerobic conditions. The anaerobic surface mud of the salt marsh provides precisely such an environment. Therefore, in addition to *Lyngbya*, many other nonheterocystous blue-green algae found on these surface muds may have a potential for nitrogen fixation on the marsh mud surface.

4 Discussions and Conclusions

An unusual feature of the seasonal pattern of salt marsh nitrogen fixation is seen in the differences between the two winters. During the first winter, fixation was reported only on the mud surface of the barren and *Spartina* zones, and it was not until late spring that fixation rates began to rise. However, during the second winter of sampling, fixation was reported from all zones and all substrates other than water. Values were similar to those reported for the previous spring. The winter of 1971–1972 was unusually mild for the Pensacola area. Weather bureau statistics

Table 4. Mean ethylene production (μM per gram dry weight per day) in three northwest Florida salt marshes.

Site Date	Raccoon Bayou Jan.–Mar. 1971	Hogtown Bayou Mar.–Apr. 1971	Horseshoe Bayou Feb. 1972
Spartina:			
Water	—	0	0
Young stem	—	0	0.1
Old stem	—	0.4	1.2
Mud	—	0.2	0.6
Juncus:			
Water	0	0	0
Young stem	0	0	0
Old stem	0	0.1	0.5
Mud	0	0.1	0.3
Barren:			
Water	0	0	0
Mud	1.4	4.8	3.6
Cyperus:			
Water	0		
Young stem	0.1		
Old stem	0.2		
Mud	0.9		

for northwest Florida indicate highly significant temperature differences between the winters of 1970–1971 and 1971–1972. In 1970–1971 average minimum temperatures were 44.1°F in December, 40.9°F in January, and 41.4°F in February. The corresponding figures for 1971–1972 were 53.2°F in December, 48.9°F in January, and 43.3°F in February. The former had 44 nights with temperatures below 40°F, while the latter had only 19. During the former winter, the average number of warm days between cold periods was 3.9 (warm days are defined as those with high

Table 5. Number of days required for resumption of acetylene reduction after exposure to constant temperature of 5°C.

Length of Cold Exposure	Days at Room Temperature				
	0	2	4	6	8
2 days	+[a]	+	+	+	+
4 days	0	0	+	+	+
6 days	0	0	+	+	+
8 days	0	0	+	+	+
10 days	0	0	0	0	+
12 days	0	0	0	0	0
14 days	0	0	0	0	0

[a] Fixation.

temperatures above 70°F and low temperatures above 40°F). During the second winter of sampling, however, the average warm days between cold periods amounted to 10.4. The greater amounts of fixation recorded during the second winter of sampling therefore might be attributed to the warmer weather of that season.

Biological nitrogen fixation evidently contributes a significant amount of available nitrogen to the salt marsh ecosystem. In soybean fields, the maximum recorded rate of ethylene production was 180 μM ethylene per gram fresh weight per day (Hardy *et al.*, 1968). The rate of fixation on old *Spartina* stems alone reached a mean of 294.9 μM ethylene per gram dry weight per day in July. During the same month in the *Spartina* zone, contribution from the mud surface was 8.9 μM ethylene per gram dry weight per day, and the young stems added an additional 30.6 μM ethylene per gram dry weight per day.

The standing crop of living *Spartina* stems in the sampling area, measured several times, averaged 1.4 kg/m^2 dry weight; old stems averaged 0.8 kg/m^2. Based on these figures, the approximate amount of potential nitrogen fixation per square meter of *Spartina* zone in July was estimated to be 190 mM of nitrogen fixed per square meter per day. Thus, during this month alone, the potential contribution from nitrogen-fixing organisms was at least 1550 kg N_2/ha. The barren zone, during its peak fixation month of July, had a potential nitrogen input of 210 kg N_2/ha. In comparison, the growing season of an alfalfa crop encompasses at least two months, and during this period approximately 250 kg N_2/ha are produced. During the growing season, a crop of soybeans fixes about 150 kg N_2/ha (Stewart, 1966).

In terms of nitrogen fixation, the salt marsh is thus an extremely rich habitat. Indeed, the greater productivity of the marshes may be closely related to the high levels of nitrogen fixation occurring there.

References

Brezonik, P. L., and C. L. Harper. 1969. Nitrogen fixation in some anoxic lacustrine environments. *Science* 164, 1277–1279.

Burkholder, P. R. 1956. Studies on the nutritive value of *Spartina* grass growing in the marsh areas of coastal Georgia. *Bull. Torrey Bot. Club* 83, 327–334.

———, and G. H. Bornside. 1957. Decomposition of marsh grass by aerobic marine bacteria. *Bull. Torrey Bot. Club* 84, 366–383.

Green, F. M. 1972. Seasonal patterns of nitrogen fixation in salt marshes of the Florida Gulf coast. 28 pp. Master's Thesis. Pensacola; Biology Department, University of West Florida.

Hardy, R. W. F., R. D. Holsten, E. K. Jackson, and R. C. Burns. 1968. The acetylene-ethylene assay for N_2 fixation: Laboratory and field evaluation. *Plant Physiol.* 43, 1185–1207.

Moul, Edwin T. 1959. Ecological observation of pools in the salt marshes of New Jersey. *Proc. Salt Marsh Conf.*, Sapelo I., Ga. 58–60.

Odum, E. P., and A. A. de la Cruz. 1967. Particulate organic detritus in a Georgia salt marsh-estuarine ecosystem. In Lauff, G. A., ed., Estuaries. AAAS Publ. 83, 383–388.

Pomeroy, L. R. 1959. Algal productivity in salt marshes of Georgia. *Limnol. Oceanogr.* 4(4), 386–397.

Postgate, J. R., ed. 1971. *The Chemistry and Biochemistry of Nitrogen fixation.* 326 pp. London: Plenum Press.

Stewart, W. D. P. 1965. Nitrogen turnover in marine and brackish habitats. I. Nitrogen fixation. *Ann. Bot. London N.S.* 29, 229–239.

——— 1966. *Nitrogen Fixation in Plants*. London: Athalone Press.

———, and M. Lex. 1970. Nitrogenase activity in the blue-green alga *Plectonema boryanum* strain 594.

Teal, John M. 1959. Energy flow in the salt marsh ecosystem. *Proc. Salt Marsh Conf.*, Sapelo I., Ga. 101–104.

——— 1962. Energy flow in the salt marsh ecosystem of Georgia. *Ecology* 43, 614–624.

U.S. Department of the Interior. Federal Water Pollution Control Administration. 1970. *Effects of Pollution on Water Quality*. Escambia River and Bay, Florida. 60 pp.

3.3 Phenology and Seasonal Modeling in Insects[1]

William E. Bradshaw

Biology Department
University of Oregon
Eugene, Oregon, 97403 USA

1 Introduction

Insects exhibit a wide variety of seasonal adaptations. Studies of seasonality in insects have generally considered whole populations rather than repeated observations of individuals. The reason for this approach is that insects, as a rule, are small, short-lived, and extremely abundant. Insects are therefore more cryptic, more ephemeral, and are potentially endowed with much greater genetic variability than most plants or vertebrates. The importance of these attributes in phenological studies cannot be overemphasized. In different seasons of the same year or during the same season of successive years, one can expect to find individuals and gene frequencies which are significantly different from those encountered in prior sample populations. Consequently, the insect phenologist may deal with greater variation in the timing of a phenophase or in the extent of population growth than a phenologist concerned with plants or vertebrates. Seasonal models of insect populations will reflect this greater variation by exhibiting greater variance or lower confidence limits in their prediction of future events than will models concerning plants or vertebrates.

2 Definition of Insect Phenophases

The following phenophases are essentially reiterations or additions to those outlined by the US/IBP Phenology Committee (French *et al.*, 1972). Whenever possible, the terms have been left the same and only the definitions modified to correspond to insect bionomics.

[1] Research supported by the University of Oregon's Office of Scientific and Scholarly Research. Thanks to Philip Lounibos and Christina Holzapfel for their help with the manuscript.

KEYWORDS: Insect; phenophase; diapause; seasonality; phenology; modeling; latitudinal effects; pesticides control; photoperiodism; development.

Arrival: Emergence of dormant individuals or appearance of the stage of the life cycle following that in which dormancy occurs; arrival of migrants or seasonal immigrants.
Departure: Observation of dead or dormant individuals or noted absence of reliably observed individuals.
Flocking: Observation of swarms or aggregations.
Molt: Molting from one stage of development to the next: hatching from the egg, molting from one larval or nymphal instar to another, molting from larva to pupa, or molt from pupa or nymph to adult.
Singing: Characteristic song or call of the species.
Courtship: Ritual behavior or observation of mating pairs.
Nesting: Construction of characteristic structure by social or solitary insects; hoarding, construction of refugia.
Eggs: Appearance of eggs.
Polymorphism: Observation of a change in frequency of phenotypes; appearance of new phenotypes, including behavioral, morphological, physiological, or reproductive states.
Voltinism: Generations per year.

The first appearance of a given phenophase is often used to score its occurrence. The great variety of insects suggests that other portions or proportions of seasonal change will frequently be significant, and insect phenophases may have the following attributes:

Onset: First observation of a phenophase.
Termination: Last observation of a phenophase.
Duration: Time from onset to termination of a phenophase.
Extent: The number or proportion of organisms in a given phenophase.
Timing: When the onset, termination, maximum extent, or some other attribute of a phenophase occurs during the year.

3 Seasonal Arrival and Departure of Insects

Arrival and departure of insects at temperate latitudes are usually the consequence of terminating or initiating diapause. Diapause is a state of developmental dormancy brought about by inactivity of neuroendocrine centers in the brain; it persists until the brain again becomes endocrinologically active, even though conditions for development may be otherwise favorable (de Wilde, 1970). Diapause ordinarily endows an individual with resistance to environmental harshness and is known to occur at virtually any stage of development. Within a species or geographic race, diapause usually occurs only at a specific stage and thus synchronizes both the development of individuals within a population and development of that population with the seasonal cycle. Most of the discussion below concerning the factors controlling the onset and termination of diapause has been condensed from the excellent considerations of Danilevskii (1965), Andrewartha (1952), Lees (1955), and Beck (1968).

Obligate diapause is evoked endogenously without regard to the environment and imposes univoltinism upon an insect. More frequently, diapause is facultative and brought about by a variety of environmental parameters, including daylength, temperature, and food quality. Facultative diapause may occur in uni-, bi-, or multivoltine insects. At temperate latitudes, insects are mainly of the "long-day" variety, meaning that long days avert or terminate diapause and short days evoke or maintain it (Lees, 1955; Danilevskii, 1965). Nutritional food (Lees, 1955, pp. 37–39; Adkisson, 1961; Bull and Adkisson, 1962; Danilevskii, 1965, pp. 79–99) and high temperatures (Danilevskii, 1965; Beck, 1968) complement the action of long-day photoperiod. Among long-day insects, diapause is usually determined in phenophases proximal to that in which diapause occurs.

Short-day insects are those in which short days avert or terminate diapause and long days promote or maintain it. Short-day insects tend to occur at southern latitudes where summer dryness and not winter cold represents the major threat to insect survival (Geyspits, 1953; Masaki, 1958). Therefore it is not surprising that among short-day species, warm temperatures complement the action of short-day rather than long-day photoperiod. At more northern latitudes, short-day evocation of diapause may occur when diapause is determined during a phenophase remote from that in which diapause takes place (Kogure, 1933).

The resumption of development among diapausing individuals involves two distinct processes: the termination of diapause, *per se*, which renders the insect competent to develop, and postdiapause morphogenesis. The distinction is extremely important for modeling seasonality since the duration of these periods varies with respect to the two predominant factors mediating the termination of diapause: temperature and photoperiod. The termination of diapause in many insects is brought about by simple chilling at temperatures well below those required for normal morphogenesis (Andrewartha, 1952; Lees, 1955, pp. 53–64). The termination of diapause may also be brought about entirely through the action of photoperiod (Beck, 1968, pp. 151–161; Bradshaw and Lounibos, 1972). The action of photoperiod is generally more pronounced at temperatures compatible with morphogenesis than at lower temperatures (Danilevskii, 1965, pp. 106–113). Once diapause is terminated, postdiapause morphogenesis continues regardless of photoperiod and is affected by the same factors that control continuous development.

4 Modeling Seasonality in Insects

Modeling of seasonality in insects has two major components: modeling arrival and departure and modeling the continuous morphological, behavioral, or physiological phenophases which occur in between. Development during favorable conditions is controlled largely by food, temperature, and humidity. Presence and availability of food in nature are difficult to quantify, but reasonable knowledge of an insect's bionomics may justify assumptions concerning correlations between climate and food availability. Davidson and Andrewartha (1948), using multiple regression analysis, were able to predict the numbers of rose-thrips likely to be present during spring and early summer, based on four sets of meteorological measurements which accounted for 76% of the interannual variability among the population. Of these parameters, they found temperature during the preceding autumn to be the most highly correlated with changes in population density. They proposed that this strong effect of autumnal temperature was due to greater growth, earlier maturation, and earlier pollen production (= thrip food) by the host plant. The important point is that they did not have to quantify food availability to make their predictions of population size. A reasonable level of observation and assumptions sufficed.

Both temperature and humidity vary widely in space and time among insect microhabitats (Chauvin, 1967, pp. 15–51). Like low temperature, low humidity tends to retard development, primarily among insects that normally live in otherwise humid habitats (Bursell, 1964). Insects are often adept at seeking out favorable microhabitats (Fraenkel and Gunn, 1961), and the availability of favorable microclimates probably correlates well with sunlight, rainfall, temperature, and evaporative index. Haufe and Burgess (1956) sought to make such a correlation. They considered tundra and taiga pools inhabited by mosquito larvae. They found that pool surface temperature (X) was related to air temperature (T), hours of sunshine (Y), and wind-chill or dry-cooling power of the atmosphere (Z):

$$X = a\left[\frac{\sqrt{Y\times 1}\left(\frac{T_1\times T_2}{2}\right)}{Z}\right]^b \tag{1}$$

where a and b are constants, T_1 the maximum daily temperature (°F) minus 32, and T_2 the minimum daily temperature (°F) minus 32. Haufe and Burgess (1956) observed that the mosquito larvae in these pools tended to have species-specific temperature preferences. They would stay at shallow, warm depths until the temperature exceeded their preferred level and then migrate downward. Therefore, whenever the surface temperature exceeded their preferred level, the effective temperature was set equal to the preferred temperature.

Once one is able to correlate standard meterological data with insect microclimates, a variety of equations exist which relate rate of development to temperature. Each equation, in a sense, represents a different model relating the duration of a phenophase to temperature. The most commonly used equation is the thermal summation formula:

$$D = \frac{b}{T-a} \tag{2}$$

where D is the duration of the phenophase, T is the temperature, and a and b are constants. The constant a is the threshold temperature below which development ceases or the developmental zero. The constant b is the thermal constant and is the number of degree-days above developmental zero required for completion of a given phenophase. Temperature data below a must be set equal to a since regressive development does not ordinarily occur.

Danilevskii (1965, pp. 223–234) found that in the cabbage moth, *Barathra brassicae*, the developmental zero was around 10°C and the thermal constant about 560–580 degree-days. As shown in Fig. 1, *B. brassicae* is univoltine near Leningrad (620 degree-days), bivoltine near Veronezh (1100 degree-days) and trivoltine in eastern Georgia (1860 degree-days). Facultative pupal diapause in all strains of *B. brassicae* is mediated by photoperiod. The critical photoperiod at a given temperature increases with latitude of origin but critical photoperiod is lowered by higher temperatures and raised by lower ones. Thus, near Leningrad, summer temperatures do not normally rise above 18°C; consequently, all larvae experience short days and result in diapausing pupae. Near Veronezh, warmer temperatures and lower critical photoperiod combine to produce long-day conditions from about May 30 to July 15. The first generation then results in developing pupae and the

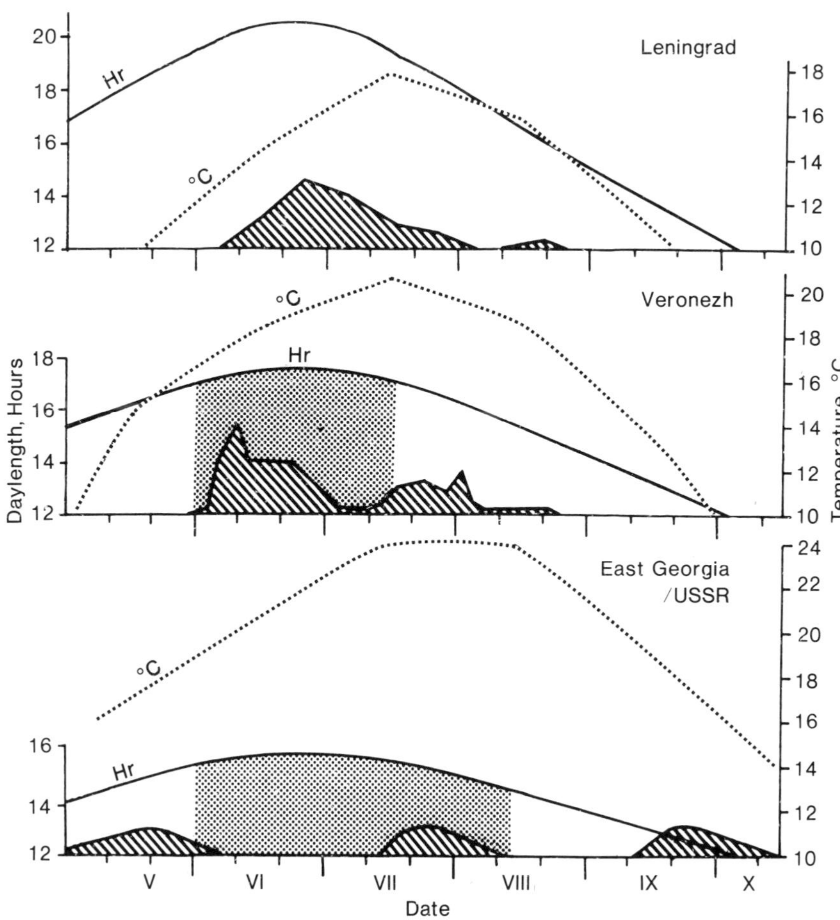

Fig. 1. Seasonal pattern in development of *Barathra brassicae* from Leningrad (60° N lat.), Veronezh (52° N lat.), and eastern Georgia (USSR) (42° N lat.). Diagonal shading: relative number of flying adults; light shading: diapause averting photoperiod. (After Danilevskii, 1965.)

second in diapausing pupae. In eastern Georgia, long-day conditions are present from about May 30 to August 15. The earliest larvae see short days; consequently, 22 to 30% of the first summer generation enters diapause. The latest larvae of the second summer generation likewise see short days and 45 to 60% of this generation enter diapause. All of the larvae of the third generation enter diapause. Some of the diapausing pupae of the first summer generation do not overwinter but emerge in the fall at about the same time as the third generation (Dolidze, 1957). This response in eastern Georgia corresponds well with that observed by Masaki (1956) in Japan at the same latitude where *B. brassicae* commences an aestival as well as an hibernal diapause. At still lower latitudes in Japan, the aestival diapause predominates. The very important point here is that a seasonal model based upon physiological experiments in the laboratory and using average temperature in the field for input data is likely to provide a good prediction of voltinism and departure under natural conditions.

Summation of degree-days does not always provide a satisfactory model for the relation between temperature and the completion of a phenophase. Other formulas based on various exponential, logistic, and catenary functions are provided by Wigglesworth (1965), Huffaker (1944), Clements (1963), and Nielsen and Evans (1960). These formulations are all based on uniform environmental temperatures, a condition which hardly ever exists in nature. Headlee (1940, 1941) has found that fluctuating temperatures may have a different effect than a constant temperature equal to the arithmetic mean of the fluctuating temperature. Cloudsley-Thomson (1953) has, however, pointed out that the arithmetic mean would be appropriate only when temperature has a linear relationship to biological rate processes, as in Equation (2). Otherwise, temperatures have to be weighted according to their nonlinear properties before summation. Thus logarithmic or other transforms of temperature data may yield more significant correlations between environmental temperature and predictability of phenophases.

Good models predicting arrival are equally as important as those describing departure. Arrival serves as the starting point of the phenological year and is largely responsible for the timing of the ensuing phenophases. Late arrival may delay the phenophases which are responsive to photoperiod so that an ordinarily bivoltine organism becomes univoltine (Danilevskii, 1965, pp. 212–221). Similarly, early arrival may result in a second generation with many nondiapausing individuals which are eliminated by winter. The timing of arrival in a given year can therefore affect the extent of population growth not only in that year, but in the subsequent one as well.

Individuals among a population may be in a variety of developmental stages when the transition from diapause-averting to diapause-promoting conditions occur. Departure at the end of the phenological year may therefore be gradual. By marked contrast, when the transition from diapause-maintaining to diapause-terminating conditions occur, individuals of a population tend to be in a uniform stage of development. Arrival can therefore be extremely abrupt, resulting in the sudden appearance of large numbers of insects. This difference in arrival and departure is illustrated by *Chaoborus crystallinus* whose larvae live in shallow palearctic ponds and diapause in the fourth larval instar beneath the winter ice

(Parma, 1969) (Fig. 2). The autumnal rise in proportion of fourth instar larvae shows that the onset of diapause takes place over a two-month period; vernal decline in the proportion of fourth instar larvae indicates the rapid, synchronous termination of diapause. Adult emergence in *Chaoborus* can also be synchronized through the action of, and not despite, fluctuating temperatures. Larvae of *C. americanus* live in shallow nearctic ponds. Diapause occurs in the fourth larval instar and is terminated through the action of feeding and long-day photoperiod (Bradshaw, 1969). Postdiapuse morphogenesis then leads to pupation and finally adult emergence. Each of these three phenophases—termination of diapause, post-diapause morphogenesis, and the duration of the pupal stage—has its own relation

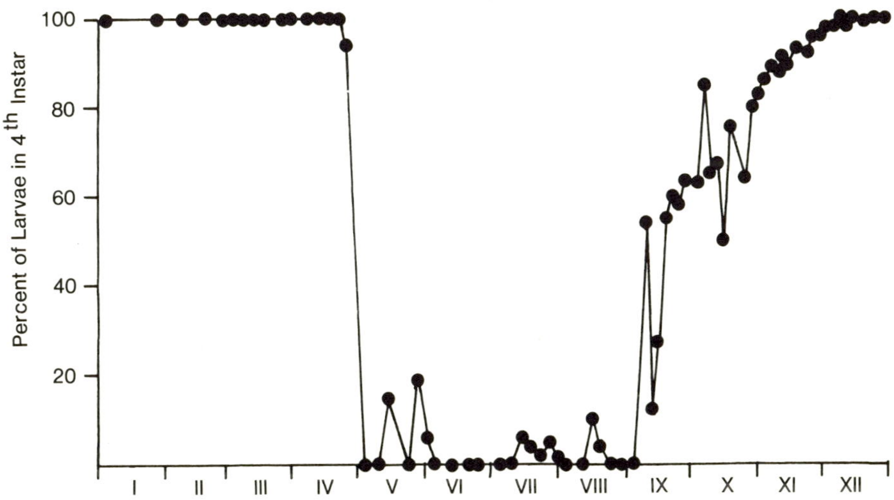

Fig. 2. Seasonal variation in the proportion of fourth instar larvae of *Chaoborus crystallinus*. (After Parma, 1969.)

to temperature as shown by the equations in Fig. 3. These equations serve as models of how each of these phenophases would progress during the spring. Cumulative percent completion of a phenophase can then be calculated as:

$$(\%\ \text{completion})_t = (\%\ \text{completion})_{t-1} + \frac{100}{\text{duration of phenophase at } T\,^\circ\text{C}} \tag{3}$$

where t is the present day, $t-1$ the preceding day, and T°C the average temperature on calendar day t. If temperatures were constant, one would expect that larvae commencing the termination of diapause 20 days apart would emerge as adults over a 20-day period. But it can be seen from Fig. 3 that under conditions of fluctuating temperature, adults emerge within a four-day period.

The most important aspect in determining the arrival of *C. americanus* adults in the above example is the termination of diapause. In the majority of insects, the termination of diapause is controlled mainly by temperature or photoperiod. These factors are considered below.

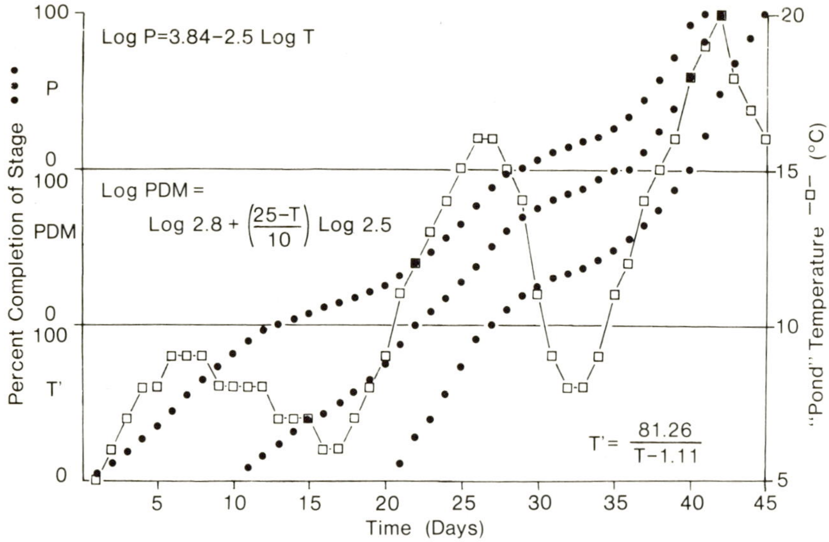

Fig. 3. Development of *Chaoborus americanus* in a fluctuating thermal environment. The duration of diapause termination (*T'*), postdiapause morphogenesis (*PDM*), and the pupal stage (*P*) are expressed as a function of temperature (*T*). Pond temperature was determined by drawing a smooth curve through temperature data from a shallow pond in southern Michigan during the spring of 1969. Three populations were presumed to initiate the termination of diapause on days 0, 10, and 20; they emerged as adults on days 40, 42, and 45, respectively (After Bradshaw, 1973).

The optimum chilling temperature for diapause termination may lie within the normal range of environmental temperatures and may be well above the lower sublethal limit (Lees, 1955, pp. 50–67; Danilevskii, 1965, pp. 1–35); Andrewartha, 1952). A plot of diapause duration as a function of temperature is therefore usually U-shaped; parabolic, elliptical, or hyperbolic functions can often be fitted to experimental data to form a model. Such a plot is frequently asymmetric about the temperature optimum, and two formulations are then required for a model: one equation for temperatures above the optimum and one for temperatures below it.

Problems concerning photoperiodically terminated diapause are various. Critical photoperiods as defined in the laboratory may be a physiologically useful index of photoperiodic response, but have several drawbacks for seasonal models: (1) In the laboratory, one usually employs instantaneous dawns and dusks, quite unlike the natural sequence of twilight. (2) There is a certain amount of reciprocity between the critical daylength and the duration of the experiment (Paris and Jenner, 1959). (3) Daylengths in nature are not static, but change with the season. (4) On a given date after the vernal equinox, daylengths are longer in the north than in the south. That insects from northern latitudes have longer critical daylengths than those from southern latitudes (Danilevskii, 1965; Beck, 1968) does not necessarily imply that the former will commence development at a later calendar date than the latter.

To account for some of these problems, I have proposed a model which predicts development of a photoperiodic mosquito in constant-temperature years with instantaneous dawns and dusks but changing photophases (Bradshaw, 1971). The mosquito *Wyeomyia smithii* is a long-day insect which diapauses in the third larval instar. The resumption of development is scored by molting to the fourth instar; the number of days required for half the population to molt is designated as IV_{50} (Bradshaw and Lounibos, 1972). IV_{50} is a function of daylength, being longest at the critical photoperiod, declining rapidly at longer daylengths, and finally reaching a minimum value at a daylength approximately 1 hr longer than the critical photoperiod. The crucial point is that, instead of asking how much development is evoked by a given daylength, one is asking how long does it take an animal to develop in response to a given photoperiod. IV_{50} as a function of several static daylengths is first determined experimentally, and date of median molting during dynamic daylengths predicted from the following model (Equations 4 to 9) [where t = day of solar year, $t-1$ = day before t, and $(IV_{50})_t = IV_{50}$ elicited by a static daylength equal to the hours of light per day on day t]:

A. Commence with t whose daylength elicits the maximum IV_{50} (4)
B. $t = t \times 1$ (i.e., add 1 day to t) (5)
C. $(\text{projected date})_t = t + IV_{50}$ (6)
D. If $(\text{projected date})_t$ earlier than $(\text{projected date})_{t-1}$, go to B (7)
D′. If $(\text{projected date})_t$ later than or equal to $(\text{projected date})_{t-1}$, go to E (8)
E. Predicted date of median molting = $(\text{projected date})_{t-1}$ (9)

For an artificial year created by an astronomic time clock at 25°C, the model successfully predicted median molting to within four days. Instead of an astronomic time clock, tables of sunrise, sunset, and twilight could be used and calendar dates of median molting in a constant-temperature year could be predicted. These calendar dates would be independent of the duration of the original static photoperiod experiments. Furthermore, they can be based on whatever astronomic time units (extent of twilight) are subsequently found to be important. Finally, the model can be used to estimate the extent to which longer critical daylengths at higher latitudes are due to real differences in environmental timing of development and to which extent they are due to the simple consequences of higher latitude.

The format employed in the model is intended to be compatible with computer programming. Continuous functions, while creating elegant models, do not always conform to the vagaries of insect physiology and climatic change. In such cases, the iterative capabilities of digital computers can be used to sum the effects of climatic variation on the progress of a phenophase, be it one controlling arrival, departure, or the duration of any phenophase in between.

5 Prospects

Levins (1969) and Cohen (1970) have formulated general models for the optimal timing of diapause. Both models require probabilities of survivorship as their input data, data which are at best difficult to obtain. An alternative approach has been pursued by Danilevskii (1965), who assumed that populations have already optimized the timing of diapause and that their seasonal development may be explained on the basis of known physiological rate processes. Danilevskii (1965) did not formulate a general model but it is clear from his work that seasonal development in insects may reliably be predicted from quantitative meteorological measurements if one knows the aspects of the environment that affect developmental rate processes. Models of photoperiodic reactions (Equations 4 to 9), when combined with those analyzing the temperature dependence of individual phenophases (Fig. 3), will eventually result in predictive models of vernal development. Models describing the timing and extent of successive phenophases during the ensuing periods of continuous development will likely evolve from multiple regression analyses of meteorological data similar to those of Davidson and Andrewartha (1948) and Haufe and Burgess (1956).

Accurate predictions of seasonality would permit the use of nonpersistent pesticides without requiring multiple application. Models predicting arrival and departure would indicate how many generations could be expected in a given year, and whether organisms found late in the season would be destined to diapause or whether they would continue developing and perish in winter without any intervention by man. Finally, if a model predicts a small outbreak of a pest species in a given year, long-term interests may be better served by not taking any control measures and permitting the pest species to maintain a healthy population of natural predators and parasites. The future of seasonal models in insects must therefore coordinate studies of insect physiology, biometeorology, and experimental population biology.

References

Adkisson, P. L. 1961. Effect of larval diet on the seasonal occurrence of dispause in the pink bollworm. *J. Econ. Entomol.* 54, 1107–1112.

Andrewartha, H. G. 1952. Diapause in relation to the ecology of insects. *Biol. Rev.* 27, 50–107.

Beck, S. D. 1968. *Insect Photoperiodism*. New York-London: Academic Press.

Bradshaw, W. E. 1969. Major environmental factors inducing the termination of larval diapause in *Chaoborus americanus* Johannsen (Diptera: Culicidae). *Biol. Bull.* 136, 2–8.

——— 1971. Photoperiodic timing of development in the pitcher-plant mosquito, *Wyeomyia smithii. Amer. Zool.* 11, 670–671.

——— 1973. Homeostasis and polymorphism in vernal development of *Chaoborus americanus. Ecology.* 54, 1247–1259.

———, and L. P. Lounibos. 1972. Photoperiodic control of development in the pitcher-plant mosquito, *Wyeomyia smithii. Can. J. Zool.* 50, 713–719.

Bull, D. L., and P. L. Adkisson. 1962. Fat content of the larval diet as a factor influencing diapause and growth rate of the pink bollworm. *Ann. Entomol. Soc. Am.* 55, 499–502.

Bursell, E. 1964. Environmental aspects: Humidity. In *The Physiology of the Insecta*, M. Rockstein, Ed. Vol. 1, pp. 323–361. New York-London: Academic Press.

Chauvin, R. 1967. *The World of an Insect*. pp. 15–51. New York: McGraw-Hill.

Clements, A. N. 1963. *The Physiology of Mosquitoes*. pp. 74–79. New York: Macmillan.

Cloudsley-Thomson, J. L. 1953. The significance of fluctuating temperatures on the physiology and ecology of insects. *Entomologist* 86, 183–189.

Cohen, D. 1970. A theoretical model for the optimal timing of diapause. *Am. Nat.* 104, 389–400.

Danilevskii, A. S. 1965. *Photoperiodism and Seasonal Development of Insects.* Edinburgh and London: Oliver and Boyd.

Davidson, J., and H. G. Andrewartha. 1948. The influence of rainfall, evaporation, and atmospheric temperature on fluctuations in the size of a natural population of *Thrips imaginis* (Thysanoptera). *J. Anim. Ecol.* 17, 200–222.

Dolidze, G. V. 1957. Study of cabbage-head insect pests in eastern Georgia. *Trans. Acad. Sci. Georgia SSR* 18, 83–90.

Fraenkel, G. S., and D. L. Gunn. 1961. *The Orientation of Animals: Kineses, Taxes, and Compass Reactions.* New York: Dover.

French, N. R., C. H. Lamoureux, H. Lieth, J. A. MacMahon, W. A. McKee, D. Parsons, F. Stearns, P. Webber, and D. Zobel. 1972. *Report of the US/IBP Phenology Committee.* Environmental Programs Coordinating Office, U.S. Component of the International Biological Program.

Geyspits, K. F. 1953. The reaction of univoltine Lepidoptera to day-length. *Entomol. Rev.* 33, 17–31.

Haufe, W. O., and L. Burgess. Development of *Aedes* (Diptera: Culicidae) at Fort Churchill, Manitoba, and prediction of dates of emergence. *Ecology* 37, 500–519.

Headlee, T. J. 1940. The relative effects on insect metabolism of temperatures derived from constant and variable sources. *J. Econ. Entomol.* 33, 361–364.

——— 1941. Further studies on the relative effects on insect metabolism of temperatures derived from constant and variable sources. *J. Econ. Entomol.* 34, 171–174.

Huffaker, C. B. 1944. The temperature relations of the immature stages of the malarial mosquito *Anopheles quadrimaculatus* Say, with a comparison of the developmental power of constant and variable temperatures in insect metabolism. *Ann. Entomol. Soc. Am.* 37, 1–27.

Kogure, M. 1933. The influence of light and temperature on certain characters of the silkworm, *Bombyx mori*. *J. Dept. Agr. Kyushu Univ.* 4, 1–93.

Lees, A. D. 1955. *The Physiology of Diapause in Arthropods.* London and New York: Cambridge University Press.

Levins, R. 1969. Dormancy as an adaptive strategy. *Symp. Soc. Exp. Biol.* 23, 1–10.

Masaki, S. 1956. The effect of temperature on the termination of pupal diapause in *Barathra brassicae* L. *Japan. J. Appl. Zool.* 21 (3), 97–107.

——— 1958. The response of a "short-day" insect to certain external factors: The induction of diapause in *Abraxis miranda* Butl. *Japan J. Appl. Entomol. Zool.* 2, 4, 285–294.

Nielsen, E. T., and D. G. Evans. 1960. Duration of the pupal stage of *Aedes taeniorhynchus* with a discussion of the velocity of development as a function of temperature. *Oikos* 11, 200–221.

Paris, O. H., and C. E. Jenner. 1958. Photoperiodic control of diapause in the pitcher-plant midge, *Metriocnemus knabi*. In *Photoperiodism and Related Phenomena in Plants and Animals*, R. B. Withrow, ed. pp. 601–624. Washington D.C.: American Association for the Advancement of Science.

Parma, S. 1969. The life cycle of *Chaoborus crystallinus* (DeGeer) (Diptera, Chaoboridae) in a Dutch pond. *Verh. Internat. Verein. Limnol.* 17, 888–894.

Wigglesworth, V. B. 1965. *Insect Physiology.* pp. 594–628. London: Methuen.

de Wilde, J. 1970. Hormones and insect diapause. In *Hormones and the Environment*, G. K. Benson and J. G. Philipps, eds. Cambridge: University Press.

3.4 Seasonality in Birds

Charles M. Weise

Zoology Department
University of Wisconsin
Milwaukee, Wisconsin, 53201 USA

1 Introduction

Seasonality in birds is most evident for physiological and behavioral events associated with reproduction, migration, and molt. These events have been intensively studied and thoroughly reviewed in the recent literature (Assenmacher, 1958; Farner, 1955, 1959, 9167; Farner and Follett, 1966; Immelmann, 1971; Lofts and Murton, 1968; Marshall, 1959; Wolfson, 1960). Therefore the present summary of the annual cycle in birds and its underlying physiological regulatory mechanism emphasizes the variations in timing attributable to weather-related phenological factors. The discussion pertains mainly to songbirds of temperate and high latitudes. Seasonality in other taxonomic groups or in birds of the sea, tropics, or deserts is mentioned only incidentally.

2 The Annual Cycle in Birds

In most birds of mid and high latitudes the energy-demanding events (phenophases) of migration, breeding, and molt are sequential and largely nonoverlapping. Figure 1 depicts the sequence in a generalized migratory songbird, based mainly on the white-crowned sparrow, *Zonotrichia leucophrys gambelii* (Farner, 1964), and the white-throated sparrow, *Z. albicollis* (Weise, 1956, 1963; Helms, 1963). In winter self-preservation is paramount, with most energy directed toward feeding and thermoregulation. As the year proceeds the overt events are, in order: decline of winter fat deposits, prebreeding molt, spring migratory fat deposition, spring migration, territorial and/or courtship behavior upon arrival in the breeding area, breeding (which itself involves a sequence of complex events), gonadal regression, postbreeding molt, autumn fattening, autumn migration, and finally deposition of

Contribution Number 8 from the University of Wisconsin–Milwaukee Field Station.

KEYWORDS: Seasonality; birds; *Zonotrichia*; *Junco*; breeding; migration; molt; fat; restlessness; refractory; timing.

winter fat and return to the winter condition. An important feature of the cycle, determinable only by experimentation, is the change in responsiveness to photoperiod indicated by the photosensitive and photorefractory periods in Fig. 1.

Of these events only the gonadal cycle, courtship, breeding, and postbreeding molt are universal. The others are highly variable among different species or populations. Some migratory birds, for example, carry out part or all of their migration without appreciable migratory fattening; some exhibit fattening in the spring but not in the fall, and so on. The prebreeding molt is absent in many birds, and where it does occur there are important species differences in the number of feathers involved and the consequent energy demands. In nonmigratory species all the events associated with migration are absent. The breeding period is variable in length, being particularly long in some nonmigratory or early-migrating species which produce two or three broods within the period. But whatever the particular combination of events a species has in its cycle, the timing of the phases is highly predictable.

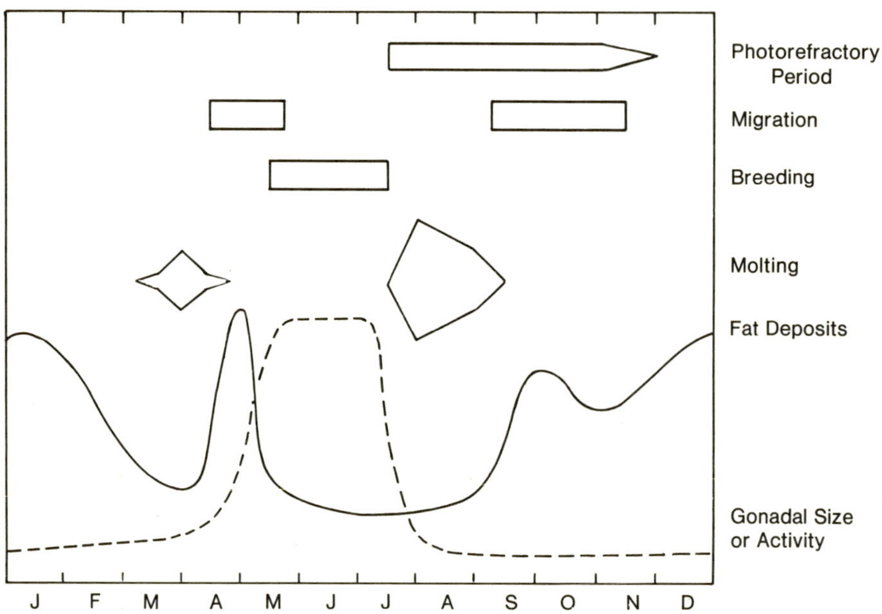

Fig. 1. Events in the annual cycle of a migratory songbird, generalized.

3 The Core Control System for the Annual Cycle

Because of the neat sequence of events in the annual cycle of birds, its precision in timing, and the relative ease with which birds can be studied, there have been many experimental explorations of the regulation of the cycle. Currently it is agreed that the events are linked and coordinated in a core control system which is fundamentally the same in all birds (Farner, 1967; Lofts and Murton, 1968).

The basic components of this system, following the scheme of Farner (1967), are: (1) external environmental information, (2) perception by eye or other receptors, (3) relay to brain centers and hypothalamus, (4) evaluation by circadian clock, (5) production of hypothalamic neuro-hormones and vascular transport to anterior pituitary gland, (6) secretion by the anterior pituitary of prolactin which is involved in migratory fattening and behavior, and gonadotropic hormones which stimulate gonadal growth, production of sex hormones and gamtogenesis, all of these leading to the attainment of the breeding condition. While most of the spring events can be accommodated in this scheme, the postbreeding events of late summer and fall cannot, and the controlling mechanisms of these are inadequately understood.

The adaptive variations in the annual cycles of different species of birds may then be traced to one or several of the components of the core control system. Birds may be stimulated by different types of environmental information, e.g., photoperiod versus rainfall, or they may be differentially sensitive to the same type, perhaps at the level of the circadian clock. The control of the pituitary by the hypothalamus may vary, resulting in different endocrine balances or hormone secretion rates. Finally, peripheral target organs may vary in their responses to pituitary or sex hormones.

Farner (1967) has recognized four timing mechanisms which operate on the core control system and are responsible for the timing of events in the individual bird (or species):

1. *Primary timing mechanisms.* These are responsible for setting the control system into operation and are dependent on such external environmental information as change in photoperiod, rainfall (in some Australian desert birds), or possibly food supply. Although there have been many claims and some evidence for endogeneity of the annual cycle at least in some birds (see Immelmann, 1971, p. 358), it is clear that in most birds increasing daylength after the winter solstice provides the primary impetus.
2. *Essential supplemental mechanisms.* These are responsible for completion or finer temporal adjustment of events initiated by the primary timer. These mechanisms are largely behavioral and are dependent on information provided by the presence of suitable habitat, presence of a mate, availability of nest site or nest materials, perception of courtship, or other social displays.
3. *Modifying mechanisms.* These are the acceleratory or inhibitory influences of environmental conditions other than the primary timer, including temperature and other weather aspects, the seasonal development of vegetation, and the seasonal development of a quantitatively or qualitatively adequate food supply.
4. *Terminal timing mechanisms.* These are the factors responsible for terminating or resetting the operation of the core control system. At present these are poorly understood and cannot be specified.

In this paper we are mostly concerned with category 3, the modifying mechanisms, although as Farner (1967) pointed out, these and the essential supplemental mechanisms are interacting and not always separable in practice.

4 Breeding

Since photoperiod is the primary timer in most nonequatorial birds, the initial development of the gonads, accessary organs, and early breeding behavior is quite precise, with modifying phenological factors playing a minor, although measurable and sometimes significant, role. Farner and Wilson (1957) found temperature to have a slight effect on the growth rate constant of the testis in white-crowned sparrows under various experimental photoperiods. Blanchard (1941) reported that development of the testis in a nonmigratory race of the white-crowned sparrow, *Z. l. nuttalli* was retarded or advanced somewhat by temperatures during the winter, i.e., during the early part of the gonadal recrudescence period. Marshall (1949) showed similar effects in several British species. In the great tit, *Parus m. major*, in Holland, Kluijver (1951) found a relation between winter and early spring temperatures and the subsequent start of breeding in later spring. This was interpreted as an influence of temperature on the early stages of development of the ovary.

At intermediate stages of development, supplemental mechanisms become essential. Neither photoperiod nor weather factors can bring the ovary to full maturity in most birds; this depends on the behavioral stimuli associated with habitat, mate, and nest materials, mentioned earlier (Hinde, 1967). However the timing of these behaviors may be indirectly influenced by weather.

Thus it can be seen that the date of attainment of the breeding condition may be influenced to some extent by modifying factors and interacting supplemental factors. At this point the role of modifying factors becomes more critical, and the date of onset of actual breeding, as expressed in the laying of eggs, may vary considerably from year to year. Nice (1937) found that the earliest laying in the song sparrow, *Melospiza melodia*, in Ohio varied from April 3 to April 23 in different years. She determined that it occurred five days after three days with temperature averaging 18.2°C, and that there was a decline of 0.87°C per day during April in the temperature threshold for laying. That is, if cold weather delayed the onset of laying, the first eggs would follow temperatures lower than those required earlier in the season. Subsequent workers have found similar effects in many other species, e.g., great tit (Kluijver, 1951), house wren, *Troglodytes aedon* (Kendeigh, 1963), and skylark, *Alauda arvensis* (Delius, 1965). Kluijver (1951) thought that low temperatures influenced only the start of laying, but it seems more likely that unfavorable weather can interrupt or interfere with the laying sequence (or other nesting activity) at any point, as argued by Kendeigh (1963).

The relationship between temperature and laying date may differ with species. In high-latitude regions most resident species and short-distance migrants begin breeding much earlier in the spring than long-distance migrants. Von Haartman (1963) found that laying dates were more variable in the former group than in the latter. This was because the early breeders, e.g., the starling, *Sturnus vulgaris*, tended to begin laying at a specific temperature threshold regardless of geographic location. Starlings in northern Finland thus began breeding significantly later than those in southern Finland. On the other hand, the late spring migrants like the spotted flycatcher, *Muscicapa striata*, tended to start breeding shortly after arriving on their breeding grounds, regardless of temperature, and egg-laying occurred only slightly later in northern than in southern Finland.

Weather conditions other than temperature have been implicated in determining the onset of breeding, but when critically tested have been found to play minor roles. Kluijver (1951) showed insignificant effects of sunshine, humidity, and rainfall in the Great Tit. Rainfall however may be much more important in tropical or desert birds (Immelmann, 1971).

Food supply is undoubtedly an important ultimate factor in the evolution of breeding seasons, but it could also serve as a proximate factor responsible for the final adjustment in the timing of laying, after the photoperiodically and behaviorally induced maturation of the ovary. Perrins (1970) argues that in some birds like the great tit the onset of laying is dependent on the female being able to find sufficient food to produce the clutch of eggs, which in this species amounts to about 94% of the female's body weight. Thus temperatures in many of the above examples might have exerted their effects indirectly by affecting the development of an insect food supply. Conversely, Kluijver (1951) found that great tits living in gardens where their food was artificially augmented started laying only insignificantly earlier than birds with natural food supplies.

Nevertheless it seems that most studies of the onset of breeding have not been designed to detect the precise mode of action of phenological factors. Temperatures or other seasonal factors could exert their effects by (1) influencing the core control system at the level of evaluation of external information; (2) influencing egg production directly by requiring more or less energy for thermoregulation, i.e. changing existence energy at the expense of productive energy; (3) influencing food supplies or foraging behavior so that total energy intake is affected; or (4) influencing territorial, courtship, or other breeding behaviors so that supplemental mechanisms are enhanced or disorganized.

The mechanisms responsible for the termination of breeding are poorly understood, and the influence of modifying phenological factors has scarcely been investigated. Although gonadal regression in most songbirds begins when daylengths are decreasing, it can be shown experimentally that such a decrease is not essential and is probably not the timing mechanism. Breeding failure may advance regression and late renesting attempts may delay it in individual birds, but how this is accomplished endocrinologically is not known. In Britain two species of thrushes, *Turdus ericetorum* and *T. merula*, were reported to have bred in 1960 for a month longer than usual (Mayer-Gross, 1964). This was attributed to the prolonged availability of an earthworm food source due to exceptionally rainy summer weather.

Following gonadal regression, most birds undergo a period when the core control system is unresponsive to long photoperiods. This refractory period dissipates naturally sometime in the fall. It has been shown experimentally that short photoperiods are required to terminate the refractory condition (Wolfson, 1959), but the possible role of modifying weather factors has not been investigated.

5 Migration

The timing of bird migration has long been famous for its precision. Many of the major facts regarding the phenology of migration have been known for a long time. By the beginning of this century it was possible to chart isochronal lines of spring arrival for many species in North America (Lincoln, 1952). Many present-day state or regional bird books show the average dates of arrival, peak movement, and departure for both spring and fall migrations for each species. A large portion of the journal *American Birds* (formerly *Audubon Field Notes*) is devoted to recording the year-to-year variations in these parameters and analyzing their causes.

As noted earlier there are great interspecific differences in migration physiology and behavior. In birds that migrate long distances or that migrate over water or through hostile climatic conditions, the development of the migratory disposition involves metabolic changes ultimately expressed in fat deposition, and behavioral changes such as the onset of migratory restlessness and the attainment of specific orientation patterns. In short-distance migrants these changes may be slight, but nevertheless it seems safe to say that in all cases the attainment of the migratory disposition involves some degree of preparation.

The timing of this preparation for migration may be approached in the same way as that for reproduction, in terms of primary, supplemental, modifying, and terminating mechanisms. The primary timer for spring migration is again photoperiod and the attainment of the migratory condition is governed by the core control system as previously described. The visible indicators of the migratory condition, fat deposition and migratory restlessness, develop rapidly at a certain time in the spring. Severe weather at this time can certainly delay the date of attainment of the migratory condition by several days (or weeks, in some early-spring migrants). This is probably a direct energetic effect. It is doubtful that favorable weather at this time can materially advance the date of attainment except possibly in some early migrating species like *Junco hyemalis*. The evidence for these statements is fragmentary and derived largely from experiments (Weise, 1956, 1963; King and Farner, 1959).

The final part of the migratory disposition is the migration flight. It is the action of modifying factors at this stage that largely determines the observed annual variations in arrival and departure dates and other migration phenomena.

The influence of weather on migration flights has been reviewed by Bagg *et al.* (1950), Lack (1960), Lowery and Newman (1966), Nisbet and Drury (1968), and Gauthreaux (1971). More recent studies have involved direct observation of migration flights by radar, portable ceilometers, and other sophisticated techniques. While there are some controversial issues the consensus is that bird migration flights are highly correlated with synoptic weather patterns rather than with simple events like temperature change. In the spring large migrations occur in the center or west side of a high-pressure air mass, where there is generally a consistent flow of air toward the north or northeast. Thus the birds are migrating on following winds. Beginning a flight in such weather also increases the probability of the birds finding favorable weather at the end of the flight (Nisbet and Drury, 1968). Large migrations may also occur in the warm sector of a low-pressure system (Bagg *et al.*, 1950). Weather systems that inhibit migration flights include the centers of low-pressure disturbances, and the east side of high-pressure cells, where winds would be adverse (Nisbet and Drury, 1968). The progress of migration through a given area in a given spring season, then, depends on the sequence of weather systems that move through the area.

The phenological importance of weather systems as opposed to simple changes in single weather factors is also illustrated by their role as ultimate factors in the selection and evolution of certain aspects of migration physiology. In the slate-colored junco (*Junco hyemalis*) an early spring migrant that winters in the United States and breeds mostly in Canada, the vernal fat deposition is timed to occur, not when the birds are departing from their wintering grounds, but after they have traveled several hundred kilometers and have reached the latitude of Wisconsin in early April (Weise, unpublished). At this time the average position of the Polar Front lies across southern Wisconsin (Bryson, 1966). Birds migrating beyond the front probably require extra energy reserves in the form of fat because of the increased likelihood of encountering cold weather and opposing winds.

The factors that terminate the spring migratory condition are almost unknown. In captive birds migratory restlessness persists through what would ordinarily be the breeding period (Weise, 1956, 1963; Helms, 1963). Possibly the physiological termination is induced by the birds' perception of a familiar home area.

The mechanisms responsible for timing the attainment of the fall migratory disposition are likewise poorly understood. Decreasing daylengths are not necessary for the development of autumn fat deposition and migratory restlessness in caged experimental birds (Weise, unpublished). Possibly the entire sequence of events from spring to fall is phased by increasing daylengths in the spring.

Once the birds have attained the fall migratory condition their flights are profoundly influenced by weather, as in spring. At this season the heaviest migration flights occur on the west sides of low-pressure systems, especially in the regions behind cold fronts (Hassler *et al.*, 1963; Lowery and Newman, 1966).

6 Molt

Molting or feather replacement is an important phenophase in birds, and its timing is as predictable as that of migration or breeding. There is agreement that the postbreeding molt, requiring 6 to 12 weeks or longer in most songbirds, must occur in a period of highly favorable weather and food supply. The evolution of breeding and migration times may in part reflect adjustments to this need (Pitelka, 1958).

The onset of the postbreeding molt is usually closely correlated with gonadal regression and the onset of photorefractoriness (except in Arctic birds where molt and breeding overlap to some extent). After a bird has been in the breeding condition for some time, these events occur automatically and are not dependent on changes in photoperiod. However the progress and termination of the molt are influenced by daylength: short days speed up the rate of molting, while artificially long days can prolong the molt by several weeks.

There is a little evidence for some influence by weather conditions (modifying factors). In the bullfinch (*Pyrrhula pyrrhula*), in England, Newton (1966) found that the initiation of molt in adults was spread over a long period from July to October and that this varied in different years. Delays in molting occurred in birds with late broods of young, and the occurrence of these was in turn related to unusually abundant food (weed seeds) in late summer due to weather conditions. The bullfinch has an atypical dependence on seeds as food for the young, and its molting schedule is probably not representative of songbirds in general. In most songbirds greater precision is probably the rule.

Many birds have a prebreeding molt, often resulting in a bright or conspicuous breeding plumage different from that worn during the rest of the year. These molts are usually partial, not involving the flight feather of the wing and tail, and they usually occur just before spring migration at a time of low fat deposits and body weights. The timing of this molt is governed by the core control system, but the endocrinologic mechanisms are practically unknown. Possibly the thyroid hormones are involved, which implies regulation by the anterior pituitary gland. In any case, modifying weather factors have been shown to cause some annual variation in the onset and progress of this molt (Blanchard, 1941; Morton and Mewaldt, 1962), warm weather tending to advance it and cold weather to retard it. However there have been very few critical studies on the phenology of molting.

References

Assenmacher, I. 1958. La mue des oiseaux et son déterminisme endocrinien. *Alauda* 26, 241–289.

Bagg, A. M., W. W. H. Gunn, D. S. Miller, J. T. Nichols, W. Smith and F. P. Wolfarth. 1950. Barometric pressure patterns and spring bird migration. *Wilson Bull.* 62, 5–19.

Blanchard, B. D. 1941. The white-crowned sparrows (*Zonotrichia leucophrys*) of the Pacific seaboard: Environment and annual cycle. *Univ. California Publ. Zool.* 46, 1–178.

Bryson, R. 1966. Air masses, streamlines, and the boreal forest. *Geogr. Bull.* 8, 228–269.

Delius, J. D. 1965. A population study of skylarks, *Alauda arvensis*. *Ibis* 107, 466–492.

Farner, D. S. 1955. The annual stimulus for migration: Experimental and physiologic aspects. In *Recent Studies in Avian Biology*, A. Wolfson, ed. pp. 198–237. Urbana: University of Illinois Press.

——— 1959. Photoperiodic control of annual gonadal cycles in birds. In *Photoperiodism*, R. B. Withrow, ed. pp. 717–750. Washington, D.C.: American Association for the Advancement of Science.

——— 1964. The photoperiodic control of reproductive cycles in birds. *Am. Sci.* 52, 137–156.

——— 1967. The control of avian reproductive cycles. *Proc. 14th Intern. Ornithol. Congr. 1966*, 107–133.

———, and B. K. Follett. 1966. Light and other environmental factors affecting avian reproduction. *J. Animal Sci.* 25, 90–118.

Farner, D. S. and A. C. Wilson. 1957. A quantitative examination of testicular growth in the white-crowned sparrow. *Biol. Bull.* 113, 254–267.
Gauthreaux, S. A. Jr. 1971. A radar and direct visual study of passerine spring migration in southern Louisiana. *Auk.* 88, 343–365.
Hassler, S. S., R. R. Graber, and F. C. Bellrose. 1963. Fall migration and weather, a radar study. *Wilson Bull.* 75, 56–77.
Helms, C. W. 1963. The annual cycle and *Zugunruhe* in birds. *Proc. 13th Intern. Ornithol. Congr. 1962*, 925–939.
Hinde, R. W. 1967. Aspects of the control of avian reproductive development within the breeding season. *Proc. 14th Intern. Ornithol. Congr. 1966*, 135–153.
Immelmann, K. 1971. Ecological aspects of periodic reproduction. In *Avian Biology*, D. S. Farner and J. R. King, eds. pp. 341–389. New York and London: Academic Press.
Kendeigh, S. C. 1963. Regulation of nesting time and distribution in the house wren. *Wilson Bull.* 75, 418–427.
King, J. R., and D. S. Farner. 1959. Premigratory changes in body weight and fat in wild and captive male white-crowned sparrows. *Condor* 61, 315–324.
Kluijver, H. N. 1951. The population ecology of the great tit, *Parus m. major* L. *Ardea* 39, 1–135.
Lack, D. 1960. The influence of weather on passerine migration: A review. *Auk.* 77, 171–209.
Lincoln, F. C. 1952. *Migration of Birds*. Garden City, N.Y.: Doubleday.
Lofts, B., and R. K. Murton. 1968. Photoperiodic and physiological adaptations regulating avian breeding cycles and their ecological significance. *J. Zool.* 155, 327–394.
Lowery, G. H., and R. J. Newman. 1966. A continent-wide view of bird migration on four nights in October. *Auk* 83, 547–586.
Marshall, A. J. 1949. Weather factors and spermatogenesis in birds. *Proc. Zool. Soc. London* 119, 711–716.
——— 1959. Internal and environmental control of breeding. *Ibis* 101, 456–478.
Mayer-Gross, H. 1964. Late nesting in Britain in 1960. *Brit. Birds* 57, 102–118.
Morton, M. L., and L. R. Mewaldt. 1962. Some effects of castration on a migratory sparrow (*Zonotrichia atricapilla*). *Physiol. Zool.* 35, 237–247.
Newton, I. 1966. The moult of the bullfinch *Pyrrhula pyrrhula*. *Ibis* 108, 41–67.
Nice, M. M. 1937. Studies in the life history of the song sparrow. I. A population study of the song sparrow. *Trans. Linn. Soc. New York* 4, 1–246.
Nisbet, I. C. T., and W. H. Drury. 1968. Short-term effects of weather on bird migration: A field study using multivariate statistics. *Animal Behaviour* 16, 496–530.
Perrins, C. M. 1970. The timing of bird's breeding seasons. *Ibis* 112, 242–255.
Pitelka, F. A. 1958. Timing of molt in Stellers jays of the Queen Charlotte Islands, British Columbia. *Condor* 60, 38–49.
Von Haartman, L. 1963. The nesting times of Finnish birds. *Proc. 13th Intern. Ornithol. Congr. 1962*, 611–619.
Weise, C. M. 1956. Nightly unrest in caged migratory sparrows under outdoor conditions. *Ecology* 37, 274–287.
——— 1963. Annual physiological cycles in captive birds of differing migratory habits. *Proc. 13th Intern. Ornithol. Congr. 1962*, 983–993.
Wolfson, A. 1959. The role of light and darkness in the regulation of spring migration and reproductive cycles in birds. In *Photoperiodism*, R. B. Withrow, ed. pp. 679–716. Washington, D.C.: American Association for the Advancement of Science.
——— 1960. Regulation of annual periodicity in the migration and reproduction of birds. *Cold Spring Harbor Symp. Quant. Biol.* 25, 507–514.

3.5 Seasonality in Mammals[1]

Michael H. Smith

Savannah River Ecology Laboratory
Aiken, South Carolina, 29801 USA
and
Department of Zoology
University of Georgia
Athens, Georgia, 30601 USA

1 Introduction

Seasonality is commonly observed among mammals. As a phenomenon it represents the integrated evolutionary and homeostatic responses of populations to a series of complex interactions of biological and physical variables. Despite this complexity, certain principles are obviously applicable to the study of seasonality. An organism must convert available resources into as many or more viable offspring than its intra- and interspecific competitors or else its genotype will decrease in relative frequency. If the various species populations are close enough to carrying capacity for the operation of some form of density-dependent feedback control, then the efficiency of energy conversion becomes an important factor in the outcome of the competitive interactions. Widescale environmental fluctuations change the functional efficiencies of organisms. Mammals, by developing homeothermy, have extended their range of peak physiological performance and tolerance and have gained certain advantages for the price of higher energy requirements for maintenance metabolism. Thus it is especially important in a discussion of mammals to recognize that seasonal changes in populational characteristics and energetics are directly related.

The fields of population ecology and bioenergetics have developed almost autonomously and have different goals. The prime objective in ecological bioenergetics is to assign caloric equivalents to ingestion, egestion, assimilation, maintenance metabolism, growth, and reproduction. The population ecologist is

[1] Preparation of the Manuscript was supported by Contract AT(38-1)-310 between the U.S. Atomic Energy Commission and the University of Georgia.

KEYWORDS: Metabolism; reproduction; survival; mammal diet; population; bioenergetics; ecology; evolution; resources; phenology; seasonality.

interested in the fluctuations in numbers of individuals as affected primarily by the difference between natality and mortality but also by immigration and emigration. Few studies have been conducted which seek to integrate information on all of the above subjects into a complete analysis for even one species of mammal during a limited time of the year (Odum *et al.*, 1962). Despite these limitations, it still seems possible to present the general picture for seasonal adaptations in mammals by carefully selecting data from a number of different species.

This discussion of seasonality will involve two major areas. First is the availability of resources, which includes both the quality and quantity of the resources and the mediating effect of the environment in determining access to them. There are many adaptations to increase accessibility or to decrease the need for certain resources at times when they are limited or absent. Understanding how these adaptations function to change the allocation of energy and population characteristics is the second objective.

2 Resource Availability

If all resources were available uniformly and in sufficient quantity throughout the year, there might be no need for seasonal adaptations. Water and food are probably the two most important resources that, when restricted in abundance, generate selection pressures for compensating adaptations. Seasonal variation in rainfall has been well documented in many areas, but the same is not true for food. Also, the real problem is in defining "available". For example, most mammals do not have rigidly restricted diets, even if they function at one trophic level (e.g., Child and von Richter, 1969), and many others are not restricted to one level (e.g., Gentry and Smith, 1968). For most mammals there are no precise data on seasonal variation in the abundance of food. Despite this lack of data, it has been argued that the abundance of certain types of food are responsible for animal cycles (Lauckhart, 1957).

There are numerous indications that the quality and type of food vary seasonally for a variety of mammal species. For example, the occurrence of forbs in the diet of mule deer is almost an order of magnitude higher during the summer than in the winter (Boeker *et al.*, 1972). Large variations in percent occurrence of the main components in the diet are common in the field vole (*Microtus agrestris*) (Hansson, 1971); lesser jerboa (*Allactoga elater*) (Pavlenko and Davletshina, 1972); bank vole (*Clethrionomys glareolus*) and yellow-necked field mouse (*Apodemus flavicollis*) (Gorecki and Gebczynska, 1962); raccoons (*Procyon lotor*), foxes (*Vulpes fulva* and *Urocyon cinereoargenteus*) and skunk (*Mephitis mephitis*) (Llewellyn and Uhler, 1952); pocket gopher (*Thomomys talpoides*) (Vaughn, 1967); shrews (*Sorex arcticus*, *S. cinereus*, and *Blarina brevicauda*) (Buckner, 1964); and lechwe (*Kobus lechee*), puku (*K. vardoni*), and waterbuck (*K. ellipsiprymnus*) (Child and von Richter, 1969). Mammals of many different taxonomic groups and functional types show seasonal shifts in dietary components. Some of the shifts in diet (e.g., *Kobus* spp.) are associated with migration caused by local flooding. Omnivores show greater seasonal dietary shifts in the types of food than herbivores.

The largest differences in diet are commonly observed between the coldest and warmest months or between the wet and dry seasons.

Considerable differences exist also in the quality of the same type of food eaten at different times during the year. Most of the supporting data for this generalization come from proximate analysis of the foods of wild ungulates and farm animals. The relationships of quality and quantity of forage utilized by black-tailed deer (*Odocoileus hemionus*), at low elevations in Alaska, to the nutritional requirements of the deer are given in Fig. 1 (*Klein*, 1965). Seasonal changes in population and

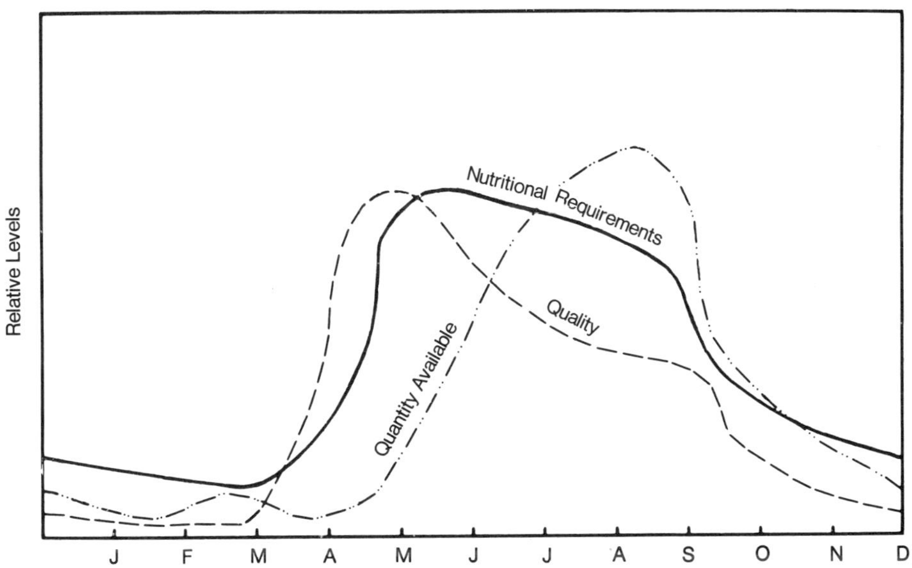

Fig. 1. Assumed relationship of quality and quantity of forage utilized by deer versus their nutritional requirements. (From Klein, 1965.)

individual characteristics vary such that maximum requirements occur at times of maximum quality or quantity. Deer increase their food intake as the quality of food declines. The factors contributing to the decline in quality are percentage protein, fat, crude fiber, ash, and concentrations of essential elements. Most of these components vary as a function of the age of the plant. For example, high nitrogen/carbohydrate ratios are associated with rapid shoot growth and low ones with root growth (Loomis, 1953). Since plant growth patterns vary with altitude and latitude, as well as the local pattern of rainfall, details of the seasonal shifts in the quality of diet for herbivorous mammals reflect local conditions.

Certain general trends in the quality of diet can be discerned in the data from northern latitudes. Maximum protein content normally occurs during the spring or early summer and the lowest in the winter (Klein, 1965; Kirkpatrick *et al.*, 1969). Digestibility follows the same trend (Eadie, 1970) (Fig. 2). The diet of *Microtus agrestis* varies in the same way as that of ungulates (Hansson, 1971), probably indicating the generality of these observations for all mammalian herbivores

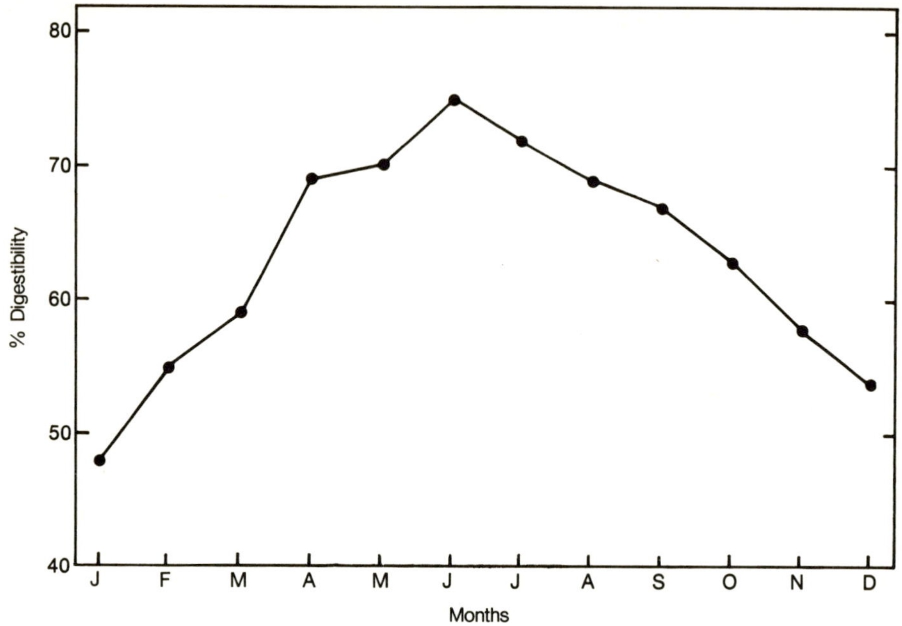

Fig. 2. Temporal changes in the digestibility of organic matter in forage from pastures stocked with sheep. (From Eadie, 1970.)

occurring at these latitudes. Data are needed on the foods of tropical species and the trace element concentrations in the diet at different times during the year.

Moisture content of food can be twice as high during the summer as in the winter (Hundley, 1959), and it varies as a function of relative humidity of the air (Schmidt-Nielsen and Schmidt-Nielsen, 1951). There are few other general measures of water availability published. Field observations show that many species of mammals do not have access to drinking water. Dew may be an important source of water for these forms. Grouping of mammals at water holes in deserts and grasslands, along with their many adaptations for water conservation (Chew, 1965), indicate the importance of water as a limiting resource, and abundance of water obviously varies seasonally at many locations. Certain biological characteristics, to be discussed later, also seem to fluctuate in accordance with the wet and dry seasons, and they will be used to illustrate how availability of water is related to seasonality in mammals.

3 Seasonal Adaptations

Metabolism

In mammals over 95% of assimilated energy is used for maintenance metabolism (Wiegert and Evans, 1967). Constant elevation of body temperatures requires a high rate of ingestion during all seasons. Many species forego a constant body temperature and show varying degrees of hypothermia. Mammals exhibit all degrees of regulation, from strict homeothermy to aestivation and deep hibernation (Kayser, 1965).

Metabolism in many nonbreeding mammals is lower in winter than in summer. The common shrew, *Sorex araneus*, has a winter metabolism that averages 76.2% of the summer values when the animals are living at 7°, 10°, 15°, 20°, 25°, 30°, or 35°C (Gebczynski, 1965). Rectal temperatures in bighorn sheep (*Ovis canadensis*) are lower (almost 2°C) during the winter than in late spring [body temperature data were not available for the summer period (Franzmann and Hebert, 1971)]. Heat production was 25% less in winter than in summer, and caloric intake was 35 to 45% less for reindeer and caribou (McEwan and Whitehead, 1970). Daily food consumption of tree squirrels (*Sciurus carolinensis* and *S. niger*) also shows the lowest values during winter (Short and Duke, 1971); the same trend holds for *Microtus arvalis* (Trojan, 1970). Water voles (*Arvicola terrestris*) have lower rates of oxygen consumption during autumn than in summer (Drozdz *et al.*, 1971). Lower winter metabolic rates might not be expected because of the tendency for metabolism to increase with decreasing temperature. However shifts in pelage insulation (Hart and Heroux, 1953), body composition (Caldwell and Connell, 1968), body temperatures (Franzmann and Hebert, 1971), and activity levels (Gebczynski, 1965) tend to decrease metabolism during the winter. Almost all of the data come from species in the temperate zone, and certain subtropical and tropical forms (e.g., *Sigmodon hispidus*) may not show decreased winter metabolism.

Hibernation represents the extreme case of reduction in metabolism during the colder winter months. Hibernation allows mammals to avoid periods of restricted food supplies or of low ambient temperatures or other physical conditions which decrease access to food. The causative factors for hibernation are well discussed by Kayser (1965), who emphasizes that temperature and relative humidity have immediate and direct effects upon hibernation. The lack of water or available moisture indirectly affects the animals by varying the moisture content of the food. If the herbaceous food remains greener and has a higher proportion of water later in the fall, animals delay going into hibernation (Shaw, 1925; Kashkarov and Lein, 1927). Availability of water is also important in causing aestivation in the cactus mouse, *Peromyscus eremicus* (Macmillen, 1965). This species lives in the dry southwestern deserts of the United States and does not have access to free water. Mammals frequently show decreased metabolism during winter but there is also occasionally a reduction during summer. The timing, degree, and duration of the reduction in metabolism and availability of food or water are probably correlated.

Body Composition

Most of the organs and physiological functions associated with metabolism show seasonal variation. For example, in the muskrat (*Ondatra zibethica*) body weight peaks during winter as do the relative weights of the heart and body fat, water, and protein. The thyroid, liver, and kidney show the smallest relative weights during the winter and largest during summer (Aleksiuk and Frohlinger, 1971). The amount of brown adipose tissue reaches a high during winter in a variety of mammals, including the red squirrel (*Tamiasciurus hudsonicus*) (Aleksiuk, 1971) and meadow vole (*Microtus pennsylvanicus*) (Didow and Hayward, 1969). Caloric value of the whole body shows the highest values in winter and lowest in summer in cotton rats (*Sigmodon hispidus*) (Golley, 1970) and shrews (*Sorex araneus* and *S. minutus*) (Myrcha, 1969). These caloric changes are affected by fat content, which is usually highest in winter; fat has a higher caloric value than protein or carbohydrate. There are major seasonal changes in the composition of the total body in almost all species of mammals. However, the data are again biased in that little is known about tropical species.

The gross modifications of the body components are also correlated with shifts in more precise physiological functions. For example, kangaroos (*Megaleia rufa* and *Macropus robustus*) have the highest electrolyte concentrations and osmolality in the urine in summer and lowest in winter (Dawson and Denny, 1969). Most indices of thyroid activity in livestock show seasonal variation even at southern latitudes (Afiefy, 1966). Vitamin A content of the liver of northern mammals shows varying degrees of seasonal fluctuation, with the magnitude of fluctuation depending upon the specific process by which the vitamin is stored (Boikova, 1971). The common pattern among different species is an increase in vitamin A with an increase in body weight.

Some seasonal patterns appear to be endogenous. Ascorbic acid concentrations of adrenals, thymus, and spleen of white laboratory rats are highest in winter and lowest in summer, even when the animals are kept under constant environmental conditions (Gabos, 1967). Synthesis of ascorbic acid is related to carbohydrate metabolism. Most of the other seasonal fluctuations in body function mentioned above are directly related to metabolism (e.g., thyroid and brown fat); many other metabolic functions not mentioned or yet studied would also be expected to show seasonal fluctuations.

The ability to achieve relatively high metabolic rates for short periods during the winter is probably selectively advantageous. This type of adaptation requires changes in the oxygen transport system. In general, erythrocyte, hematocrit, and hemoglobin levels tend to decrease with increasing temperature from winter to summer (Veiga *et al.*, 1963) but not in all cases (McEwan and Whitehead, 1969). Seasonal trends in these variables are observable in large farm animals and rodents living in relatively exposed environments (Sealander, 1962). However, small mammals living in the relatively more constant subnivean environment in the extreme north do not show variations in hemoglobin concentration or hematocrit (Sealander, 1966). Other observed changes in the blood may be related to seasonal variations in renal function and diet; these include significant spring to summer decreases in calcium, urea, creatinine, and uric acid and concurrent increases of glucose and potassium in brown bears (*Ursus arctos*) (Halloran and Pearson, 1972). Blood is a sensitive indicator of environmental fluctuations since it is quite variable in composition and reflects both short- and long-term stresses.

Energy conservation is extremely important to mammals, and fur decreases thermal losses to the environment. At low temperature mammals increase the insulation value of the hair by changing its thickness, as well as increase the amount of body fat. The relative seasonal changes in the insulation value of hair range from 12 (varying hare, *Lepus americanus*) to 52% (black bear, *Euarctos americanus*) of the winter values in nine species of Arctic and north temperate zone mammals (Hart, 1956). Larger mammals show the greatest absolute changes in the thickness of the fur. Most mammals do not molt during the winter nor during peak reproductive times when energy requirements for maintenance are highest (Ling, 1970). Pelage growth and replacement are very seasonal and are usually coordinated in respect to season by the relative length of the day.

Mammals appear to be capable of attaining large, short-term increases in metabolic rate during winter, and they have higher energy reserves at this time. However, they do not maintain a higher average metabolism during the colder months. Greater insulation, decreased function for certain organs, lower activity levels, and an appropriate reproductive cycle all contribute to the lower energy requirements during the winter period. Limited food is very likely to occur during winter in certain circumstances (Smith, 1971), and the population dynamics and physiology of mammals reflect this situation.

Reproduction

Much is known about patterns of reproduction in mammals (Asdell, 1946). From an evolutionary point of view, reproduction is the most important function of a species, even though it represents only a small portion of its total energy budget (Wiegert and Evans, 1967). During pregnancy the energy requirements of females increase slightly, but during lactation there is a dramatic increase (Brody,

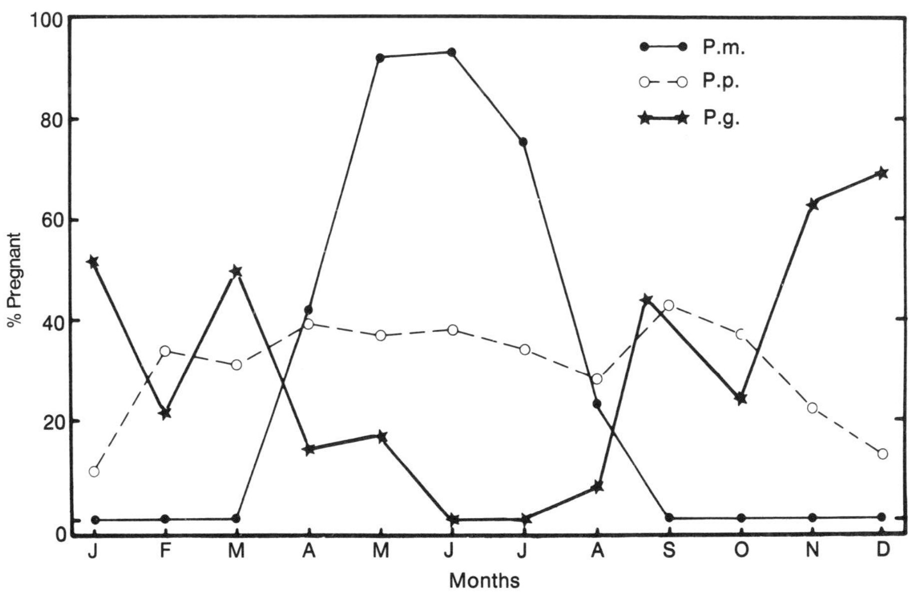

Fig. 3. Seasonal changes in the percentage of pregnant females in *Peromyscus maniculatus* from a northern locality characterized by extreme temperature fluctuations, *P. gossypinus* from a southern locality with a mild climate, and *P. polionotus* from an intermediate location and climate. (From Pournelle, 1952; Davenport, 1964; Brown, 1966, respectively.)

1945; Barnett and Little, 1965). Energy requirement of the common vole (*Microtus arvalis*), which averages 11 kcal per day in nonreproducing females, rises to 16 kcal per day maximum during pregnancy and 42 kcal per day during the last days of lactation (Migula, 1969). Males spend a minimal amount of energy on the production of reproductive tissue, but probably invest considerable energy into searching activities during the breeding season. Energetic equivalents for these activities have not been calculated for males, but they are probably relatively small compared to the energetic cost of reproduction in females. Lactation must be a very critical stage in the life history of a species and selection for appropriate timing severe in regards to season. It is not surprising then that the most obvious adaptations in mammalian reproduction have been to increase the efficiency of the process (Asdell, 1966).

The number and variety of local reproductive cycles are almost overwhelming—partly because reproductive characteristics vary with altitude and latitude (Smith and McGinnis, 1968). There is also variation in the cycle of a population at one location from year to year (Jameson, 1953). Nevertheless, certain trends are apparent. For example, breeding seasons are shorter in the north and young are born during the summer months, e.g., *Peromyscus maniculatus* (Brown, 1966) (see Fig. 3). Proceeding south, the breeding season is lengthened until animals breed throughout the year with a slight depression during the winter months, e.g., *P. polionotus* (Davenport, 1964). Further south there is a shift of the cycle to the point

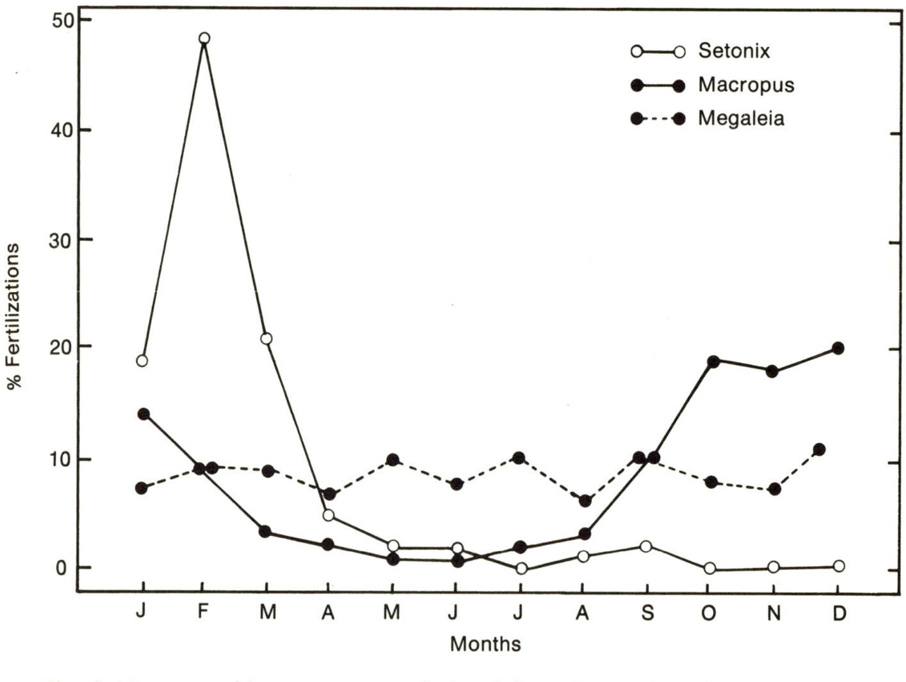

Fig. 4. Mean monthly percentages calculated from the total number of fertilizations observed during one year in three marsupials: *Setonix brachyurus*, *Macropus canguru*, and *Megaleia rufa*. (Modified from Sharman *et al.*, 1966.)

where most young are born during winter, and the hot summer months are characterized by reproductive quiescence, e.g., *P. gossypinus* (Pournelle, 1952). These trends are not clearly apparent across widely separated taxonomic units or large habitat diversity.

The seasonal pattern of rainfall may be more important than that of temperature in determining reproductive cycles of some neotropical species of rodents (Fleming, 1971) and for diprotodont marsupials (Sharman *et al.*, 1966) (see Fig. 4). The diprotodont marsupials (*Setonix* and *Macropus*), living in the coastal areas of southern Australia, are adapted to breed at a time that will insure that most of the young become semi-independent in the spring, when there is a flush of green vegetation. *Megaleia* inhabit areas where prolonged droughts occur often and rainfall is irregular. Its breeding cycle insures that there will always be some young to take advantage of improved forage conditions after rainfall. There are obvious seasonal reproductive patterns in many mammals but there are also aseasonal forms. Mammals exhibit almost all of the possible combinations of breeding strategies within the limits set by their anatomy (Asdell, 1966).

Reproduction is confined to periods when mammals are regulating their body temperatures (Fig. 5). Weights of the reproductive organs are lowest during hibernation (Kayser, 1965) and during the nonbreeding season in nonhibernators (Wildhagen, 1953; Lord, 1963). Reproductive organ weights are probably also influenced by seasonal changes in density (Christian and Davis, 1964) and most

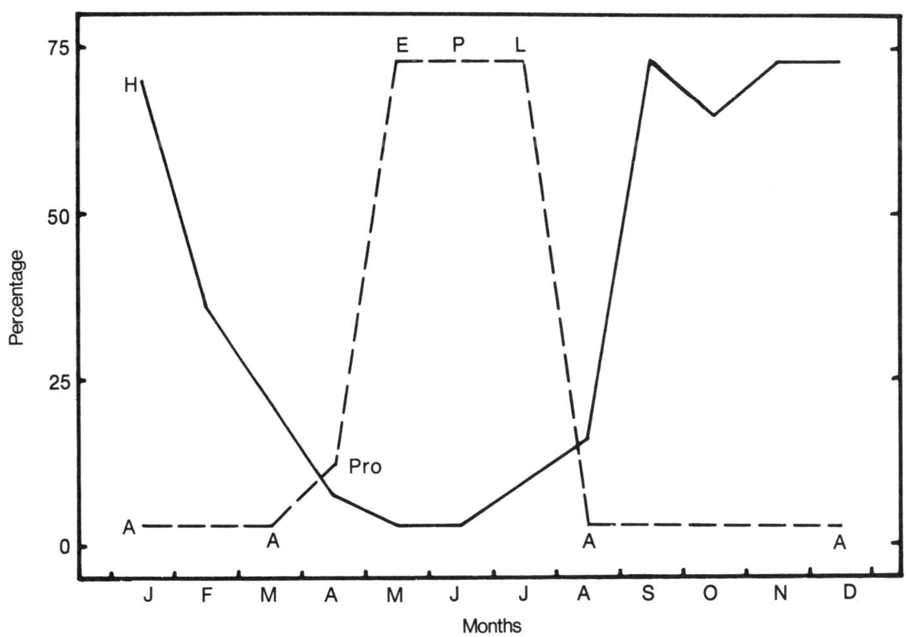

Fig. 5. Reproductive cycle in relation to the tendency to hibernate (*H*) in the 13-lined ground squirrel as measured by the percentage of days spent in hibernation each month. The reproductive cycle is represented by the broken line. *A* = anestrus; *Pro* = proestrus; *E* = estrus; *P* = pregnancy; *L* = lactation. (From Foster *et al.*, 1939.)

short-lived species show dramatic changes in number during one annual cycle (Lord, 1963; Fleming, 1971). Age structure of the populations also changes at the same time (Gliwicz *et al.*, 1968). Size-specific differences in reproductive rates are commonly encountered in small mammals (Caldwell and Gentry, 1965); within one species larger females have larger litters than smaller females. Many of the changes in body composition in the short-lived species of mammals may be due to shifts in age structure and not a direct result of environmental fluctuations.

Survival

The other important process influencing age structure is mortality. Most of the data on survivorship are not given by season. The question of seasonal variation is compounded by differences in survival rates of different cohorts born during separate seasons (e.g., *Arvicola terrestris*) (Stoddart, 1971). In the water vole those animals born in the latter part of the breeding season have the poorest survivorship. Survival of cottontails (*Sylvilagus floridanus*) is best when they are born during June and poorest if they are born before or after this part of the cycle (Lord, 1961).

Food shortages associated with unfavorable weather can directly cause death

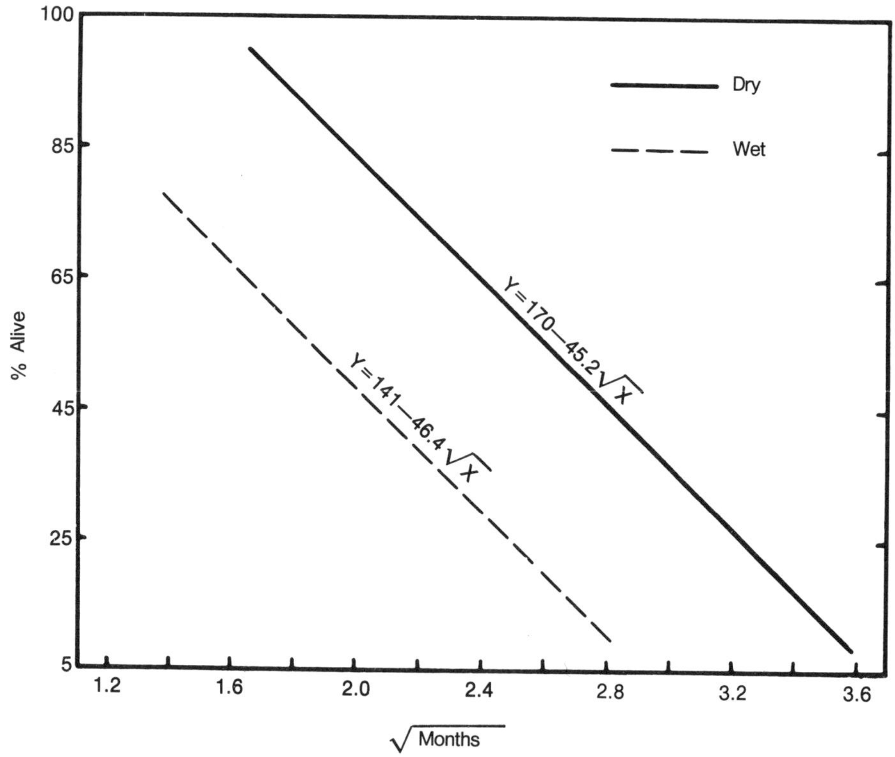

Fig. 6. Survival of the spiny rat, *Proechimys semispinosus*, born during the wet or dry season. Y equals the percentage alive and X equals the square root of the number of months after birth. (Modified from Fleming, 1971.)

in prairie deermice (*Peromyscus maniculatus*) (Howard, 1949). Survival is poorest in this species and in *Microtus agrestis* during spring, while *Clethrionomys glareolus* shows annual differences in its seasonal survival pattern (Chitty and Phipps, 1966). Man shows a highly seasonal pattern of stillbirths with maximum occurrence in April, May and June (Slatis and Decloux, 1967).

The relationship between survivorship and periods of limited resources is at best circumstantial and in some cases confusing. Survival of the spiny rat (*Proechimys semispinosus*) is best during the dry season and poorest in the wet season (Fleming, 1971) (see Fig. 6). Actually the rate of survival is probably constant after the early part of their life. Food and water should probably be more limiting in the dry season. In most cases, including this one, the cause of death is not known because mortality is seldom directly observed. Seasonal shifts in survivorship do occur in mammals, but more data are needed on the relationship between fluctuations in survival and important environmental variables to provide a better understanding of these processes.

Acknowledgments

I would like to thank Alan Fink and John Gentry for criticizing the original version of this paper.

References

Afiefy, M. M. 1966. Seasonal variations in thyroxine and iodine contents in relation to fertility in buffaloes and cattle. *Egyptian Vet. Med. J.* 8, 73–100.

Aleksiuk, M. 1971. Seasonal dynamics of brown adipose tissue function in the red squirrel (*Tamiasciurus hudsonicus*). *Comp. Biochem. Physiol.* 38a, 723–731.

———, and A. Frohlinger. 1971. Seasonal metabolic organization in the muskrat (*Ondatra zibethica*). 1. Changes in growth, thyroid activity, brown adipose tissue and organ weights in nature. *Can. J. Zool.* 49, 1143–1154.

Asdell, S. A. 1946. *Patterns of Mammalian Reproduction.* Ithaca, N.Y.: Comstock Publishing Co.

——— 1966. Evolutionary trends in physiology of reproduction. *Symp. Zool. Soc. London* 15, 1–13.

Barnett, S. A., and M. J. Little. 1965. Maternal performance in mice at −3°C: Food consumption and fertility. *Proc. Royal Soc.* B162, 492–501.

Boeker, E. L., V. E. Scott, H. G. Reynolds, and B. A. Donaldson. 1972. Seasonal food habits of mule deer in southwestern New Mexico. *J. Wildlife Management* 36, 56–63.

Boikova, F. I. 1971. Seasonal variations in vitamin "A" content in the liver of certain mammals of the far north. *Sov. J. Ecol.* 2, 77–81.

Brody, S. 1945. *Bioenergetics and Growth.* New York: Reinhold.

Brown, L. N. 1961. Reproduction of *Peromyscus maniculatus* in the Laramie Basin of Wyoming. *Am. Midland Naturalist*, 76, 183–189.

Buckner, C. H. 1964. Metabolism, food capacity, and feeding behavior in four species of shrews. *Can. J. Zool.* 42, 259–279.

Caldwell, L. D., and C. E. Connell. 1968. A précis on energetics of the old-field mouse. *Ecology* 49, 542–548.

Caldwell, L. D., and J. B. Gentry. 1965. Natality in *Peromyscus polionotus* populations. *Am. Midland Naturalist* 74, 168–175.

Chew, R. M. 1965. Water metabolism of mammals. In *Physiological Mammalogy*, W. V. Mayer and R. G. Van Gelder, ed. Vol. II, pp. 43–178. New York: Academic Press.

Child, G., and W. von Richter. 1969. Observations on ecology and behaviour of Lechwe, Puku and Waterbuck along the Chobe River, Botswana. *Z. Säugetierkunde*, 34.

Chitty, D., and E. Phipps. 1966. Seasonal changes in survival in mixed populations of two species of vole. *J. Animal Ecol.* 35, 313–331.

Christian, J. J., and D. E. Davis. 1964. Endocrines, behavior and populations. *Science* 146, 1550–1560.

Davenport, L. B., Jr. 1964. Structure of two *Peromyscus polionotus* populations in old-field ecosystems at the AEC Savannah River Plant. *J. Mammal.* 45, 95–113.

Dawson, T. J., and M. J. S. Denny. 1969. Seasonal variation in the plasma and urine electrolyte concentration of the arid zone kangaroos *Megaleia rufa* and *Macropus robustus*. *Australian J. Zool.* 777–784.

Didow, L. A., and J. S. Hayward. 1969. Seasonal variations in the mass and composition of brown adipose tissue in the meadow vole, *Microtus pennsylvanicus*. *Can. J. Zool.* 47, 547–555.

Drozdz, A., A. Gorecki, W. Grodzinskı, and J. Pelikan. 1971. Biogenergetics of water voles (*Arvicola terrestris* L.) from southern Moravia. *Ann. Zool. Fennici* 8, 97–103.

Eadie, J. 1970. Sheep production and pastoral resources. *Symp. Ecol. Soc. Brit.* 10, 7–24.

Fleming, T. H. 1971. Population ecology of three species of neotropical rodents, *Misc. Publ.* No. 143. 77 pp. Ann Arbor: Museum of Zoology, University of Michigan.

Foster, M. A., R. C. Foster, and R. K. Meyer. 1939. Hibernation and endocrines. *Endocrinology* 24, 603–612.

Franzmann, A. W., and D. M. Hebert. 1971. Variation of rectal temperature in bighorn sheep. *J. Wildlife Management* 35, 488–494.

Gabos, M. 1967. Seasonal variations of ascorbic acid content in some organs of the white rat. *Rev. Roumaine Biol.* 12, 87–90.

Gebczynski, M. 1965. Seasonal and age changes in the metabolism and activity of *Sorex araneus* Linnaeus 1758. *Acta Theriol.* 10, 303–331.

Gentry, J. B., and M. H. Smith. 1968. Food habits and burrow associates of *Peromyscus polionotus*. *J. Mammal.* 49, 562–565.

Gliwicz, J., R. Andrzejewski, G. Bujalska, and K. Petrusewicz. 1968. Productivity investigation of an island population of *Clethrionomys glareolus* (Schreber, 1780). I. Dynamics of cohorts. *Acta Theriol.* 8, 401–413.

Golley, F. B. 1969–1970. Caloric value of cotton rats (*Sigmodon hispidus* Say and Ord). In *Energy Flow Through Small Mammal Populations*, K. Petrusewicz and L. Ryszkowski, ed. pp. 143–147. Warszawa: Polish Scientific Publishers.

Gorecki, A., and Z. Gebczynska. 1962. Food conditions for small rodents in a deciduous forest. *Acta Theriol.* 6, 275–295.

Halloran, D. W., and A. M. Pearson. 1972. Blood chemistry of the brown bear (*Ursus arctos*) from southwestern Yukon Territory, Canada. *Can. J. Zool.* 50, 827–833.

Hansson, L. 1971. Habitat, food and population dynamics of the field vole *Microtus agrestis* (L.) in south Sweden. *Viltrevy*, *Swed. Wildlife* 8, 267–378.

Hart, J. S. 1956. Seasonal changes in insulation of the fur. *Can. J. Zool.* 34, 53–57.

———, and O. Heroux. 1953. A comparison of some seasonal and temperature-induced changes in *Peromyscus*: Cold resistance, metabolism, and pelage insulation. *Can. J. Zool.* 31, 528–534.

Howard, W. E. 1949. Dispersal, amount of inbreeding, and longevity in a local population of prairie deermice on the George Reserve, southern Michigan. Contributions to the Laboratory of Vertebrate Biology. 43, 1–52.

Hundley, L. R. 1959. Available nutrients in selected deer-browse species growing on different soils. *J. Wildlife Management* 23, 81–90.

Jameson, E. W., Jr. 1953. Reproduction of deer mice (*Peromyscus maniculatus* and P. *boylei*) in the Sierra Nevada, California. *J. Mammal.* 34, 44–58.

Kashkarov, D., and L. Lein. 1927. The yellow ground squirrel of Turkestan *Cynomys fulvius oxianus* Thomas. *Ecology* 8, 63–72.

Kayser, C. 1965. Hibernation. In *Physiological Mammalogy*, W. V. Mayer and R. G. Van Gelder, ed. Vol. II, pp. 179–296. New York: Academic Press.

Kirkpatrick, R. L., J. P. Fontenot, and R. F. Harlow. 1969. Seasonal changes in rumen chemical components as related to forages consumed by white-tailed deer of the southeast. *Trans. N. Am. Wildlife Natural Resources Conf.* 34, 229–238.

Klein, D. R. 1965. Ecology of deer range in Alaska. *Ecol. Monograph* 35, 259–284.

Lauckhart, J. 1957. Animal cycles and food. *J. Wildlife Management* 21, 230–234.

Ling, J. K. 1970. Pelage and molting in wild mammals with special reference to aquatic forms. *Quart. Rev. Biol.* 45, 16–54.

Llewellyn, L. M., and F. M. Uhler. 1952. The foods of fur animals of the Patuxent Research Refuge, Maryland. *Am. Midland Naturalist* 48, 193–203.

Loomis, W. E. 1953. *Growth and Differentiation in Plants.* Ames: Iowa State College Press.

Lord, R. D., Jr. 1961. Mortality rates of cottontail rabbits. *J. Wildlife Management* 25, 35–40.

——— 1963. *The Cottontail Rabbit in Illinois.* Carbondale: Southern Illinois University Press.

Macmillen, R. E. 1965. Aestivation in the cactus mouse, *Peromyscus eremicus. Comp. Biochem. Physiol.* 16, 227–248.

McEwan, E. H., and P. E. Whitehead. 1969. Changes in the blood constituents of reindeer and caribou occurring with age. *Can. J. Zool.* 47, 557–562.

——— 1970. Seasonal changes in the energy and nitrogen intake in reindeer and caribou. *Can. J. Zool.* 48, 905–913.

Migula, P. 1969. Bioenergetics of pregnancy and lactation in European common vole. *Acta Theriol.* 14, 167–179.

Myrcha, A. 1969. Seasonal changes in caloric value, body water and fat in some shrews. *Acta Theriol.* 14, 211–227.

Odum, E. P., C. E. Connell, and L. B. Davenport, Jr. 1962. Population energy flow of three primary consumer components of old-field ecosystems. *Ecology* 43, 88–96.

Pavlenko, T. A., and A. G. Davletshina. 1972. Nutrition of lesser jerboa in the Fergana Valley. *Sov. J. Ecol.* 2, 69–71.

Pournelle, G. H. 1952. Reproduction and early postnatal development of the cotton mouse, *Peromyscus gossypinus gossypinus. J. Mammal.* 33, 1–20.

Schmidt-Nielsen, B., and K. Schmidt-Nielsen. 1951. A complete account of the water metabolism in kangaroo rats and an experimental verification. *J. Cellular Comp. Physiol.* 38, 165–181.

Sealander, J. A. 1962. Seasonal changes in blood values of deer mice and other small mammals. *Ecology* 43, 107–119.

——— 1966. Seasonal variations in hemoglobin and hematocrit values in the northern red-backed mouse *Clethrionomys rutilus dawsoni* (Merriam), in interior Alaska. *Can. J. Zool.* 44, 213–225.

Sharman, G. B., J. H. Calaby, and W. E. Poole. 1966. Patterns of reproduction in female diprotodont marsupials. In *Comparative Biology of Reproduction in Mammals*, I. W. Rowlands, ed. pp. 205–232. New York: Academic Press.

Shaw, W. T. 1925. Duration of aestivation and hibernation of the Columbian ground squirrel (*Citellus columbianus*) and sex relation to the same. *Ecology* 6, 75–81.

Short, H. L., and W. B. Duke. 1971. Seasonal food consumption and body weights of captive tree squirrels. *J. Wildlife Management* 35, 435–439.

Slatis, H. M., and R. J. Decloux. 1967. Seasonal variation in stillbirth frequencies. *Human Biol.* 39, 284–294.

Smith, M. H. 1971. Food as a limiting factor in the population ecology of *Peromyscus polionotus* (Wagner). *Ann. Zool. Fennici* 8, 109–112.

———, and J. T. McGinnis. 1968. Relationships of latitude, altitude, and body size to litter size and mean annual production of offspring in *Peromyscus*. *Res. Popul. Ecol.* 10, 115–126.

Stoddart, D. M. 1971. Breeding and survival in a population of water voles. *J. Animal Ecol.* 40, 487–494.

Trojan, P. 1969–1970. Energy flow through a population of *Microtus arvalis* (Pall.) in an agrocoenosis during a period of mass occurrence. *In Energy Flow through Small Mammal Populations*, K. Petrusewicz and L. Ryszkowski, ed. pp. 267–279. Warszawa: Polish Scientific Publishers.

Vaughn, T. A. 1967. Food habits of the northern pocket gopher on shortgrass prairie. *Am. Midland Naturalist* 77, 176–189.

Veiga, J. S., E. Ghion, and C. A. C. Aggio. 1963. Aspectos fisiológicos associados com a adaptacão das bovinos nas regioes tropicais e subtropicais. *Arguiv. Escola Vet.* 15, 167–204.

Wiegert, R. G., and F. C. Evans. 1967. Investigations of secondary productivity in grasslands. In *Secondary Productivity of Terrestrial Ecosystems*, K. Petrusewicz, ed. Vol. II, pp. 499–518. Warzawa: Panstwowe Wydawnictwo Naukowe.

Wildhagen, A. 1953. *On the Reproduction of Voles and Lemming in Norway*. Oslo: Arbeidernes Aktietrykkeri.

3.6 Seasonality of Decomposers[1]

Barry A. Katz[2] and Helmut Lieth

Department of Botany
University of North Carolina
Chapel Hill, North Carolina, 27514 USA

1 Introduction

The largest portion of the biomass, once produced, is remineralized by decomposers after a shorter or longer time. In strongly seasonal climates, where large amounts of litter are produced within a short period, the activity of the decomposers is also seasonal. At the present, too little is known about the decomposers to say how they act in endogenous seasonal pulses. We summarize what is known about decomposer phenology in general, with special emphasis on the soil fungi.

2 Input into the Decomposer System

The input of biomass into the decomposer system in the ground is prodigious. Table 1 compares litter input in different climatic regions of the world. Roughly two-thirds of the total annual litter fall from above ground is leaf litter. Deposition varies, in time, from one climatic zone to another and, within a particular biome, according to the habit of the species producing litter (Fig. 1).

This dead biomass (and dead roots) are deposited from land vegetation primarily on top of the soil and in the upper few decimeters of the mineral soil. Many significant turnover processes must therefore occur within these upper soil layers. Lieth (1963) calculated that the soil-litter decomposing system is responsible for no less than 60% of the total carbon dioxide that a forest returns to the atmosphere.

Also the soil-litter system is the reservoir for 70% of the carbon dioxide retained as carbon in recently fixed or readily available biomass (excluding carbon in the form of petroleum, coal, etc.) Most of the nitrogen, phosphorus, sulfur, etc., originally bound in the biomass enters the producer and decomposer systems;

[1] Contribution Number 88 from the Eastern Deciduous Forest Biome, US/IBP.
[2] Present address: IPC-KFA, 517 Jülich 1, W. Germany.

KEYWORDS: Seasonality; decomposer; fungi; forest litter; summergreen forest; North Carolina; review paper; basidiomycetes; ascomycetes; fungi imperfecti.

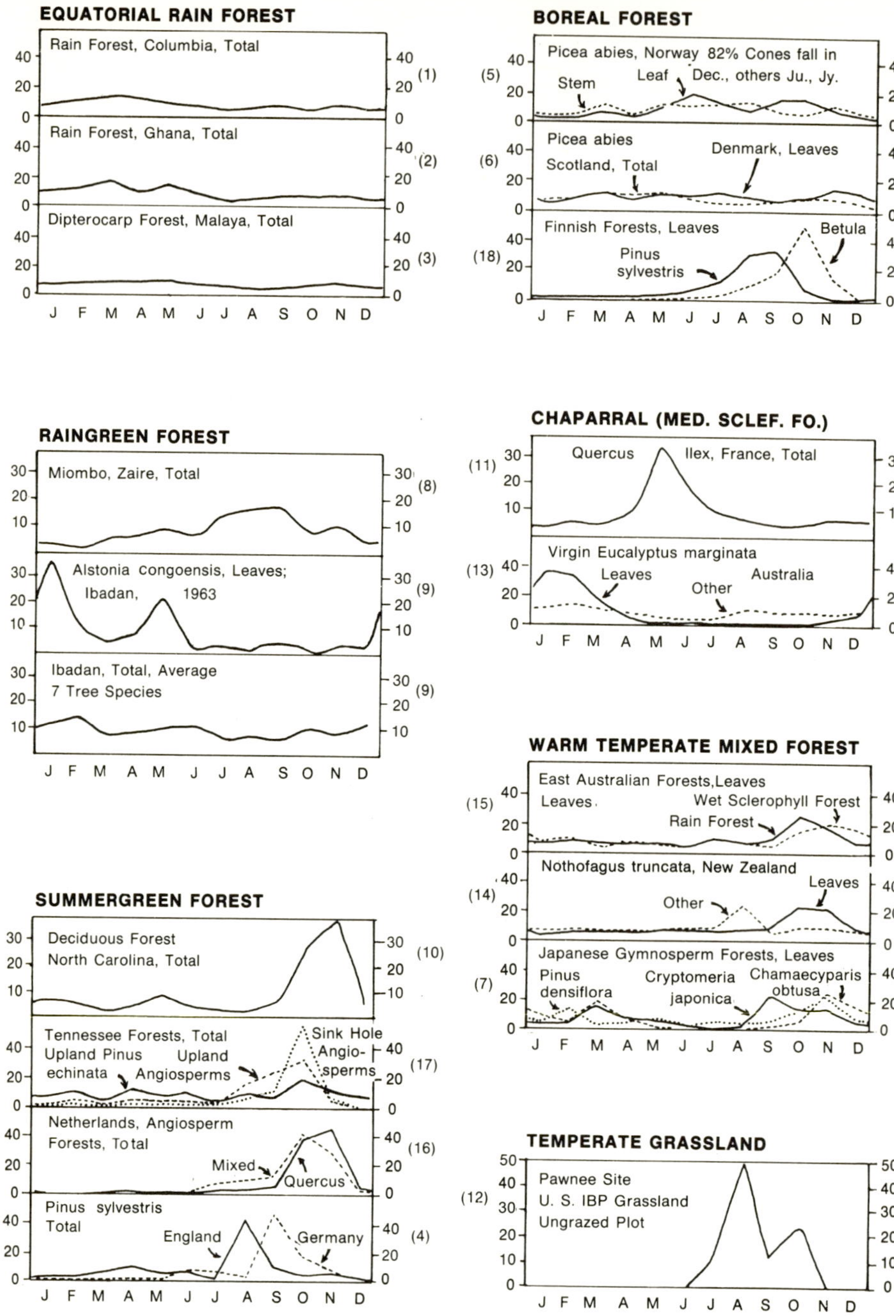

Fig. 1. Monthly plant litter production in various biomes. *Abscissa*, months; *ordinate*, monthly litter production in percent of yearly total. Distribution of litterfall is strongly seasonal in temperate and boreal deciduous forests and dry seasonal woodlands in the tropics.

The reference numbers indicate the following source papers. Papers 1 to 7 and 13 to 18 are adapted from Bray and Gorham (1964, pp. 134, 135 and 136). (These citations may not

Table 1. Annual litter production in four major climatic zones.[a]

	Leaves		Other		Total	
	No. Regions Averaged	Metric Tons/Ha	No. Regions Averaged	Metric Tons/Ha	No. Regions Averaged	Metric Tons/Ha
Arctic-alpine	1	0.7	1	0.4	3	1.0
Cool temperate	15	2.5	10	0.9	22	3.5
Warm temperate	8	3.6	5	1.9	7	5.5
Equatorial	2	6.8	1	3.5	4	10.9

[a] Updated table after Bray and Gorham (1964).

the continued growth of high-level primary producer systems depends upon the decomposer system for many of its essential inorganic mineral nutrients.

Microorganisms may depend on the seasonality of environmental forces or their seasonally different performance may affect the environment for plants and animals. Possible interactions include physical influences on the biological activity, range and stability of temperature and moisture, biomass deposition, seasonal mineral fallout in precipitation and overall availability of dissolved minerals (both harmful and beneficial), the diffusion properties of substances in relation to a given microhabitat under a given set of conditions, exposure to sunlight; biological influences on biological activity (products of primary mineral weathering, solubility of mineral salts due to free organic chelations (such as citric acid), and selective uptake by living organisms), pH, redox potential, presence of exoenzymes, growth promoters and growth inhibitors, and concentration of dissolved oxygen and carbon dioxide.

The methodological drawbacks revealed in the attempt to distinguish microbial species, taxonomically and functionally, is surely a cause of the slow advancement of soil microbial ecology, in general, and its phenology, in particular. Nevertheless, there are a number of seasonality-bound fluctuations reported in the literature. In Table 2 we have compiled the various components of the decomposer systems for which seasonality is reported. In spite of this evidence, some investigators (Cutler *et al.*, 1923; Parkinson and Balasooriya, 1969) claim that nonseasonal and diurnal variations in physical parameters are as influential as yearly seasonal changes. Cutler's protozoan samples did, in fact, show nearly as much diurnal variation as yearly variation, while Parkinson could show no difference between fungal populations collected simultaneously from selected plots and populations

be included in the literature list). 1—Jenny *et al.* (1949). 2—Nye (1961). 3—Mitchell (1936) 4—Kendrick (1959), and Danckelmann (1887). 5—Mork (1942). 6—Wright (1957), Bornebusch (1937). 7—Osmasa and Mori (1937). 8—Malaisse and Malaisse-Mousset (1970). 9—Madge (1965). 10—Wells, *et al.* (1972). 11—Rapp (1971). 12—Calculated from Morris and Brunner (1971) in N. R. French, ed. (1971). 13—Hatch (1955). 14—Miller and Hurst (1957). 15—Webb (personal communication) to Bray and Gorham (1964). 16—Witkamp and van der Drift (1961). 17—Olson (personal communication) to Bray and Gorham (1964). 18—Viro (1955).

Table 2. Summary of major groups of organisms in the soil-litter system.

1	2	3	4	5
Group	Reference to Seasonality	Population Estimates	Feeding Habits	Comments
Bacteria and actinomycetes	Witkamp, 1966; Eggleton, 1938; Waksman, 1916	1. $2–9\times10^9$/g. soil. 2. 4000 lb live wt/acre for 6 in. depth[a] (~ 4.5 tons/ha for 15-cm depth) 3. 336–700 kg/ha.[c]	Primary decomposers, secondary decomposers, autotrophs, pathogens, predators	The same method of estimating population as No. 1 in col. 3 yields 2×10^9 for human feces. This points up the importance of bacteria in soil.
Fungi	Kendrick & Burgess, 1962, Hodgkiss & Harvey, 1960	336 kg/ha[c]	Primary decomposers, secondary decomposers, pathogens, predators	Ubiquitous, important primary and secondary decomposers.
Algae	Not considered	$1–300\times10^6$/g [a]	Autotrophs, saprotrophs	
Protozoans	Cutler *et al.*, 1923	$10^3–10^7$/g soil [a]	Primarily predators of bacteria	May consume $1.5–9\times10^{14}$ bacteria/m²/year which is 15–85 times the standing crop of readily available bacteria.
Nematoda	Coleman, 1971	$3.3–300\times10^5/m^2$ [b]	Plant parasites, predators of bacteria & algae or larger organisms	Affect numbers of primary producers and decomposers but have little direct decomposition significance.
Enchytraeidae (small worms)	O'Connor, 1967	$3–5\times10^5/m^2$ [a]	Mycophagic; diet little known	
Lumbricidae	Satchell, 1967; Evans & Guild, 1947	$5–250\ g/m^2$ [a]	Litter consumers	Effects on decomposers much publicized.
Arthropoda (ex. Acari & Collembola)	Seasonality not considered in this paper	$2\times10^5/yd^2$ [a] ($\sim 2.5\times10^5/m^2$)	Mycophagic, detritophagic, predators, coprophagic	Numbers and mass vary with species.
Acari	Celecia & Smythe, personal communication 1972; Wallwork, 1959	$1.6\times10^5/m^2$ [a]	Predators, detritophagic, mycophagic	Aid primary decomposition indirectly by nutrient and organismal transport and humus formation.
Collembola	Smythe & Celecia, personal communication 1972; Glasgow, 1939; Metz, 1967	$7–500\times10^2/m^2$ [a]	Mycophagic and detritophagic	As above

[a] Burges and Raw (1967); [b] author(s) cited in column 2 [c] Volobuev (1964).

collected year-round from the same plot in some soil horizons. Ultimately, we must reconcile the environmental complex with the successional nature of the process.

We restrict ourselves to the discussion of the first two groups in Table 2; for details about the other groups, we refer to the literature listed in column 2 of the table. Seasonality and phenology will be discussed in two aspects: seasonal quantity of individual members and seasonal changes in species composition.

3 Seasonal Changes in Quantity. Bacteria, Actinomycetes, and Fungi: Methodological Problems

Early in the twentieth century Waksman (1916) systematically applied Koch's dilution plating technique to soil samples taken from a meadow, a garden, an orchard, and a forest. He took monthly samples from prescribed depths and measured the moisture, nitrogen, and carbon content of the soils. Waksman concluded that bacterial numbers were not governed by the above-mentioned parameters, and that the time of maximum bacterial populations varied with each soil.

Koch's method was originally designed to obtain pure cultures, and requires that the experimenter provide a nutritionally satisfying, asceptic culture medium to insure colonization. The weakness of soil dilution plating stems from the fact that, of the entire range of soil microorganisms, only a selected few will be suited to the limitations of media, temperature, and atmospheric composition of choice. The method does not insure that the organisms of this nutritional spectrum will be enumerated, as there are competitive effects that bias the technique to an even more limited range.

It is lamentable that, as of the beginning of 1973, no physical technique or no medium will allow soil microbiologists to isolate members of the class Basidiomycetes, which include the primary cellulolytic and ligninolytic fungi, except under the most fortuitous conditions.

The vast majority of the quantitative literature is based on experimentation involving the serial dilution plate method or a modification of it. Eggleton (1938) studied the physical conditions and microbial populations of English pasture soils. Plots were treated with all combinations of manuring, irrigation, or lack of treatment. Analysis of variance and regression indicated that soil moisture content was a statistically significant factor in soil population, whereas temperature was not. Eggleton ventured the opinion that moisture controlled by affecting the herbage and only indirectly influenced soil bacteria.

The work of Meiklejohn (1957) illustrates another study in comparative soil microbial analysis, and points out another weakness of the serial dilution plate method. Working in Kenya, she attempted to analyze the seasonality of bacteria and actinomycete populations of three adjacent vegetation types: a pasture, tall grassland, and an acacia wood. In addition to the dilution plate method, she estimated populations by direct observation (Parkinson, 1972). The dilution plate method consistently gave low bacterial densities at low dilution ratios and higher bacterial densities at higher dilution ratios in all three of the series she did using different culture media. Direct observation also confirmed that the unusually high actinomycete populations she reported (1955) were due to enormous numbers of actinomycete spores in soil, which she attributed to drought that preceded the experimental period.

The population figures derived from dilution plates assume that each colony on agar develops from a single microorganism; however, there were changes in the numbers of individuals initiating colonies, which are related to the dispersion and fragmentation of soil particles and bacterial clumps in suspension.

The most rigorous mathematical analysis of errors in dilution plating can be found in the work of Hornby (1969). Other important references for understanding the limitations of agar plate culturing from natural substrates are Rao (1959), Wastie (1961), and Dwivedi and Garrett (1968).

Population estimates by direct observation methods suffer from equally serious drawbacks: the similarities in many microorganisms of their active, inactive, and dead individuals do not allow for estimation of the current active population; and the indistinguishability of the vegetative state among many members within the same class of microorganisms prohibits enumeration of most of the visible organisms to even their genera. For progress in this method see Perfil'ev and Gabe (1969), Casida (1969), and Parkinson (1972).

A promising approach to studying soil bacteria comes from the Russian investigators Perfil'ev and Gabe (1969). They developed thin-walled, rectangular capillary tubes which allow the investigator to examine the organisms that develop within the capillaries under powerful oil immersion optics. Series of capillaries are placed in the soil for varying periods of time to insure equilibration with the surrounding soil. The success of these tubes in moist muddy habitats is indicated by the fact that it was possible to detect the stratification of microorganisms according to oxidation potential and/or oxygen tension within the capillaries as indicated by oxidation-related color stratifications of the minerals within the capillary tubes. Organisms, colonial arrangements, and unusual feeding habits that have never been seen before have been recorded using these tubes. A micromanipulator has been adapted for use within the capillaries while viewing under oil immersion. Casida (1969) developed another technique that allows for the investigation of bacteria using high quality optics, but the use of capillaries is consistent with the normal capillary environment of the soil.

Other investigators have attempted to sample the decomposer population by other methods which yield greater taxonomic and ecological information than the serial dilution plate method.

The spectrum of techniques available to investigate fungi are reviewed in Booth (1971) and in Parkinson (1972). Many valuable techniques that are mentioned in these sourcebooks are not considered here because no phenological research has been conducted using these techniques.

In an effort to isolate the more fastidious soil fungi that are poor competitors on standard media, pure or unusual substrates are introduced into the soil-litter system as bait; horns, hair, skin, dung, cellophane, and shrimp exoskeletons are but a few of the more common baits (Koch, 1972; Luttrell, 1967). Mabee and Garner (Chapter 3.7) employed the cellulose baiting technique described by Luttrell (1967).

Kendrick and Burges (1962) made a detailed study of the seasonality of colonization, sporulation, and stratification of pine litter (*Pinus sylvestris*). A variety of procedures, including media culturing, moist chambers, partial sterilization, and paraffin sectioning, were used in an effort to assess all of the fungi. This was necessary because most of the important pine needle inhabitants do not grow readily in culture. The organic horizon was divided into separate *L*, F_1, and F_2 layers. Distinctions between interior and exterior colonizers were possible. Temperature and moisture were recorded and thought to have remained within tolerable limits throughout the study, and were not considered significant factors in deter-

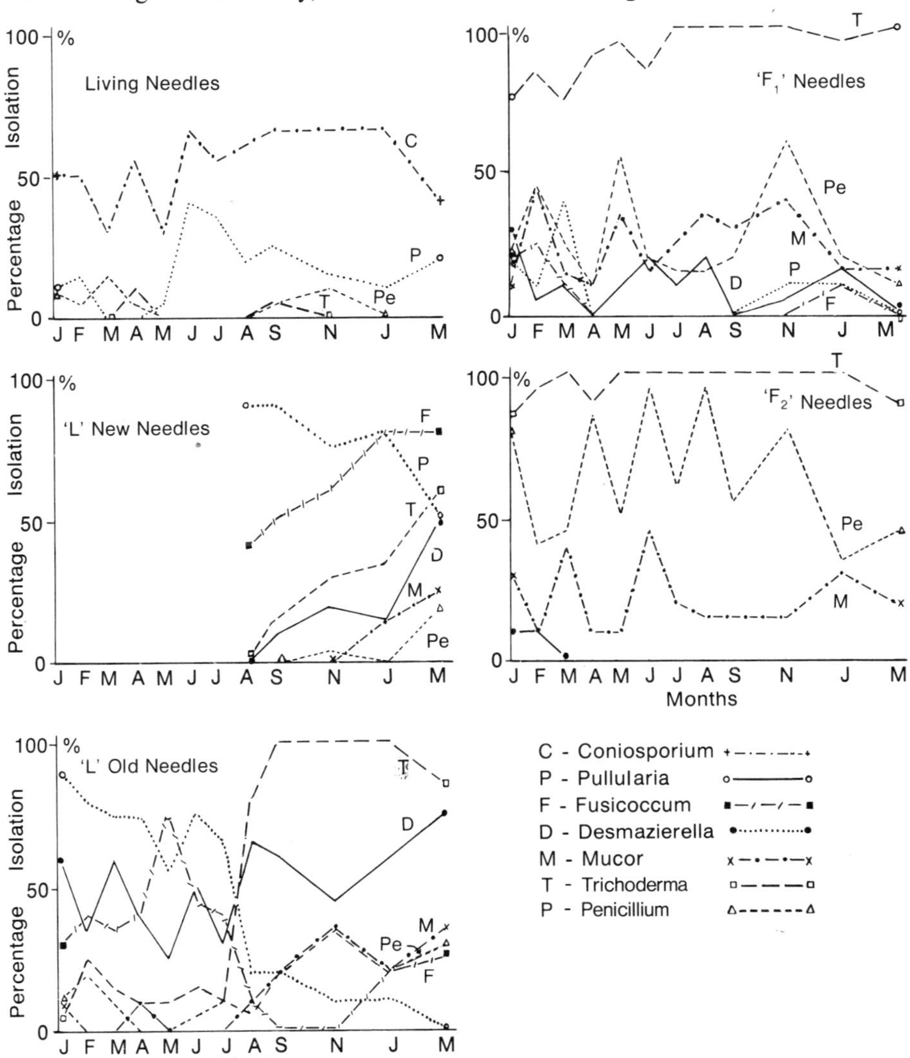

Fig. 2. Percentage of occurrence of fungi most frequently isolated from washed needles of different layers of the top organic layer in a pine forest. Strong seasonality is present in only a few genera. (Adapted from Kendrick and Burges, 1962.)

mining the seasonality. The monthly isolation patterns of different fungi from washed pine needles obtained in this study are presented in Fig. 2.

4 Biochemically Related Population Studies

Studies on the overall biochemical activities of microorganisms have also been undertaken. Ross (1965) attempted to measure the seasonal uptake of oxygen and the enzyme hydrolysis of sucrose and starch in 12 environmentally different New Zealand pasturelands. There was no regular seasonal pattern of oxygen uptake even when samples were amended with glucose to eliminate substrate as a limiting factor. Reducing sugars tended to be more abundant in the soil during late spring and early summer; however, starch hydrolysis rates were randomly distributed and

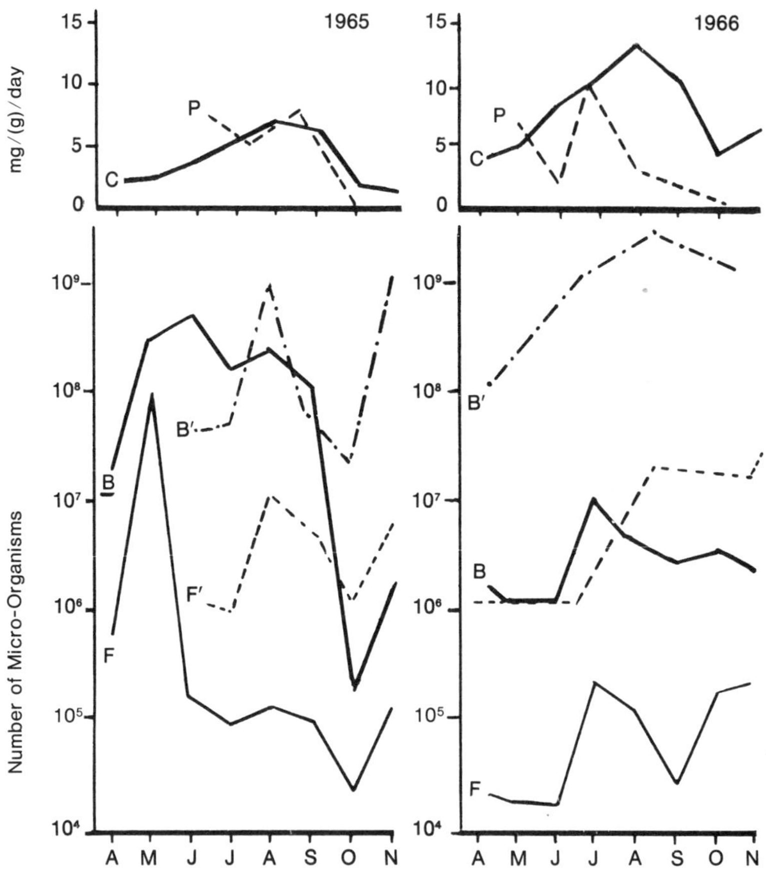

Fig. 3. Dynamics of the microbiological activity of soil of the Strzeleckie Meadows 1965 and 1966. (Adapted from Jakubczyk, 1970, p. 133.) B = bacteria number, F = fungi number per milliliter of soil, B' = bacteria number, and F' = fungi number per gram of dead plant material, C = decomposition rate of cellulose in milligrams per gram per day, P = decomposition rate of dead plant material in milligrams per day.

sucrose was maximally hydrolyzed in winter. Data were not repeatable for two years. None of the measured properties was significantly correlated with the number of days of drought before the last rainfall preceding sampling or with other climatic parameters. Ross cites, however, a report from the Soviet Union (Kozlov, 1962) in which regions of more contrasting seasons showed maximal rates of sucrose hydrolysis in a seasonal pattern.

Jakubczyk (1970), in Poland, has examined the seasonal populations of bacteria and fungi as well as the decomposition rates of cellulose and dead plant material (Fig. 3). Her data reveal some correlation between seasonal patterns of the populations and degradation activities. However, the peaks of microbial activity and the decomposition of organic matter seem to vary throughout the year with the habitat.

In Holland, Witkamp and van der Drift (1961) compared the production and breakdown of litter in mull and mor forest soils in relation to changes in soil moisture, temperature, microbial population, litter-feeding fauna, rates of breakdown of the main leaf components, and the carbon dioxide concentration in the soil atmosphere. Litter decomposition was strongly seasonal with the timing and biomass input affected by precipitation patterns. The rates of litter accumulation and breakdown varied through the year in such a way that the breakdown rate of mull litter was initially faster than that of mor, but was slower later (Fig. 4). The individual species components of the mull leaf litter reflect this trend. Rate of cellulose decomposition was measured by recording the weight loss of filter paper placed in the litter; it was also assumed to be the principle component of microbial activity as measured by soil carbon dioxide evolution. Bacteria populations were estimated solely by serial dilution plates, while fungal populations were estimated by dilution plates and mycelium development evaluated by Cholodny slides. All of these measurements, and others, showed seasonal variations.

They found the following yearly pattern: Maximum litter fall occurred in autumn when temperatures were still about 10°C. Mycelial growth and accumula-

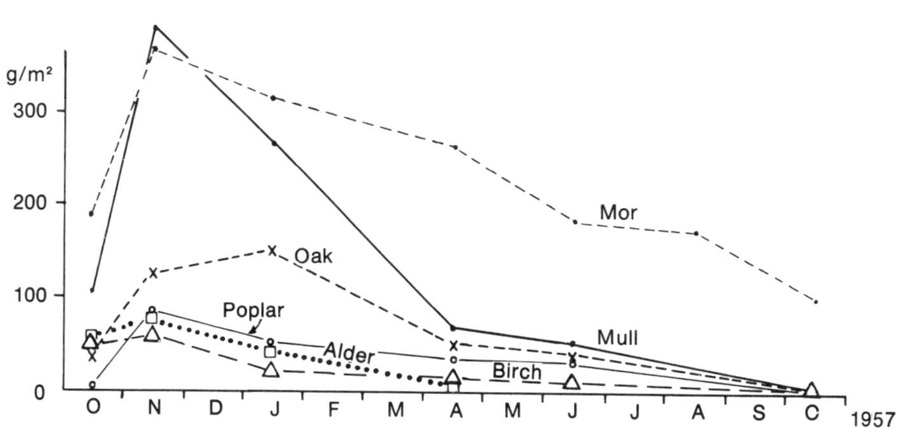

Fig. 4. Amount of litter falling on a cleaned surface of mor and mull soil. The litter of the mull site is broken down into the major contributing species, showing the delay of oak litterfall. (Adapted from Witkamp and Van der Drift, 1961).

tion were rapid as was bacterial growth. Colder temperatures of the next few months brought about a reduction in carbon dioxide evolution; and the numbers of fungal propagules (reflected in dilution plates counts) and accumulation of mycelium reached their maxima. Microbial activity was lowest in the coldest months.

The rate of weight loss of mull leaves was greatest during the December-March period. The authors suggested that this was due to leaching and the superficial attack by great numbers of enchytraeids and larger burrowing saprophagous animals whose numbers all reached their maximum in this period.

As temperatures rose in May, bacterial and fungal growth resumed and soil animals decreased. The increased decomposition of cellulose and wood is reflected in the rise in carbon dioxide evolution. Activity in the litter slowed down during dry spells in early summer, but the moister mineral horizons were less affected. Late summer rains caused maximum microbial activities and stimulated invertebrate activity. Thus the newly fallen leaves of the autumn were met by a vigorous soil biota.

They concluded that high temperatures and favorable moisture stimulate populations and activities. Seasonal climatic changes affect horizons differently: deeper soil is affected by temperature and the top 2 cm is governed by moisture as well.

In his later studies at Oak Ridge, Tennessee, Witkamp (1963, 1966) pointed out the importance of frequent measurements of microbial populations, carbon dioxide evolution from soil, temperature, and moisture. Modifying his basic 1963 methodology, he was able to show that microbial respiration, measured as carbon dioxide evolution from soil, was significantly correlated to the carbon-nitrogen ratio of the leaf species, temperature, bacterial population, moisture content of the leaves, and age of the litter (1966). He presented the following multiple regression formula:

$$C = 46.50 \times 3.22\, T + 26.86 \sqrt{M/D} + 11.39 \log B - 0.64\, W \qquad (1)$$

where respiration is measured as C = carbon dioxide evolved in microliters per hour per gram; temperature is T; the square root of moisture content is g moist wt/g dry wt = M/D; bacterial colony counts are B = numbers of colonies from 10^{-6}g air-dry wt litter; and litter age is W = weeks of exposure. The transformation $\sqrt{M/D}$ was used instead of M/D to emphasize the influence of low moisture contents.

A comparison of the measured respiration rate and the calculated rate shows a reasonable fit (Fig. 5). The rate of weight loss for the leaf species was correlated ($P < 0.01$) to the carbon-nitrogen ratio. The weight loss was also correlated to the combined mean annual colony counts of bacteria and fungi. In order to equalize the values of colony counts of bacteria and fungi, a factor of 153 times the number of fungal colonies was used in the following formula. This was decided on the assumption that total colonies of bacteria and total colonies of fungi represent equal breakdown potentials (there were 153 times as many bacterial colonies as fungal colonies). The regression formula is

$$W = 39.38 + 0.47\,(B + 153\, F) \qquad (2)$$

in which W = percent loss of weight per year, B = numbers of bacterial colonies from 10^{-6}g dry-wt litter, and F = numbers of colonies from 10^{-6}g dry-wt litter.

The formula assumes that litter would lose about 40% of its weight per year to leaching, by rain autolysis, and removal of solids by gravity and soil fauna.

Witkamp concluded from these results that carbon-nitrogen ratios of litter leaves, more than the type of forest stand, affects decomposition rate, microbial density, and that, in the fairly humid experimental site location, temperature is a dominant influence over moisture on the rate of decomposition.

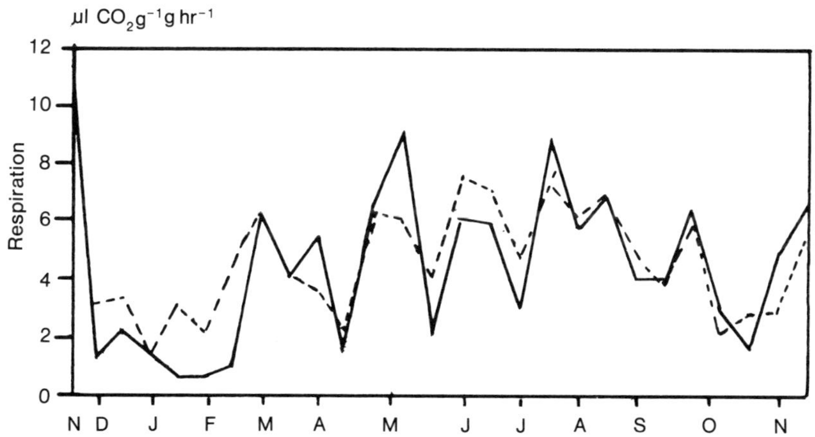

Fig. 5. Mean calculated (- - - - - -) and measured (———) respiration rates of redbud, oak, and pine litter in three forest stands for November 1960–November 1961. (From Witkamp, 1966.) The extremely low values measured from December through February were stated by the author as possible errors, but there is a reasonable similarity between model and measured values. The average monthly respiration rates show the expected high during the summer months.

His data also suggest that year-round dilution plating does reflect the overall microbial population, even though it favors only organisms that are nutritionally and competitively able to grow on agar plates. He offers his methods as a procedure that could lead to understanding of habitats, climates, and decomposition if they are repeated in diverse environmental sites.

5 Seasonality of Species Composition-Decomposer Phenology

Seasonal differences in the presence of decomposer organisms at one particular place are reported in several studies and found in our own work. We report here the following examples:

Using standard substrate isolation methods for myxomycetes, Pendergrass and Frederick attained the following seasonal variations among the 36 species isolated (personal communication, 1972):

Seasonal (Total: 20)[1]	Nonseasonal (Total: 16)
Arcyria carnea	*Arcyria cinerea*
A. globosa	*Comatricha elegans*
A. pomiformis	*C. fimbriata*
Comatricha suksdorfii	*C. laxa*
Diderma rugosum	*C. lurida*
Didymium megalosporum	*C. nigra*
Macbrideola cornea	*Cribraria microcarpa*
M. scintillans	*C. minutissima*
Metatrichia vesparium	*C. violacea*
Perichaena corticalis	*Didymium clavus*
P. minor	*D. quamulosum*
P. vermicularis	*Echinostelium minutum*
Physarum galbeum	*Lamproderma scintillans*
P. leucophaeum	*Licea operculata*
P. notabile	*L. pedicellata*
P. nudum	*Perichaena chrysosperma*
P. nutans	
P. viride	
Stemonitis virginiensis	
Trichia floriformis	

Of the 20 seasonal species, the following graph illustrates their overwhelming abundance during the summer (Fig. 6*a*). Only two of 36 species did not appear in summer, but 12 or more species were absent in each of the other seasons. There were interesting relationships between seasonality and the diversity of bark types on which the 36 species were found (Fig. 6*b*, *c*). Of the 19 species of myxomycetes showing seasonality for which complete data were given, 13 were restricted to a single species of bark, while only one of the 16 nonseasonal myxomycetes was as restricted.

In spring 1972, the senior author began a study of the seasonality of the litter fungi in a mixed pine-hardwood forest in the Chapel Hill, North Carolina area, adapting Parkinson's (1972) soil washing technique to chopped leaf litter, which could then be subjected to a series of selective media. The selective agents were commercial fungicides. A variety of temperature and light regimes used during culturing were also compared. Hopes of ascertaining the origin of the high numbers of colonies of Penicillia and other culture "weeds" that dominate similar species lists was satisfied.

The importance of culturing the fruiting bodies developing at the research site is twofold: culture characteristics of sterile mycelia could then be compared to a series of known cultures of higher fungi for identification: frequencies of isolations could sometimes be related to periods of high spore load. Thus the real importance of colony numbers could be assessed.

In an effort to intensively collect the fungi of a single area, the sampling plot

[1] This material was presented by Dr. Frederick at the Minneapolis meeting. Full publication is expected soon in:

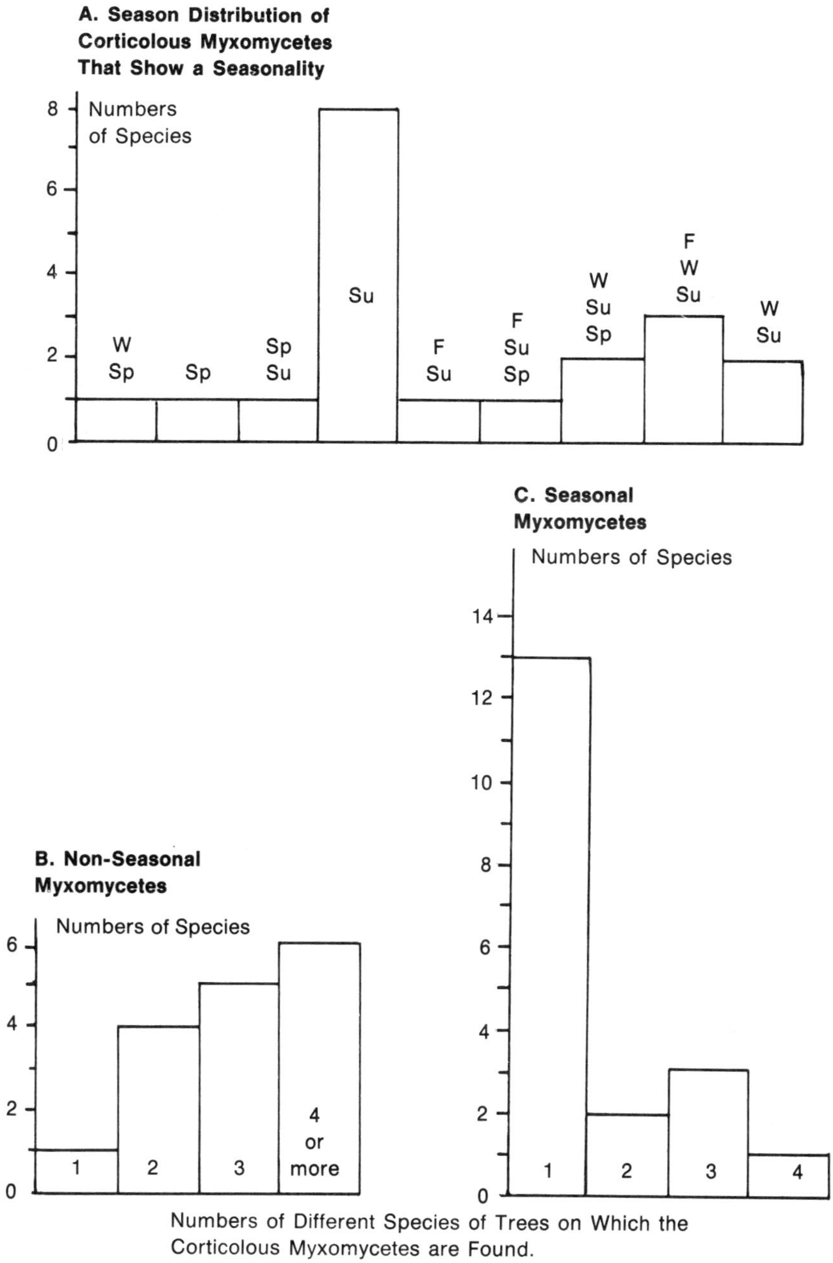

Fig. 6. Seasonality analysis of the growth of corticulous Myxomycetes in central Georgia. These data were presented during the symposium by Pendergrass and Frederick.[2] W = Winter, Sp = Spring, Su = Summer, F = Fall.

[2] Pendergrass, Levester and Lafayette Frederick. 1973. Myxomycetes from bark of living trees in Panola Mountain State Park, Georgia. Manuscript in preparation, to be submitted for publication in Castanea. Ku, Chyuan L. and Lafayette Frederick. 1973. Studies on Myxomycetes occurring on bark of living trees in Atlanta, Georgia. Manuscript in preparation, to be submitted to Bulletin of the Georgia Academy of Science.

was deliberately kept small—2 sq m. Collections of litter were made approximately once every two weeks; and about 500 litter fragments were divided into two groups, one of which was washed and the other left unwashed. Half of each of the two groups was then subjected to another physical or chemical treatment so that all combinations of treatments were available for species analysis, and chemical and physical methods could be assessed for their selective abilities. Details of procedures and results are in preparation and will be published elsewhere.

In addition to isolation methods and seasonality, the decomposition abilities of the isolates were assessed. The ability of fungi to decompose cellulose and lignin and the possession of peroxidase were examined as outlined by Desmoulins (1967) and Blaich (1971). We regard these properties as the primary tools of the microbes for initiating plant biomass decomposition. Therefore, identifying these abilities should accompany the analysis of decomposer species lists. In addition to these, we hope to characterize those species best able to degrade microbial polysacchorides. Once the likely decomposer niches of these organisms are revealed, the highly productive organisms may find useful applications in tertiary sewage treatment or even in the fermentation of substrates like newsprint. Table 3 is a partially completed list of the organisms found through one year on the 2-sq m plot.

Investigations into the seasonality of spore discharge by wood decay fungi suggest that they are frequently coordinated with some physical factor. Hodgkiss and Harvey (1969) monitored the spore release of seven Pyrenomycete species growing on wood using a Hirst spore trap. The authors' graphs illustrating the sporulation patterns are presented in Fig. 7.

Six of the seven species showed positive correlations of the volume of spore discharge to rainfall or humidity, while the seventh, *Eutypa acharii*, showed a correlation of monthly averages to number of hours of incident light. None of the species showed correlation to temperature as long as it remained above freezing.

The seasonal appearance of many species of mushrooms is well-recognized and a part of the folk culture in many lands. Details of the seasonality of basidiophores and the fruiting of the Discomycetes may be found in the works of Dennis (1968), Hinitikka (1969), and others.

Both Feher (1933) and Mishustin (1954) have indicated that there is a correlation between latitude and density of bacterial colonization in the soil. Very few references to this work exist in the literature in English, although such an ecologically significant relationship deserves close attention.

In recent papers Hirata (1968, 1972) supports the thesis of a relationship between the presence of certain parasitic fungi and the great climatic areas differing in strength of seasonality. He analyzed the distribution pattern of the Meliolineae and the Erysiphaceae, and their numbers of host species. According to Hirata (personal communication) and Blumer (1967), the Erysiphaceae are most strongly represented in the northern hemisphere, especially in Europe. Of the 13 genera mentioned by Hirata, five are endemic to extra topical climates with strong seasonality, one genus is from tropical areas with the description of Blumer indicating that they occur especially in tropical climates with seasonality, and seven are global in distribution. Hirata's detailed analysis of the family ranges and host

Month, 1972/1973:	Apr.	May[a]	June[a]	July Summer	Aug.	Sept.	Oct.	Nov.	Dec.	Jan. Winter	Feb.	Mar.
Total Isolations per month	909	94	69	508	313	1555	1065	1056	1022	1033	1482	1619
Sterile hyphae	14	13	1		4	14	4	29	70	29	33	17
Unidentified imperfects	4	2	4	0	1	3	2	9	3	2	14	13
Bacteria and yeasts	62	2	2	35	30	185	142	217	187	259	234	273
No growth in culture	103	—	—	47	34	266	59	30	84	132	102	110
Phycomycetes	133	3	2	175	111	628	253	220	364	303	566	605
Coemansia spp.				5		2			1	1		
Fungi imperfecti:												
Moniliales, Moniliaceae												
A—Nonseasonal												
Trichoderma viride	144	11	5	94	25	173	160	151	32	29	187	280
Penicillium spp.	153	6	7	81	37	38	295	229	63	132	230	233
Verticillium spp.	12	15	7	40	17	56	34	27	19	25	52	7
Geotrichum sp.[b]	4		1				7	8		5	1	5
Chloridium sp.	3			1		1	3			1	4	
Cladosporium spp.	34	3	8	2		15	7	69	66	46	36	32
Scolebobasidium constrictum				3	1	1	1			1	5	
Cephalosporium spp.	1				3	2	1	2		2	3	1
Monocillium humicola	3	2	6	1	3	2	1				5	
B—Seasonal												
Microsporium sp.	1											2
Arthrinium sp.	1											
Phialophora sp.		1										
Polyscytalum sp.		1										
Trichothecium sp.		2										
Cephalosporiopsis sp.		1										
Diplosporium sp.	1											
Fusidium sp.	2											

Table 3 *continued*

Month 1972/73:	Apr.	May[a]	June[a]	July	Aug.	Sept.	Oct.	Nov.	Dec.	Jan.	Feb.	Mar.
				Summer						Winter		
Sporotrichum sp.	3						1					
Tritirachium sp.	3	2				1		4				1
Botrytis sp.	30	1			3						1	
Stachybotrys sp.		1		2								
Gonatobotryum sp.			1									
Cephalosporium serrae					1							
Virgaria nigra			1									
Diplorhinotrichum sp.					1							
Umbellopsis			1									
Verticladium sp.			2									
Stemphylium sp.					1							
Gliomastix sp.					1		1					
Geniculosporium sp.			1					1				
Nodulisporium sp.			2			1						
Beauveria sp.				1		1						
Fusarium sp.			1	1				1		1		
Diheterospora chlamydosporia	1				1	1						
Geotrichum candidum	1		1			1		2				
Helicosporium sp.		2			1	1						
Oidiodendron sp. 1	1	5			1	6						
Oidiodendron sp. 2				1	1							
Chloridium chlamydosporis	2			4		1						
Sporothrix sp. 2								1				
Amblyosporium sp.							1					
Gilmaniella humicola						1						
Sporothrix sp. 1							1	1				
Calcarisporium arbuscula						1	1					
Monacrosporium sp.						3						
Aspergillus sp.	0	2				8						
Monochaete sp.	1						1	1				
Bipolaris sp.							2					
Trichoderma album							2			1		
Hyalodendron sp.					1	13	4	5	2		20	
[illegible]			1			1	1		1	2		

Cephalosporium sp. No. 459											1	
Rhinocladiella sp.										1		
Paecilomyces sp.											1	
Gliocladium sp.											1	
Menisporella sp.									1		2	
Spadicoides sp.	1									1		
Sclerotium sp.	1	1				1					1	
Chaetopsina sp.						1	4	1	1			1
Epicoccum nigra	9	2						1	2		4	1
Oedocephalum sp.	4	1				1	4		1		4	2
Alternaria spp.	111	1				1	1	20	27	30	7	14
Tuberculariaceae												
unidentified two species			3		1			1				
Sphaeropsidales												
Total (17) "Species"	8	4	1			5	32	18	24	18	1	3
Melanconiales												
unidentified one species.									1			
Pestalozzia spp.	38	2	3	3	8	47	33	6	25	38	57	35
Sexual forms:												
Ascomycetes												
Dasyscypha sp.	2											
Chaetomium sp.											1	
Xylaria sp.	1											
Basidiomycetes												
Stromatoscypha fimbriata	1											
Total (24) "Species"	5	1	1	1	15	4	3	5	4	5	3	9
Unidentified Basidiomycete No. 293						1						
Unidentified Basidiomycete No. 534												1

[a] May and June, 1972, data were collected by direct observation only. No special isolation methods were used. Only separate hyphae were scored.
[b] These fungi produce arthrospores, and no clamp connections were present, however, they are thought to be Basidiomycetes.

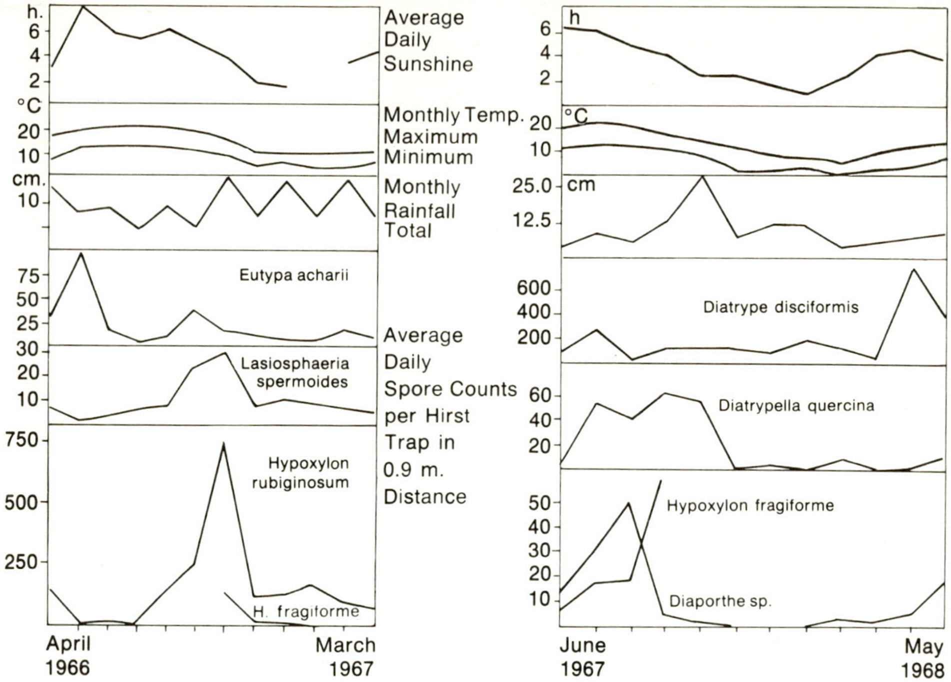

Fig. 7. Seasonal variations in spore discharge of seven wood decaying ascomycetes. The insolation, temperature, and precipitation data recorded during the spore counting periods is shown in the upper part of the figure. Wind velocity through the Hirst spore traps ranged 4.8–24.1 km/hr, but only 4.8–8.0 km/hr during 15 of the total 26 months. Spore discharge is markedly seasonal. (From Hodgkiss and Harvey, 1969, pp. 357, 358.)

Table 4a. Geographic range of genera of Erysiphaceae.[a]

Genera with Limited Range	Tropical	Extra-Tropical
Cystotheca		X
Typulochaeta		X
Medusophaera		X
Salmonomyces		X
Pleochaeta		X
Brasiliomyces	X[b]	
	8%	38%
Genera with Global Range	54% of Erysiph. genera	
Erysiphe *Microsphaera*		
Leveillula *Uncinula*		
Sphaerotheca *Phyllactinia*		
Podosphaera		

[a] From Hirata (1968) and Blumer (1967).
[b] The tropical regional hosts are deciduous.

Table 4b. Numbers of host families and host species of Erysiphaceae and their ratios according to geographic distribution.[a]

Climatic Zone distributions of the Families with Species Hosts to Erysiphaceae	No. of Families with Species Hosts to Erysiphaceae	Total No. of Species in All Families	No. of Host Species from These Families	% of Species Hosts to Erysiphaceae
Predominantly Seasonal: northern; northern and southern; northern and tropical; northern, southern, and tropical	91	124,430	7,225	5.8
Predominantly Nonseasonal: tropical; southern; tropical and southern	31	32,888	118	0.35

[a] From Hirata (1972).

ranges of this group indicates that the group has 98% of its hosts in the northern hemisphere (Table 4).

Such distribution patterns are unexpected, especially in view of the opinion generally held among mycologists that the overall ranges of the major groups of the fungi are not restricted among the readily inhabitable land masses.

6 Seasonality of Mycorrhizal Fungi

The seasonality of mycorrhizal fungi can be inferred from its close relationship to the producer component. Seasonal growth is likely to be related to growth changes of the roots of the producer organism; activity may be related to the seasonal changes in nutrient input and the activities of the primary decomposers (most mycorrhizal fungi do not possess a cellulolytic ability); and sporulation has not been seriously monitored. Harley (1969) offers some information about the problems of work with mycorrhiza.

7 Modeling Fungal Growth and Seasonality

Modeling fungal growth and seasonality of the decomposers under natural conditions is hardly possible, given the state of methodology and information now at hand. Paul and Domsch (1972) have offered a model of the nitrification process in soil. Waggoner, Horsfall, and Lukens[3] have pioneered in the modeling of fungal

[3] Waggoner, P. E., Horsfall, J. G., and Lukens, R. J. 1972. EPIMAY, a simulator of southern corn leaf blight. The Connecticut Agricultural Experimental Station, New Haven.

pathogens of grain crops so that epidemics can be predicted from obtainable physical data. The incorporation of general and specific seasonal data and models into ecosystems models is still far in the future.

Acknowledgment

The senior author wishes to thank Dr. H. Lieth for the opportunity to collaborate with him on this paper. He would also like to acknowledge his gratitude to Mr. and Mrs. David Moses and Walter J. Urban for their help through the years.

References

Blaich, R. 1971. Beiträge zur Physiologie holzabbauender Basidiomyceten. 113 pp. Thesis, Ruhr Universität Bochum.

Blumer, S. 1967. *Echte Mehltaupilze (Erysiphaceae)*. 346 pp. Jena: Fischer Verlag.

Booth, C. 1971. *Methods in Microbiology*. Vol. 4. 795 pp. London: Academic Press.

Bray, J. R., and E. Gorham. 1964. Litter production in forests. In *Advances in Ecological Research*, J. B. Cragg, ed. Vol. 2, pp. 101–158. New York: Academic Press.

Burges, A., and F. Raw. 1967. *Soil Biology*. 530 pp. London: Academic Press.

Casida, L. E., Jr. 1969. Observations of microorganisms in soil and other natural habitats. *Appl. Microbiol.* 18. 1065–1071.

Coleman, D. C. 1971. Numbers and biomass of soil nematodes of two South Carolina old fields. *Am. Midland Naturalist.* 85, 262–265.

Cutler, D. W., M. A. Lettuce, M. Crump, and H. Sandon. 1923. A quantitative investigation of the bacterial and protozoan population of the soil, with an account of the protozoan fauna. *Phil. Trans. Roy. Soc.* 211, 317–350.

Dennis, R. W. G. 1968. *British Ascomycetes*. 455 pp. Lehre, Germany: J. Cramer.

Desmoulins, V. 1967. Proposition d'un processus de detection des phenoloxidases et peroxidases chez les macrophytes. *Planta* 76, 129–137.

Dwivedi, R. S., and S. D. Garrett. 1968. Fungal competition in agar plate colonization from soil inocula. *Trans. Brit. Mycol. Soc.* 51, 95–101.

Evans, A. C., and W. J. L. Guild. 1947. Studies on the relationships between earthworms and soil fertility. *Ann. Appl. Biol.* 34, 307–330.

Eggleton, W. G. E. 1938. The influence of environmental factors and numbers of soil microorganisms. *Soil Sci.* 46, 351–363.

Feher, D. 1933. *Untersuchungen über die Mikrobiologie des Waldbodens*. 272 pp. Berlin: Julius Springer.

French, N. R., ed. 1971. Preliminary analysis of structure and function in grasslands. *Range Sci. Dept. Sci. Ser. No. 10*. 387 pp. Ft. Collins: Colorado State University.

Glasgow, J. P. 1939. A population study of subterranean soil collembola. *J. Anim. Col.* 8, 323–353.

Harley, J. L. 1969. *The Biology of Mycorrhiza*. 2nd ed., 334 pp. London: Leonard Hill.

Hirata, K. 1968. Notes on host range and geographic distribution of the powdery mildews. *Trans. Mycol. Japan.* 9, 73–88.

——— 1972. Notes on the host range and geographic distribution of the powdery mildew fungi. IV. *Trans. Mycol. Soc. Japan.* 13, 1–21.

Hodgkiss, I. J., and R. Harvey. 1969. Spore discharge rhythms in pyrenomycetes. VI.

The effects of climatic factors on seasonal and diurnal periodicities. *Trans. Brit. Mycol. Soc.* 52, 355–363.
Hornby, D. 1969. Gravimetrical and mycological investigations of soil suspensions in the soil dilution plate technique. *J. Appl. Bacteriol.* 32, 244–258.
Hinitikka, V. 1969. Studics in the genus *Mycena* in Finland. *Karstenia* 6–7, 77–87.
Jakubczyk, H. 1970. Dynamics of microbiological activity in some soils of meadow communities. In *Methods of Study in Soil Ecology*, J. Phillipson, ed. pp. 131–135. Geneva: UNESCO.
Kendrick, W. B. and A. Burges. 1962. Biological aspects of the decay of *Pinus sylvestris* leaf litter. *Nova Hedwigia* 4, 313–344.
Koch, W. J. 1972. *Fungi in the Laboratory*. 291 pp. Chapel Hill: University of North Carolina Student Stores.
Kozlov, K. A. 1962. Study of the biological activity of soils of eastern Siberia. *Pochvovedenie* 4, 40–47.
Lieth, H. 1963. The role of vegetation in the carbon dioxide content of the atmosphere. *J. Geophys. Res.* 68, 3887–3898.
Luttrell, E. S. 1967. A strip bait for studying the growth of fungi in soil and aerial habitats. *Phytopathology* 57, 1266–1267.
Malaisse, F., and M. Malaisse-Mousset. 1970. The phenology of defoliation. *Bull. Soc. Roy. Bot. Belg.* 103, 115–124.
Madge, D. S. 1965. Leaf fall and litter disappearance in a tropical forest. *Pedobiologia* 5, 272–288.
Meiklejohn, J. 1957. Numbers of bacteria and actinomycetes in Kenya soil. *J. Soil Sci.* 8, 240–247.
——— 1955. The effect of bush burning on the microflora of a Kenya upland soil. *J. Soil Sci.* 6, 111.
Metz, L. J., and M. H. Farrier. 1967. Acarine associated with decomposing forest litter in the North Carolina Piedmont. In *Proc. 2nd Intern. Congr. Acariology*, pp. 43–52.
Mishustin, E. N. 1954. The law of zonality and studies of microbial associations in the soil. *Usp. J. Sov. Biol.* 37, 1.
Morris, M. S., and J. D. Brunner. 1971. Primary productivity of the fescue grassland in western Montana. *US-IBP Grasslang Biome Tech. Rept. 113.* 74 pp. Ft. Collins: Colorado State University (cited from Sims and Singh, 1971).
O'Connor, F. B. 1967. The Enchytraeidae. In *Soil Biology*, N. A. Burges and F. Raw, eds. pp. 213–257. London: Academic Press.
Parkinson, D. 1972. Methods for studying the ecology of soil micro-organisms. *IBP Handbook No. 19.* Oxford: Blackwell Scientific Publications.
———, and I. Balasooriya. 1969. Studies on fungi in pinewood soil. IV. Seasonal and spatial variations in the fungal population. *Rev. Ecol. Biol. Sol.* 6, 147–153.
Paul, W., and K. H. Domsch. 1972. Ein Mathematisches Modell für den Nitrifikationsprozess im Boden. *Arch. Mikrobiol.* 87, 77–92.
Perfil'ev, B. V., and D. R. Gabe. 1969. *Capillary Methods of Investigating Micro-organisms.* 627 pp. Toronto: University of Toronto Press.
Rao, A. S. 1959. A comparative study of competitive saprobic ability in 12 root-infecting fungi by an agar plate method. *Trans. Brit. Mycol. Soc.* 42, 97–111.
Rapp, M. 1971. Cycle de la matière organique et des éléments minéraux dans quelques écosystèmes méditerranéens. *Extrait PBI Recherche Cooperative sur programme du C.N.R.S. No. 40.* 184 pp. Paris: Editions C.N.R.S.
Ross, D. J. 1965. A seasonal study of O_2 uptake of some pasture soils and activities of enzyme hydrolysis of sucrose and starch. *J. Soil Sci* 16, 73–85.

Sims, P. L., and J. S. Singh. 1971. Herbage dynamics and net primary production in certain ungrazed and grazed grasslands in North America. p. 29–124. In *Range Sci. Dept. Sci. Ses. No. 10*, N. R. French, ed. pp. 29–124. Ft. Collins: Colorado State University.

Satchell, J. E. 1967. Lumbricidae. In *Soil Biology*, N. A. Burges and F. Raw, ed. pp. 259–322. London: Academic Press.

Stotsky, G. 1972. Activity, ecology, and population dynamics of microorganisms in soil. *Crit. Rev. Microbiol.* 2, 59–137.

Volobuev, V. R. 1964. Ecology of soils. Jerusalem. Israel Program for Scientific Translation.

Waksman, S. A. 1916. Bacterial numbers in soils at different depths and in different seasons of the year. *Soil Sci.* 1, 363–380.

Wallwork, J. A. 1959. The distribution and dynamics of some forest soil mites. *Ecology* 40, 557–563.

Wastie, R. L. 1961. Factors affecting competitive saprobic colonization of the agar plate by various root-infecting fungi. *Trans. Brit. Mycol. Soc.* 44, 145–159.

Wells, C. A., D. Whigham, and H. Lieth. 1972. Investigation of mineral nutrient cycling in an upland piedmont forest. *J. Elisha Mitchell Sci. Soc.* 88, 66–78.

Witkamp, M. 1963. Microbial populations of leaf litter in relation to environmental factors and decomposition. *Ecology* 44, 370–377.

——— 1966. Decomposition of leaf litter in relation to environment, microflora, and microbial respiration. *Ecology* 47, 194–201.

———, and J. Van der Drift. 1961. Breakdown of forest litter in relation to environmental factors. *Plant Soil* 15, 295–311.

3.7 Seasonal Variations of Soil Fungi Isolated from the Rhizosphere of *Liriodendron tulipifera* L.[1]

Harold F. Mabee

Biology Department
East Tennessee State University
Johnson City, Tennessee 37601

and J. H. B. Garner

Office of Criteria Development
Environmental Protection Agency
Research Triangle Park, North Carolina, 27711 USA

1 Introduction

Seasonal variation exists among the variety of organisms within an ecosystem, but few studies show such variation among soil fungi, except for obvious structures like mushrooms. Soil fungi that are active during a particular season may not be active during other seasons. When unable to compete with other fungi or to adapt to soil conditions, many soil fungi form spores or other resistant structures that are capable of remaining dormant until conditions necessary for germination are present. To better understand the role of fungi in the soil, the presence or absence of seasonal variation should be determined. The present study was initiated to determine whether seasonal variation occurred among the fungi isolated from the soil surrounding two tulip poplars (*Liriodendron tulipifera* L., also called tulip tree and yellow poplar; see Chapters 4.4 and 5.2).

2 Materials and Methods

The study area was located in the Cherokee National Forest near Johnson City, Tennessee. The two tulip poplars were 14.6 m apart and situated on a south-facing

[1] Study conducted with the facilities of the Biology Department, East Tennessee State University, Johnson City, Tennessee.

KEYWORDS: Soil fungi; seasonality; decomposers; strip-bait; rhizosphere; populations; *Liriodendron tulipifera*; spores; active fungi; inactive fungi.

slope. Rosebay rhododendron (*Rhododendron maximum* L.) and hemlock (*Tsuga canadensis* [L.] Carr.) made up the greater part of the nearby vegetation. The leaf litter was dominated by pine leaves from Virginia pine (*Pinus virginiana* Mill.) and hemlock leaves (*T. canadensis*).

The strip-bait method of Luttrell (1967) was used to isolate fungi from the soil. The advantage of this method is that the baits adopt the existing soil conditions, making them selective for fungi that are actively growing in the soil. The strip-baits were made by soaking filter paper in one of two different solutions, 2% malt and dextrose-peptone-yeast extract (DPY) solution of Papavizas and Davey (1959). Paper disks 12.7 mm wide were punched from the filter paper. The disks from each medium were stacked in fives. A piece of electrician's tape 19.1 mm wide and 17.9 cm long was punched with five holes 6.4 mm wide and 2.5 cm apart and then stretched on a board with the adhesive side up. One stack of five disks was placed over each of the five holes. A second tape with corresponding holes was placed on top of the first, and the two adhesive sides of the tapes were sealed together with the stacks of disks in between.

The two kinds of strip-baits were punch-coded to identify the medium used in their construction. All of the strip-baits were autoclaved and kept in a sterile container until taken to the site of the tulip trees. The strip-baits were placed in pairs in the soil near both trees. Holes were made in the ground with an iron rod, 17.8 cm deep and 7.6 cm apart; one bait was then inserted into each hole. Each pair consisted of two strip-baits, one of each kind.

The strip-baits were placed four times during 1968. The first series of baits was put out February 24, 1968 and collected March 4, 1968. The second group was set out March 26, 1968 and collected April 5, 1968. The third set was inserted into the ground June 11, 1968 and collected June 21, 1968. The last strip-baits were set out August 29, 1968 and collected September 16, 1968. The strip-baits around both trees were placed in the same location throughout the study.

Three pairs of strip-baits were inserted in the soil 1.5 cm from the base of each tree. The first of each three pairs was placed in the soil on a slope that was slightly above the bases of the trees. The second pair of strip-baits was placed in the soil on the same level as the tree bases. The third pair of baits was inserted in a lower slope below the bases of the trees.

Following each collection of strip-baits, the tapes were pulled apart. The middle disk was removed from each stack with sterile forceps. Each middle disk was placed in a separate Petri dish. The 2% malt-soaked disks were placed on 2% malt agar; those disks soaked in DPY were placed on DPY agar. The Petri dishes were incubated at room temperature. When growth occurred on any disk, it was isolated to a new culture dish and retransferred until bacteria free. Since only the middle disk of each stalk was used, dormant spores from inactive fungi were excluded; active fungi, which had to grow through two outer disks to reach the center disk, were isolated. Identifications of species was done where possible; however, only the genera are reported here.

Soil temperatures and pH readings were taken for each period the strip-baits were placed. A LaMotte Soil Test Kit was used to determine the pH values of the soil. A standard laboratory thermometer was calibrated and used to obtain soil temperatures. Soil samples were analyzed in the laboratory at the end of the study to determine the moisture-holding capacity, the percent of organic matter, the pore volume, and the texture of the soil. Moisture-holding capacity was determined using a method similar to that described by Gates (1949). Percent of organic matter and the pore volume of the soil were measured using the methods described by Lutz and Chandler (1955). The texture of the soil was determined by the hydrometer method as described by Day (1955).

3 Results

From a total of 261 isolations, 23 genera were identified. Seasonal growth variations were distinguished among five of the genera isolated.

Table 1 shows the genera isolated during 1968, giving number of isolations and percent of total isolations made during each season. Five genera of fungi were isolated from the first group of strip-baits which were set out between February

Table 1. Genera isolated during the four seasons of 1968.

Genera	Number of Isolations in			
	Winter	Spring	Summer	Autumn
Penicillium	4	24	17	10
Trichoderma	1	2	24	31
Phoma	2	3		
Scopulariopsis	1	1		
Sartorya	1			
Mucor		3	7	16
Sepedonium		3	13	20
Verticillium		1	3	8
Cladosporium		3		
Phomopsis		2		
Yeasts		2		
Absidia		1		
Alternaria		1		
Chaetophoma		1		
Dendrophoma		1		
Haplographium		1		
Aspergillus			3	
Paecilomyces			2	3
Acrocylindrium			1	
Chloridium			1	
Cladosporium			1	4
Eurotium			1	
Haplographium			1	1
Nigrospora			1	3
Mycelia sterilia		1	1	1
Total isolates	9	50	76	97

24 and March 4, 1968. *Penicillium* was isolated four times. The soil from the slopes near both trees had a 6.1 to 6.4 pH range. During the first period of isolations, the average soil temperature was 0°C. The ground was covered by snow, and the soil surface was frozen.

Sixteen genera were isolated during the second period between March 26 and April 5, 1968. *Penicillium* represented 48.0% of the total genera isolated during this period. *Cladosporium*, *Mucor*, *Phoma*, and *Sepedonium*, each represented 6.0% of the total genera isolated during this period. Soil pH range during this period, 4.0 to 4.6, was considerably below the winter readings. The soil temperature during this spring period averaged 13°C.

Fourteen genera were isolated from strip-baits set out during the period between June 11, and June 21, 1968. *Trichoderma*, *Penicillium*, *Sepedonium*, and *Mucor* represented 31.6, 22.4, 17.1, and 9.2%, respectively, of the total genera isolated during the period. The soil pH values of the upper, middle, and lower slopes near both trees ranged from 4.4 to 4.8. Soil temperatures averaged 21°C during this period.

The fourth sample period, August 29 to September 16, 1968, yielded 10 genera. *Trichoderma*, *Sepedonium*, and *Mucor* represented 32.0, 20.6, and 16.5% respectively, of the total genera isolated. *Penicillium*, which was quite prominent during the spring, now represented 10.3% of the genera isolated. The soil from the slopes near both trees had a 4.6 to 5.5 pH range. The average soil temperature was 18°C.

Laboratory analysis revealed that the moisture-holding capacity of the soil near the tulip trees was 75.3% water. The percent of organic matter was 30.3%. The pore volume of the soil tested was 82.5%. The texture of the soil was 80% sand, 18% silt, and 2% clay, and was classified as a loamy sand using a soil characteristic table compiled by the Soil Survey Staff (1951).

4 Discussions

Seasonal variation of fungi in the soil was not observed during the 19-month study of Novak and Whittingham (1968). The soil dilution plate method (Warcup, 1950) was employed in their studies for isolating soil fungi. Brian (1960) mentions that most colonies of fungi that are isolated by soil dilution plates develop from spores or other resistant fungal structures which are capable of remaining dormant until development of conditions necessary for germination. Dilution plates for isolation of soil fungi appear to be inadequate for studying their seasonal variation. This method not only isolates soil fungi actively growing during the season while the dilutions are made, but also isolates fungi from spores that are awaiting conditions favorable for germination.

Luttrell's strip-bait method used in this study adopts the soil conditions which exist when the strip-baits are placed into the ground. A fungus must grow through two outer disks before reaching the middle disk, and this eliminates the chance of mixing inactive fungi with active forms. Fungi which are not active during the period when baits are placed out are eliminated from the culture. The soil fungi which grew through the paper disks used to construct the strip-baits not only showed seasonal variation, but also demonstrated that they were capable of developing on another substrate such as the cellulose disks.

Pronounced seasonal variations were observed in our studies with respect to the number of genera isolated and the total number of isolates (Table 2). Seasonal variation in the number of genera isolated indicates changing composition of the active soil fungi. Variation in the total number of isolates indicates changes in the density of active populations. A number of possible explanations might account for the observed seasonal variations. It is known that the pH of the soil is usually higher in cool, moist weather (Russels, 1961). Increases in soil moisture may result in a proportional drop in fungal population (Orpurt and Curtis, 1957). Saturation of the soil observed during the winter period possibly resulted in unfavourable anaerobic conditions. Temperature variations may also influence the growth and activity of specific microflora, because seasonal fluctuations in pH values are related to changes in soil moisture and temperature (Bowser and Leat, 1958).

Table 2. Seasonal variations in the number of genera isolated, total number of isolations, average soil temperatures, and observed range of soil pH values.

Season	No. of Genera Isolated	Total No. of Isolates	Ave. Soil Temp. (°C)	pH Range
Winter	5	9	0	6.1–6.4
Spring	16	50	13	4.0–4.6
Summer	14	76	21	4.4–4.8
Autumn	10	97	18	4.6–5.5

Only *Trichoderma* and *Penicillium* were isolated in each of the four periods (Table 1). *Penicillium* was most prominent in the winter and spring, representing 44.4 and 47.0%, respectively. *Trichoderma*, in contrast, showed the lowest numbers of genera during the same periods, representing only 11.1% in the winter and 4.0% in the spring. During the summer period, both *Trichoderma* and *Penicillium* were quite prominent. However, *Trichoderma* represented 31.6% of the isolations, while *Penicillium*, which had been more abundant during the winter and spring periods, now represented 22.4% of the isolations. Isolations made during the autumn period showed *Trichoderma* representing 32.0% and *Penicillium* representing 10.3% of total isolations. Favorable growth conditions for *Trichoderma* appeared to be the reverse of those for *Penicillium*.

Sepedonium, *Mucor*, and *Verticillium* also showed seasonal growth variations. All were absent during winter isolations. During the spring period, *Mucor* and *Sepedonium* each represented 6.0% of the isolations, and *Verticillium* represented 2.0% for the same period. The percent of isolations for *Mucor*, *Sepedonium*, and *Verticillium* during the summer period were 9.2, 17.1, and 4.0% respectively. During the autumn, isolations *Mucor*, *Sepedonium*, and *Verticillium* represented 16.5, 20.6, and 8.3%, respectively. Each of these genera increased in isolations from the spring to autumn periods. During the autumn period, both *Sepedonium* and *Mucor* were isolated more frequently than *Penicillium*.

This study shows that seasonal variations of fungal populations do exist in the soil. The fungi isolated were mostly Deuteromycetes, many of which are known cellulose users.

References

Bowser, W. E., and J. N. Leat. 1958. Seasonal pH fluctuations in a grey wooded soil. *Can. J. Microbiol.* 3(2), 351–358.

Brian, P. W. 1960. Antagonistic and competitive mechanisms limiting survival and activity of fungi in soil. In *The Ecology of Soil Fungi*, D. Parkinson and J. S. Waid, ed. pp. 115 to 129. Liverpool, England: Liverpool University Press.

Day, P. R. 1965. Particle fraction and particle-size analysis. In *Method of Soil Analysis*, C. A. Black, ed. Part I, pp. 545–567. Madison, Wis.: American Society of Soil Agronomists.

Gates, F. C. 1949. Soil water content. In *Field Manual of Plant Ecology*. pp. 87–88. New York: McGraw-Hill.

Luttrell, E. S. 1967. A strip bait for studying the growth of fungi in soil and aerial habitats. *Phytopathology* 57, 1266–1267.

Lutz, H. J., and R. F. Chandler, Jr. 1955. *Forest Soils*. 514 pp. New York: John Wiley.

Novak, R. O., and W. F. Whittingham. 1968. Soil and litter microfungi of a maple, elm, ash floodplain community. *Mycologia* 60, 776–787.

Orpurt, P. A., and J. T. Curtis. 1957. Soil microfungi in relation to the prairie continuum in Wisconsin. *Ecology* 38, 628–633.

Papavizas, G. C., and C. B. Davey. 1959. Evaluations of various media and antimicrobial agents for isolation of soil fungi. *Soil Sci.* 88, 112–117.

Russels, E. W. 1961. *Soil Conditions and Plant Growth*. 9th ed., 688 pp. London: Longmans, Green.

Soil Survey Staff. 1951. Soil characteristic table. In *U.S. Dept. Agr. Soil Survey Manual*. Handbook No. 18. p. 209. Agr. Res. Admin., U.S. Dept. Agr.

Warcup, J. H. 1950. The soil-plate method for isolation of fungi from soil. *Nature* 166, 177.

3.8 Periods of Activity of Water Molds in a North Carolina Lake

John C. Clausz

Department of Biology
St. Andrews Presbyterian College
Laurinburg, North Carolina, 28352 USA

1 Introduction

Although many of the earliest reports on water molds dealt primarily with taxonomy and morphology, the works of Humphrey (1893), Peterson (1910), and Lund (1934) directed attention to the occurrence of these organisms and to their role in the habitat. They observed the fungi growing on a variety of organic remains such as dead fish, frogs, insects, and plant matter. More frequently, these fungi are not directly observed growing in the body of water. However, when hemp seeds or other baits are added to water samples brought into the lab, water molds are soon observed growing on the baits.

The practice of placing baits in habitats and later retrieving the infested baits takes advantage of the saprophytic nature of the water molds and yields information as to their activity. In the reports of Kanouse (1926), Crooks (1937), Waterhouse (1942), Hohnk and Bock (1954), Perrott (1960), and Cooke and Bartsch (1960), this "trapping technique" was used to learn which species of water molds were present and active.

Relative to the activities and importance of aquatic fungi in aquatic ecosystems, Weston (1941) commented that these fungi occur "as saprophytes in manifold capacities unceasingly active in the essential degradation of complex materials; and as parasites ever preying inconspicuously on the major groups of plants and animals important in hydrobiology." There does not appear to be a major group of plants or animals that escapes some measure of destruction by the aquatic fungi (Weston, 1941). Chytrid epidemics have been reported on diatoms and dynoflagellates by Canter and Lund (1948, 1951, 1953) and Paterson (1960).

Sparrow (1960) is replete with citations of plant and animal hosts for species of aquatic fungi. The association of saprolegniaceous species with diseased and moribund fish and amphibians has been reported since the Middle Ages (Sparrow,

---

KEYWORDS: Saprophyte; parasite; aquatic fungi; seasonality; viability; activity index; stenophagic; euryphagic; oospore; zoospore; propagule; germination; survival; oxygen concentration; *Achlya*; *Saprolegnia*; *Dictyuchus*; *Aphanomyces*.

1968). The primary species associated with fish kills is *Saprolegnia parasitica*, but species of *Achlya*, *Aphanomyces*, and *Dictyuchus* also have been reported on diseased or wounded fish (Tiffney, 1939a, 1939b; Shanor and Saslow, 1944; Vishniac and Nigrelli, 1957; Scott and O'Bier, 1962). However, actual proof of the parasitic nature of these fungi on healthy, naturally occurring fish has yet to be presented. Fuller (1971) described the consistent association of *Saprolegnia* with ulcerative dermal necrosis of salmonid fish but has not been able to establish the primary parasitic role in this case.

However, for two crustacean species the primary parasitic role of three species of the Saprolegniaceae has been established (Unestam, 1965; Hubschman and Schmitt, 1969). The parasite, *Aphanomyces astacus*, causes 100% death of attacked populations of the crayfish (*Astacus astacus*), which is a delicacy and of great economic importance to Finland and Sweden (Unestam, 1965, 1966). Such devastation on a herbiverous population has a dramatic effect on the aquatic ecosystem because the aquatic plants begin to choke the system.

Thus the water molds, and particularly the Saprolegniaceae, play a dual role in an ecosystem. They are active as saprobes in mineral and nutrient cycling processes and several species are effective and destructive parasites.

A water mold growing on a petiole or an insect has some hyphae ramifying through the nutritive source and other hyphae growing out into the surrounding water. After a while sporangia begin to form on these external hyphae. A model describing the mechanism by which a combination of factors would change a vegetative hyphal tip into a reproductive hyphal tip awaits the definition of conditions necessary for sporulation. See also Couch (1932).

Nevertheless a phase of assimilative growth is necessary for sporulation. If spore viability is short, one could assume that the presence of spores in a water sample indicated vegetative activity of that fungus in the very recent past. However, in his consideration of dispersal by aquatic fungi, Ingold (1971) states that there is little definite information concerning the lifespan of the spore.

The purpose of this investigation is to ascertain the viability of zoospores for possible use as an index to fungal activity, and to discern patterns of seasonal activity in Muskrat Pond, a pond in the North Carolina Botanical Garden.

2 Phenology of Aquatic Fungi in the Literature

Spring and fall maxima of water molds have been observed by Chaudhuri (1947), Dick and Newby (1961), Perrott (1960), and Suzuki (1960a, b, c, 1961a, 1961b), Dayal and Tandon (1962), Roberts (1963), Dayal and Ji (1966), and Srivastava (1967) indicated that in their experience water molds were most abundant in the winter months. Hughes (1962), from literature studies and his own research, observed that patterns of seasonal periodicity exhibited by water molds divide the species into two groups. Species of Saprolegniaceae producing excentric oospores have exhibited no particular seasonal patterns of occurrence, whereas those species producing centric or subcentric oospores have occurred in greater abundance usually in one or another season. The reports of Coker (1923), Coker and Matthews (1937), and Forbes (1935) on water molds seem to support the hypothesis of Hughes. Hughes noted that the species reported by Coker to occur more frequently at one or another time of the year were those forms producing centric or subcentric oospores, and those species found by Coker to occur during the year with no great fluctuations in abundance produced excentric oospores. Coker's isolation included about an equal number of species producing both types of oospores. Forbes' (1935) data, though, showed that the marked periodicity he observed (high frequency in winter, absence in summer) was based on a preponderance (85%) of species producing centric or subcentric oospores. Milanez (1967) also presented evidence supporting Hughes, but noted that the periodicity he saw was not as conspicuous as the literature would lead one to believe.

In the phenology section of his review of the ecology of the freshwater fungi, Sparrow (1968) states that for some species temperature may control fungal activity indirectly by reducing competition for substrata: "This has been proven by common practice over many years when certain so-called 'rare' groups of fungi have been induced to appear under laboratory conditions on natural substrata at low temperatures (3–5°C), where competition from coarser organisma, thriving at higher temperatures, is eliminated."

3 Effects of Dissolved Oxygen and Temperature on Viability of Zoospores

The mere presence of fungus propagules at a place in the water is not necessarily a significant indicator of active growth and reproduction of the fungus in that place, for spores of many fungi are known to be able to retain viability under environmental conditions not conducive to growth (Sussman, 1968). After imposition of conditions conducive to growth, the same propagules might then germinate and grow.

Experiments were performed to determine the effect of low oxygen concentrations and temperature on viability of zoospores and thus to provide some insight as to any value for survival the zoospores might have. A detailed description of the experimental procedure is given in Clausz (1970). Results were taken for three different temperatures, anaerobic as well as aerobic environments, at four time intervals after introduction of zoospore suspension into incubation vessels.

At all temperatures the greatest reduction in viability of zoospores for any single period of measurement occurred at the beginning of the experiment in the first six hours (Table 1). By the end of 12 hr, at all temperatures viability reached a lower level in aerated water than in oxygen-deficient water. After 12 hr at 15°C, only 13% of the zoospores in aerated water were viable, while at the same time zoospores in oxygen-deficient water showed 40% viability. A similar pattern of loss of viability was noted at the other experimental temperatures. The most striking loss appeared in the 25°C aerated suspension where viability had become reduced to zero after 12 hr. In all instances, at the end of 12hr, viability of zoospores suspended in oxygen-deficient water was higher than in aerated water. For both treatments retention of viability seemed to be inversely related to temperature 12 hr after incubation started. Similar results were obtained in repeat experiments.

Table 1. Effect of temperature and dissolved oxygen on viability of zoospores of *Achlya* sp.

		Temperature					
		15°C		20°C		25°C	
Hours	vs/s + %v[a]	Air	N_2	Air	N_2	Air	N_2
0	vs/s	130	154	146	148	153	147
	% v	100	100	100	100	100	100
6	vs/s	74	63	26	59	6	50
	% v	57	39	18	40	4	34
12	vs/s	17	65	5	33	0	19
	% v	13	40	3	22	0	13
18	vs/s	3	46	2	14	0	15
	% v	2	28	1	10	0	10
24	vs/s	0	18	3	6	1	10
	% v	0	12	2	5	1	7
ppm 0_2		8.3	1.5	8.0	1.2	7.1	1.3

[a] vs/s viable spores per sample. % v percent viable spores based on viable spores per sample at time zero.

In aerated water, zoospores at the warmest temperature suffered the greatest loss of viability during the first 12 hr. After 24 hr the effect of temperature was no longer apparent in aerated water. In oxygen-deficient water, a similar pattern was evident, but after 24 hr a possible effect of temperature was still apparent. The lower temperatures appeared to prolong viability.

In the only other experiment of similar design, Willoughby (1962) observed a die off of spores, but not to such a large an extent as reported here. Of the total number of saprolegniaceous spores in the original sample, only 27% and 38% were alive after 24 hr at laboratory temperature. Under a very different set of circumstances, Unestam (1966) and H. T. Fuller (personal communication) have described how zoospores may have longer term survival value. Unestam (1969) reports that viable spores were obtained from a spore suspension stored at 2°C for two months.

In the past, isolation of saprolegniaceous fungi from samples has not been taken to indicate growth and sporulation at the time of isolation. This observation has been a serious impediment to meaningful insights pertaining to the degree of importance of these organisms in their role as decomposers in ecosystems. Indications are that zoospores are short lived. Since zoosporogenesis is preceded by a period of vegetative activity, the number of spores in a sample is probably related to the amount of growth and reproductive activity occurring at or just prior to the time of collection.

4 Seasonal Patterns of Activity in Muskrat Pond

Counting Technique Description

Descriptions of methods used to obtain information in ecological studies of aquatic fungi have been presented by Willoughby (1962), Collins and Willoughby (1962), Willoughby and Collins (1966), Ji and Dayal (1966), and Dick (1966). While all of these techniques have produced numerical data, the method evolved in this study has reduced some of the disadvantages of those already published.

Each phase of the agar isolation technique used in this work was tested to determine its effect of survival of zoospores (Clausz, 1970). Antibiotics proved useful in the isolating technique of Fuller and Poyton (1964), and their use in this study at concentrations which reduce bacterial, but not fungal, growth did not affect detrimentally the recovery of zoospores. Experimentation showed that corn meal agar (CMA) with antibiotics was as good as other media tested: CMA with dextrose, peptone-yeast extract-glucose; and yeast extract-glucose-soluble starch.

Experiments designed to test survival of zoospores subjected to the isolation technique revealed that occasionally more than 40% of the spores centrifuged and embedded in agar did not appear in resulting cultures. However, in some instances all spores tested were retrieved and grew. Variations of this magnitude are presented in the works of Collins and Willoughby (1962) and Ji and Dayal (1966), where between 92 and 67% of the spores survived their techniques. Apparently, all of the quantitative techniques devised thus far for studying water mold populations impose limitations on the counts obtained since some spores do not survive the technique.

Table 2. Species of the Saprolegniaceae collected at the Botanical Garden Lake.

Achlya americana Humphrey
A. colorata Pringsheim
A. diffusa Harvey *ex* Johnson
A. klebsiana Pieters
A. orion Coker and Couch
Achlya sp.

Aphanomyces laevis deBary
Aphanomyces sp.

Dictyuchus sterile Coker

Leptolegnia caudata deBary
Leptolegnia sp.

Saprolegnia delica/diclina
S. ferax (Gruith.) Thuret.
S. hypogyna (Pringsheim) deBary
Saprolegnia sp.

In this study, saprolegniaceous propagules were removed from samples of lake water by continuous flow centrifugation. A Sorvall SS-3 automatic superspeed centrifuge equipped with the Szent-Gyorgi and Blum 4-tube continuous flow system was used. A flow rate of 80 to 100 ml/min at 12,000 × *g* was sufficient to collect all saprolegniaceous spores in the 1-liter samples obtained from the lake. For further details on methods see Clausz (1970).

Many apparently sterile isolates of *Saprolegnia* and *Leptolegnia* were found at the Botanical Garden Lake. Their identifications to species was facilitated, however, by growing hemp seed cultures in an unilluminated, 10 to 15°C cold room as suggested by Couch (1932) and Szaniszlo (1965). Coker (1923), Coker and Matthews (1937), Johnson (1956), Scott (1961), and Seymour (1966) were the monographic references used for identification. Species isolated and identified are listed in Table 2; some of the original isolates are shown in Fig. 1.

Patterns of Seasonal Activity

Muskrat Pond is an 8-acre impoundment which was filled in the winter of 1965–1966. The lake is located on the Mason Farm portion of the North Carolina Botanical Garden in Chapel Hill (Fig. 2). Terres (1969) and Meyer (1971) described the lake and the surrounding vegetation. The locations of the six sampling stations are seen in Fig. 2. The six sampling stations included all the habitat types present in the lake. Station 1 was located at the deepest part of the lake, 4.2 m. No attached benthic plants were growing at Stations 2 and 3, 2.0 and 1.5 m, respectively. Station 2 was located where the bottom was steeply sloping away from the shore, whereas Station 3 was positioned in a gently sloping portion of the lake. Stations 4 to 6 were located in the shallow portion of the lake, where species of *Chara* and *Najas* were growing on the bottom. Station 6, being closest to shore on the wooded side of the lake, contained a great deal of detrital plant material. At each sampling station weekly measurements of water temperature and dissolved oxygen content were made with a YSI model 51 oxygen meter at 0.5-m intervals from the surface to the bottom. One set of weekly measurements always coincided with the regular monthly sampling of the water. Clausz (1970) has shown that in the summer thermal isolation of the upper and lower portions of the lake leads to the development of an oxygen-deficient zone in the deeper portions. The shallow portions of the lake exhibit temperature profiles which reflect the prevailing atmospheric conditions.

At monthly intervals water samples were taken at 0.5-m intervals of depth at each station except Station 1 where samples were taken at 1-m intervals. Subsurface samples were collected with a 2-liter Van Dorm water sampler from which 1 liter was brought to the laboratory for analysis. No more than 1.5 hr elapsed between collection and embedding in agar.

Spring and fall are the seasons for maximal fungal activity (Fig. 3). During these periods species of *Saprolegnia* and *Leptolegnia* were more active than others. Species of *Achlya* were active mainly in late summer. The seasonal occurrence of all other species listed in Table 2 is shown in Fig. 4. *Saprolegnia delica/diclina* and *Saprolegnia* sp. were active primarily from spring through fall; whereas *Lepto-*

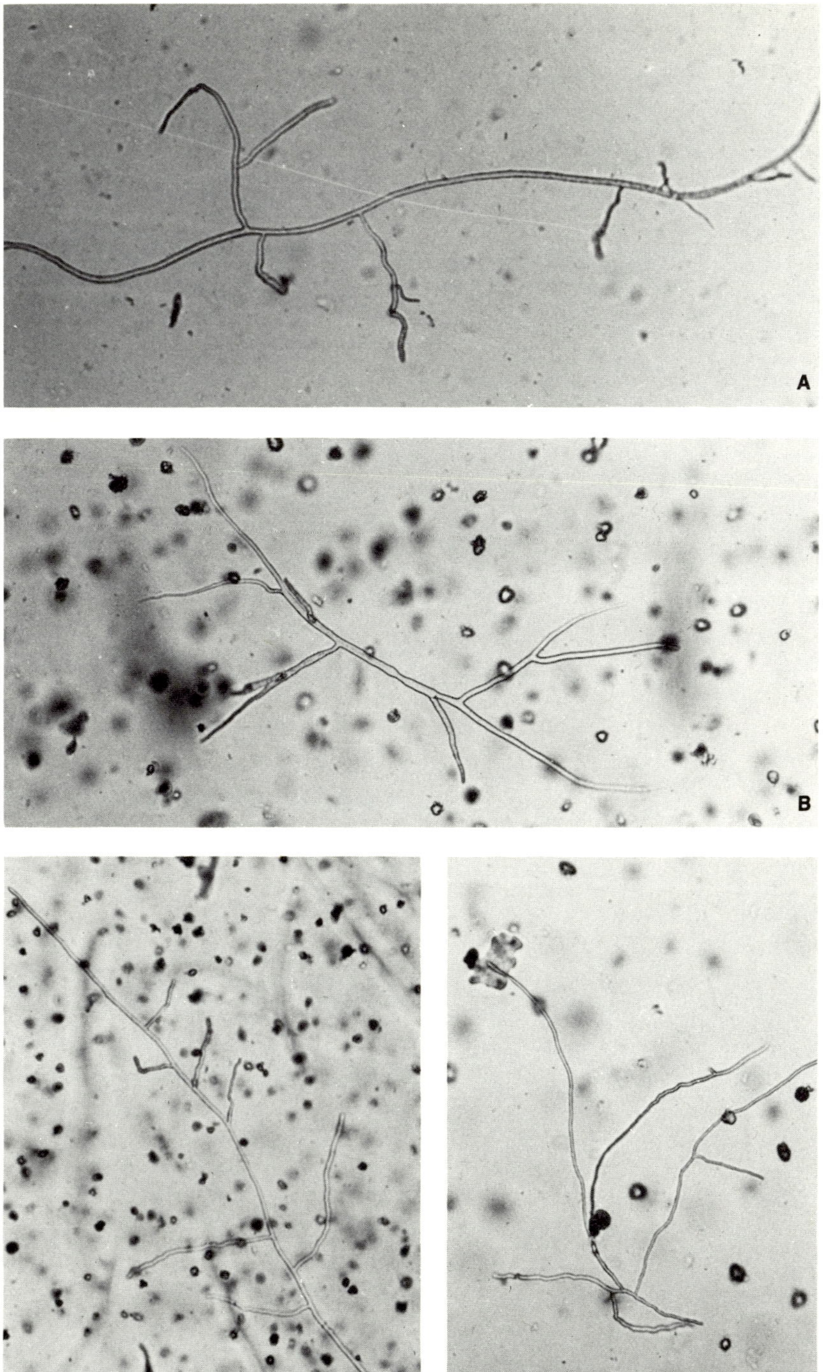

Fig. 1. Germlings of species of water molds growing in corn meal agar with antibiotics. *A. Achlya* sp. *B. Dictyuchus* sp. *C. Saprolegnia* sp. *D. Aphanomyces* sp. All figures × 100.

legnia caudata and *Leptolegnia* sp. were active primarily from fall through spring.

Since most of the *Achlya* sp. cultures that have reproduced sexually were determined to be *A. klebsiana*, the data shown in Fig. 3 for *Achlya* sp. may be representative of *A. klebsiana* also. *Achlya* sp. was active during the hot and warm months, July–October 1967, and July 1968. (Fig. 4.) *Achlya* sp. and *A. klebsiana*

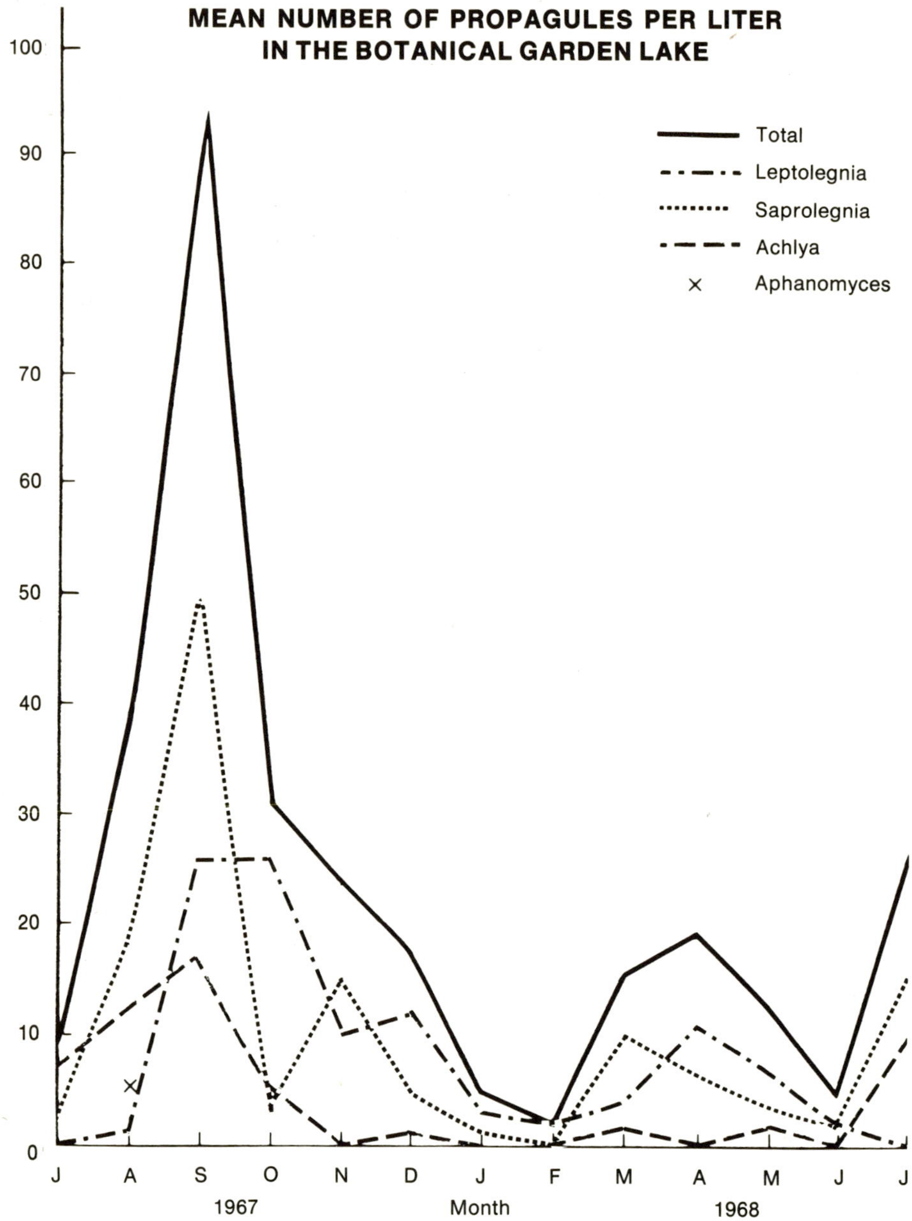

Fig. 3. Mean number of propagules per liter in the Botanical Garden Lake.

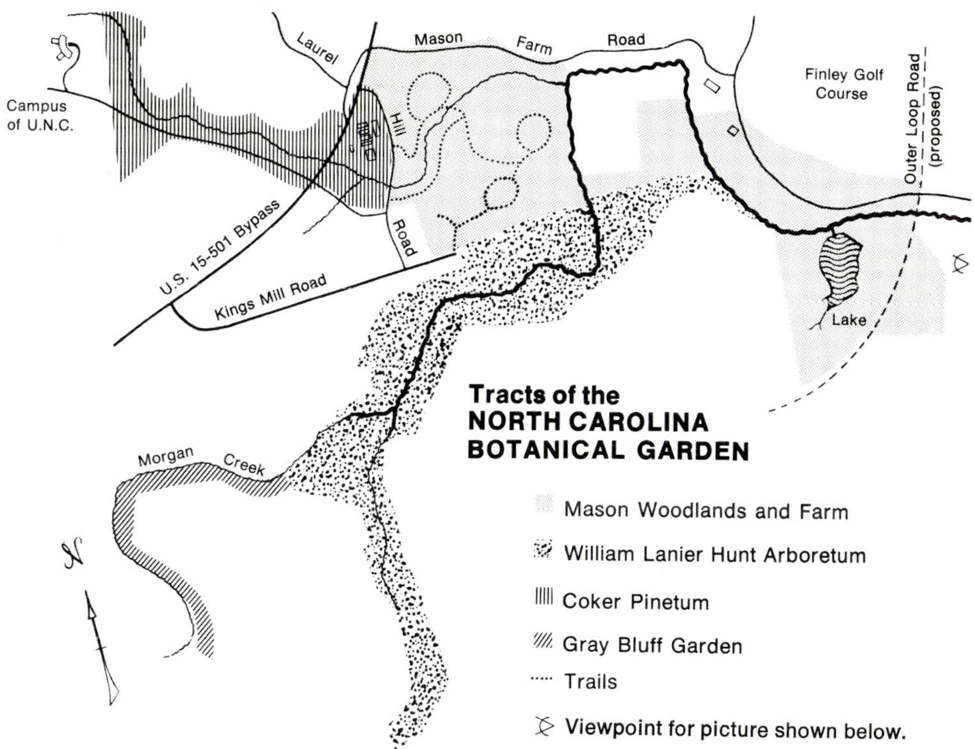

Fig. 2. *A.* Location of the Muskrat Pond in the North Carolina Botanical Garden, *B.* Position of six sampling stations in the lake.

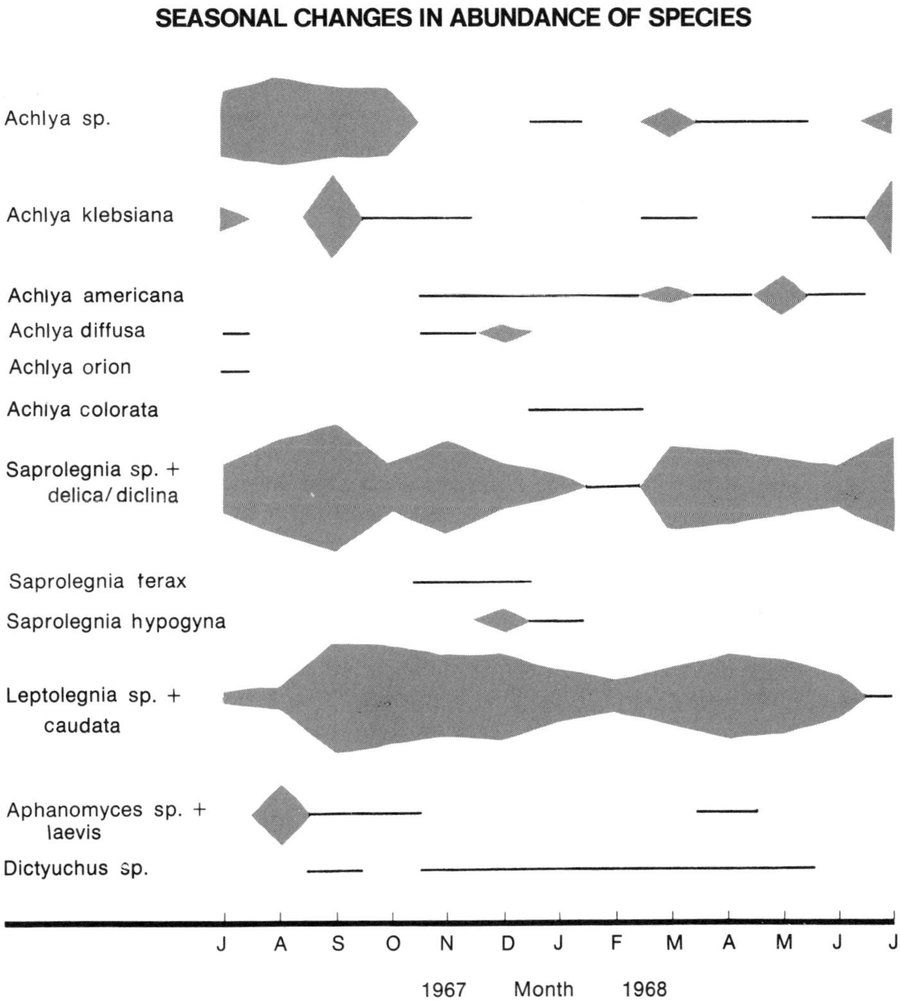

Fig. 4. Seasonal changes in abundance of fungal species listed in Table 2 in the Botanical Garden Lake.

were both absent or present in low numbers during the winter and spring months. In contrast, *Achlya americana* is active during the cooler months of November through May. *A. americana* was conspicuously absent from samples taken in the summer months. A similar period of activity during the colder months is exhibited by *Achlya colorata.*

Aphanomyces sp. and *A. laevis* were most active in August, the same month that numerous tadpoles of *Rana* sp. were observed in Muskrat Pond (Terres, 1969). These tadpoles supported luxuriant growth of *Aphanomyces* sp.

Except for sterile strains of *Dictyuchus*, the other species were collected infrequently and were few in number (Fig. 4).

How might these periodicity patterns arise? Seasonal variations in climate might affect periodicity through its control on oospore germination. Three species collected at Muskrat Pond—*Achlya colorata*, *Saprolegnia ferax*, and *S. hypogyna*—were active only in the winter months. It is possible that low temperatures were required to trigger germination of oospores for these species. Sussman (1968) noted temperature as an important factor in breaking dormancy. However, the activity pattern for *Saprolegnia delica/diclina* and *Leptolegnia caudata* reveals activity throughout the year with periods of minimal activity. The reduced activity was probably the result of a temperature effect on metabolism, the high or low temperature reducing rates of growth or development.

If species of fungi could be characterized as stenophagic or euryphagic, it might well be shown by experimentation that the occurrence and abundance of stenophagic species coincide with availability of the food favoring metabolism and that little but the rate of their development might be affected by environmental conditions. Even if conditions optimal for growth of stenophagic forms were present, though, the fungi would not develop in the absence of suitable substratum. The development of euryphagic species, not highly selective as to energy source, would be influenced more by the existing environmental circumstances, and unless physical conditions were limiting, would grow within limits imposed only by the paucity or profusion of food materials.

Since substrata used in the present study were intentionally selective, little can be said regarding the range of substrata utilizable by the species investigated. However, certain observations of some species were interesting. Propagules of *Aphanomyces laevis* and *Aphanomyces* sp. were collected in greatest abundance in August, the same month when numerous dead tadpoles of *Rana* sp. were observed with luxuriant vegetative growths of *Aphanomyces* sp. Propagules of these species were collected in lower numbers in September, October, and April. It is possible that the remains of the tadpole carcasses continued to support the fungus to lesser degrees in September and October, but in April it is more likely that eggs of *Rana* or other similar organisms provided the food source. Frequently frog eggs have been seen in the spring supporting growth of this water mold, which seems to have stenophagic leanings. *Saprolegnia delica/diclina* or *Leptolegnia caudata* seems to have been more heavily influenced by physical conditions than by availability of certain substrata. These two fungi occurred in moderate numbers throughout the year, with *Saprolegnia* reduced only in the very warm months. The occurrence and abundance of these species appear to fit a pattern to be expected of euryphagic species. Whether the periodicity of a species is controlled directly by climate acting on the fungus or indirectly by regulating abundance of food is a problem requiring more study. It is difficult to actually observe a fungus growing in a lake.

References

Brock, T. D. 1966. *Principles of Microbial Ecology*. 306 pp. Englewood Cliffs, N.J.: Prentice-Hall.

Canter, H. M., and J. W. G. Lund. 1948. Studies on plankton parasites. I. Fluctuations in the numbers of *Asterionella formosa* Hass. in relation to fungal epidemics. *New Phytologist* 47, 238–261.

——— 1951. Studies on plankton parasites. III. Examples of the interaction between parasitism and other factors determining the growth of diatoms. *Ann. Bot.* (*London*) (*N.S.*) 15, 359–371.

——— 1953. Studies on plankton parasites. II. The parasitism of diatoms with special reference to the lakes in The English Lake District. *Brit. Mycol. Soc. Trans.* 36, 13–37.

Chaudhuri, H., *et al.* 1947. *A Handbook of Indian Water Moulds.* Part I, 70 pp. India: University of Punjab.

Clausz, J. C. 1970. Distribution of propagules of species of the saprolegniaceae in the North Carolina Botanical Garden Lake and some possible controlling factors. Ph.D. Thesis. Chapel Hill: University of North Carolina.

Coker, W. C. 1923. *The Saprolegniaceae of the Eastern United States and Canada, with Notes on Other Water Molds.* 201 pp. Chapel Hill: University of North Carolina Press.

———, and V. D. Matthews. 1937. Blastocladiales, Monoblepharidales, Saprolegniales. *N. A. Flora* 2, 1–76.

Collins, V. G., and L. G. Willoughby. 1962. The distribution of bacteria and fungal spores in Blelham Tarn with particular reference to an experimental overturn. *Arch. Mikrobiol.* 43, 294–307.

Cooke, W. B., and A. F. Bartsch. 1960. Aquatic fungi in some Ohio streams. *Ohio J. Sci.* 60, 144–148.

Couch, J. N. 1932. The development of the sexual organs in *Leptolegnia caudata*. *Am. J. Bot.* 19, 584–599.

Crooks, K. M. 1937. Studies on Australian Phycomycetes. *Proc. Roy. Soc. Victoria* (*N.S.*) 49, 206–233.

Dayal, R., and T. Ji. 1966. The occurrence and distribution of aquatic fungi in certain ponds of Varanasi. *Hydrobiologia* 27, 548–558.

———, and R. N. Tandon. 1962. Ecological studies of some aquatic Phycomycetes. *Hydrobiologia* 20, 121–127.

Dick, M. W. 1966. The Saprolegniaceae of the environs of Blelham Tarn: Sampling techniques and estimation of propagula numbers. *J. Gen. Microbiol.* 42, 257–282.

———, and H. V. Newby. 1961. The occurrence and distribution of Saprolegniaceae in certain soils of south-east England. I. Occurrence. *J. Ecol.* 49, 403–419.

Forbes, E. J. 1935. Water moulds of the Manchester district. *Mem. Proc. Manchester Lit. Phil. Soc.* 79, 1–11.

Fuller, H. T. 1971. *Saprolegnia* and salmon disease in Ireland. Abstr. of papers. First International Mycological Congress. Exter, England.

Fuller, M. S., and R. C. Poyton. 1964. A new technique for the isolation of aquatic fungi. *BioScience* 14, 45–46.

Hawker, L. 1966. Environmental influences on reproduction. In *The Fungi*, G. C. Ainesworth and A. S. Sussman, ed. Vol. 2, pp. 435–469. New York: Academic Press.

Hohnk, W., and K. J. Bock. 1954. Ein Beitrag zur Ökologie der saprophytischen Wasserpilze. *Veröffentl. Inst. Meeresforsch. Bremerhaven* 3, 9–26.

Hubschman, J. H., and J. A. Schmitt. 1969. Primary mycosis in shrimp larvae. *J. Invertebrate Pathol.* 13, 351–357.

Hughes, G. C. 1962. Seasonal periodicity of the Saprolegniaceae in the south-eastern United States. *Brit. Mycol. Soc. Trans.* 45, 519–531.

Humphrey, J. E. 1893. The Saprolegniaceae of the United States, with notes on other species. *Trans. Am. Phil. Soc.* 17, 63–148.

Ingold, C. T. 1971. *Fungal Spores, Their Liberation and Dispersal.* 302 pp. Oxford, England: Clarendon Press.

Ji. T., and R. Dayal. 1966. The occurrence and distribution of reproductive spores of Saprolegniales in certain ponds of Varanasi. *Nova Hedwigia* 12, 509–517.

Johnson, T. W., Jr. 1956. *The Genus Achlya: Morphology and Taxonomy.* 180 pp. Ann Arbor: University of Michigan Press.

Kanouse, B. B. 1926. On the distribution of water molds, with notes on the occurrence in Michigan of members of the Leptomitaceae and Blastocladiaceae. *Papers Mich. Acad. Sci.* 5, 105–114.

Lund, A. 1934. Studies on Danish freshwater Phycomycetes and notes on their occurrence particularly relative to hydrogen ion concentration of the water. *Mem. Acad. Roy. Sci. Danemark, Sect. Sci.* 6 (9), 1–97.

Meyer, K. A. 1971. Seasonal fluctation of phytoplankton composition, diversity and production in a fresh water lake. *J. Elisha Mitchell Sci. Soc.* 87, 127–138.

Milanez, A. I. 1967. A study of the aquatic 'Phycomycetes' of the Gull Lake area in Michigan. *Diss. Abst.* 28, No. 6.

Paterson, R. A. 1960. Infestation of chytridiaceous fungi on phytoplankton in relation to certain environmental factors. *Ecology* 41, 416–424.

Perrott, P. E. 1960. The ecology of some aquatic Phycomycetes. *Brit. Mycol. Soc. Trans.* 43, 19–30.

Peterson, H. E. 1910. An account of Danish freshwater-Phycomycetes, with biological and systematic remarks. *Ann. Mycol.* 8, 494–560.

Roberts, R. E. 1963. A study of the distribution of certain members of the Saprolegniales. *Brit. Mycol. Soc. Trans.* 46, 213–224.

Scott, W. W. 1961. A monograph of the genus *Aphanomyces. Virginia Agr. Exp. Sta. Tech. Bull.* 151, 1–95.

———, and O'Bier. 1962. Aquatic fungi associated with diseased fish and fish eggs. *Progr. Fish-Cult.* 24, 3–15.

Seymour, R. L. 1966. A revision of the genus *Saprolegnia.* Ph.D. Thesis. Blacksburg: Virginia Polytechnic Institute.

Shanor, L., and H. B. Saslow. 1944. *Aphanomyces* as a fish parasite. *Mycologia* 36, 413–415.

Sparrow, F. D. 1960. *Aquatic Phycomycetes.* 2nd ed. 1187 pp. Ann Arbor: University of Michigan Press.

——— 1968. Ecology of the freshwater fungi. In *The Fungi*, G. C. Ainsworth and A. S. Sussman, ed. Vol. 3, pp. 41–93. New York: Academic Press.

Srivastava, G. C. 1967. Ecological studies on some aquatic fungi of Gorakhpur, India. *Hydrobiologia* 30, 385–404.

Sussman, A. S. 1968. Longevity and survivability of fungi. In *The Fungi*, G. C. Ainsworth and A. S. Sussman, ed. Vol. 3, pp. 447–486. New York: Academic Press.

Suzuki, S. 1960a. Seasonal variation in the amount of zoospores of aquatic Phycomycetes in Lake Shinseiko. *Bot. Mag.* (*Tokyo*) 73, 483–486.

——— 1960b. The seasonal variation of aquatic fungi in Senshunike pond. *Japan. J. Limnol.* 21, 271–278.

——— 1960c. Ecological studies on the aquatic fungi in the Arakawa River. *Japan. J. Limnol.* 21, 25–31.

——— 1961a. Some considerations on anaerobic life of aquatic fungi in lake bottom. *Japan. J. Ecol.* 11, 219–221.

——— 1961b. The seasonal changes of aquatic fungi in the lake bottom of Lake Nakanuma. *Bot. Mag.* (*Tokyo*) 74, 30–33.

Szaniszlo, P. J. 1965. A study on the effect of light and temperature on the formation of oogonia and oospores in *Saprolegnia ferax. J. Elisha Mitchell Sci.* 81, 10–15.

Terres, J. K. 1969. *From Laurel Hill to Silers Bog*. 227 pp. New York: Alfred A. Knopf.

Tiffney, W. N. 1939a. The identity of certain species of the Saprolegniaceae parasitic to fish. *J. Elisha Mitchell Sci. Soc.* 55, 134–151.

——— 1939b. The host range of *Saprolegnia parasitica*. *Mycologia* 31, 310–321.

Unestam, T. 1965. Studies on the crayfish plague fungus *Aphanomyces astaci*. I. Some factors affecting growth *in vitro*. *Physiol. Plant.* 18, 483–505.

——— 1966. Studies on the crayfish plague fungus *Aphanomyces astaci*. II. Factors affecting zoospores and zoospore production. *Physiol. Plant.* 19, 1110–1119.

——— 1969. On the physiology of zoospore production in *Aphanomyces astaci*. *Physiol Plant.* 22, 236–245.

Vishniac, H. S., and R. F. Nigrelli. 1957. The ability of Saprolegniaceae to parasitize platyfish. *Zoologia* 42, 131–134.

Waterhouse, G. M. 1942. Some water molds of the Hogsmill River collected from 1937–1939. *Brit. Mycol. Trans.* 25, 315–325.

Weston, W. H. 1941. The role of aquatic fungi in hydrobiology. In *A Symposium on Hydrobiology*. pp. 129–151. Madison: University of Wisconsin Press.

Willoughby, L. G. 1962. The occurrence and distribution of reproductive spores of Saprolegniales in fresh water. *J. Ecol.* 50, 733–759.

———, and V. G. Collins. 1966. A study of the distribution of fungal spores and bacteria in Blelham Tarn and its associated streams. *Nova Hedwigia* 12, 150–171.

PART 4

Representative Biome Studies

Introduction

Helmut Lieth

Botany Department
University of North Carolina
Chapel Hill, North Carolina, 27514, USA

The map on the inside back cover of this book shows that most regions of the world have some degree of seasonality. The reasons for such seasonality, the fluctuations in the environmental parameters, as well as the response of the biota, vary greatly. Part 4 includes studies that may serve as examples for major biomes, such as tundra (Chapter 4.1), desert (Chapter 4.2), grassland (Chapter 4.3), deciduous forest (Chapter 4.4), chaparral (Chapter 4.5), woodland (Chapter 4.6), and tropical rain forest (Chapter 4.7). Some important biomes are omitted (e.g., savanna) because the projected contributors could not participate.

While taxonomic considerations vary greatly from one continent to another, phenological principles are the same everywhere; environmental conditions generally force the sequence of events within and between trophic levels. Especially interesting is the seasonality shown in biomes with environmental extremes. Studies from these areas point to a phenological approach for systems analysis in almost any ecosystem.

This section contains examples from North and South America, Europe, and Africa. Phenological studies are also available from Asia and Australia. The papers of Hopp (Chapter 2.1), Mooney *et al.* (Chapter 4.5), and Wielgolaski (Chapter 4.1) include information and literature from these continents. We hope the realization that phenological principles and problems apply worldwide will result in a permanent, international phenological working group.

Contributors to this section emphasize different methods and trophic levels. Most authors examine the primary producers; however, the papers of Malaisse (Chapter 4.6) and Frankie *et al.* (Chapter 4.7) include much information on the consumers. See Chapter 3.8 for reference to aquatic phenology. Studies of decomposers are generally unavailable outside the deciduous forest (see Chapters 3.6 and 3.7) and grassland biomes.

As more information is available from a biome, attempts to model are being made (Chapter 4.3 on grassland development and Chapter 6.1 on agriculture; see also Part 5 on modeling in general). Seasonal flow of energy is modeled in the following components (in descending frequency); habitat, green plants, consumers, and decomposers. The area for new phenological research is vast. We hope that the detailed case studies presented here will reinforce interest in phenological system models for whole ecosystems and biome regions.

4.1 Phenological Studies in Tundra

Frans-Emil Wielgolaski

Norwegian IBP
Botanical Laboratory
University of Oslo
Blindern — Oslo 3, Norway

1 Introduction

A comprehensive monograph on plant phenology in Arctic tundra was presented by Sørensen (1941). Many climatic data, as well as discussions of life forms of tundra flowering plant species, are included in his analysis of phenological seasons in northeast Greenland. Phenological studies in Arctic areas are being performed (e.g., Webber, personal communication 1972) in the Baffin Island area. Many papers on phenology and related topics are available, e.g., from the Rocky Mountains Alpine tundra, which has been studied botanically since early in this century (Cooper, 1908; Clements, 1920; Cox, 1933). Bliss (1956) compared plant development in microenvironments of Arctic and Alpine tundras. Billings and Bliss (1959) studied an Alpine snowbank and its effects on plant development. Mooney and Billings (1961) carried out comparative studies of physiological ecology and phenological development on *Oxyria digyna* from many tundra areas, both in Arctic and Alpine regions; they have correlated the carbohydrate cycle with phenological development in several Alpine species (1960). Holway and Ward (1963) considered the retarding influence of snow cover and meltwater on phenology in the Rocky Mountains, and suggested possible correlations between plant phenology and environmental data (1965). Snow cover was considered to be the primary factor influencing phenology, with soil temperature also important, while slope and exposure were distinctly secondary in the high Rocky Mountain area.

Phenological studies are carried out in the tundra projects of the International Biological Program (IBP) in many countries. Few results are available yet from these studies, but some preliminary values from Devon Island in the Canadian high Arctic are given by Muc (1972) and Svoboda (1972). Phenological observa-

KEYWORDS: Tundra phenology; Arctic tundra; Alpine tundra; phenology; transplant garden; ecophysiological phenology; *Vaccinium vitis-idaea*; *Salix herbacea*; *Polygonum viviparum*; snow bank effect.

tions are necessary to relate the production of an ecosystem to the same developmental stages at various sites. Use of a calendar time scale alone often leads to misinterpretation of data, i.e., in investigating variation of nutrient use in an organism over time without considering the phenology of the organism.

2 Phenological Studies in the IBP Tundra Projects

Within the IBP tundra biome studies, data from high Arctic sites, as in northern Canada, may be compared with data, e.g., from Austrian Alpine tundra. Even if the vegetation in the Alps might be classified as tundra (caused by low temperatures), the daylength is quite different. This may be one reason for developmental rates other than those found in the Arctic. Having the phenological phases recorded for all major species, the biomass, mineral nutrient content, carbohydrates, etc., may be compared at a specific stage of the life cycle in various countries. All these plant measurements at various phenological phases, as well as at each of the phenological phases itself, will be related with environmental data.

In the international tundra working groups for primary production and decomposition, standardized phenological studies are performed. The decomposition group has submitted special types of cellulose strips to all tundra sites with detailed instructions on how they should be used in the soil and handled after specific intervals. To be sure that the methods of chemical analysis do not differ among countries, all samples are sent to one country for analysis.

The primary production group uses what might be called "the international pea" for phenological comparisons among sites, nationally and internationally. Seeds of the same variety of pea (*Pisum sativum*) taken from a single source are supplied to all sites and sown in the same type of soil and watered and fertilized in the same manner during the growing season at all sites. Time of germination, bud, flower, etc., is observed. In addition, height, number of leaves, and leaf areas are measured nondestructively at short intervals. In the North American IBP tundra sites, additional species are used for comparison in transplant gardens, e.g., a few graminoids.

The phenology of natural vegetation is studied, some aspects in detail, in many of the international IBP tundra sites. Ten to 25 specimens of the most important species (mostly vascular plants) are marked, and the phenological stages are recorded for each plant during the entire season; shrubs are studied in various seasons. When possible, number of new leaves per shoot, leaf areas, and sometimes leaf lengths of green and nongreen parts, height of plants, number of inflorescent stems, etc., are measured at specific intervals.

Studies of phenology in animals are less developed internationally. However, in North America some observations are made, mostly on birds, but a few also on arthropods. Arrival and departure dates of migrating birds are noted. Dates of display, molt, singing, nesting, complete clutch, egg hatches, and nest leaving are recorded for some species. At Point Barrow and Prudhoe Bay in Alaska and at Devon Island in Canada, the seasonal changes in weight, diet, and microhabitat for a few bird species are studied on a daily basis.

3 Results from the Alpine Tundra Sites of the Norwegian IBP

A few results from the Norwegian Alpine tundra-sites at Hardangervidda, about 200 m above the tree line and 500 m away from each other, reveal phenological differences over a short distance. One of the sites is a dry oligotrophic lichen heath with coarse sandy soil and only a few millimeters of humus. Another site is an eutrophic meadow with about 10 cm humus, flooded in spring, but later somewhat drier; a third site is an eutrophic, very wet meadow on organic soil about 1 m deep.

At all sites low perennial vegetation is dominant: dwarf shrubs less than 10 cm high, perennial herbaceous vascular plants of similar height, and mosses and fruticose lichens, the latter especially at the lichen heath. The yearly mean temperature is about -3°C, but maximum air temperature close to the soil may be above 30°C in the summer months (seldom above 20°C in 2-m height); still, it might happen to snow even during the summer months. The yearly precipitation is about 700 to 1000 mm, with the least in the early summer months, sometimes causing drought at the coarse sandy site. The snow-free period is from early June to mid-September at the lichen heath, and snow melts about two weeks later at the wet meadow.

Among all the species studied *Vaccinium vitis-idaea* is one of the most common at many of the Norwegian sites. At the dry lichen heath the new leaf emergence for this species took place in late June, about one week earlier than at the driest meadow site, because of earlier snow melting. The same time difference was found for flower bud and flower drop of the species. Webber (personal communication 1972), however, found that late snow melting may delay the sequence of spring phenophases, but such delay was reduced in the onset of fall phenophases.

The height of the plants was the same at the two Norwegian sites, both at leaf emergence and at the time of the first visible bud. At the time of flower drop, however, in late July, the length (and the weight) of the new shoots at the meadow was about twice the length (and weight) of the shoots at the lichen heath, possibly because of the dryness at the latter site. This indicates that other factors may counteract the height-inhibiting effect of long-lasting snow cover in tundra areas reported by, e.g., Billings and Bliss (1959) and Holway and Ward (1965). According to Bock and Webber (personal communication 1972), a reduced height-growth rate in tundra snow bank plots in early season may be followed by faster growth rate in these plots for some species.

In the studies on *Vaccinium vitis-idaea* in Norway, if increased precipitation occurred before the growing period was ended, the rate of height increase at the dry lichen heath might be even faster than at the meadow. The heights of the same species were considerably greater at the various phenological stages studied at a site in the sub-Alpine birch forest than at the sites above the tree line. Thus the relationship between environmental factors and the dates of phenological stages of tundra plants, as well as plant heights at these stages, seems to be rather complicated.

Phenological data of *Vaccinium vitis-idaea* from the Norwegian sites can, in IBP, be compared to similar data from five sites in Alaska, ranging from 180 to

1200 m above sea level, from boreal forest to Alpine tundra. This may increase the possibility of understanding the growth pattern of the species in relation to the environment.

One observation which possibly is closely linked to the temperature factor is that almost no flowers occur on the *Vaccinium* plants above the tree line in the Norwegian IBP studies, except close to rocks, and then on the southern side. In extremely warm summers a few flowers might also be found on exposed areas. Further studies are necessary to determine if the climate in the current year is most important, or if flower frequency is more dependent on conditions in the preceding year, as indicated by Sørensen (1941) and mentioned by Holway and Ward (1965).

Two other plant species found at several tundra sites internationally and at most of the Norwegian sites, too, are the dwarf shrub *Salix herbacea* and the herb *Polygonum viviparum*. As seen in Fig. 1 the reproductive phases of the early flowering species *Salix herbacea* are found much earlier at the lichen heath than at the two meadows (especially the wet meadow with late snow melting) because of the early snow-free ground at the lichen heath. *Salix herbacea* shows almost the same phenological differences between the sites in fall as in spring. The flowering period seemed, however, to be shortest at the wet meadow. The reason may be high temperature during late June, the main flowering period for *Salix herbacea* at that site.

The first flower bud for both *Ranunculus acris* and *Polygonum viviparum* was seen somewhat earlier at the wet meadow than at the dry meadow (Fig. 1). One possible cause is that dry conditions at the dry meadow in the previous year retarded initiation of new flower buds. The other phenological phases of the same species were later at the wet meadow than at the dry meadow (Fig. 1), possibly because of generally lower soil temperatures at the wet meadow. The flowering of *Polygonum viviparum* occurred at the dry meadow at the same time as found for similar vegetation types in the US/IBP Alpine tundra studies in Rocky Mountains (Ebert, personal communication 1972), while the first mature fruit was noted one to two weeks later at the Norwegian dry meadow than in the United States. At flowering, the height of the plants and the number of leaves were similar at dry sites in both countries. The height increased at both places with increasing soil water, except for the wettest plot at Rocky Mountains where possibly aeration was poor.

Some phenological studies are also done on tundra animals in Norway. For food intake by wild reindeer during different times of the year, fruticose lichens which grow on windswept areas are the most important winter forage plant, while different species of *Salix* are important as food for the reindeer during early summer (Thompson, personal communication 1971). Later in summer, herbs are the most important forage plants. The reindeer turn, however, to lichens again as their main diet even before new snow covers the herb areas. The numbers of reindeer, small mammals and invertebrates are counted at different times of the year at Hardangervidda, Norway. Similar phenological studies are also done in other tundra areas.

From all examples mentioned, obviously a strong need exists for linking the phenological viewpoint to the measurements of assimilation, respiration, biomass, and nutrient levels found at various times of the year at the tundra sites. An understanding of the functioning of the ecosystem and its parts is only possible when the phenological stages of the various species are identified, and the assimilation and nutrient status are measured at various times of the season.

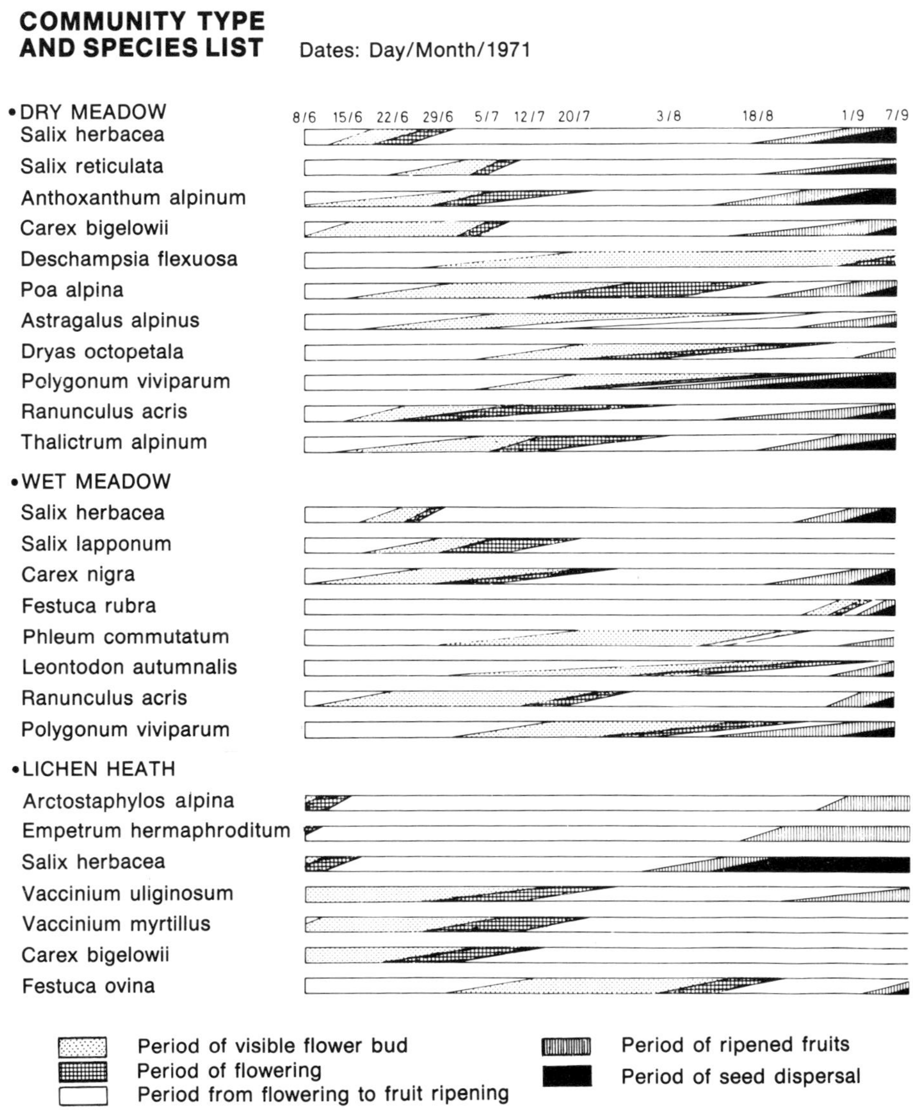

Fig. 1. Reproductive development at Alpine tundra sites at Hardangervidda, southern Norway, 1971.

References

Billings, W. D., and L. C. Bliss. 1959. An alpine snowbank and its effects on vegetation, plant development and productivity. *Ecology* 40, 388–397.

Bliss, L. C. 1956. A comparison of plant development in microenvironments of arctic and alpine tundras. *Ecol. Monograph* 26, 303–337.

Clements, F. E. 1920. Plant indicators: The relation of plant communities to process and practices. *Carnegie Inst. Wash. Publ. 290*, 1–388.

Cooper, W. S. 1908. Alpine vegetation in the vicinity of Long's Peak, Colorado. *Bot. Gaz.* 45, 319-337.

Cox, C. F. 1933. Alpine plant succession on James Peak, Colorado. *Ecol. Monograph* 3, 300–372.

Holway, J. G., and R. T. Ward. 1963. Snow and meltwater effects in an alpine area of Colorado. *Am. Midland Naturalist* 60, 189–197.

——— 1965. Phenology of alpine plants in northern Colorado. *Ecology* 46, 73–83.

Mooney, H. A., and W. D. Billings. 1960. The annual carbohydrate cycle of alpine plants as related to growth. *Am. J. Bot.* 47, 594–598.

——— 1961. Comparative physiological ecology of arctic and alpine populations of *Oxyria digyna. Ecol. Monograph* 31, 1–29.

Muc, M. 1972. Vascular plant production in the sedge meadows of the Truelove lowland. In *Devon Island IBP Project, High Arctic Ecosystem, Project Rept. 1970–71*, L. C. Bliss, ed. pp. 113–145. Edmonton, Canada: Department of Botany, University of Alberta.

Sørensen, T. 1941. Temperature relations and phenology of the northeast Greenland flowering plants. *Medd. Grønland* 125(9), 1–305.

Svoboda, J. 1972. Vascular plant productivity studies of raised beach ridges (semi-polar desert) in the Truelove lowland. In *Devon Island IBP Project, High Arctic Ecosystem, Project Rept. 1970–71*, L. C. Bliss, ed. pp. 146–184. Edmonton, Canada: Department of Botany, University of Alberta.

4.2 Phenological Studies in the Mojave Desert at Rock Valley (Nevada Test Site)[1]

Thomas L. Ackerman and Sam A. Bamberg

Laboratory of Nuclear Medicine and Radiation Biology
University of California, Los Angeles
Mercury, Nevada, 89023 USA

1 Introduction

Desert regions have an environment characterized by low moisture that is irregularly distributed in space and time, temperatures with large diurnal and seasonal ranges; and periodic strong winds. Plants in deserts have adapted to these conditions by a rapid response when conditions are favorable and to long periods of dormancy or inactivity under poor conditions. Thus they have a phenology that is closely related to rainfall and temperature and, in particular, to soil moisture. Our purpose is to determine the influence of environmental factors on phenology of desert shrubs in an area of the Mojave Desert.

Phenology of desert plants was studied at Rock Valley in the northern Mojave Desert (Shreve, 1942), starting in 1968 and continuing to the present. Rock Valley is one of many sites selected to relate phenology over a wide area of the Nevada Test Site (Wallace and Romney, 1972). In 1971 Rock Valley was set up as validation site as a part of the IBP Desert Biome effort.

Specific studies on phenology in desert plant species or on plant phenophases are rare. Oechel *et al.* (1972) discussed node formation and elongation of shoots in *Larrea tridentata* in the Colorado Desert where there is no "dormant" period and plants respond to periods of high water-tissue potential at any time of year. Tueller *et al.* (1972) have presented a numerical analysis of phenology in Great Basin shrubs and grasses based on a technique suggested by West and Wein (1971). Other phenological work at the Nevada Test Site has been done by Beatley (1967, 1969).

[1] Research supported by Contract AT(04-1) GEN-12, U.S. Atomic Energy Commission; and as a part of the US/IBP Desert Biome, supported in part by National Science Foundation Grant No. GB-15886.

KEYWORDS: Phenology; Mojave Desert; US/IBP Desert Biome; desert shrubs; annuals; air-soil temperatures; rainfall; soil moisture; dormancy; flowering; species response; productivity.

2 Location and Characteristics of the Study Area

Rock Valley is located in the southwest portion of the Nevada Test Site of the U.S. Atomic Energy Commission; it is 130 km northwest of Las Vegas, Nevada, and 20 km west of Mercury, Nevada. The valley is more or less surrounded by low mountains of the Specter Range (1400 m) and Skull Mountain (1830 m), and drains to the southwest into Amargosa Valley. The elevation of the study area is 1006 to 1037 m with a northwest-facing slope of 1% to 3%. Soils of the study area are derived mostly from highly calcareous alluvium from the northwest side of the Specter Range, which is limestone and dolomite of middle Cambrian age, and partly from volcanic basalt and tuff of Tertiary age (Sargent and Stewart, 1971). The soil surface has a well-developed desert pavement that is underlain by a massive 30- to 60-cm deep caliche hardpan layer which stops root penetration (Romney *et al.*, in preparation).

The northern Mojave Desert has a more "continental" climate with greater extremes of temperature than in its more southern part. Temperatures range from −15°C (5°F) in winter to 47°C (117°F) as a summer maximum.

In this desert the plant responses are largely determined by the amount and intensity of rainfall and favorable temperatures for growth. The rains here are most

Table 1. Plant species measured for phenology and vegetation characteristics at Rock Valley. (From Bamberg *et al.*, in preparation)

Species	Desert Affinity	Density, no./ha	Relative Abundance, %	Relative Plant Cover, %
Acamptopappus shockleyi (goldenhead)	Mojave	37	0.4	0.2
Ambrosia dumosa (white bur sage)	Mojave	2561	28.8	15.9
Atriplex confertifolia (shadscale)	Great Basin (mostly)	103	1.2	1.1
Coleogyne ramosissima (blackbrush)	Transitional	20	0.2	0.5
Ephedra nevadensis (mormon tea)	Both	754	8.5	7.9
Ceratoides lanata (winterfat)	Great Basin (mostly)	449	5.0	1.3
Grayia spinosa (spiny hop sage)	Great Basin (mostly)	1149	12.9	6.0
Krameria parvifolia (range ratany)	Mojave	1417	15.9	18.7
Larrea tridentata (creosote bush)	Mojave	959	10.8	19.6
Lycium andersonii (wolfberry)	Mojave	725	8.1	18.4
Lycium pallidum (box-thorn)	Mojave	530	6.0	10.3
Total for 11 species			97.8	99.8
Total for 10 species			97.4	99.6

predictable from November to March, but the amount and time in this period vary from year to year. Rains at this time are widespread and of relatively light intensity and mostly penetrate the soil, producing little runoff. Summer thunderstorms are infrequent and local and of such intensity that much runoff is produced. The average yearly rainfall for eight years (1964 to 1971) at Rock Valley is 13.9 cm (5.47 in.).

Vegetation in the area is a *Larrea-Ambrosia* association with floristic elements of both the Mojave and Great Basin deserts present (Table 1). *Coleogyne* is a transitional plant in that its distribution is intermediate between both deserts (Beatley, 1969), and occurs little in the Rock Valley site. The vegetation of the area is rarely over 1 m high, and consists of widely spaced individuals or clumps of plants.

3 Methods

Phenological and environmental measurements were made once a week in the spring and early summer when plants are most active, and every two to four weeks the rest of the year. For this study observations were made for 11 species (Table 1). In 1971 observations were made on 10 species only; *Acamptopappus* was eliminated. Three recent nomenclature changes are used here: *Ambrosia dumosa* (Gray) Payne was *Franseria dumosa* Gray, *Larrea tridentata* (Ses and Moc. ex DC.) Cov., (Beatley, 1971) was *L. divaricata* Cav. and *Eurotia lanata* (Pursh) Moq. has just been changed to *Ceratoides lanata* (Pursh) J. T. Howell, (Howell, 1971). Nomenclature of the species follows Munz and Keck (1959) and Munz (1968), except for *Ceratoides* and *Larrea*. Recorded for each species was the date of the "first" observed events of leaf bud, leaf, flower bud, flower, fruit, seed fall, leaf fall, and dormancy. Within the time of these events there is a great variation from shrub to shrub. Dormancy is difficult to determine for some deciduous shrubs which retain a few leaves for a short while after becoming dormant.

Environmental data collected at time of each visit included rainfall, using a totaling rain gauge and maximum and minimum air temperatures using Taylor instruments at 30 cm above ground measured since the previous phenologial observation. Soil temperature was measured only for 1970 and 1971 at time of each visit at 15-, 30-, and 45-cm depths, using Fenwall KA31L4 thermistors. These temperature measurements indicate only trends and do not show the range needed for better interpretation. A computer program to better analyze these data is planned. Given in this report are only inferences from those data.

4 Results and Discussions

Seasonal Effects on Phenology and Productivity

The phenological and environmental data for Rock Valley for the years 1968 to 1970 are given in Wallace and Romney (1972, pp. 284–289). The data for 1971 are presented in Figs. 1 to 4.

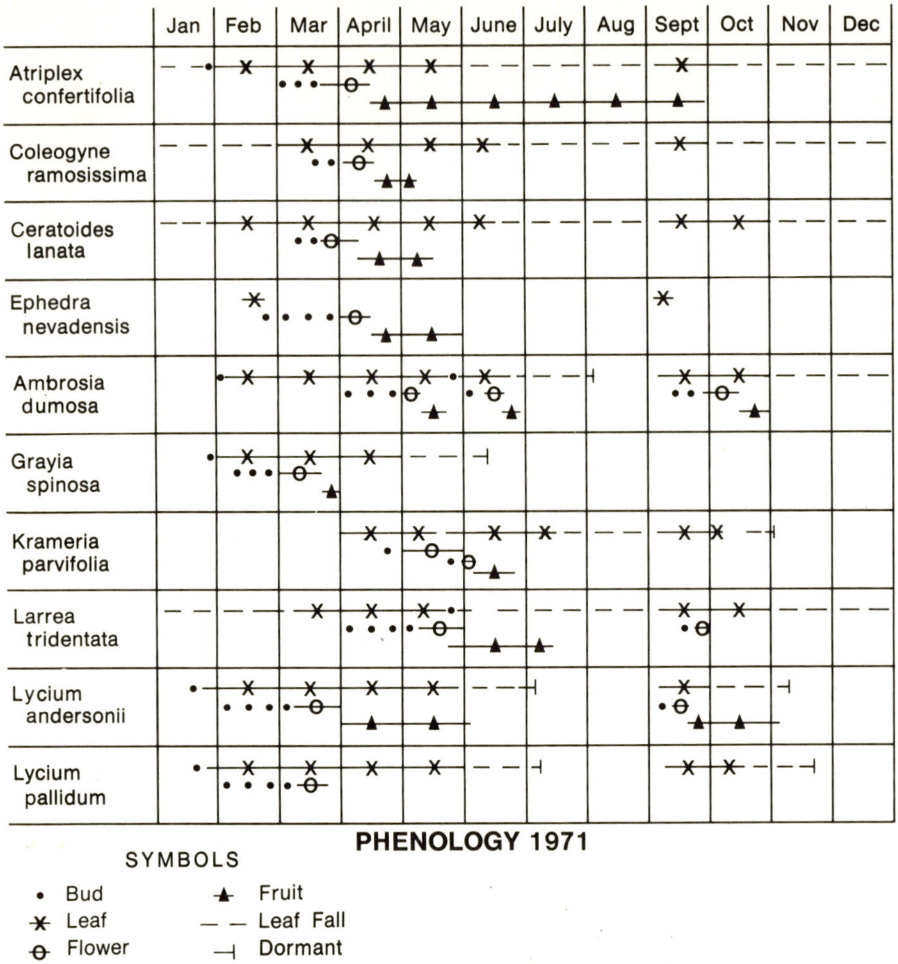

Fig. 1. Phenology 1971.

Comparison of rainfall available for the spring season and summer rain events (Table 2) and biomass production (Table 3) for the four years shows that 1969, with its higher rainfall, had a somewhat better production This was also borne out by field observations, although some species had the same production as in other years. Production was the least in 1971 when there was the lowest rainfall for the period under consideration. A summary of seasonal effects on phenology is given here.

Before the spring season of 1968 precipitation totaling 8.7 cm (3.5 in.) fell, so that shrubs responded with a fair amount of new growth. In addition, fruit production in 1968 was relatively high for *Lycium andersonii*, medium for *Larrea*, and low for *Ephedra*, *Ambrosia*, and *Krameria*. Little rain fell after February, so there was probably insufficient moisture for later flowering species, such as *Ambrosia* and *Krameria*, to produce much flowering and fruiting. The many annuals, which

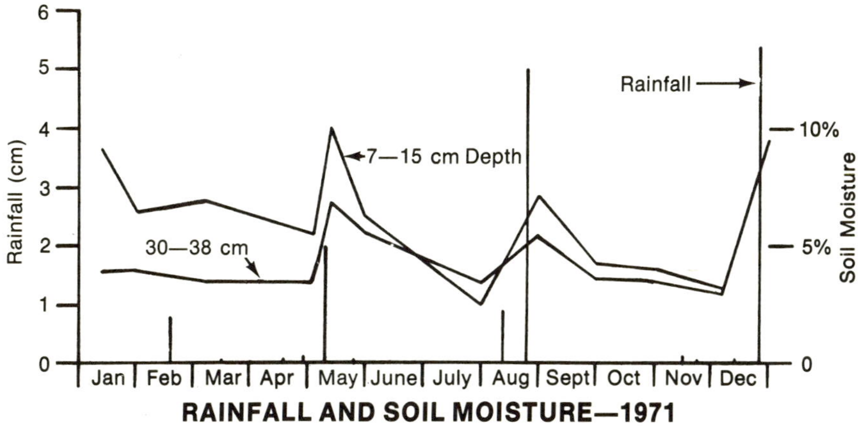

Fig. 2. Rainfall and soil moisture 1971.

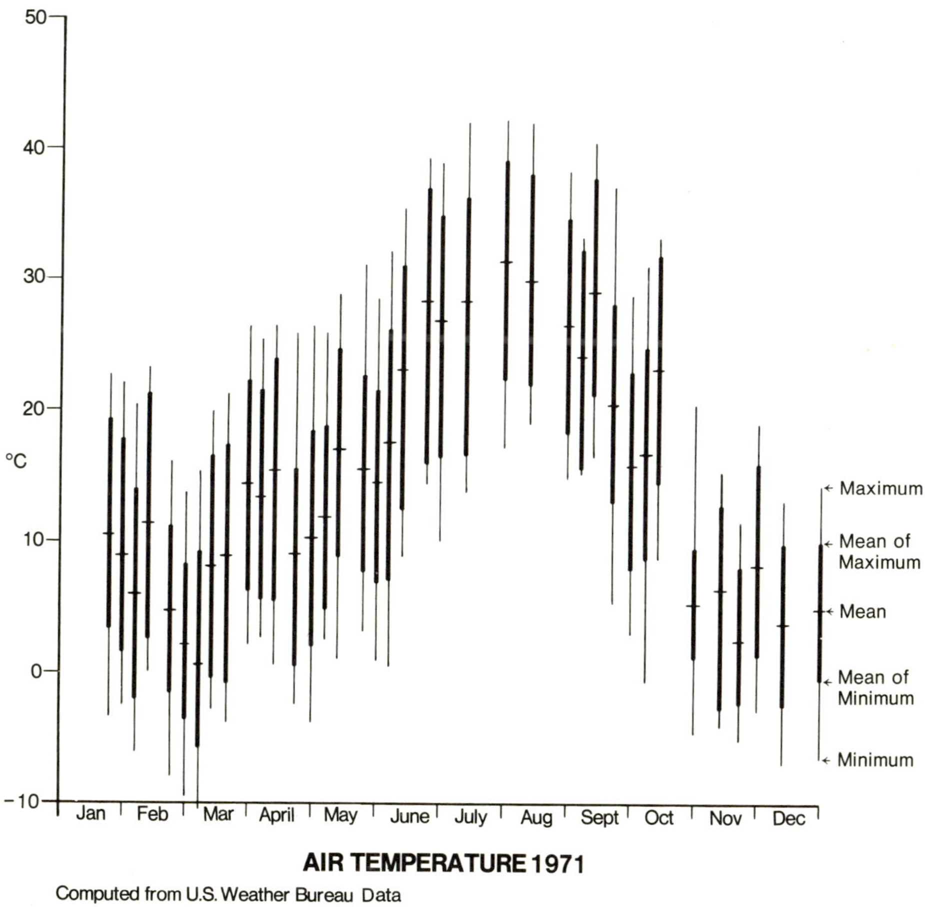

Fig. 3. Air temperature 1971.

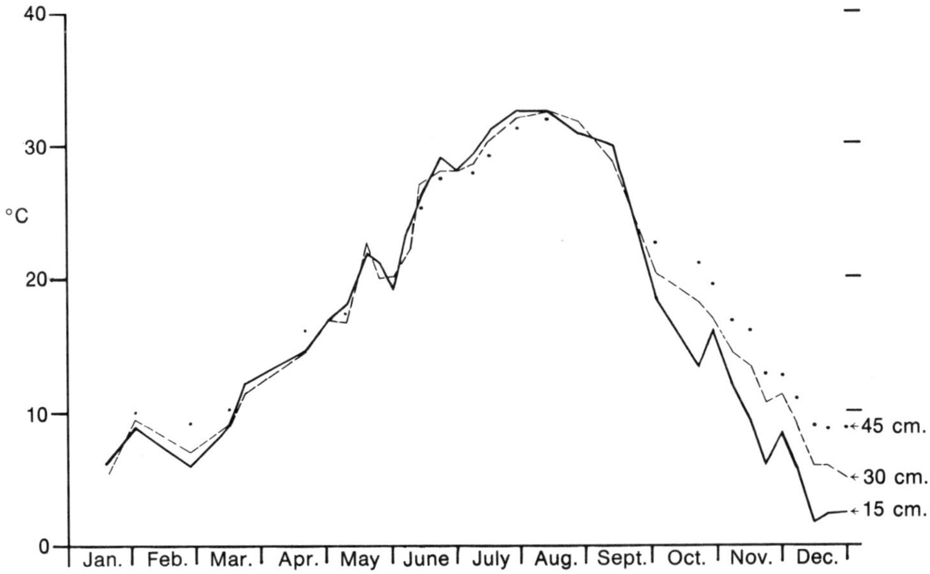

Fig. 4. Soil temperature 1971.

germinated after rains in August and November 1967, probably had good survival after the late February rains. Annuals had a production of 260 kg/ha in 1968 (Wallace and Romney, 1972) compared to 4.8 kg/ha for 1971.

In 1969 an unusually high amount of rain, 21.3 cm (8.4 in.), fell between mid-January and mid-March, resulting in a high density of annuals of medium height, probably due to a late germination in early February. New growth and flowering

Table 2. Rainfall at Rock Valley on a seasonal basis.

Time interval	Year	Amount in cm	Equivalent in in.
Late November to mid-February	1968	8.8	3.45
July		5.2	2.06
Total for year		15.7	6.17
Mid-January to mid-March	1969	21.3	8.40
Total for year		25.0	9.86
November	1970	1.9	0.76
Mid-February to early March		6.4	2.50
Total for year		11.7	4.60
Late November to December	1971	4.6	1.80
August		2.5–5.1	1–2
Total for year		13.5	5.29
Average rainfall for 4 years		16.5	6.48

Table 3. Plant biomass of Rock Valley.

	New leaves plus new stems, kg/ha			New leaves, kg/ha		
	1969[a]	1970[a]	1971[b]	1968[c]	1970[a]	1971[b]
Ambrosia dumosa		26.4	15.0	19.4	21.6	13.2
Grayia spinosa	11.7	11.7	5.7	4.3	9.4	5.4
Krameria parvifolia		109.9	115.7	72.0		96.6
Larrea tridentata			54.3	158.6	73.6	1.0
Lycium andersonii	117.8	53.9	42.8	89.7		41.9
Lycium pallidum	9.4	4.2	2.0	1.7		2.0
Annuals, whole plant			4.8	260.0		

[a] From Romney, unpublished.
[b] From Bamberg *et al.*, in preparation.
[c] From Wallace and Romney, 1972.

on shrubs was the highest during the years of record, and some shrubs almost doubled in biomass. *Ephedra* produced an unusually heavy crop of staminate and pistillale cones. Most plants became dormant later than in other years of our record.

After a rain of 1.9 cm (0.7 in.) in November 1969, a few annuals germinated and after rains from mid-February to March, 1970, totaling 6.3 cm (2.5 in.), many very small annuals germinated; thus two sizes of annuals were produced in this season. Most shrubs also responded to the February rains but produced very little new growth. A temperature of -6°C (21°F) on the night of April 28 froze shrubs at lower elevation of the study area and killed some new stems and all flower buds of *Ambrosia*, *Krameria*, and *Acamptopappus*; and some stems of *Grayia*. No fruit was produced by *Ephedra*, *Lycium andersonii*, and only a few by *Lycium pallidum*.

Before the spring growth season of 1971 there were two rains in late November through December 1970, totaling 4.5 cm (1.8 in.). Both *Lycium* species broke dormancy during temperatures of 25° to 29°C (77° to 83°F) on January 18, the earliest recorded for the four years, however, shrubs produced little growth and few flowers and fruit. This was especially true for *Ambrosia*, *Krameria*, and *Lycium andersonii*. Most flower buds on *Ambrosia* and *Krameria* withered and others set fruit only after a May 8 rainfall of 2.0 cm (0.79 in.). *Lycium pallidum* produced moderate amounts of flowers, but was so heavily infested by an insect larvae, which ate approximately 40 to 80% of leaves and most flower buds, that practically no fruit was set and stem production was low. *Ephedra* produced a heavy set of cones in Rock Valley and also in Mercury Valley which had rains totaling only 1.2 cm (0.48 in.). Reasons for this heavy production are not known, since in most years, except for 1969, *Ephedra* produced very few cones.

1971 was the first year phenology was recorded for annuals as a part of the IBP effort. Annuals germinated after winter rains but were inactive during subfreezing night temperatures and developed only shallow root systems. During the unusually high temperatures in January the top 7.6 cm (3 in.) of soil quickly dried out and probably resulted in high transpiration rates in the annuals. Even though these temperatures were favorable for growth, annuals were inactive or withered and died. After a mid-February rain of 0.8 cm (0.3 in.), stems of annuals under shrubs in the more favorable habitats elongated and produced new leaves. These eventually flowered and set fruit. Most of the annuals in the more exposed sites did not survive, as indicated by a survival rate of only 35% for the Desert Biome site.

Since soil moisture was so limited, phenology for any single species of annual varied tremendously from microsite to microsite. The same species was flowering, setting fruit, and dying at the same time at different sites because of differences in the microclimate.

Phenological observations made on annuals in the northern Mojave Desert near the northern end of Death Valley and Saline and Eureka valleys indicate that both productivity and survival are best after adequate spring moisture; in fact, annuals may survive into mid June with late spring rains.

Effects of Winter and Early Spring Rains

Rains during the winter and early spring are most predictable and widespread. Rains in late November may or may not produce visible growth. There was no plant response after rains in late November 1967, although plant response the next spring was good. After rains in late November 1970, however, leaf buds appeared on both *Lycium* species and *Acamptopappus*, but remained closed until temperatures became favorable in January. The amount of rain during the winter or early spring is probably more important than the time it comes during this period, as winter rains have their effect the following spring. Because this is principally a winter rainfall area, these widespread, gentle rains are most important for plant growth and productivity.

Low moisture or low temperatures may result in a delayed start of the growth season, i.e., phenology of each species is determined by the right moisture and temperature range. Usually from late November through early January the night temperatures in the area are near or below freezing, so moisture from rains is stored in the soil until favorable conditions permit bud burst and leafing out. For example, in 1970 leaf bud initiation was delayed until late February when rains first came; also in 1969 leaf formation was delayed until late February after a mean temperature of 5.6° C (42° F) as compared to 10°C (50°F) for a comparable period during 1968.

Rains from March to June are rare in Rock Valley; however, observations (Bamberg, unpublished) in other parts of the Mojave Desert indicate that rains at this period produce renewed growth and flowering in some species.

Effects of Summer Rains

Summer rains are very infrequent and of such intensity that most water runs off and little moisture penetrates the soil to become available for plant growth except in drainages. The effects of these rains vary, depending again on amounts, penetration in soil, and subsequent temperatures. For example, after two rains in July 1968 totaling 5.2 cm (2.1 in.), all the species under observation except *Grayia* produced new leaves and stems. *Ceratoides*, *Ambrosia*, *Krameria*, and *Larrea* flowered and fruited sparsely in the wetter drainages. One species, *Acamptopappus*, responded with only a few leaves.

In 1971 a late August thunderstorm varied from 2.5 to 4.9 cm (1.0 to 1.9 in.) over the IBP site. The effects of this rain on plant growth and productivity were less than the one in 1968. Following this rain, germination was observed in two species of summer annuals and in a few *Larrea* and *Ambrosia*. Late summer rains in other years have permitted massive germination of *Larrea*, *Ambrosia*, and *Atriplex* in other parts of the Mojave Desert.

Actual growth is usually for a short two- to three-week period after summer rains. Plants quickly become inactive and then deciduous species start dropping leaves. High evaporation and transpiration rates of the shrubs during daily air temperatures of 35° to 46°C (95° to 115°F) during August 1968 were probably responsible for the short growth period. By early September 1968 deciduous shrubs started to drop leaves, indicating that soil moisture was probably depleted.

In July and August 1970, several rains of about 1 cm each failed to bring plants out of dormancy or inactivity. Observations in the summers of 1969 and 1972 indicate that summer rains of less than 2 cm had little effect on shrubs. The data suggest that summer rains probably have to be over 2.0 cm (0.7 in.) before they have any appreciable effect on shrubs.

Differential Species Response

Each species has a temperature range in which it will grow if enough soil moisture is present. Species that begin growth the earliest, usually in late January or February when minimum temperatures are around 3°C (37°F) and maximum temperatures are about 20°C (68°F) are *Atriplex*, *Acamptopappus*, *Ceratoides*, *Grayia* and both *Lycium* species. Both *Lycium* species break dormancy the earliest of any plant when night temperatures are around freezing and maximum day temperatures average 15°C (60°F).

In December 1967, *Acamptopappus* was the only species to produce a few leaves after the late November rains, and after a winter inactive period renewed growth in the spring. Summer rains also produced few leaves in this species, showing little tolerance for high temperatures.

In June and July, as soil moisture falls below 4% and air and soil temperatures increase above 30°C (86°F), *Grayia* and both *Lycium* spp. are the first species to start shedding leaves and become dormant. Wallace and Romney (1972), in the laboratory, found that *Grayia* had much better root and stem growth at 16°C than at 21°C and 28°C root temperatures. *Lycium andersonii* grew well from 16° to 28°C, indicating that this species has a wide temperature tolerance. *Coleogyne* grew better at 21°C than at either 16° or 28°C, showing a narrow temperature tolerance. They found that *Ceratoides* and *Atriplex* respond better to lower soil temperatures, but also that both species grow best when root temperatures are 16° to 24°C and air temperature is 32°C—i.e., with a large differential between air and soil temperatures—but grow less when both air and soil temperatures are the same. These conditions occur in early spring with increasing air temperatures while soil temperatures are still low from the winter period.

Krameria is the last to come out of dormancy in the spring when maximum air temperatures are around 27°C (80°F) and minimum air temperatures around 6°C (43°F), and is never dormant in the summer no matter how dry the soil becomes. Wallace and Romney (1972) found that *Krameria* is still active when soil moisture is below 3%. This species maintains active photosynthesis under natural conditions during the summer at temperatures of 45°C (113°F) and plant tissue water potential of −73 bars (Bamberg and Wallace, unpublished). *Larrea* can also carry on photosynthesis at 50°C (122°F) when soil moisture is not critical, i.e., less than −40 bars.

Plants which grow and flower during the summer with adequate moisture include *Larrea*, *Ambrosia*, and *Krameria*. *Larrea* is the last to flower in late spring, and all three species continue activity throughout the summer as long as moisture is adequate.

Dormancy

For this study, plants which lose all their leaves and exhibit no activity are considered dormant. In contrast, those plants which never lose all their leaves throughout the year are considered evergreens. Evergreens are termed inactive when they exhibit no obvious growth during the summer with low soil moisture and high temperature periods, as well as during the winter.

Dormancy in desert shrubs is a complex phenomenon involving physiological changes, drought, and temperature responses. As the growing season progresses in early summer, soil moisture is depleted by rapid evaporation and transpiration during increased air and soil temperatures. At this time deciduous shrubs start to shed their leaves and soon go dormant, so that an interaction of high temperatures and low soil moisture probably results in dormancy in the summer.

This interaction of high temperature and low soil moisture as a cause of dormancy is illustrated by events following a heavy summer rain when all dormant or inactive species leaf out, except *Grayia*. Shrubs rapidly put out new growth; then after a short period start shedding their leaves and become dormant again when soil moisture levels decrease below 4%. However, daytime summer temperatures of 35° to 46°C (95° to 115°F) are probably too extreme for such species as *Lycium*, *Ceratoides*, *Ephedra*, *Coleogyne*, and *Atriplex* to remain active, except possibly during morning hours between 5 and 9 AM when temperatures are lower. This is indicated by our glasshouse studies where these species became inactive and did not grow, even when watered, at air temperatures of around 40°C (104°F). When moved to lower air temperatures 31°C (88°F), they again produced new growth.

Water temperature of summer rains is much lower than daytime air temperatures, and percolating water as well as evaporation lowers soil temperatures so that plant growth can resume. As mentioned previously, Wallace and Romney (1972) found that *Atriplex* and *Ceratoides* grow best when root temperatures are lower than air temperatures. This is the situation found both after a summer rain and in early spring. From our watering experiments on the desert, a continuous soil temperature recorder shows that irrigation water in the summer does lower soil temperature for a short time. In Rock Valley, however, soil temperature is measured only every two weeks during the summer, and this effect of lowered soil temperatures is not evident, since in the interim the soil temperatures have again increased. The increasing soil temperature and decreasing soil moisture probably account for the growth ceasing only two to three weeks after a summer rain.

Grayia always becomes dormant when daytime air temperatures are from 35° to 40°C (95° to 104°F) and does not come out of dormancy, even with summer rains, until the following spring. Wallace and Romney (1972) found that *Grayia* will break dormancy only after a low temperature period, or after an application of the hormone gibberellin. Dormancy in *Grayia* is induced by an internal physiological mechanism triggered by either heat or photoperiod. This dormancy may explain the restriction of *Grayia* to the northern Mojave Desert and Great Basin where there is a chill period in the winter to break dormancy. *Krameria* only becomes dormant in the fall when temperatures fall to around 4°C (40°F) at night.

Besides having a dormant period in the summer, some species have a later dormant period during the winter. In 1968 *Ambrosia* was green until the end of December when temperatures of −15°C (5°F) froze all leaves. From plots in other areas we know that *Lycium andersonii* leaves will freeze and the shrubs do not leaf out soon again, even with adequate soil moisture.

Observations on dormancy in other species showed *Acamptopappus* went through the winter of 1967 with no frost damage. *Ephedra* has no leaves so it is hard to tell when it is physiologically active, but the plant grows new stems after summer rains and has a stem color change from green to brown during periods of inactivity.

Shrubs that never have become dormant in our area and put out new growth after spring, summer, and fall rains are *Larrea*, *Coleogyne*, *Ceratoides* and *Atriplex*. During the dry summer with high temperatures these shrubs lose most of their leaves and keep only those at the very ends of the branches. During this period they are probably inactive physiologically. In extreme conditions during dry years either ends of stems or complete stems die in these species.

Flowering and Fruiting

Shrubs which have flowered only in the spring in Rock Valley and in other areas where we have plots, and are probably determined by photoperiodism, are *Ephedra*, *Coleogyne*, *Atriplex*, and *Grayia*. Species that seem to be able to flower anytime during the spring, summer, or fall, if they get enough rain, are *Larrea*, *Ceratoides*, and *Ambrosia*. *Krameria* will flower anytime during its shorter growing period. *Lycium andersonii* has flowered both in the spring and fall.

The amount of moisture in the soil at the start of the spring growth period determines initial flowering response. Lack of moisture at this time results in some species producing few or no flowers. For example, in 1970 *Lycium andersonii* produced no flowers and *Ephedra* no cones, while *Lycium pallidum* had a few flowers and fruit. Again in the spring of 1971 most shrubs produced no flowers or fruit, although two species, *Lycium andersonii* and *Ambrosia*, had a few flowers and fruit. In contrast, adequate early season rains in January and February 1969 allowed all shrub species to flower and set abundant fruits.

Moisture later in the spring growing season can result in either renewed growth or flowering with fruit set. In 1971, after a May rain of 2.2 cm (0.86 in.), several shrubs responded, *Ambrosia* flowered again and set fruit, including some individual plants that were barren earlier; *Krameria* reflowered and produced fruit for the first time; and *Lycium pallidum* produced a few flowers but no fruit.

Acknowledgments

The authors thank Dr. Evan Romney for the use of data collected under the broad phenological observation studies at the Nevada Test Site and for unpublished productivity figures at Rock Valley. James Childress provided soil moisture data for the entire period and environmental data for part of 1971. Bernardo Maza developed a computer program for weather data summarization in connection with this study.

References

Beatley, Janice C. 1967. Survival of winter annuals in the northern Mojave Desert. *Ecology* 48, 745–750.

——— 1969. Vascular plants of the Nevada Test Site, Nellis Air Force Range, and Ash Meadows (northern Mojave and southern Great Basin deserts, south-central Nevada). *US AEC Rept. UCLA No. 12–705*. 122 pp.

——— 1971. Vascular plants of Ash Meadows, Nevada. *US AEC Rept. UCLA No. 12–845*. 59 pp.

Howell, J. T. 1971. A new name for 'winter fat.' *Wasmann J. Biol.* 29, 105.

Munz, P. A., and D. D. Keck. 1959. *A California Flora*. 1681 pp. Berkeley: University of California Press.

——— 1968. *Supplement to a California Flora*. 224 pp. Berkeley: University of California Press.

Oechel, W. C., B. R. Strain, and W. R. Odening. 1972. Tissue water potential, photosynthesis, 14_C-labeled photosynthate utilization, and growth in the desert shrub, *Larrea divaricata* Cav. *Ecol. Monograph.* 42(2), 127–141.

Sargent, K. A., and J. H. Stewart. 1971. *Geologic Map of the Specter Range NW Quadrangle Nye County, Nevada. Map GQ-884.* Washington, D.C.: U.S. Geological Survey.

Shreve, F. 1942. The desert vegetation of North America. *Bot. Rev.* 8, 195–246.

Tueller, P. T., A. D. Bruner, and J. B. Davis. 1972. Ecology of Hot Creek Valley—Vegetation and soil response to underground detonation. *Univ. Nevada Agr. Exp. Sta. Rept. NVO-409–1.* 131 pp.

Wallace, A., and E. Romney, 1972. Radioecology and ecophysiology of desert plants at the Nevada Test Site. *TID-25954*. 439 pp. Oak Ridge, Tenn. US AEC Technical Information Center.

West, N. F., and R. W. Wein. 1971. A plant phenological index technique. *BioScience* 21, 116–117.

4.3 Phenological Studies and Modeling in Grasslands[1]

Norman French and Ronald H. Sauer

Natural Resource Ecology Laboratory
Colorado State University
Ft. Collins, Colorado 80521, USA

1 Grassland Types

Energy in the form of solar radiation is the basic driving force of the vegetation system. A small fraction of this energy is utilized in photosynthesis. Much is utilized in changing the state of water in the system, that is, in evapotranspiration. Since evapotranspiration depends upon temperature and daylength, there is a gradient of evapotranspiration decreasing from south to north in the contiguous 48 United States with some minor modifications according to changes in elevation.

In the eastern United States precipitation exceeds the potential evapotranspiration, while much of the western United States has a deficit. The line where excess becomes deficit approximates the western boundary of the tallgrass prairie (Fig. 1). Subtle interactions of climate and the control of vegetation are much less apparent although equally important. We seek a reliable key to these interactions in order to construct better phenological models that can be incorporated into the grassland ecosystem models.

The International Biological Program is utilizing the variability in this complex pattern of interaction in an effort to better understand ecosystem functions. The research design of the US-IBP Grassland Biome includes detailed investigations in a variety of grassland types (Fig. 1). These characteristic types have different species composition, climate, and seasonality. They vary in growth response and quantity of production.

Grasslands in general are characterized by relatively dry growing seasons and/or high evapotranspiration. The North American prairie extends eastward in mid-continent, as do dry westerly winds that prevail during the summer season (Borchert, 1950). Nonetheless, the tallgrass prairie extending eastward is characterized by

[1] Supported by National Science Foundation grants GB 13096 and GB 31862X to the Grassland Biome US/IBP.

KEYWORDS: Phenology; phenophases; model; grassland; driving variables; IBP-Grassland; North America.

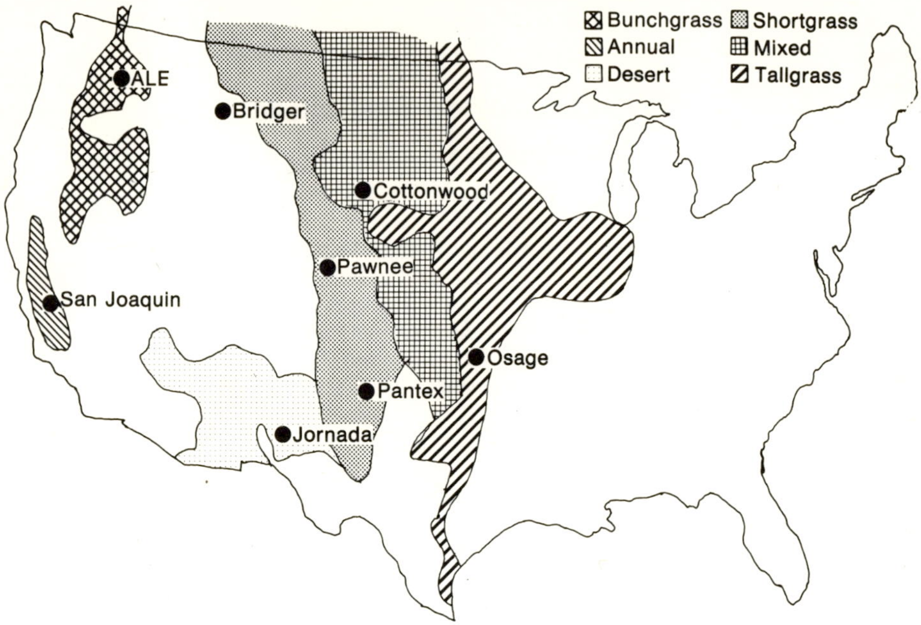

Fig. 1. Outline map of the United States showing major grassland types and the sites of US/IBP grassland studies.

the highest total precipitation (90 cm), generally well distributed during the growing season. Westward in the mixed prairie zone the total precipitation is less (40 cm). Shortgrass is characterized by even less precipitation (30 cm). Much of the moisture in all of this prairie region in the summer months arrives in the form of thunder showers, and is erratically distributed.

2 The Role of Phenology in Comparative Grassland Studies

This mosaic of climatic parameters has selected community types which differ in their climatic responses. In the tallgrass prairie, the biomass of the live vegetation is maximum early in the season. Above-ground plant biomass is far less in the shortgrass prairie, even though the plants may grow for a longer period of the potential growing season. Growth in some desert grassland, dependent on summer rains, begins late and reaches a maximum at a later date than in the other grasslands. Thus there are patterns of seasonality that influence growth and production of vegetation in these grassland types; comparison across grassland types requires a key for meaningful study of progress during the growing season in different locations. Such a key is the relationship between abiotic and biotic variables. We want to demonstrate this with a few facts concerning the vegetation period and some functional reactions and interactions among the abiotic compartments of the ecosystem.

Initiation of the growing season varies according to the climatic type. The earliest initiation of growth occurs in the annual grassland, followed by the Pacific Northwest bunchgrass type, then progressing from south to north in the continental prairies, and beginning last in high mountain grassland.

The length of the growing season, defined as the period when a 15-day moving average of temperature is above 15°C, decreases from south to north in the region of the tallgrass to shortgrass prairies. The southernmost sites in this region have the longest growing season (270 to 340 days). The Mediterranean climate of the annual grassland provides a very long potential growing season, but has limited moisture.

The progression of temperature and moisture conditions influences vegetative growth (Coupland, 1950). In the tallgrass prairie, with long growing season and ample moisture, net primary production continues through at least 90% of the potential growing season. This is also generally true for the other mid-continent prairie sites. The main difference between sites is the total biomass obtained. In other types of grassland, moisture becomes limiting earlier so that primary production rate reaches a peak toward the middle of the growing season and then declines. Variations in the erratic precipitation regime during any particular season are readily reflected in productivity at the drier sites (Lewis, 1971, p. 339). Thus the shortgrass prairie and desert grassland react as a pulsed system, the vegetative growth responding to increments of rainfall when and if they occur (Rauzi and Dobrenz, 1970).

Decomposition of organic materials in grassland sites appears predictable as a function of soil water and temperature. At the tallgrass prairie sites decomposition proceeds rapidly and may be nearly complete within 60 days. At the shortgrass prairie sites, decomposition proceeds at a comparable rate when conditions are optimal (Reuss, 1971, p. 136). When soil water in the surface layer is depleted, however, decomposition ceases. With a pulse of rainfall, the rate increases again.

Consumers often select plants according to developmental stage as well as for species. Herbivores may select young, rapidly growing material, and switch to different species as the stems mature and cure. Birds may depend on seeds, or may select insects which change their dietary preference according to stage of development. Thus the ramifications of phenology reverberate throughout the ecosystem.

3 Phenological Records in Grasslands

Initial efforts in the US/IBP Grassland Biome program included recording phenophases of major plant species for each quadrant that was sampled for determination of above-ground biomass. The results proved difficult to interpret because of infrequent observations, but showed interesting comparisons for the same species on grazed and ungrazed pastures. Grazed pastures show an accelerated development, probably due to increased isolation and the resulting higher surface temperatures during the early part of the growing season.

Current records of phenology include an estimate of the percentage of the plants in the phenophase in question to help evaluate intraspecific variation and phenophase progression. Observations were recorded three times each week. Sample results from natural shortgrass prairie in the spring of 1972 (Fig. 2) show three species with somewhat different developmental patterns: two early (represented by *Oxytropis sericea* and *Festuca octoflora*) and one late (*Sitanion hystrix*).

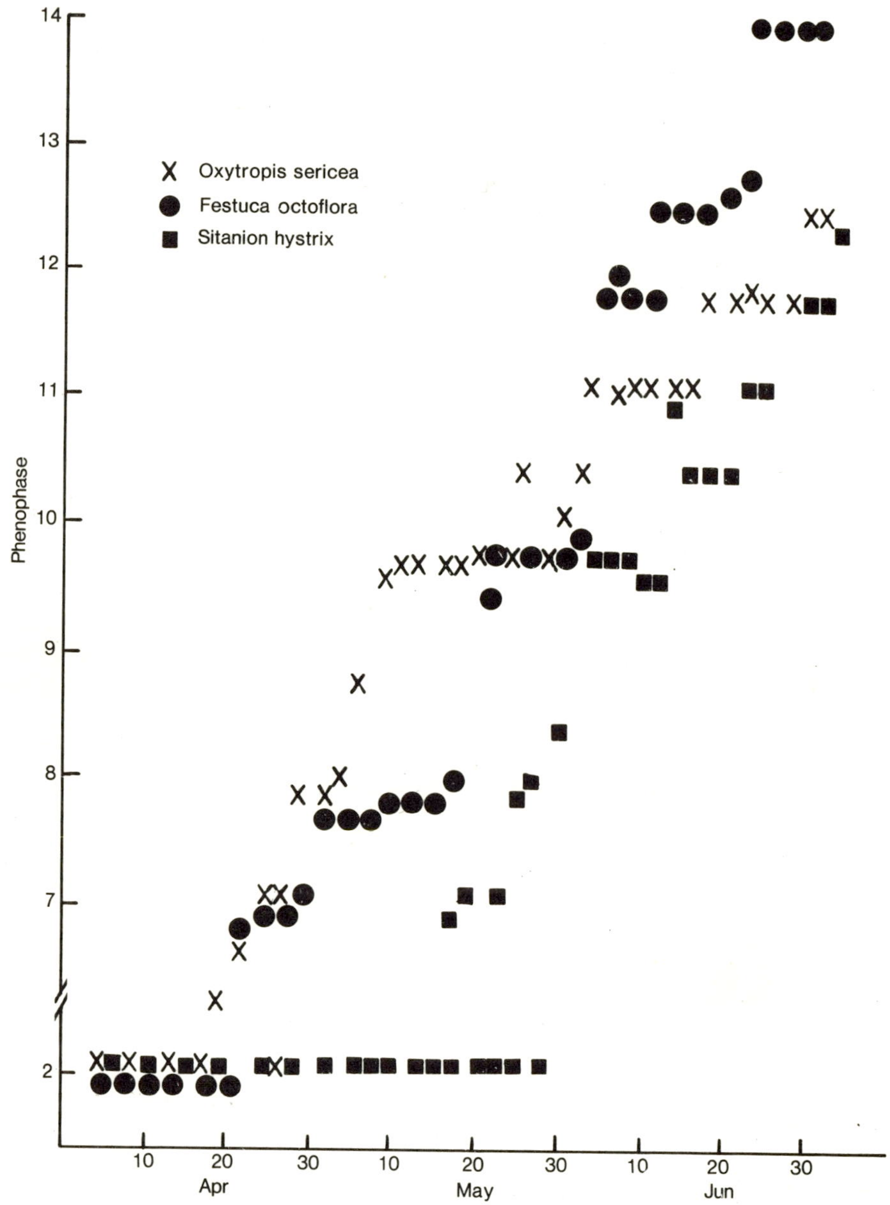

Fig. 2. Phenophase progression of three selected species of the shortgrass prairie (Pawnee site). Phenophases: *2* = early vegetative growth, *7* = bud stage, *10* = full bloom, *12* = ripe seed; for other phenophase codes see Table 2.

The majority of the plants that flower in the spring months follow the early pattern. Other species which flower later in the season are not represented here.

4 Biotic Response to Abiotic Variables

The influence of abiotic parameters on the developmental regime of the three groups of plants represented in Fig. 2 was investigated by multiple regression analysis. Each of the three sets of data in Fig. 2 was tested for correlation with precipitation, soil water, evaporation, soil temperature at −5 cm, soil temperature at −20 cm, and daylength. Precipitation was expressed as the cumulative rainfall through the season. Soil water was in centimeters, also expressed as a cumulative function. Evaporation was cumulative loss from an open pan. Temperatures were cumulative degrees above 40° F (4°C) using daily maximum temperatures. Daylength was the duration in minutes of the period between sunrise and sunset.

Multiple regression analysis indicated that the development (Table 2) of the early developing species (*Oxytropis sericea*) was best correlated with soil water and daylength. All the significant variation in the progression of phenophases was explained by these two variables, resulting in an r^2 value of 0.9729. In the other species showing an early but somewhat different rate of development (*Festuca octoflora*), the significant variability was explained by one variable, the soil temperature at −5 cm. The resulting r^2 value was 0.9754. In the series that showed a slower rate of development (*Sitanion hystrix*), the progression of phenophases was most highly correlated with the soil temperature at −20 cm and with soil water, resulting in a r^2 value of 0.9506.

The simple correlations among the different variables, both dependent and independent, are shown in Table 1. Variable number 1 (*Oxytropis sericea* phase) has highest correlation coefficients with variables 5 (soil water) and 10 (daylength). Similarly, variable 2 (*Festuca octoflora*) is highly correlated with variables, 8 (soil temperature −5 cm), 7 (air temperature), and 9 (evaporation, soil temperature −20 cm). Variable 3 (*Sitanion hystrix*) shows high correlation coefficients with variables 9 (soil temperature −20 cm), 7 (air temperature), and 8 (soil temperature −5 cm). However, there is such a high correlation between certain independent variables that the separate consideration of simple correlation is of questionable value. For example, variables 8 and 9 (the two soil temperatures) are very highly correlated with 7 (air temperature). Similarly, variable 6 (evaporation) is very highly correlated with variables 7, 8, and 9 (air temperature and the two soil temperatures).

Daylength, soil water, and soil temperature are the driving variables most closely associated with phenological development of plants on the shortgrass prairie. Additional refinements may provide more precise interpretation of such correlations. For example, may air moisture be a more accurate indicator of plant development than is pan evaporation? Furthermore, analysis conducted on results in different seasons, which would follow a different pattern of water and temperature regime, may clarify the roles of these factors for different species. Such basic information is essential for model development.

Table 1. Correlation coefficients between development of three plant species and six microclimatic variables.

Variable Name	Variable Number	1	2	3	4	5	6	7	8	9	10
Oxytropis sericea	1	1.000	0.932	0.852	0.805	0.980	0.962	0.941	0.948	0.927	0.978
Festuca octoflora	2		1.000	0.960	0.907	0.948	0.986	0.987	0.988	0.986	0.914
Sitanion hystrix	3			1.000	0.914	0.882	0.953	0.963	0.960	0.968	0.837
Precipitation	4				1.000	0.874	0.899	0.927	0.923	0.939	0.770
Soil Water	5					1.000	0.971	0.960	0.966	0.949	0.971
Evaporation	6						1.000	0.996	0.997	0.991	0.948
Air temperature	7							1.000	1.000	0.999	0.925
Soil temperature (−5 cm)	8								1.000	0.997	0.933
Soil temperature (−20 cm)	9									1.000	0.905
Daylength	10										1.000

5 Phenological Modeling of Grasslands

The phenological model developed in the grassland biome uses recent weather patterns simulated in the model to regulate the progression of phenological change from early vegetative through flowering and back to early vegetative phenophases, as defined in Table 2. Four suffrutescent perennial species and a cactus are modeled: *Bouteloua gracilis*, *Agropyron smithii*, *Sphaeralcea coccinea*, *Artemisia frigida*, and *Opuntia polyacantha*. Because flowering time varies with weather (Blaisdell, 1958; Sauer, 1971), the model was initially developed with flowering time a function of recent weather patterns rather than a function of a chronological time sequence. For example, in a seven-year period, *Stipa comata* had a range of 44 days in the appearance of flowering stalks (Blaisdell, 1958). The timing of the vegetative phenophases was simulated for use in the consumer models as an indication of digestibility. Earlier stages are more palatable. The yearly differences in weather patterns, generated stochastically by the abiotic section of the grassland model, now produce reasonable variations in the progression of phenophases through the growing season. A graph from the simulation model for *Boutelona gracilis* is shown in Fig. 3. The biomass did not reach a high value in 1972, perhaps because the 1972 weather was different than the simulated weather patterns driving the model.

Table 2. Definitions of 14 phenophases in the grassland model (Fig. 3 and 4).

1. Pre-emergence growth/winter dormancy
2. First visible growth
3. First leaves fully expanded
4. Middle leaves fully visible
5. First leaves senescent; middle leaves fully expanded
6. Late leaves fully expanded
7. Developing floral buds; middle-late vegetative
8. Mature floral buds; late vegetative
9. Floral buds and open flowers
10. Buds, flowers, and green fruit
11. Buds, flowers, green fruit, and ripe fruit
12. Green fruit and ripe fruit
13. Ripe fruit and dispersing seeds
14. Flowering-induced dormancy

The change from one vegetative stage to another is a function of a 20-day moving average of the product of insolation, maximum daily air temperature, and soil water condition. The 20-day moving average is used to reflect plant response to recent weather patterns. Such a delayed response seems more reasonable than a response to daily changes in weather. Temperature has been shown to influence plant development rate (Blaisdell, 1958; Sauer, 1971), presumably through its effect on metabolic rate. Daily air temperature maxima, rather than the average of the daily maximum and minimum temperatures, are used in the model. The importance of including insolation in the function was suggested by the correlation between lilac flowering time and a function of insolation (Caprio, 1971).

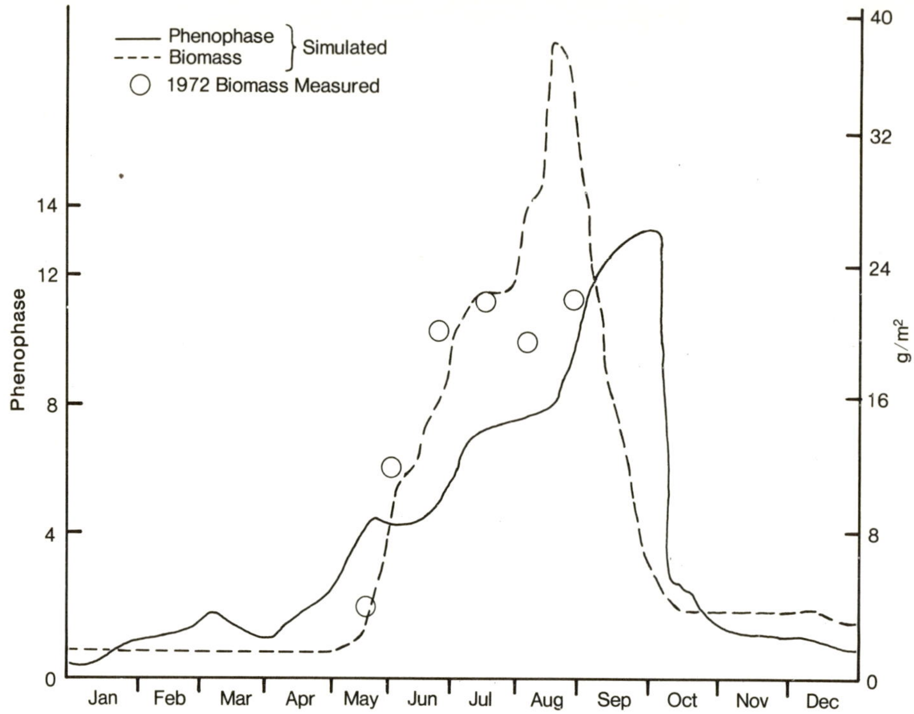

Fig. 3. Computer simulation for a one-year period of biomass dynamics and phenophase development of *Bouteloua gracilis*. Circles indicate measured biomass of this species in two areas in 1972.

When plant development in the model has progressed into the first reproductive stage (phenophase greater than 7.0), control of phenological progression is transferred to a different function. A cumulative sum, starting when the plant begins its reproductive phases, of the daily product of maximum temperature and insolation is used to determine the timing of the various reproductive phenological stages (Fig. 3). If the species has a particular photoperiod requirement, flowering is delayed until the necessary daylength has occurred. Once a plant has attained a reproductive mode, it continues on to seed maturation and dispersal. Hence a cumulative sum and not a moving average is used in the progression of the reproductive phenological stages. Flowering phenophases asymptotically approach stage 14.0, flowering-induced dormancy. End of season dormancy may or may not be induced by the flowering and fruiting process; this term is considered descriptive, not explanatory. If seeds have failed to mature by the onset of winter temperatures, the fruits are aborted and the plants return to phenophase 1.0.

Flowering-induced dormancy normally is broken when the 20-day moving average of the product of soil water conditions and daily maximum temperature decreases below a value that is preset for each species. Thus growth is resumed, assuming that conditions permit photosynthetic activity, when cool moist conditions again occur, as in the fall or winter. This approach to breaking dormancy and the progression of vegetative and reproductive phenophases is not intended to be mechanistic but rather a means of simulating biological response to abiotic variables. More realism is needed in this area.

The absence of growth during the winter months, here called winter dormancy, is simulated by preventing growth if soil temperatures are less than a minimum temperature set for each species. The bulk of a suffrutescent perennial species, when dormant, is below ground, and thus soil temperature of the rhizosphere is used as a control. Early season growth in an early spring can thus be simulated, as well as the cessation of growth in the event of a late winter or early spring snow.

The simulated development of a cool season grass, driven by the data recorded in 1972 for temperature, insolation, and soil water, compared with observed development of the cool season grass *Sitanion hystrix* and the annual grass *Festuca octoflora*, is shown in Fig. 4. The comparison of the simulation with the real data shows a two- to three-month discrepancy for the early development and a slightly early prediction for phenophases 10 to 12. *Festuca octoflora* completed its cycle in 1972 and, with continued favorable growing conditions, produced new vegetative growth. Such information will be used to refine further the parameters of the model.

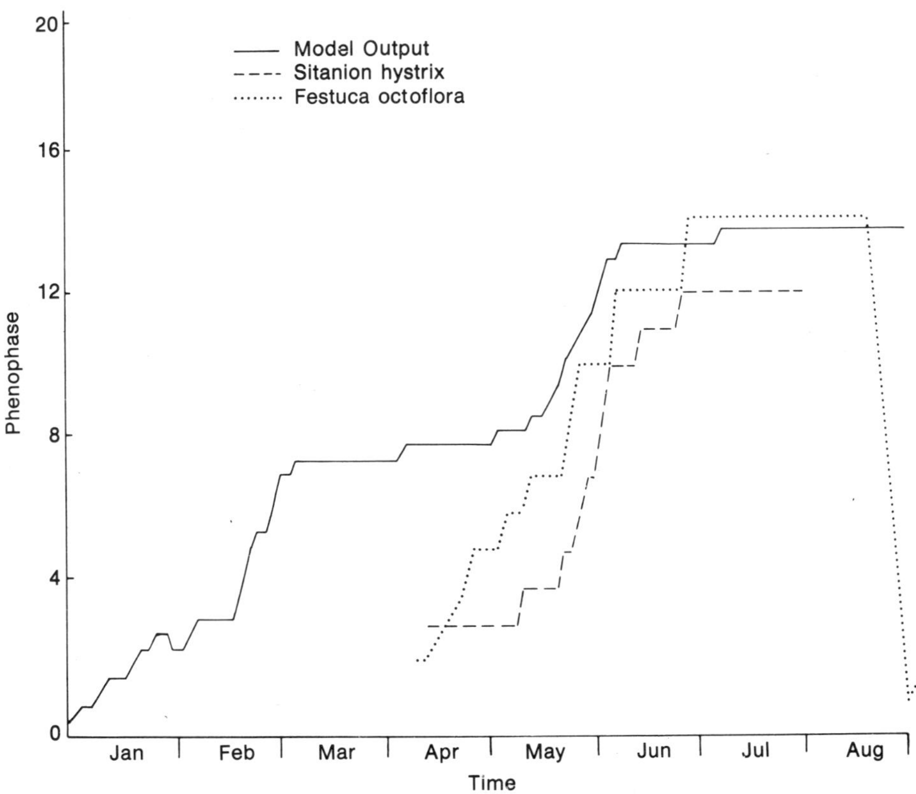

Fig. 4. The phenology of a cool season grass simulated in the model compared to observed values of the cool season grass *Sitanion hystrix* and the annual grass *Festuca octoflora*.

It is then necessary to obtain driving variable data and phenophase data from another season or other locations to validate the refined model.

6 Discussion and Conclusions

At the present time there are at least three general objectives being pursued by investigators of phenology: comparison of climate during different seasons, mapping regional changes in climatic conditions, and correlation of plant development with meteorological or other environmental variables. The first two objectives require records by date of plant (or animal) developmental stages; the third further requires accurate and concomitant records of environmental parameters and plant (or animal) developmental stages. Only through an understanding of the joint linkages between biotic events and microclimate can the driving variables (e.g., precipitation, solar insolation) be coupled reasonably to ecosystem function.

Efforts to model the grassland ecosystem have demonstrated that field studies and mathematical modeling must proceed together with constant communication, as in an efficient servomechanism. Analysis of field data provides information on the coupling of system events and directs attention to data requirements from field observations. In this manner model development can proceed toward more adequate resemblance to reality.

To advance phenological modeling, it is essential to determine the relationship between phenological events of different organisms and the sequence of environmental conditions that must precede the events. If we know these, we know the minimum set of measurements needed to understand ecosystem function.

References

Blaisdel, J. P. 1958. Seasonal development and yield of native plants on the upper Snake River plains and their relation to certain climatic factors. *US Dept. Agr. Tech. Bull. No. 1190.*

Borchert, J. R. 1950. The climate of the central North American grassland. *Ann. Assoc. Am. Geogr.* 40, 1–39.

Caprio, J. M. 1971. The solar-thermal unit theory in relation to plant development and potential evapotranspiration. *Montana Agr. Exp. Sta. Circ. 215.*

Coupland, R. T. 1950. Ecology of mixed prairie in Canada. *Ecol. Monograph* 20, 271–315.

Lewis, J. K. 1971. The grassland biome: A synthesis of structure and function, 1970. In *Preliminary Analysis of Structure and Function in Grasslands. Range Science Series No. 10,* N. R. French, ed. pp. 317–387. Ft. Collins: Colorado State University.

Rauzi, F., and A. K. Dobrenz. 1970. Seasonal variation of chlorophyll in western wheatgrass and blue gramma. *J. Range Mgt.* 23, 372–373.

Reuss, J. O. 1971. Decomposer and nitrogen cycling investigations in the grassland biome. In *Preliminary Analysis of Structure and Function in Grasslands. Range Science Series No. 10,* N. R. French, ed. pp. 133–146. Fort Collins: Colorado State University.

Sauer, R. H. 1971. Adaptations to heat and drought in the *Clarkia unguiculata* complex. p. 145. Ph.D. Dissertation. Los Angeles: University of California.

4.4 Phenodynamics of Production in a Mesic Deciduous Forest[1,2]

Fred G. Taylor, Jr.

Environmental Sciences Division
Oak Ridge National Laboratory
Oak Ridge, Tennessee 37830, USA

1 Introduction

Climatologists and agrometeorologists have predicted phenological events in many ways, e.g., to select optimum seeding dates (Hopkins, 1938) or to plan agricultural practices, particularly in Europe (Molga, 1962; Schnelle, 1955), as well as in crop pest control to maximize production (Hopp *et al.*, 1969; Wallin, 1967; Woods, 1969). In terms of economic investment, quantitative phenology (phenometry) has been applied successfully to predict harvest dates, and to schedule agricultural practices, insuring both quality and an even flow of produce to food processors (Bomalaski, 1948; Brown, 1952; Katz, 1952; Newman *et al.*, 1967; Seaton, 1955). In defining the beginning, duration, and end of seasons phenological observations may be better indices of the bioclimatic character of local areas than recordings made by mechanical instruments (Hopkins, 1918, 1938). Specific plants and selected phenological events might be useful as indicators on which to key major land management practices and productivity studies (Caprio, 1966; Lieth and Radford, 1971). The International Biological Program (IBP) has provided an impetus for incorporating phenological considerations into basic ecological studies such as analyses of ecosystems (Bliss, 1967).

[1] Contribution No. 49 from the Eastern Deciduous Forest Biome, US/IBP.

[2] Research supported in part by the Eastern Deciduous Forest Biome, US/IBP, funded by the National Science Foundation under Interagency Agreement Ag-100, 40-193-69 with the U.S. Atomic Energy Commission, Oak Ridge National Laboratory.

KEYWORDS: Phenometry; primary production; phenodynamics; flowering; phenogram; light transmission; canopy; thermal environment; heat summation; threshold; phenophase; phenological spectra; growth analysis; biomass; indicator species; population; community; forest ecosystem; meteorology.

2 The Oak Ridge Site of IBP

The phenological observations reported here are from the Oak Ridge IBP site in Roane County in eastern Tennessee (approximately 84° 18′ W longitude and 35° 54′ N latitude) on lands acquired in 1942 by the U.S. federal government. Since acquisition as farmlands, most of the forests on the Atomic Energy Commission's Oak Ridge Reservation have remained relatively undisturbed except for some recent forest management practices (Curlin, 1965). Approximately 15,175 ha, now mostly forest, have long been utilized as sites for ecological research (Auerbach *et al.*, 1964). Access to the area is denied to the general public and the reservation affords maximum protection for long-range ecological studies. Communities, populations, or individual species selected for phenological observations can provide data year after year (Taylor, 1969).

The general area was formerly classified as an oak-chestnut association (Braun, 1950) within the Ridge and Valley Physiographic Province (Fenneman, 1938). As individual chestnut trees were eliminated by the chestnut blight, suppressed or codominant species (oaks and hickories) replaced them as major canopy components (Woods and Shanks, 1959). Tulip or yellow poplar (*Liriodendron tulipifera* L.) is abundant near the base of ridge and slope habitats in secondary communities. A tulip poplar-dominated forest ecosystem has been the subject of numerous experiments (Auerbach *et al.*, 1964; Olson, 1965; Peters *et al.*, 1969; Reichle *et al.*, 1973; Waller and Olson, 1967). Data presented here are indicative of the phenodynamics within the major forest association (oak-hickory) and certain subcommunities (including escaped or naturalized species) as influenced by meteorological parameters. Phenological observations, where complemented with quantitative analyses, can reveal the importance of short-term phenological events to functional processes within the forest ecosystem.

3 Phenology of Flowering Within an Oak-Hickory Forest Association

Vegetative growth of some plants diminishes with the advent of flowering. Among selected species (indicators), the dates of first flowering reported by regional networks of observers form the basis of maps portraying the arrival of the growing seasons and offering guides to agricultural practices (Caprio, 1966; Lieth and Radford, 1971). Records of dates of first flowering for major species at the Oak Ridge Site have been observed since 1963 (Taylor, 1969). These records are summarized in a phenogram (Fig. 1) illustrating the mean date of first flower between 1963 and 1970. The figure shows that phenological development can be examined for particular communities or for various taxa. An early peak in the number of species flowering, followed by a decline and a subsequent increase in early summer, has been widely observed (Harper, 1906).

The first major peak (week 16) is composed of woodland species, mainly herbaceous, while the second or early summer peak (week 23) is dominated by species indigenous to thickets, roadsides, or openings within the forest.

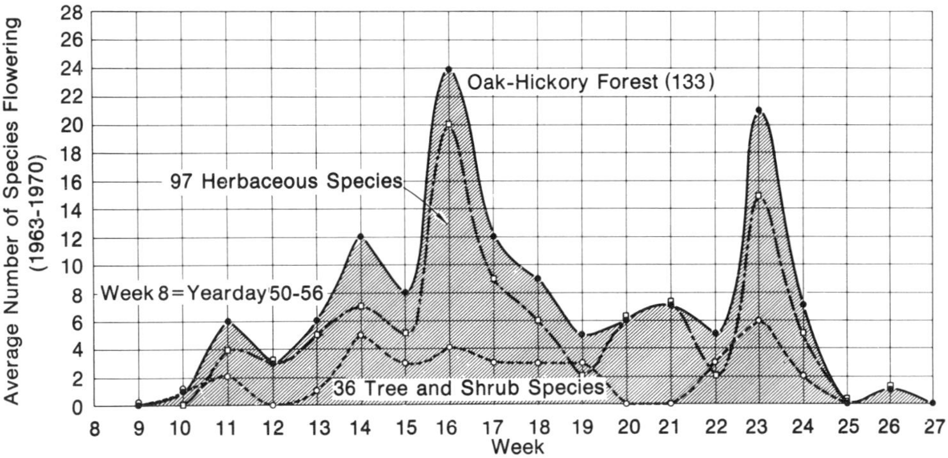

Fig. 1. Phenogram of the oak-hickory forest association depicting flowering seasons of moist and dry habitats, distinguished by spring and early summer floras. Data represent mean date of first flower between 1963 and 1970. Richness of the understory is noted by the number of herbaceous species observed. Number in parenthesis indicates the total number of species observed.

The transmission of incident radiation to ground level is reduced 40% following the spring peak in flowering, corresponding to 75% leaf emergence of the dominant canopy species of lower slope forests (Fig. 2). At the time of total leaf emergence (week 24) light transmission is reduced to 11%. These data are comparable to results in agricultural experiments (Yocum *et al.*, 1964).

The second peak flowering season (week 23) occurs after the canopy has closed.

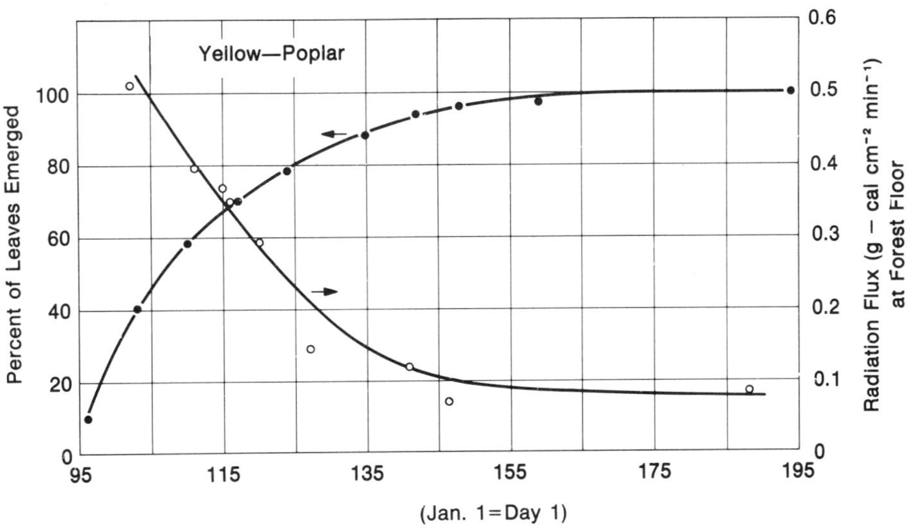

Fig. 2. Cumulative percentage of maximum leaf emergence for *Liriodendron tulipifera* during the 1970 growing season, and concomitant light extinction on the forest floor.

However, this floristic component in the development of the forest ecosystem is composed of species found in more open habitats (roadsides and forest edges) where light transmission is not so significantly reduced. Early canopy development by dominant tree species and subsequent reduction in light transmission to the forest floor might alter the photosynthetic efficiency of understory species. Consequently, the completion of the life cycle phases might result in reduced propagules (seeds), jeopardizing the continued existence of certain species within the ground layer.

4 Influence of Meteorological Factors

Variability in date of first flower within a species (Table 1) is due, in part, to variation in one or more meteorological parameters. Mean monthly temperatures during early 1970 were below the monthly mean values for the 10-year period between 1961 and 1970 (Fig. 3) but were generally higher during the remainder of the year.

The relationship between the thermal environment and the phase of development of certain plant species has been utilized by agriculturists with varying success (Baker and Brooks, 1944; Bomalaski, 1948; Brown, 1952; Hopkins, 1938; Katz, 1952; Newman *et al.*, 1967; Reaumur, 1735; Seaton, 1955) to characterize the developmental phases of plant from trends in the thermal environment. In thermal summation techniques (degree-days or heat units) threshold temperatures have been utilized to predict plant development within homogeneous systems such as corn (Scott, 1946) and peas (Katz, 1952). The development of the heat unit concept and subsequent modifications have been discussed in detail by Wang (1960).

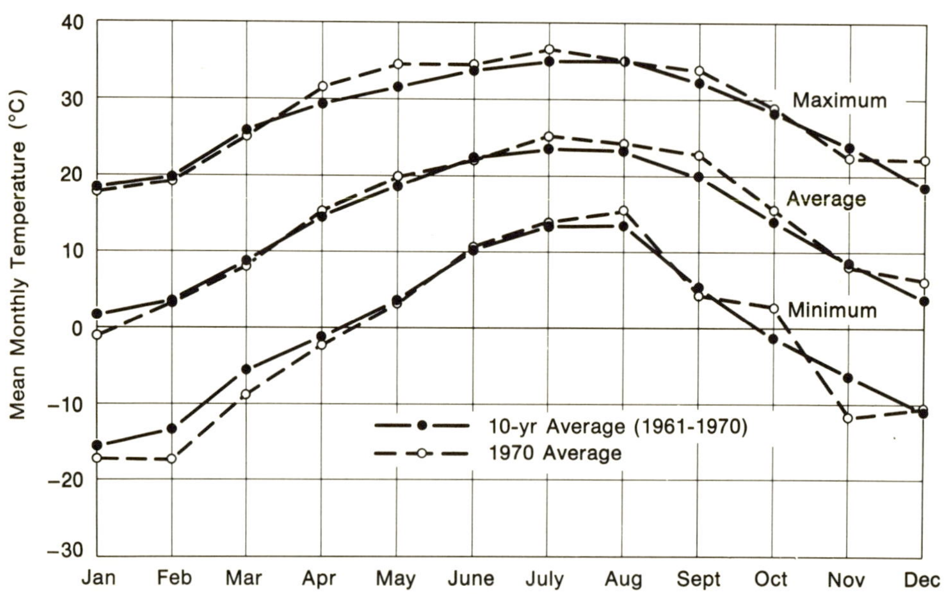

Fig. 3. Monthly temperature averages (maximum, mean, and minimum) for the 10-year period (1961–1970) as compared with averages for 1970.

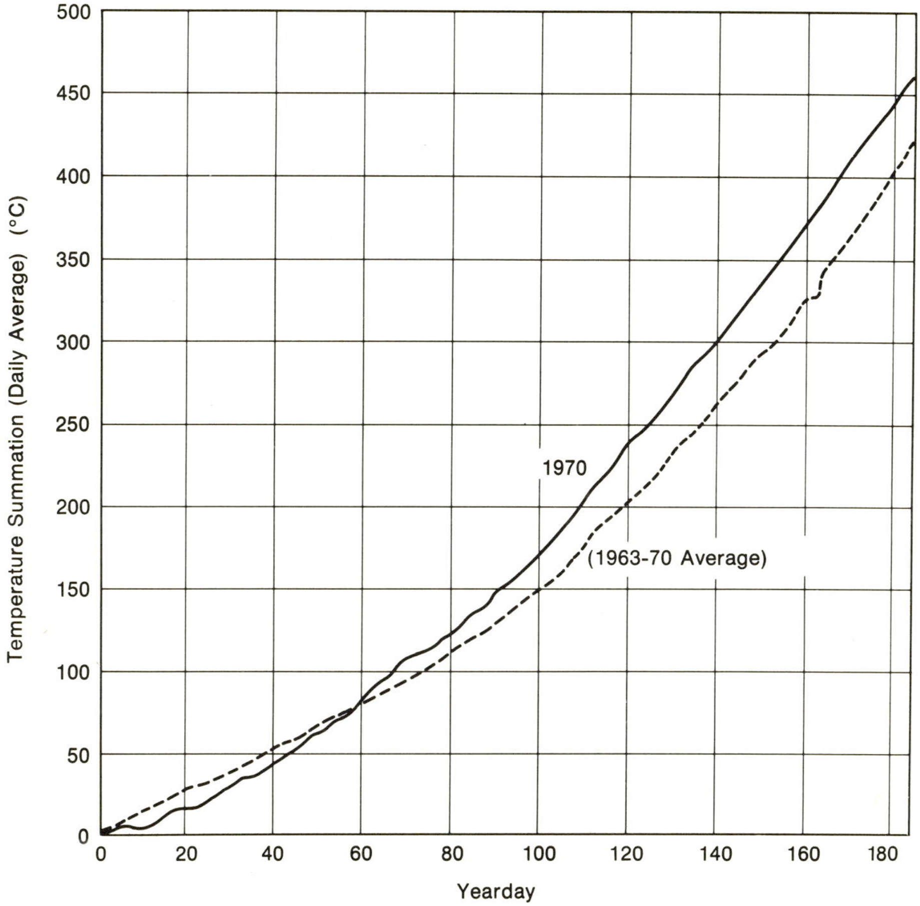

Fig. 4. Comparison of the thermal environment within the oak-hickory forest association between 1963 and 1970 and 1970 by temperature summations of the mean daily temperature.

Thermal considerations have been shown to be important to the phenological development within managed forests or natural ecosystems (Jackson, 1966). Temperature summations or degree-days have been used to depict differences in thermal environments between successive years (Bomalaski, 1948). Within the heterogeneous structure of a forest ecosystem, a common temperature threshold has little validity for analytical techniques because not all species respond strongly to the same temperature regimes.

If a diagram depicting flowering within the oak-hickory forest during 1970 were superimposed upon the phenogram illustrating mean date of flowering between 1963 and 1970 (Fig. 1), the early spring peak would be advanced by seven days, to week 15, whereas the early summer flora would be delayed seven days, to week 24. Monthly average temperatures (Fig. 3) for 1970, compared for these periods, offer little explanation for the variability. However, characterizing the thermal environment by the summation of daily mean temperature (Fig. 4) shows

Table 1. Flowering period of species on the Oak Ridge site between 1963 and 1970 arranged chronologically by date of first flower. (Nomenclature after Fernald, 1950.)

1 *Acer saccharinum**
2 *Acer rubrum**
3 *Lamium amplexicaule*
4 *Obolaria virginica*
5 *Cardamine hirsuta*
6 *Holosteum umbellatum*
7 *Ulmus americana**
8 *Draba brachycarpa*
9 *Draba verna*
10 *Hepatica americana*
11 *Stellaria media*
12 *Viola kitaibeliana*
13 *Poncirus trifoliata**
14 *Taraxacum officinale*
15 *Anemonella thalictroides*
16 *Oxalis stricta*
17 *Amelanchier arborea**
18 *Dentaria laciniata*
19 *Sassafras albidum**
20 *Antennaria plantaginifolia*
21 *Cercis canadensis**
22 *Houstonia caerulea*
23 *Lamium purpureum*
24 *Prunus hortulana**
25 *Sanguinaria canadensis*
26 *Capsella bursa-pastoris*
27 *Duchesnea indica*
28 *Salix nigra**
29 *Viola rotundifolia*
30 *Viola hirsutula*
31 *Fragaria virginiana*
32 *Aesculus octandra**
33 *Aesculus pavia**
34 *Dentaria heterophylla*
35 *Trifolium repens*
36 *Cornus florida**

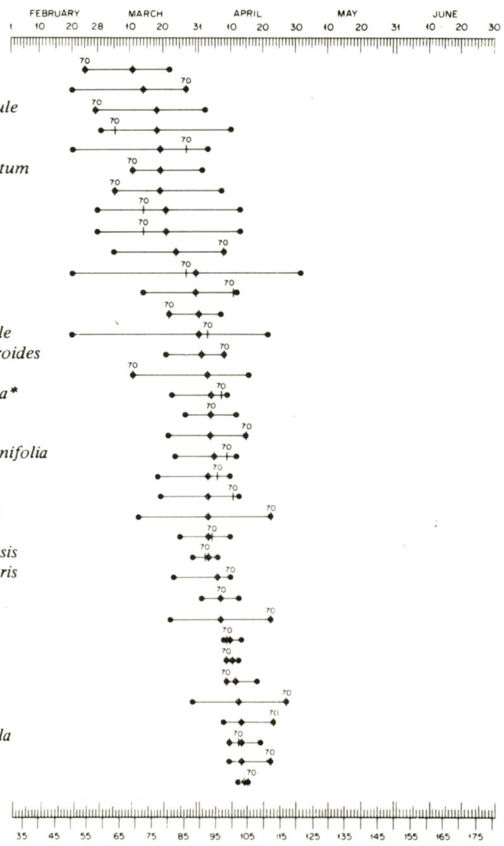

37 *Amsonia tabernaemontana*
38 *Pyrus communis**
39 *Viola canadensis*
40 *Geranium carolinianum*
41 *Lithospermum canescens*
42 *Myosotis verna*
43 *Potentilla canadensis*
44 *Trillium luteum*
45 *Aquilegia canadensis*
46 *Malus pumila**
47 *Uvularia perfoliata*
48 *Orchis spectabilis*
49 *Prunus serotina**
50 *Trillium erectum*
51 *Viola hastata*
52 *Phlox divaricata*
53 *Pinus virginiana**
54 *Viola sagittata*
55 *Viola septemloba*
56 *Nasturtium officinale*
57 *Oxalis violacea*
58 *Podophyllum peltatum*
59 *Raphanus raphanistrum*
60 *Tiarella cordifolia*
61 *Geranium maculatum*
62 *Ranunculus bulbosus*
63 *Rubus trivialis*
64 *Galium aparine*
65 *Rosa carolina**
66 *Actaea pachypoda*
67 *Morus rubra**
68 *Sedum ternatum*
69 *Salvia lyrata*
70 *Robinia pseudo-acacia**
71 *Smilacina racemosa*
72 *Sisyrinchium mucronatum*

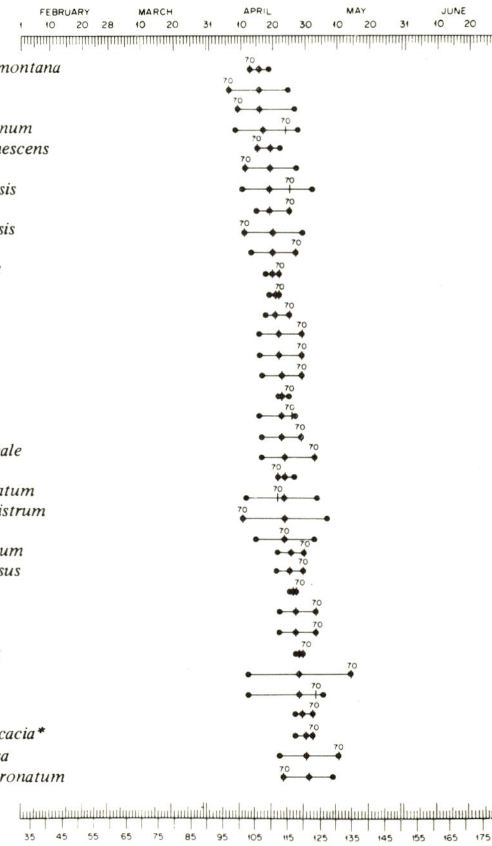

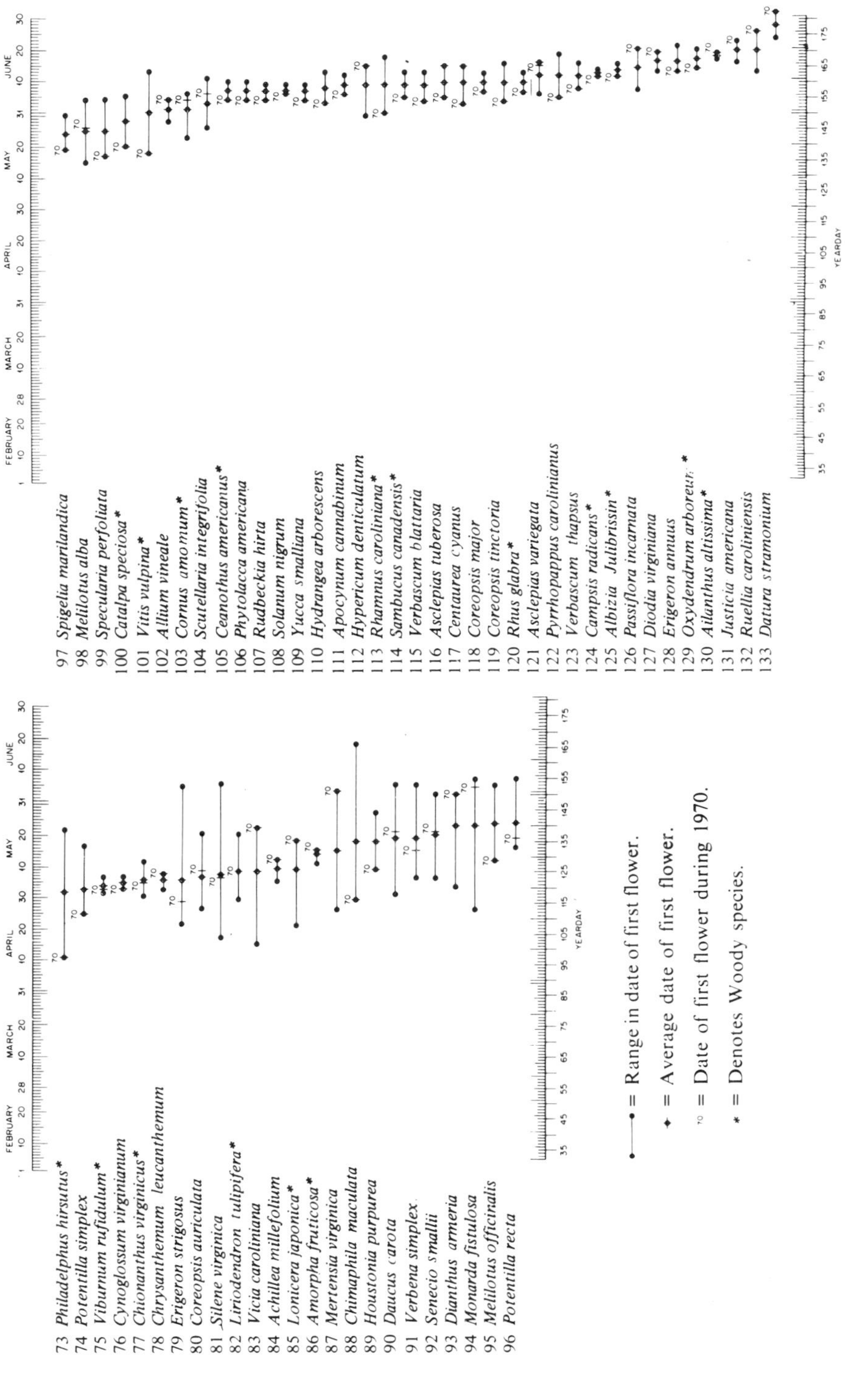
FEBRUARY MARCH APRIL MAY JUNE
YEARDAY
73 Philadelphus hirsutus*
74 Potentilla simplex
75 Viburnum rufidulum*
76 Cynoglossum virginianum
77 Chionanthus virginicus*
78 Chrysanthemum leucanthemum
79 Erigeron strigosus
80 Coreopsis auriculata
81 Silene virginica
82 Liriodendron tulipifera*
83 Vicia caroliniana
84 Achillea millefolium
85 Lonicera japonica*
86 Amorpha fruticosa*
87 Mertensia virginica
88 Chimaphila maculata
89 Houstonia purpurea
90 Daucus carota
91 Verbena simplex
92 Senecio smallii
93 Dianthus armeria
94 Monarda fistulosa
95 Melilotus officinalis
96 Potentilla recta
97 Spigelia marilandica
98 Melilotus alba
99 Specularia perfoliata
100 Catalpa speciosa*
101 Vitis vulpina*
102 Allium vineale
103 Cornus amomum*
104 Scutellaria integrifolia
105 Ceanothus americanus*
106 Phytolacca americana
107 Rudbeckia hirta
108 Solanum nigrum
109 Yucca smalliana
110 Hydrangea arborescens
111 Apocynum cannabinum
112 Hypericum denticulatum
113 Rhamnus caroliniana*
114 Sambucus canadensis*
115 Verbascum blattaria
116 Asclepias tuberosa
117 Centaurea cyanus
118 Coreopsis major
119 Coreopsis tinctoria
120 Rhus glabra*
121 Asclepias variegata
122 Pyrrhopappus carolinianus
123 Verbascum thapsus
124 Campsis radicans*
125 Albizia Julibrissin*
126 Passiflora incarnata
127 Diodia virginiana
128 Erigeron annuus
129 Oxydendrum arboreum*
130 Ailanthus altissima*
131 Justicia americana
132 Ruellia caroliniensis
133 Datura stramonium
= Range in date of first flower.
= Average date of first flower.
70 = Date of first flower during 1970.
* = Denotes Woody species.

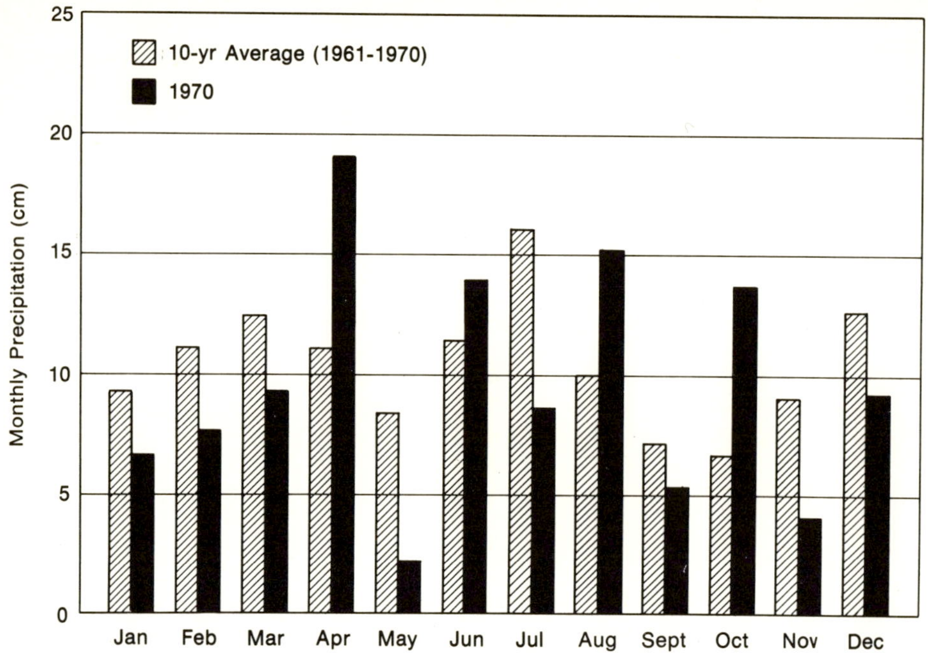

Fig. 5. Comparison of total precipitation during 1970 with the expected quantity based on 10-year average between 1961–1970. A deficit exists for 8 of the 12 months during 1970. Note particularly the deficit preceding the early summer flora (May).

that the thermal climate at yearday 112 (week 16), the mean peak flowering period between 1963 and 1970, is attained nine days earlier in 1970. This suggests that the advancement in flowering during the spring of 1970 was strongly related to the thermal climate within the forest ecosystem. Likewise, the temperature sum during 1970 for the early summer component was also attained earlier (by 14 days) than the same temperature sum during the mean peak flowering period between 1963 and 1970 (yearday 161). Flowering was delayed seven days in 1970, which suggests that other factors were more important than temperature, or that optimum thermal regimes were exceeded, thus inhibiting the initiation of flowering.

The early summer flora (week 23) includes species indigenous to more open, dry habitats where moisture may have been a limiting factor during 1970. There was a moisture deficit (approximately 10%) in comparison with the 10-year mean. During the first half of 1970, four of six months received below normal precipitation (Fig. 5). Especially noticeable was the deficit in May which was probably critical for plant species occupying drier sites. This deficit occurred prior to the early summer peak in number of species flowering and could be responsible for the delay in flowering season noted previously.

The relationship between the temperature sum and date of first flower (yearday) provides significant predictability between species (Fig. 6). Silver and red maple are the first species to begin reproductive phases, at a period when the mean daily air temperature is approximately 10°C. The significant correlation ($r^2 = 0.99$)

suggests that observations of mean date of first flower (Table 1) spanning several years are sufficient to estimate advent of the reproductive phase between many species. Thermal or temperature summations (Fig. 4) utilizing the mean daily temperatures would provide a more precise estimate of flowering date from temperature summations (*TS*) such that

$$TS_i = \sum_{k=1}^{i} adt_k \tag{1}$$

where adt_k is the average daily temperature, $\left[\dfrac{\text{sum of mean hourly temperatures}}{\text{N}}\right]$.

Since the correlation between species is highly significant, the predicted date of first flower (yearday) may indicate minimum or threshold thermal regimes necessary to initiate the reproductive phase (Fig. 6).

Moisture, light, and temperature interactions are all known to enhance or delay timing of certain phenological events (Daubenmire, 1959). Departure from normal environmental conditions can alter the appearance of various phenological events within forest ecosystems as indicated by the response of flowering to differences in

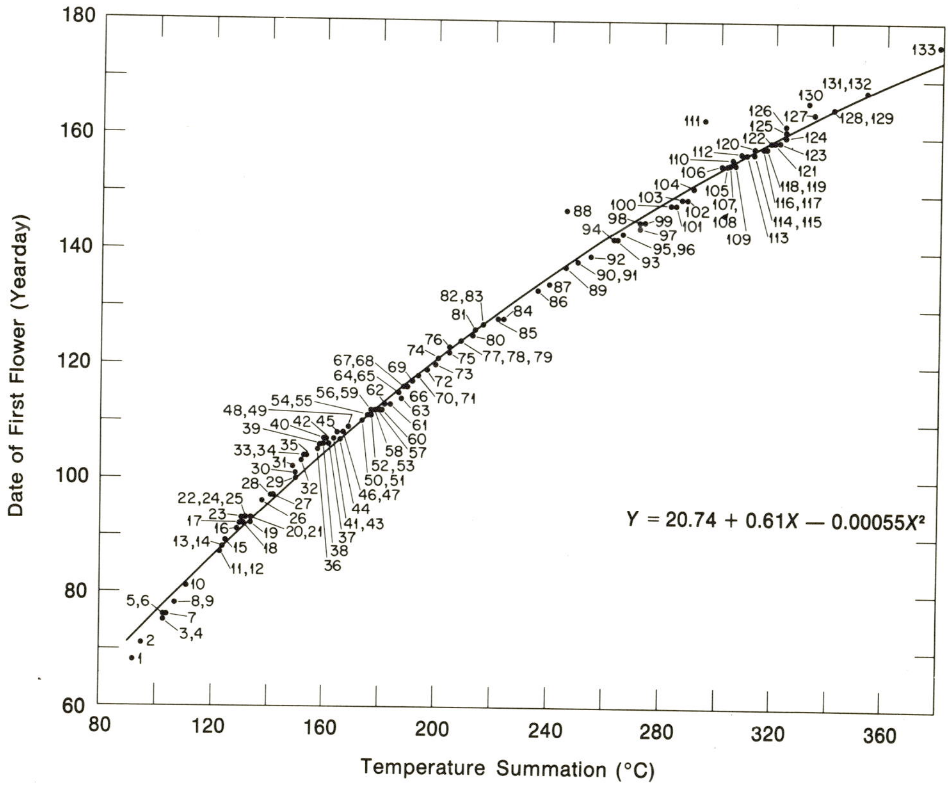

Fig. 6. Relationship between mean date of first flower and mean temperature summation (1963–1970). Numbers correspond to species numbers in Table 1.

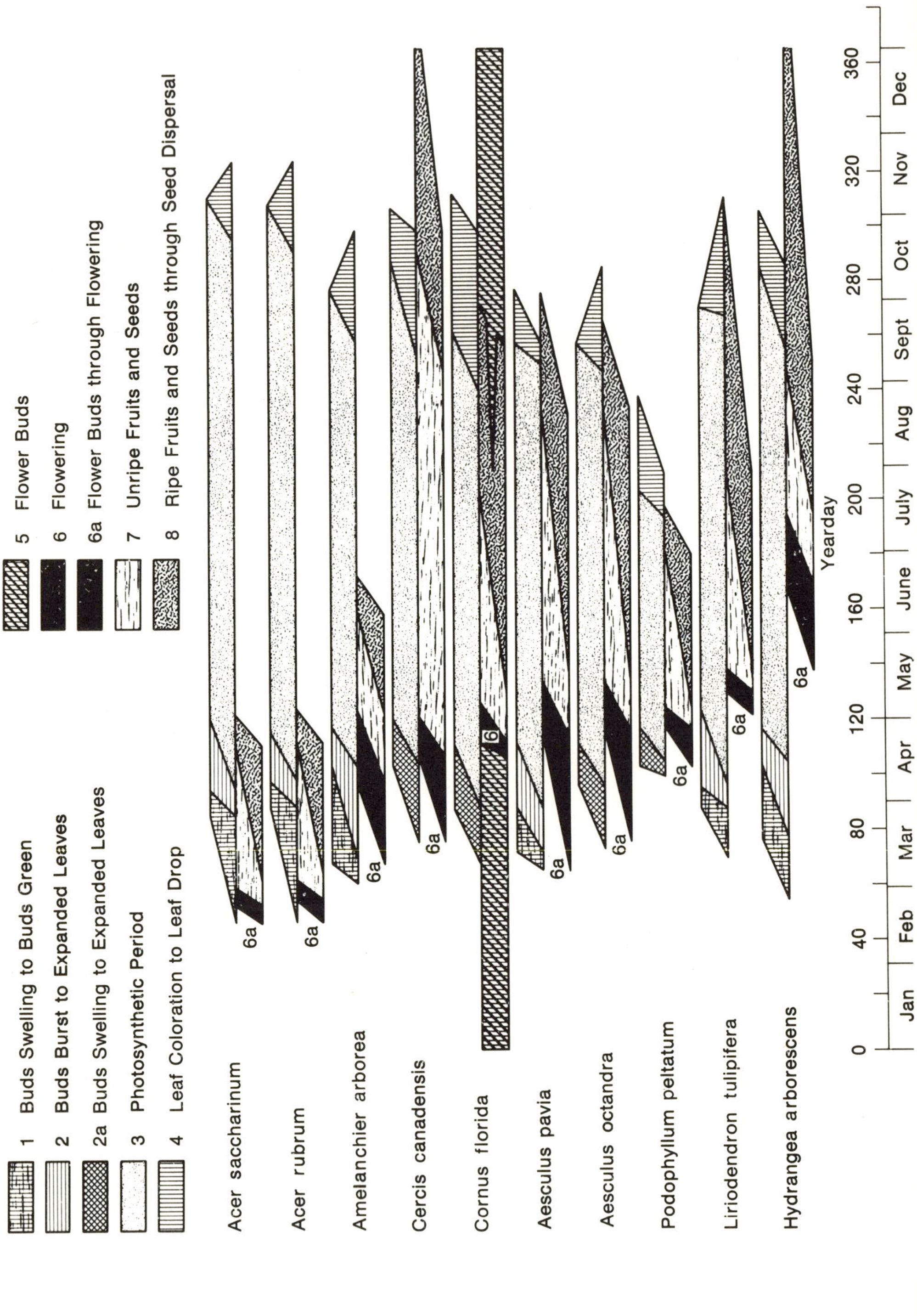
1 Buds Swelling to Buds Green
2 Buds Burst to Expanded Leaves
2a Buds Swelling to Expanded Leaves
3 Photosynthetic Period
4 Leaf Coloration to Leaf Drop
5 Flower Buds
6 Flowering
6a Flower Buds through Flowering
7 Unripe Fruits and Seeds
8 Ripe Fruits and Seeds through Seed Dispersal
Acer saccharinum
Acer rubrum
Amelanchier arborea
Cercis canadensis
Cornus florida
Aesculus pavia
Aesculus octandra
Podophyllum peltatum
Liriodendron tulipifera
Hydrangea arborescens
6a
6
0
40
80
120
160
200
240
280
320
360
Yearday
Jan
Feb
Mar
Apr
May
June
July
Aug
Sept
Oct
Nov
Dec

temperature and moisture. Deviations in micrometeorological parameters, in turn, influence dynamic characteristics of ecosystems (e.g., availability of food base to secondary producers, pollination, and rates of accumulation and turnover of standing crop).

5 Relationship of Events Within Species

In addition to responses to environmental factors in determining the growing season, it is essential to know how each phenological event within a season is related to each adjacent phase, especially for studies for productivity. Within the eastern deciduous forest biome 10 species were selected for detailed observations for the construction of phenological spectra (Fig. 7). Those selected represent the first woody species to initiate growth (red maple), the dominant canopy species of slope forests (yellow-poplar), and selected shrubs and herbaceous species within areas under intensive study. Particularly noticeable (Fig. 7) is the almost simultaneous development of the reproductive phase of shrub species most characteristic of local forest ecosystems (dogwood, redbud, and shadbush or serviceberry). This series of phenological events precedes canopy development of one of the codominant species (*Liriodendron*).

6 Phenology and Production of *Podophyllum peltatum*

An important ecological consideration in the classification of forest types or associations has been the presence or absence of binding genera or species (Braun, 1950), especially within the herbaceous layer. Although this stratum of the forest contributes little biomass to the total primary production of the ecosystem, it may well provide insight into the interactions between individual species of communities and their environment (Kubicek, 1970). The importance of some phenological events to the total productivity of single species (Kreh, 1965) suggests that each component within a community could be analyzed in a similar manner. Such studies of growth analysis would permit the construction of more precise models of production within communities or ecosystems.

Within the mesic yellow poplar-dominated slope forest at Oak Ridge, biomass and productivity of the herbaceous layer have been estimated by indirect (allometric) methods (Christofolini, 1970). Data were obtained during one growing season and revealed no relationship to phenological development and environmental interactions. By comparison, the herbaceous layer within an oak-hornbeam forest in Europe was evaluated phenodynamically (Kubicek, 1970) during two growing seasons, and variations in monthly production were directly related to

---

Fig. 7. Phenological spectra of species characteristic of secondary forest communities dominated by *Liriodendron tulipifera* within the oak-hickory association. These spectra indicate the relationships of phenological events within species and compare events between species and delineate the growing season.

climatic conditions. Such analyses provide explanations for variations in total productivity within communities from one season to the next.

Growth analysis data are presented for a single species (*Podophyllum peltatum* L.) at the Oak Ridge Site to exemplify the role of phenology in studies of the total productivity, including the herbaceous layer. Mayapple is used as an example because of its social occurrence and rapid phenological development in the spring. As one of the most conspicuous herbaceous species to initiate spring growth within the eastern deciduous forest, mayapple is a potential indicator species for regional studies. During the first three years of growth the nonreproductive individuals manifest a vertical shoot growth (Fig. 8) with developmental forms distinct from older reproductive plants (Holm, 1899). Older plants annually produce a single leaf from underground, horizontal shoots (Fig. 8). Plants within mature mayapple clones or populations exhibit an almost simultaneous response in phenological

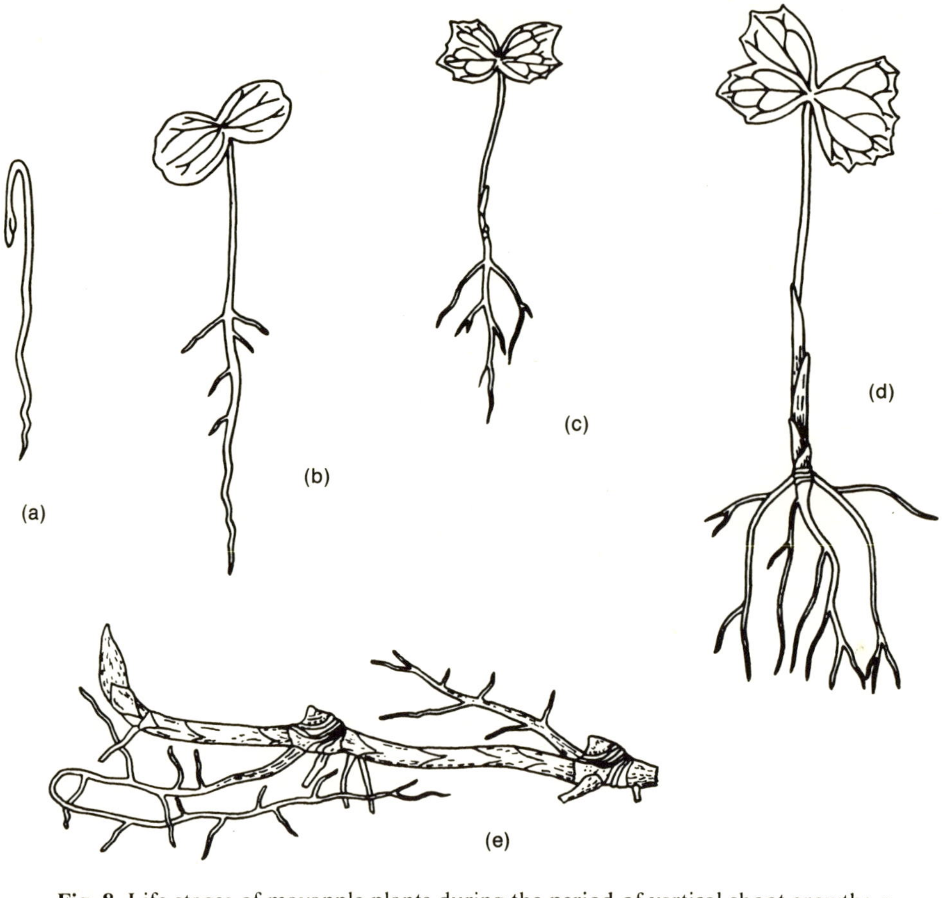

Fig. 8. Life stages of mayapple plants during the period of vertical shoot growth: a = germination, b = first-year plant with cotyledonary leaves, c = second-year seedlings, d = third-year plant depicting final vertical growth of the shoot, e = horizontal shoot (rhizome) and roots developed from a lateral bud of a three-year-old plant, illustrating an additional three years growth including the terminal bud. (Modified from Holm, 1899.)

development. The phenophases of individual plants are depicted in Fig. 9. The time from leaf emergence through the litter to date of first flowering is approximately three weeks. Distinct populations or clones of mayapple occur frequently within slope or mesic (northerly exposed) forests. Within the *Liriodendron* forest study area, mayapple is best established in social units on low, moist sites. Being partially clonal with moderate longevity, the perennial populations vary in area, appearing elliptical to circular in shape. Clone size among 21 populations averaged 21.04 m^2, ranging from 0.7 to 176.6 m^2. Plant density within this array varied from 10 to 683 with an average of 167 plants per population.

Regression analysis of computed area to number of plants permits an estimate

Fig. 9. Phenophases of aerial plants from the underground horizontal shoot. (*a*) Leaf emergence through the litter, (*b*) unfolding of leaves, (*c*) leaf expansion of nonflowering plant, (*d*) leaflet expansion and flower bud of reproductive plant, (*e*) a plant in flower.

of population density ($r^2 = 0.88$), providing that the size of the clone is known. The resulting equation, $\log_{10} Y = 1.12 \times 0.85 (\log_{10} X)$, assumes the configuration of the population to be circular. Clone area (X) is computed from diameter measurements of the population. Within the broad definition of the oak-hickory association, density per square meter is less than 1, whereas the density within the average clone or population is approximately $8/m^2$. Excavation of the rhizome system of a clone revealed that the average age of viable underground shoots was six years with obvious decay of older segments, and that the population had increased in radial growth at the rate of 1 dm per year. This particular population had therefore occupied the site for approximately 45 years. Phenodynamic studies of specific populations spanning many years could provide data depicting the impact of seasonal or annual climatic variations on production.

Beginning with leaf emergence (yearday 99) plants were harvested at each of eight life phases. Plants were compartmentalized into leaf, stem, flower or fruit, and rhizome-root for biomass estimates. Estimates of rhizome-root biomass during the leaf emergence phase (yearday 99-102) were considered in standing crop, such that subsequent estimates would reflect changes throughout the life cycle. Biomass increased for each component with the progression of phenological development (Fig. 10), reaching maximum dry weights during the phase of unripe seeds and fruits, prior to the onset of senescence (yearday 162). At that time, rhizome-root (1970 increment) showed an increase of 21% in dry weight above the standing crop estimate at leaf emergence. Leaf and above-ground stem (petiole) accounted for

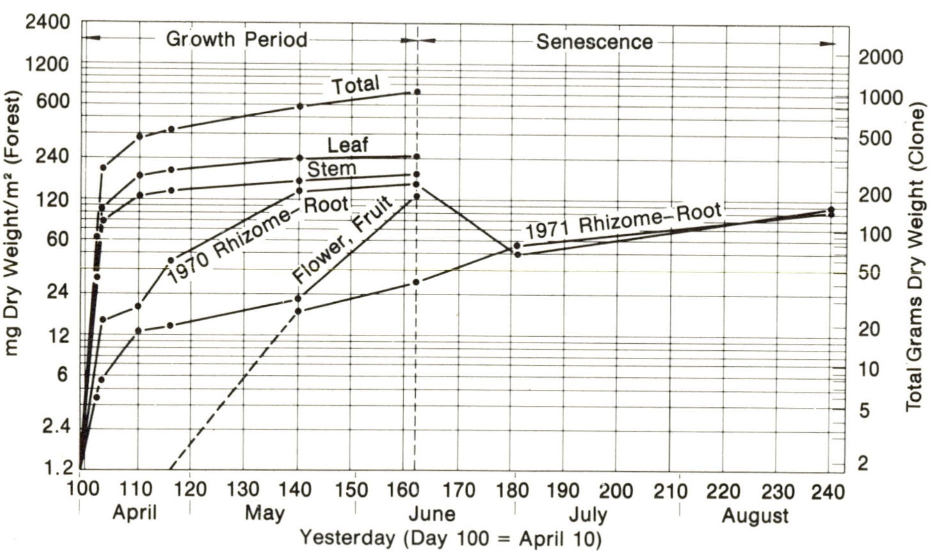

Fig. 10. Biomass in mayapple, partitioned into obvious components. Phenophases correspond to the following time intervals: Yeardays 99–102 (leaf emergence), 102–103 (leaves expanding), 99–116 (flower buds), 116–123 (flowering), 123–178 (unripe seeds and fruits), 178–196 (ripe seeds and fruits), 196–240 (leaf drop and seed dispersal). The vegetative or growth phase is between yearday 99 and 162. Color change of leaves begins on yearday 162 and terminates with total leaf drop, day 240.

57%, whereas unripe fruits and seeds represented 17% of total dry weight. The underground rhizome system grows terminally, producing an overwintering apical bud (Fig. 8*e*). The 1971 underground shoot was initiated during the reproductive phase (yearday 116-123), and showed an increase of 4% in dry weight prior to senescence of aerial parts. At the onset of senescence (yearday 162), leaves began to change color, followed by leaf and fruit drop, and subsequent decreases in dry weight. Seeds are distributed by insects and small mammals feeding on ripe fruits. Further increases in the increment to the underground shoot system (1971) were accompanied by decreases in dry weight of the previous season's growth (1970), indicating that the rhizome functions as a storage organ for subsequent growth processes. By the end of the growth period (yearday 240), each increment in rhizome-root (1970 and 1971) accounted for 25% of the biomass, with the other 50% consisting of fallen aerial parts (leaf, fruit, and seeds). Biomass is summarized (Fig. 10) per unit area within the oak-hickory forest (plant density $< 1/m^2$) or on the basis of the average mayapple population within the *Liriodendron* forest study site (plant density is $8/m^2$).

Productivity estimates per unit forest area or average population (clone) within the intensive study area (Fig. 11), reflect growth above the standing crop of rhizome-root (1970) at leaf emergence. Productivity is maximum during the reproductive (flowering) phase, yearday 103–116, decreasing rapidly through fruit development and leaf coloration phases. The decrease in production between yeardays 163 and 181 represents a loss due to the consumption of ripe fruits by animals and insects. As indicated in biomass data, unripe fruits and seeds account for as much as 17% of the total dry weight. The phenological spectrum above the productivity curve in Fig. 11 illustrates the relationship of various events to changes in productivity.

7 Comments

The concurrent phenological development of the understory flora (herbaceous) coincides with the development of dominant and codominant canopy species (Fig. 1) within the deciduous forest ecosystem. Many phenological events are correlated with changes in functional processes within the ecosystem (e.g., photosynthesis and respiration). Arrival of host-specific phytophagous insects and secondary consumers might well be predicted for certain species. Availability of the food base among some species often coincides with the arrival of specific animals, especially migrant birds.

Phenodynamics of production within ecosystems at different locations represents the interactions of meteorology with growth processes. Similarities in phenological development between sites provide benchmarks for timing of productivity, as in profiles across regional boundaries. Associated with the phenological development of species or communities are changes in biomass and rates of productivity. Such short-term phenological events represent input into the litter layer and subsequently affect the mineral cycle within the ecosystems.

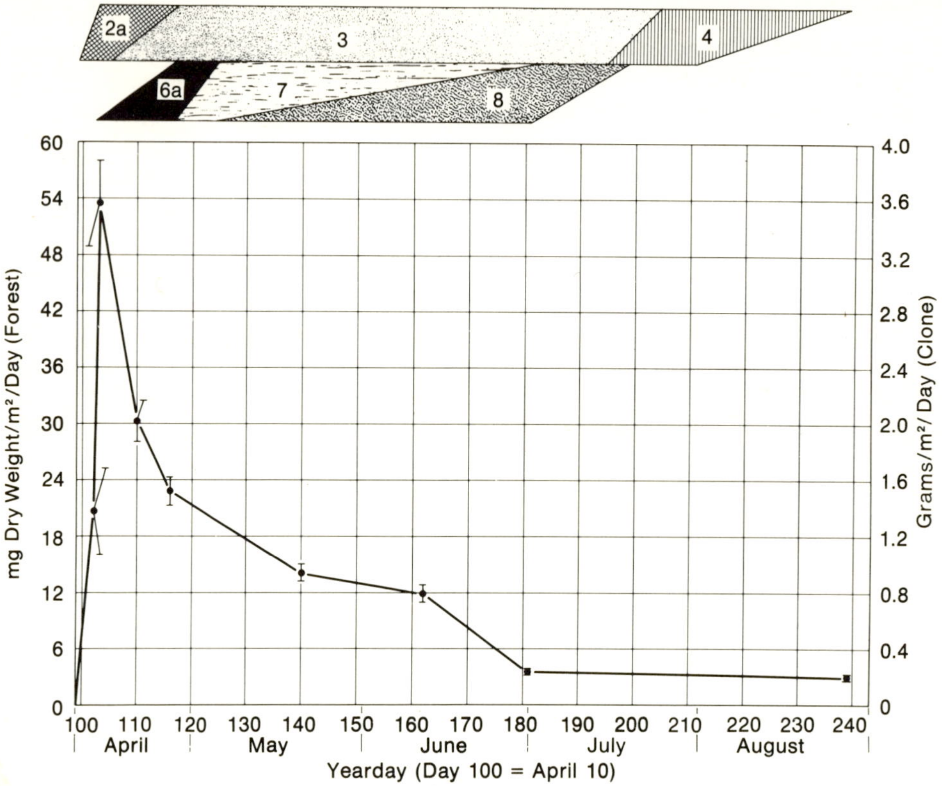

Fig. 11. Seasonality in productivity by mayapple, *Podophyllum peltatum*. As indicated by the spectral strip above the curve (numbers represent the same phenophases as identified in Fig. 7), productivity is maximum during the reproductive phase. Reduced productivity by fruit consumption of small mammals and insects coincides with the period of ripe fruits and seeds to seed dispersal. Brackets denote ± 1 S.E.

Acknowledgments

Appreciation for assistance is gratefully extended to B. Hutchison (National Oceanic and Atmospheric Administration) and R. Sayrs (Oak Ridge National Laboratory), who provided assistance in the procurement and summarization of meteorological data, and especially to L. M. Stubbs (Oak Ridge National Laboratory) for field assistance.

References

Auerbach, S. I., J. S. Olson, and H. D. Waller. 1964. Landscape investigations using caesium-137. *Nature* 201, 761–764.

Baker, G. A., and R. M. Brooks. 1944. Climate in relation to deciduous fruit production in California. III. Effect of temperature on number of days from full bloom to harvest of apricot and prune fruits. *Proc. Am. Soc. Hort. Sc.* 45, 95–104.

Bliss, L. C. 1967. Phenology program of the IBP. *BioScience* 17, 712–714.

Bomalaski, H. H. 1948. Growing degree days. How to apply this unit to measure maturity of crops. *Food Packer* 29, 51–69.

Braun, E. Lucy. 1950. *Deciduous Forests of Eastern North America.* 596 pp. Philadelphia: Blakiston.

Brown, D. S. 1952. Climate in relation to deciduous fruit production in California. V. The use of temperature records to predict the time of harvest of apricots. *Proc. Am. Soc. Hort. Sci.* 60, 197–203.

Caprio, J. M. 1966. Patterns of plant development in the western United States. *Montana Agr. Exp. Sta. Bull. 607.* 42 pp.

Cristofolini, G. 1970. Biomassa e producttivita dello strato erbaceo di un ecosistema forestale. *Giorn. Bot. Ital.* 104, 1–34.

Curlin, J. W. 1965. Forest management plan, AEC Oak Ridge Reservation. *U.S.-AEC Rept. ORNL-TM-1317.*

Daubenmire, R. F. 1959. *Plants and Environments.* 422 pp. New York: John Wiley.

Fenneman, N. M. 1938. *Physiography of Eastern United States.* 534 pp. New York: McGraw-Hill.

Fernald, M. L. 1950. *Gray's Manual of Botany.* 8th ed., 1632 pp. New York: American Book Company.

Harper, R. 1906. A phytogeographical sketch of the Altamaha Grit region of the coastal plain of Georgia. *Ann. New York Acad. Sci.* 17, 1–414.

Holm, T. 1899. *Podophyllum peltatum.* A morphological study. *Bot. Gaz.* 27, 419–433.

Hopkins, A. D. 1918. Periodical events and natural laws. A guide to agricultural research and practice. *Monthly Weather Rev. Supplement No. 9.*

——— 1938. Bioclimatics—A science of life and climate relations. *U.S. Dep. Agri. Misc. Publ. No. 280.*

Hopp, R. J., M. T. Vittum, and N. L. Canfield. 1969. Instructions for phenological observations: Persian lilac. *Vermont Agr. Exp. Sta. Pamphlet 36.*

Jackson, M. T. 1966. Effects of microclimate on spring flowering phenology. *Ecology* 47, 407–415.

Katz. Y. H. 1952. The relationship between heat unit accumulation and planting and harvesting of canning peas. *Agron. J.* 44, 74–78.

Kreh, H. 1965. Untersuchungen über den Aufbau und die Stoffproducktion eines Sonnenblumenbestandes. *Zulassungsarbeit zur wissenschaftlichen Prüfung für das höhere Lehramt.* Stuttgart: Tech. Hochsch.

Kubicek, F. 1970. Production and phenology of the herb layer in an oak-hornbeam forest. *Biologia* 25 (10), 651–666.

Lieth, H. 1970. Phenology in productivity studies. In *Ecological Studies* Vol. 1, *Analysis of temperate forest ecosystems*, D. E. Reichle ed. pp. 30–46. Berlin, Heidelberg, New York: Springer-Verlag.

——— 1971. The phenological viewpoint in productivity studies. In *Productivity of Forest Ecosystems*, P. Duvigneaud, ed. pp. 71–84. Paris: UNESCO.

———, and J. S. Radford. 1971. Phenology, resource management, and synagraphic computer mapping. *BioScience* 21, 62–70.

Molga, M. 1962. Outline of agrometeorological problems. II. In *Agriculture Meteorology.* Published for National Science Foundation and the Department of Agriculture by Centralny Instytut Informacji Naukowo-Technicznej I. Ekonomicznej, Warszawa.

Newman, J. E., W. C. Cooper, W. Reuther, G. A. Cahoon, and A. Peynado. 1967. Orange fruit maturity and net heat accumulations. In *Ground Level Climatology*, R. H. Shaw, ed. pp. 127–147. AAAS Publ. No. 86. Washington, D.C.: American Association for the Advancement of Science.

Olson, J. S. 1965. Equations for cesium transfer in a *Liriodendron* forest. *Health Phys.* 11, 1385–1399.

Peters, L. N., J. S. Olson, and R. M. Anderson. 1969. Eight-year summary of plant soil data on ^{137}Cs in a tagged Appalachian forest dominated by *Liriodendron tulipifera. U.S. AEC Rept. ORNL-TM-2737.*

Reaumur, R. A. F. 1735. Thermometric observations made at Paris during the year 1735, compared to those made below the Equator on the Isle of Mauritius, at Algiers, and on a few of our American Islands. *Paris Memoirs*, Academy of Science.

Reichle, D. E., B. E. Dinger, N. T. Edwards, W. F. Harris, and P. Sollins. 1973. Carbon flow and storage in a forest ecosystem. In *Carbon and the Biosphere*, G. Woodwell and E. Pecan, ed. Brookhaven Symposia in Biology. *AEC Conf.* 720510.

Schnelle, F. 1955. *Pflanzen-Phänologie*. 299 pp. Leipzig: Akademische Verlagsgesellschaft Geest und Portig K. G.

Scott, G. C. 1946. Quality control in canning peas and corn. *Food Packer* 27, 40–42.

Seaton, H. L. 1955. Scheduling plantings and predicting harvest maturities for processing vegetables. *Food Technol.* 9, 202–209.

Taylor, F. G. 1969. Phenological records of vascular plants at Oak Ridge, Tennessee. *ORNL-IBP-69-1* 46 pp.

Waller, H. D., and J. S. Olson. 1967. Prompt transfers of cesium-137 to the soils of a tagged *Liriodendron* forest. *Ecology* 48, 15–25.

Wallin, J. R. 1967. Ground level climate in relation to forecasting plant diseases. In *Ground Level Climatology*, R. H. Shaw, ed. pp. 149–163. AAAS Publ. No. 86. Washington, D.C.: American Association for the Advancement of Science.

Wang, J. Y. 1960. A critique of the heat unit approach to plant response studies. *Ecology* 41, 785–790.

Woods, C. 1969. Spore trap serves as "early warning system" for leaf blight of sweet corn. *Sunshine State Agr. Res. Rept.* 14, 9–12.

Woods, F. W., and R. E. Shanks. 1959. Natural replacement of chestnut by other species in the Great Smoky Mountains National Park. *Ecology* 40, 349–361.

Yocum, C. S., L. H. Allen, and E. R. Lemon. 1964. Photosynthesis under field conditions. VI. Solar radiation balance and photosynthetic efficiency. *Agron. J.* 56, 249–253.

4.5 Plant Development in Mediterranean Climates[1]

Harold A. Mooney and David J. Parsons[2]

Department of Biological Sciences
Stanford University, Palo Alto, California 94305, USA

and Jochen Kummerow[3]

Laboratório de Botánica
Universidad Católica de Chile
Santiago, Chile

1 Introduction

The summer dry, winter wet climates (Mediterranean) of the world are located in five disjunct areas which generally lie on the west sides of continents between latitudes of 30° to 40° (Fig. 1). All of these areas have vegetations dominated by evergreen sclerophyllous shrubs and trees, and with annuals and herbaceous perennials as additional important vegetation components (Specht, 1969; Mooney and Dunn, 1970; di Castri and Mooney, 1973).

The environmental limitations to plant growth in this climatic type can be illustrated by the seasonal course of climatic elements for a station in southern California (Fig. 2). During the drought period in summer, temperatures are high and evaporative stress is great. In winter, temperatures are cold, but they are generally not cold enough to disrupt photosynthetic systems in evergreen plants. However, there can be cold injury to nonhardened or young tissue.

The most favorable time for most metabolic activity is in late spring when temperatures have become warm, but prior to limitations of summer drought (Fig. 3). The summer drought may deplete the available moisture in the soil profile

[1] Contribution supported by NSF Grant GB 27151.

[2] Present address: U.S.D.I., National Park Service, Sequoia and Kings Canyon National Parks, Three Rivers, California 93271, USA.

[3] Present address: Department of Botany, San Diego State University, San Diego, California 92115, USA.

KEYWORDS: Mediterranean climate; chaparral; California; Chile; sclerophyll; drought; shrub; phenolics; predation; phenology; resource division; carbon balance; carbon apportionment.

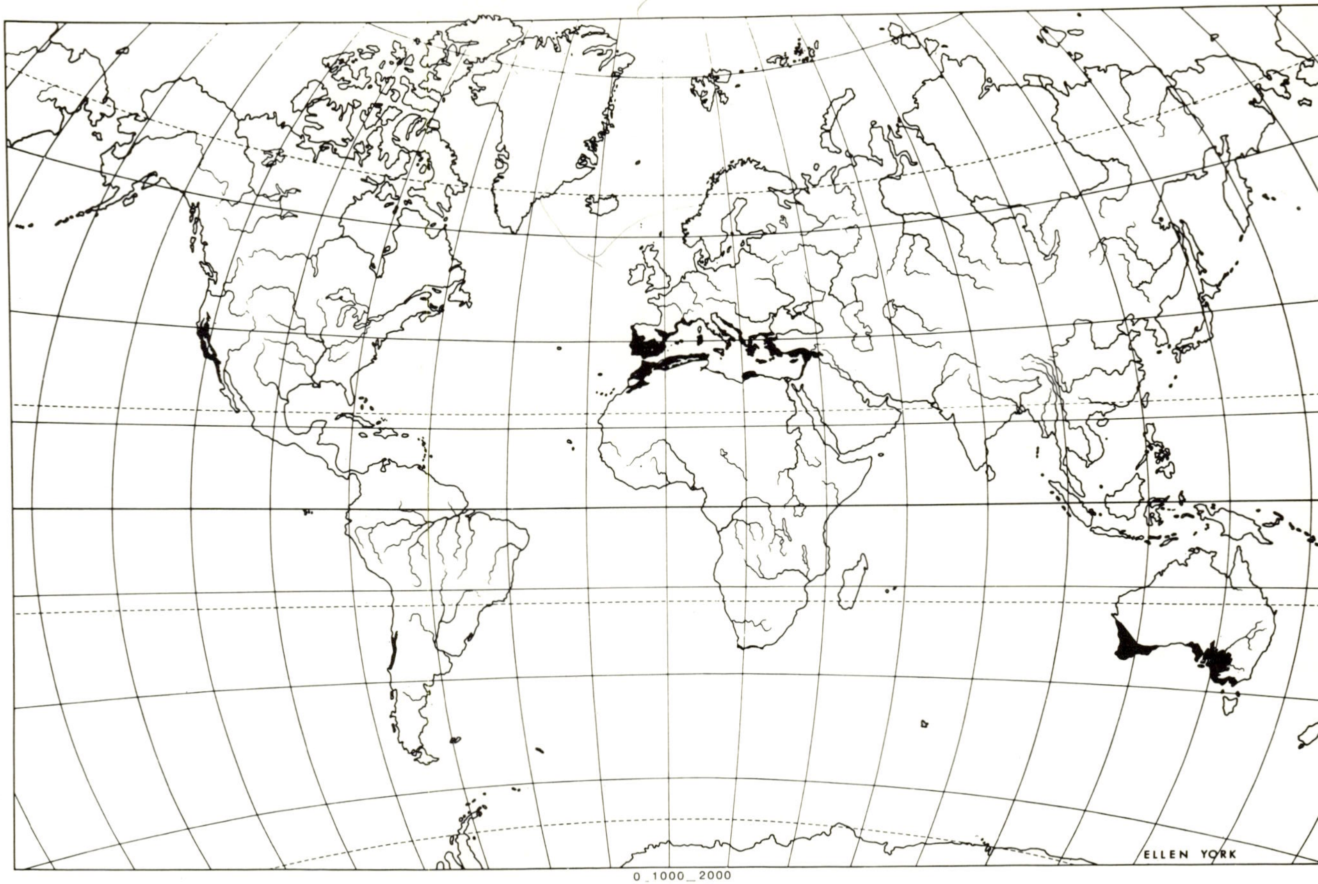

Fig. 1. Distribution of the Mediterranean climates of the world.

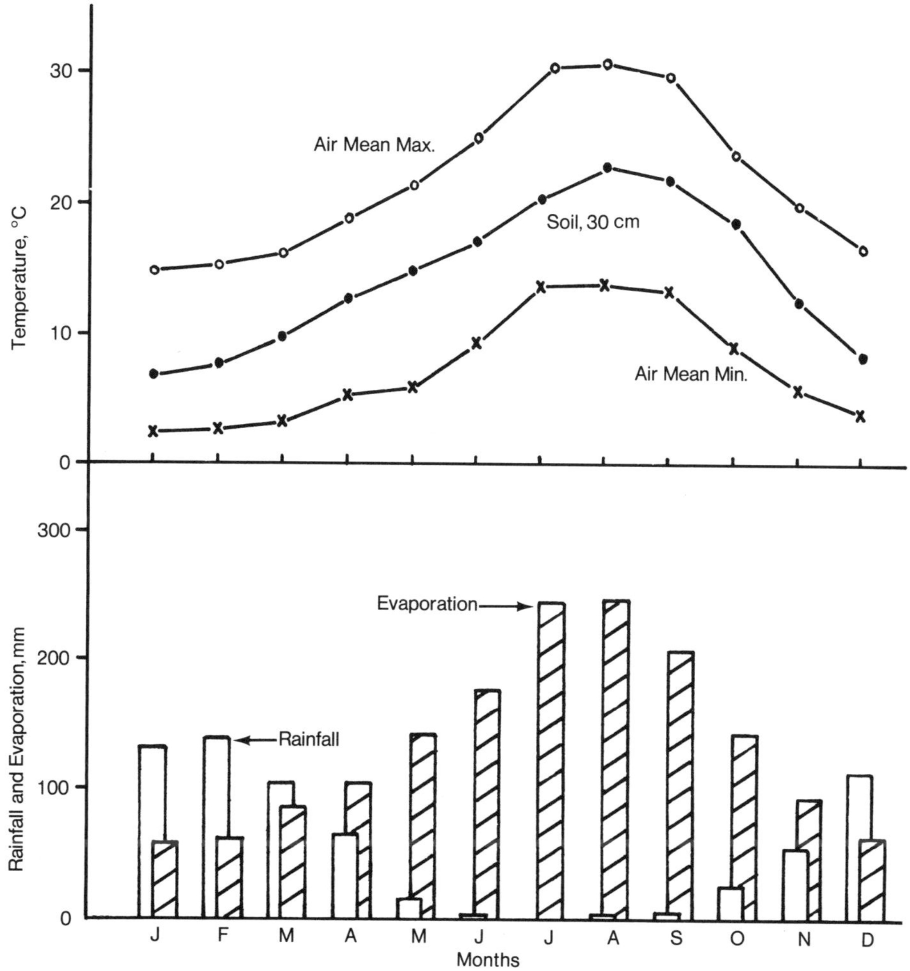

Fig. 2. The Mediterranean climate of a chaparral habitat in the San Gabriel Mountains of southern California. Data for Tanbark Flat (34°12′N, 117°46′W, 853-m elevation). Rainfall records for a 32-year period and evaporation for 25 years (Hill, 1963). Air temperature data for eight years (Qashu and Zinke, 1964).

(Fig. 4), particularly in the upper levels. Deep in the soil, moisture may be available throughout the entire year, particularly in wet years. Deep-rooted evergreen perennials can utilize this moisture to maintain a year-round positive carbon gain. The upper levels of the soil are desertlike, however, and there is often no available moisture for periods up to six months. Shallow-rooted plants occupying this habitat either must have drought dormancy mechanisms or be able to evade the drought as seeds, as do the many annuals found there. Drought duration can vary widely from year to year; hence, so can productivity. The unpredictability of drought intensity and duration has had a profound effect on the evolution of adaptive responses to this unique climatic type (Mooney and Dunn, 1970).

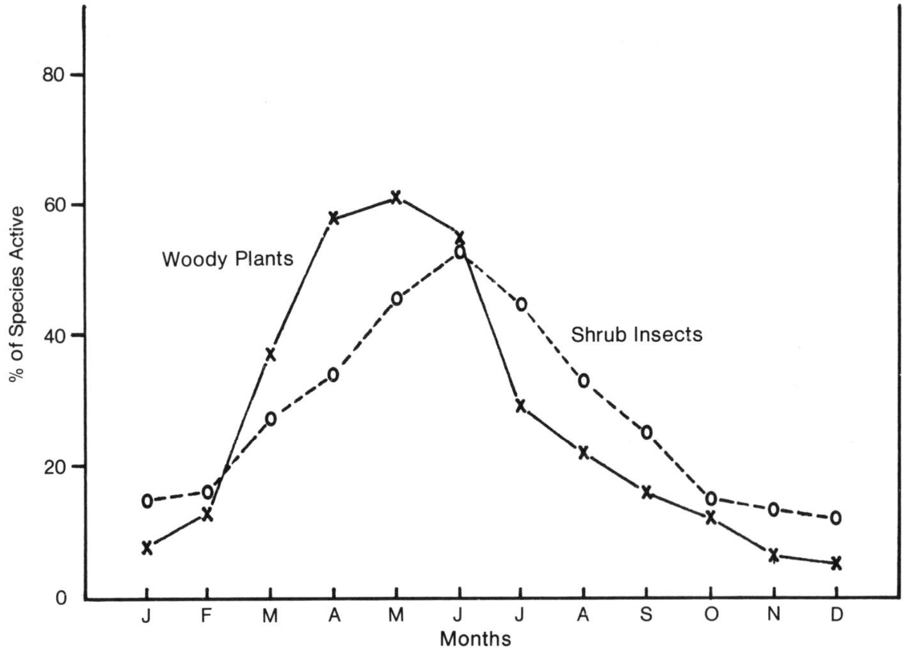

Fig. 3. Seasonal activity of woody plants and of shrub-related insects in the California chaparral. Data for plants represent the percentage of species of the total woody flora of the San Dimas Experimental Forest (101 species) flowering on the given date (Mooney and Parsons, 1973). The insect data indicate the number of species collected on a given date as a percent of the total collected during the year. The insects were all collected in association with the characteristic chaparral shrub *Heteromeles arbutifolia* in the Palo Alto, California, area (Swan, 1952). These two regions are geographically separated in California but they both are within Mediterranean climates.

The linkage between climate and plant development is generally similar in the world's various Mediterranean climates (Fig. 5), particularly in the Mediterranean Basin, California, and Chile, where the climates are most analogous. Seasonal flowering activity of the sclerophyll-dominated vegetation of Australia is also mostly in the spring; however, there is a sharp drop in the long dry summer. Yet vegetative growth of Australian shrubs may continue throughout the drought period—uncharacteristic of other Mediterranean climatic regions where growth is predominantly during spring (Specht, 1969). In the South African region of the sclerophyllous fynbosch vegetation, flowering activity is more evenly distributed throughout the year, as is rainfall.

2 Plant Development as the Timing of Carbon Apportionment

We now consider more specifically the environmental and biological characteristics of this climatic type which limit developmental events to certain periods.

Plants of any environment have certain basic functions to perform in order to survive and reproduce (Fig. 6). They must produce and maintain (and protect from predators) their systems for gaining and transporting carbon, as well as water and minerals. Furthermore, they must produce, maintain, protect, and disseminate propagules. All of these functions require energy, which is stored and transported as carbon compounds.

Generally, all of these processes do not progress at equal rates. Evidently there are time periods when one or more of the processes will not proceed either because of environmental limitations or because it may not be adaptive for a particular event to proceed; e.g., flowering at a time when pollinators are not present.

Plants probably cannot apportion all of their carbon to all life events simultaneously. There must be an economy of carbon when there will be periods of gain and storage and then periods of rapid growth. Also there must be periods when more growth is directed toward one system than another.

In order to explore how, when, and for what functions plants apportion their carbon *Heteromeles arbutifolia*, a California evergreen shrub adapted to the Mediterranean climate, will be considered as an example. This plant has the capacity to

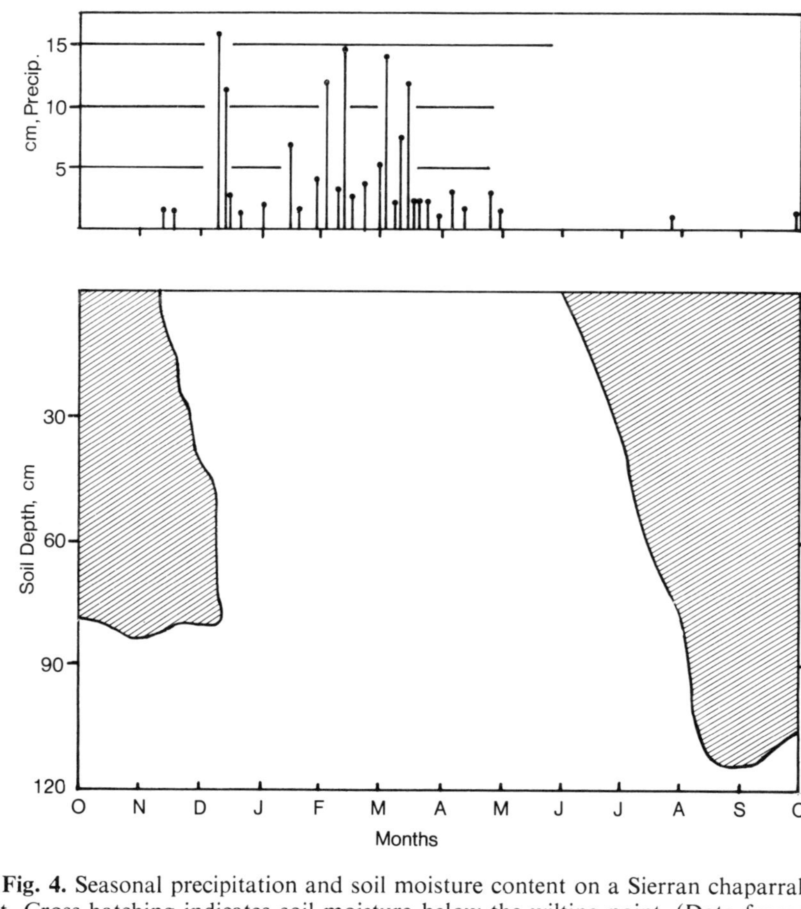

Fig. 4. Seasonal precipitation and soil moisture content on a Sierran chaparral study plot. Cross-hatching indicates soil moisture below the wilting point. (Data from Rowe, 1948.)

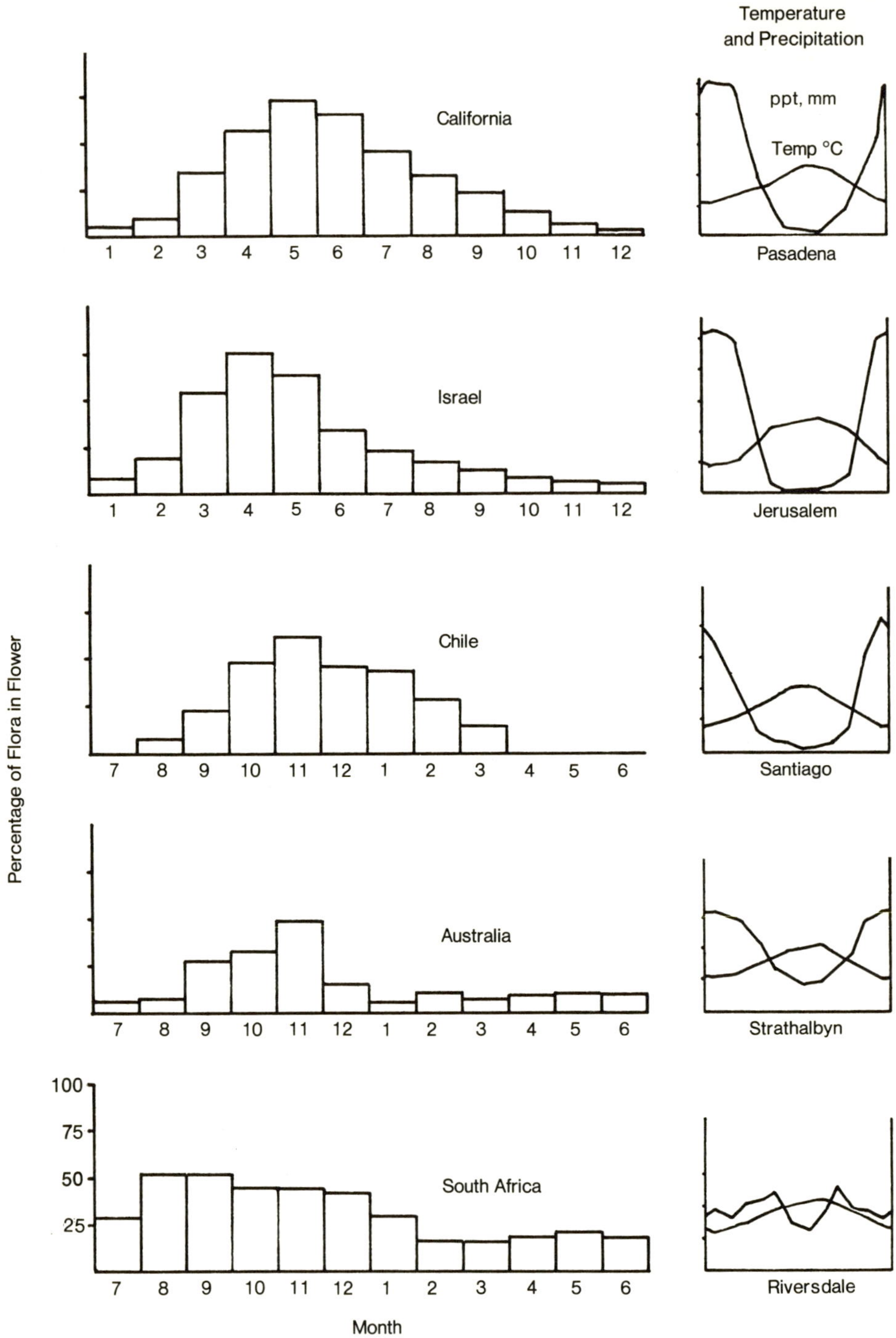
Temperature
and Precipitation
California
ppt, mm
Temp °C
Pasadena
Israel
Jerusalem
Chile
Santiago
Australia
Strathalbyn
South Africa
Riversdale
Percentage of Flora in Flower
100
75
50
25
Month

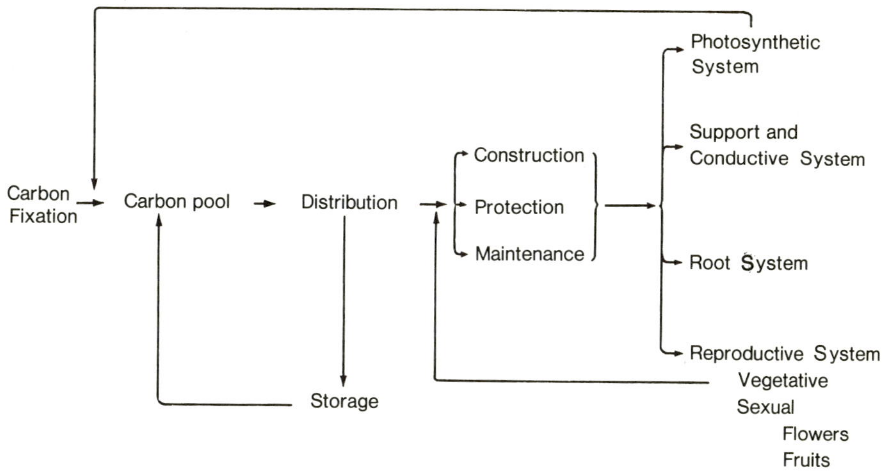

Fig. 6. The uses of carbon by plants.

fix carbon throughout the entire year (Fig. 7). Peak activity is during winter and spring when water is nonlimiting. However, photosynthesis will normally continue throughout the summer unless the drought is unusually severe.

The carbon gained is used for growth during late spring and early summer. Not until near the end of stem growth do flowering and, subsequently, fruiting begin. Exactly when growth commences in spring is probably related, in part, to when temperatures become favorable, since new growth on many shrubby sclerophylls can be damaged by cold. In *Heteromeles*, at least, the end of the stem growth period is evidently related more to carbon demands for reproductive functions in the plant than it is to climatic events. This can be seen by comparing the stem growth activity of reproductive adults versus nonreproductive plants growing in the same environment (Fig. 8).

It is during the period of the lowest carbon gains that there are the highest requirements for carbon for growth and reproduction (Fig. 7). These relationships would indicate a possible shortage for other functions at that time. Leaves of *Heteromeles* contain large amounts of phenolic compounds, including tannins. Such compounds have been strongly implicated in predator protection relationships. It can be seen that these compounds are produced in abundance only during periods when stem growth is not occurring.

Fig. 5. Seasonal flowering activity of plants growing in the five Mediterranean climatic regions of the world. Climatic data are for the nearest station to the phenological observations (from Walter and Lieth, 1960–1967). The California phenological data are for the flora of the San Dimas Experimental Forest, southern California (Mooney and Parsons, 1973); the data for Israel for the entire flora of the region (Zohary, 1962); climatic data are for Jerusalem, a representative station. Chile phenological observations are for shrubs only and are derived from personal observations and local floras for the Santiago region; Australian observations are from Specht and Rayson (1957); South Africa data are from Adamson and Salter (1950) and Muir (1929).

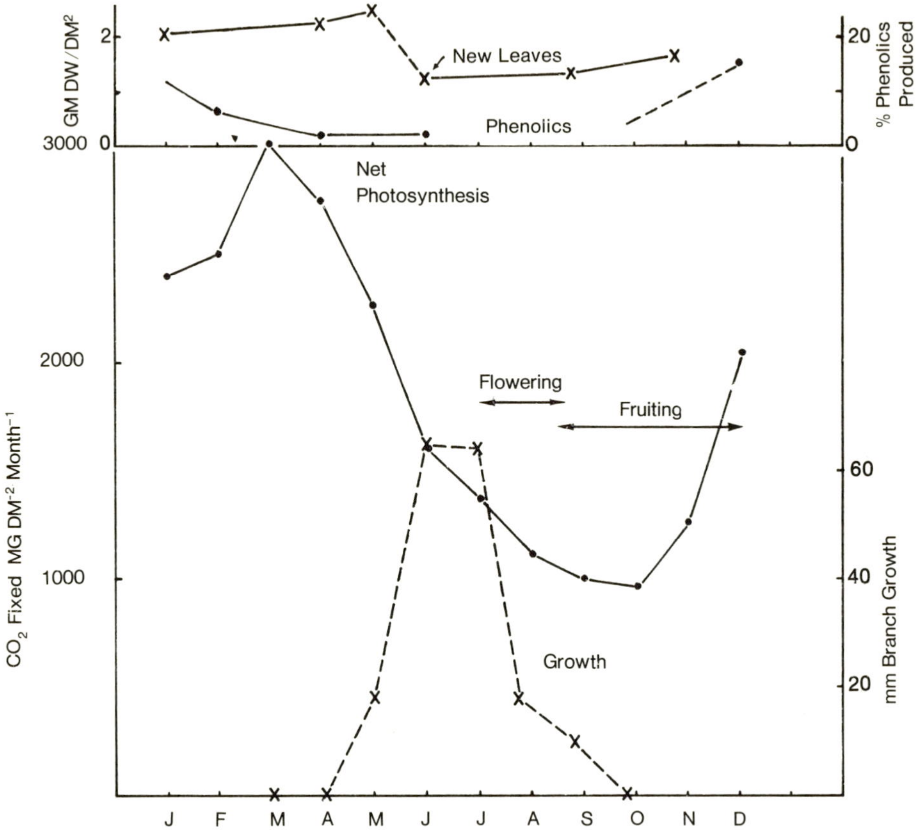

Fig. 7. The seasonal carbon gaining capacity and selected carbon-use patterns of the California sclerophyll shrub *Heteromeles arbutifolia*. Photosynthesis values are calculated for a plant growing in a native environment in northern California. Growth and phenological measurements taken from plants of a local population growing under a natural environmental regime in Palo Alto, California, in an experimental garden. The phenolics produced are as a percent of the total $^{14}CO_2$ fixed at the given time period on potted plants growing under Palo Alto climatic conditions (all of the above, unpublished data). Leaf density measurements are for plants growing in the southern California chaparral (data from Song, 1970).

Leaves of sclerophylls continue to increase in density throughout the year subsequent to their formation in late spring (Fig. 7) The denser leaves are not only more tolerant of desiccation injury but are presumably more resistant to predation. If carbon were channeled directly into sclerification immediately upon leaf formation and prior to the drought the plants would be protected from the start. This does not occur, however, perhaps because of the carbon limitations mentioned above. The results of these events are that during early summer there are leaves which are low in density and low in phenolics. Thus they have a low resistance to both drought and predators. These relationships are probably a tradeoff between the competitive benefits of rapid channeling of carbon into developing a full new canopy against the possible loss of new growth by not developing environmental protection systems concomitantly.

We are just beginning to understand some of the environmental and biological limitations on the timing of carbon allocation by *Heteromeles*. Of course, not all sclerophylls of Mediterranean climate have identical strategies of growth. A comparison of growth, flowering, and fruiting of a number of Californian and Chilean sclerophyll shrubs and trees illustrates these differences (Fig. 9). Plants of some species, such as the Californian *Ceanothus leucodermis*, manage to flower, fruit, and produce leaves at the same time; others delay fruiting at least until after stem growth, as does *Heteromeles*. The morphological and phenological bases for these differences need to be established. In order for reproduction to occur simultaneously with growth, these events either have to happen at a slower rate than if growth were occurring alone or else they must occur at the expense of reserves (cf. Chapter 4.4).

The period of stem growth is generally more synchronous between species in most communities than is reproduction. This is probably because the timing of canopy development is highly competitive between species. On the other hand, given allowance for adequate reserves and the consequences that result from storage in delaying other functions, these plants may compete for entirely different pollinators and fruit dispersers. Furthermore, they may be attempting to evade different classes of herbivores. Thus there may be quite dissimilar developmental patterns between species. This assumes, as seems to be true in Mediterranean climates, that pollinators, dispersers, and predators are not all limited climatically to a very brief time period.

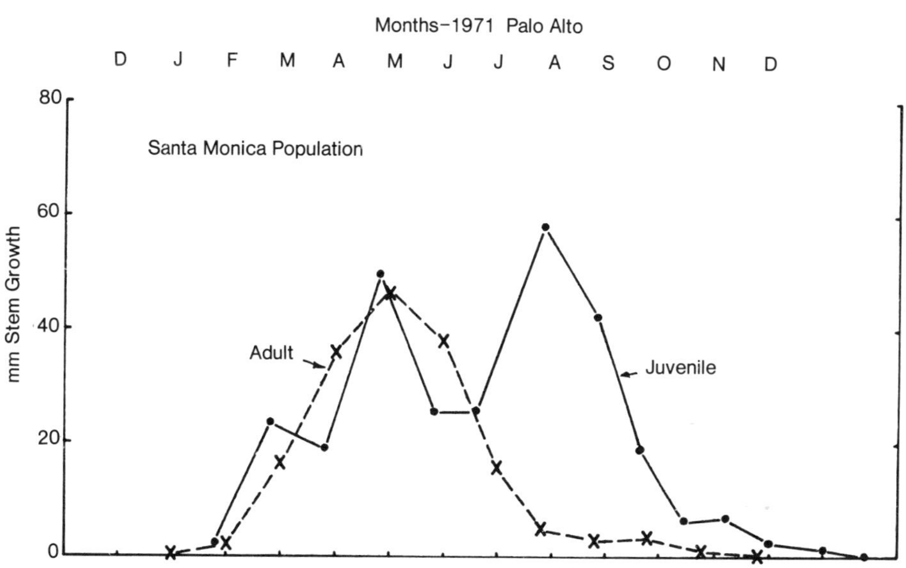

Fig. 8. Stem growth of reproductive adults and juvenile plants of *Heteromeles arbutifolia* growing in an experimental garden in Palo Alto, California, These plants were watered throughout the year. The origin of these plants was the Santa Monica Mountains in southern California. Growth was measured on vegetative branches on all plants. Reproductive parts in this species are located on the terminal portions of stems.

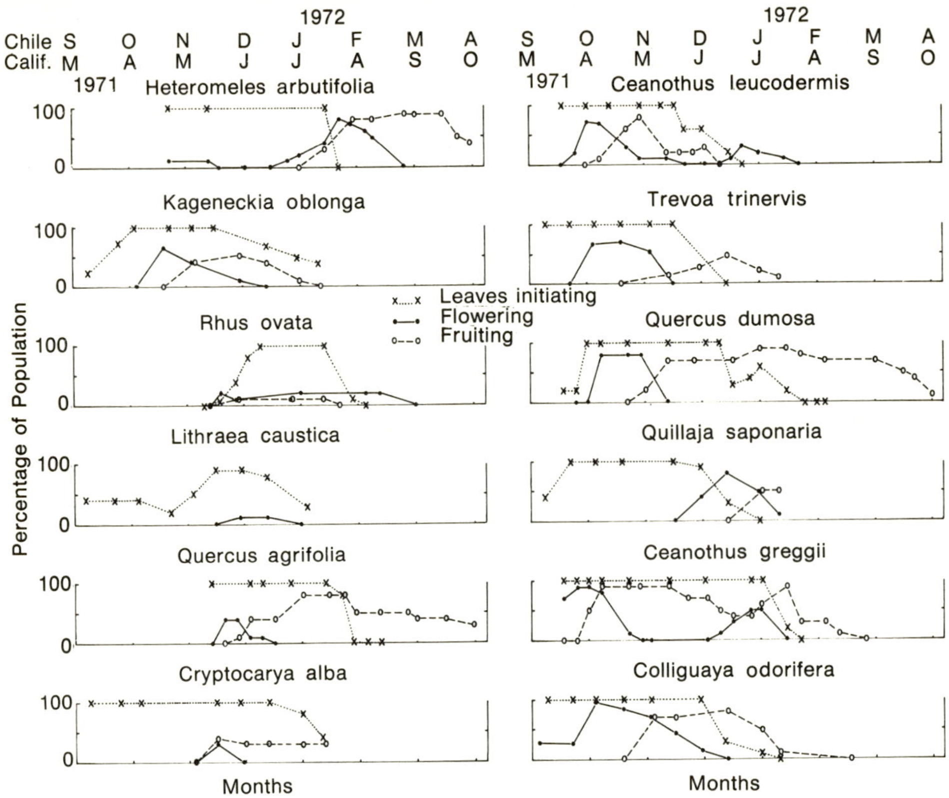

Fig. 9. Phenological development of shrubs and trees of the mediterranean climatic regions of California and Chile growing in their native environments. The plants are paired top to bottom for general morphological equivalence; e.g., *Heteromeles* is a California shrub which has a similar appearance and ecology to *Kageneckia* from Chile. *Rhus*, *Quercus*, and *Ceanothus* are all Californian, and *Lithraea*, *Cryptocarya*, *Trevoa*, *Quillaja*, and *Colliguaya* are all from Chile. Numbers represent the percent of 10 plants in a given phenological state (unpublished data).

Flowering activity of the different California chaparral shrub species in a given habitat may be staggered through a six-month period starting in late winter (Fig. 10). Of the 12 species indicated in Fig. 10, 10 are pollinated by polylectic insects (a couple also have specialist pollination and one is wind pollinated). Many of these pollinators are active year-round (A. Moldenke, personal communication). These shrubs may evolutionarily divide a common pool of pollinators by their phenological behavior.

We hope to gather considerably more information on the phenological and physiological responses of plants in Mediterranean climates so that we will be able to say for any given species or growth form how and when it gains and utilizes its carbon for competition, reproduction, and protection. Obviously, detailed information is needed, not only on microclimate and physiology of a given species, but also on the phenology of associated competitors, pollinators, and predators. A

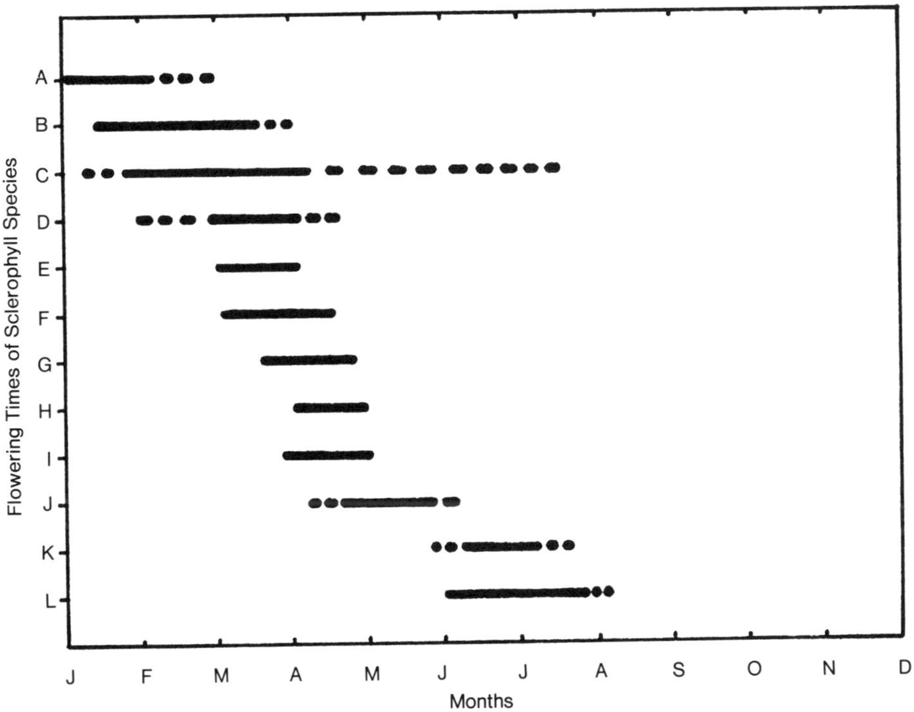

Fig. 10. Flowering times of sclerophyllous shrubs in the Santa Monica Mountains, Southern California (data for 1932 season, Bauer, 1936). *A; Arctostaphylos glandulosa; B*: *Ceanothus macrocarpus; C*: *Dendromecon rigida; D*: *Ceanothus spinosus; E*: *Rhamnus crocea; F*: *Cercocarpus betuloides; G*: *Rhus ovata; H*: *Quercus dumosa; I*: *Prunus ilicifolia; J*: *Adenostoma fasciculatum; K*: *Heteromeles arbutifolia; L*: *Rhus laurina.*

portion of this information is coming from the International Biological Program. Background is provided by *Ecological Studies*, Vol. 1 (Reichle, 1970) and Vol. 7 (di Castri and Mooney, 1973).

3 The Phenological Program of the US/IBP Mediterranean Scrub Project

One component of the US/IBP Program is a study of the origin and structure of ecosystems. This project is comparing the structural aspects of ecosystems in Mediterranean homoclimates in California and Chile (as well as desert climates of the United States and Argentina) to determine if equal environment leads to equal structural and functional ecosystems. The timing of biological events in these matched systems is being documented in a number of ways. For example, the general linkage between macroclimate and plant development in these areas is being followed on a regional basis by assessing responses of selected species along environmental gradients. Populations of 10 individuals of each species are scored for such phenological events as leaves initiating, flower bud, flowering, fruiting, fruits dispersing, and leaves drying or falling. For any date, then, the percent of the population in a given phenological state can be noted (see representative data of Fig. 9). These semiquantitative data are easily obtained, and can be applied to many species and stations to give wide regional as well as floristic coverage.

The phenological development of plants can be viewed in terms of the pulsing of energy sources into the ecosystem for consumers. Every new event, e.g., leafing out, flowering, or leaf fall, corresponds to the initiation of availability of a food source for a different group of consumers. Our program is attempting to quantify these relationships for shrubs to see how comparable they are between our matched ecosystems. The basic program establishes the dry weights of various plant parts (old leaves, new leaves, flowers, fruit) for a given sample period for a standard sample branch. The number of sample branches per square meter of shrub is determined, as is the percent coverage of the species in the community. From these data the quantity of a particular plant component can be calculated for the entire community at any given time. We are also beginning to obtain data on the energy content of these components, as well as the content of such important nutrients as carbohydrate, nitrogen, and phosphorous. Additionally, the secondary chemistry of certain species is being followed to relate energy content to actual energy availability to consumers. Additional projects are quantifying the seasonal behavior of various groups of consumers.

These data should provide the necessary links between climate and plant species, as well as community phenological response.

Acknowledgment

This study is a contribution of the United States and Chile to the International Biological Program. We thank Adrianna Hoffmann and M. Beauchamp for collection of some of the data utilized here.

References

Adamson, R. J., and T. N. Salter. 1950. *Flora of the Cape Peninsula*. Capetown: Juta and Co.

Bauer, H. 1936. Moisture relations in the chaparral of the Santa Monica Mountains, California. *Ecol. Monograph* 6, 409–454.

di Castri, F., and H. A. Mooney. eds. 1973. Mediterranean Type Ecosystems. In *Ecological Studies*. Vol. 7. New York: Springer-Verlag.

Hill, L. W. 1963. *The San Dimas Experimental Forest*. Pacific Southwest Forest Range Exp. Sta. 25 pp.

Mooney, H. A., and E. L. Dunn. 1970. Convergent evolution of Mediterranean climate evergreen sclerophyll shrubs. *Evolution* 24, 292–303.

Mooney, H. A., and D. J. Parsons. 1973. Structure and function of the California chaparral —An example from San Dimas. In *Ecological Studies*, F. di Castri and H. Mooney, ed. Vol. 7. New York: Springer-Verlag.

Muir, J. 1929. The vegetation of the Riverdale area, Cape Province. *Bot. Surv. So. Afr. No. 13* 81 pp.

Qashu, H., and P. Zinke. 1964. The influence of vegetation on soil thermal regime at the San Dimas lysimeters. *Soil Sci. Soc. Am. Proc.* 28, 703–706.

Reichle, D. E. Ed. 1970. Analysis of temperate forest ecosystems. In *Ecological Studies.* Vol. 1. New York: Springer-Verlag.

Reimann, L. F. 1959. Mountain Temperatures. Data from the San Dimas Experimental Forest. *Pac. Southwest Forest Exp. Sta. Misc. Paper 36.*

Rowe, P. B. 1948. *Influence of Woodland and Chaparral on Water and Soil in Central California.* California Forest Range Exp. Sta. 70 pp.

Song, L. C., Jr. 1970. Some studies of the seasonal nature of leaves of selected plants in the sclerophyll vegetations of the California chaparral and the Chilean matorral. Unpublished M.S. Thesis. Los Angeles: University of California.

Specht, R. L. 1969. A comparison of the sclerophyllous vegetation characteristic of Mediterranean type climates in France, California, and southern Australia. I. Structure, morphology, and succession. *Australian J. Bot.* 17, 277–292.

———, and P. Rayson. 1957. Dark Island Heath (Ninety-Mile Plain, South Australia). I. Definition of the ecosystem. *Australian J. Bot.* 5, 52–85.

Swan, L. W. 1952. The ecology of the arthropods associated with *Photinia arbutifolia.* 322 pp. Unpublished Ph.D. Thesis. Stanford, Calif.: Stanford University.

Walter, H., and H. Lieth. 1960–1967. *Klimadiagramm-Weltatlas.* Jena: VEB Gustav Fischer Verlag.

Zohary, M. 1962. *Plant Life of Palestine.* 262 pp. New York: Ronald Press.

4.6 Phenology of the Zambezian Woodland Area with Emphasis on the Miombo Ecosystem

François P. Malaisse

Laboratoire de Botanique et de Climatologie
Université Nationale du Zaïre
Lubumbashi
Zaïre, Africa

1 Introduction

Open woodland is the dominant vegetation formation class over 12% of Africa, or 3.8×10^6 km^2 (Fig. 1), occurring only in the Sudano-Zambezian region. This woodland area is in two parts: a narrow strip north of the equator related to the Sudanian Domain and a wide distribution southward, corresponding closely to the Zambezian Domain (White, 1965).

Within this formation class, several major ecosystems may be distinguished. Aquatic communities represented are natural lakes (Moero, Bangweolo, Malawi); man-made lakes (Nzilo, Lufira, Koni, Lubumbashi); permanent or temporary running waters, including streams and rivers (Upper Lualaba, Lufira, Kafue, Zambeze); and, finally, "dambos," which are temporarily flooded savannas located in shallow depressions or drainage canals. In Upper Katanga (now Shaba) we distinguish four main vegetation types: steppe, savanna (open), woodland (miombo), and forest; this last has two divisions, riparian forest (mushitu) and dry evergreen forest (muhulu) (Malaisse, 1973).

Until now little has been known about the phenology of the Zambezian woodland area. This paper compiles existing and newly gathered phenological information about the miombo ecosystem, including the water bodies found within this vegetation type.

KEYWORDS: Phenology; seasonality; woodland; miombo; Kongo, Africa; plants; flowering; birds; insects; decomposer activity; precipitation effects; Zaïre

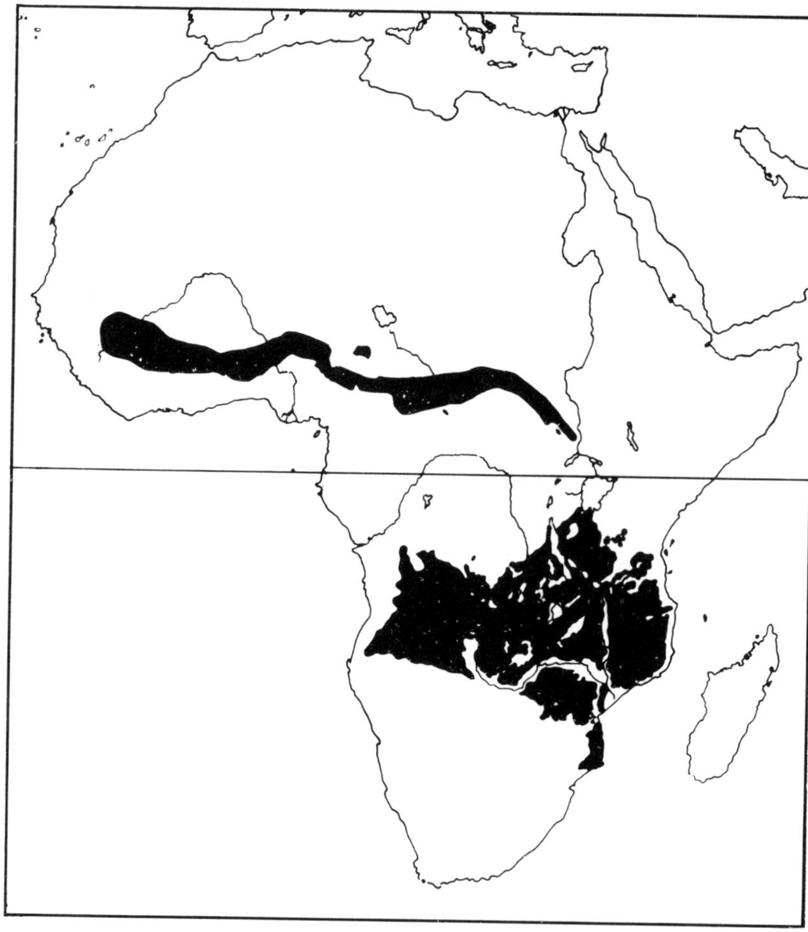

Fig. 1. Areas of Africa where woodland is the dominant vegetation type. (After A.E.T.F.A.T. vegetation map, 1959.)

2 Ecoclimatic Environment

The macroclimate of Lubumbashi (formerly Elisabethville) is characterized by one rainy season (November to March), one dry season (May to September), and two transition months (October and April). July and August are always without rainfall. Yearly mean total rainfall is about 1270 mm, with a range of 895 to 1551 mm for 1918 to 1970. Some rare showers exceed 100 mm (160 mm in 2 hr 40 min in 1923), half of them are 5 to 50 mm, and the rest are under 5 mm. The rainy season usually lasts 118 days. The mean yearly temperature is about 20° C. Temperature is lowest at the beginning of the dry season, from mid-May to mid-July. The minimum is generally measured around 6 AM, and may occasionally fall to the freezing point. September, October, and sometimes November are the warmest months, with an average maximum daily temperature of 31° to 33°C. The dry season has two distinct periods: a cold dry season (May to July) and a warm dry season (August to October). During the last period, heating of the soil induces the austral spring characterized by the flowering and flushing of trees before the beginning of rain. Temperature amplitude is low in rainy seasons and high in dry, cold seasons. The relative humidity of the air follows the rainfall pattern, with a minimum in October and a maximum in February.

The mean yearly solar radiation is 16.8×10^9 kcal/ha/yr, of which 7.75×10^9 kcal/ha/yr is between 400 and 700 mμ. The minimum average daily radiation occurs during the rainy season due to the high cloudiness (394 cal/cm^2/day in December) and increases progressively until May. Values for June and July remain somewhat low. The maximum occurs during the warm dry season (533 cal/cm^2/day in September), the sky's clarity overriding the additional effects of atmospheric opacity due to bush fires and the decrease in incoming solar radiation (Bernard, 1950). Average daily hours of sunshine are lowest in December and maximal in July; the annual daily mean is 7.67 hr. The main climatic data are tabulated in Table 1.

3 Terrestrial Communities

The Miombo Ecosystem

Miombo represents 98% of the combined forest and woodland area, and 90% of the area around Lubumbashi. Miombo is a tropical, dry, primarily deciduous woodland. In Upper Shaba, *Brachystegia* spp. and *Julbernardia paniculata* (Benth.) Troupin are the common dominant trees. Fire is a very important factor in the miombo ecosystem; miombo is generally regarded as secondary growth, largely occurring after destruction of dry evergreen forest (Fanshawe, 1971). The seasonal rainfall pattern causes striking changes in miombo appearance; the principal dry and wet aspects are shown in Plate 1.

Repeated fires late in the year produce in this area wooded savanna and finally grass savanna vegetation, while total absence of fire induces muhulu (dry evergreen forest). Fires destroy a part of the above-ground organs of the herb layer frequently and sometimes annually. Heat intensity reached in these fires depends on when they occur during the dry season and on the quantity of grass, tree litter, etc., available. Late fires often destroy the leaves at heights of 10 to 12 m.

A second typical characteristic of miombo is the presence of large termite mounds, sparsely distributed (3 to 5/ha). They reach 8 m in height and 15 m in width, and may cover 4.3 to 7.8% of the miombo (Sys, 1957). Termite mounds have a high clay/silt content, a high mineral content, and an alkaline pH, and are rich in bases and humus. They also have a higher moisture content than the surrounding woodland (Fanshawe, 1971). Consequently, termitaria vegetation is quite different from that of the surrounding miombo.

Table 1. Climatic data for the Kasapa Miombo, as recorded on the station located at the study site.

	Daily Total Radiation cal/cm²	Mean Daily Sunshine, 1/10 hr.	Temperature, °C Mean Max.	Mean	Mean Min.	Extreme Temp., °C Max.	Min.	Mean Daily ampl., °C	Average Rainfall, mm	No. Rainy Days	Rel. Hum., %	Piche Evap., ml/day
Jan.	423	48.6	28.1	22.2	16.3	36.4	12.4	11.8	**299.1**	**25**	**89**	1.8
Feb.	414	48.5	28.2	22.4	16.6	31.0	13.5	11.9	**247.4**	22	**90**	1.8
Mar.	442	58.6	28.5	22.1	15.6	31.7	10.3	12.8	**174.9**	18	**89**	1.9
April	472	82.8	29.3	20.8	12.3	33.0	6.0	16.9	36.5	7	82	2.5
May	478	99.6	28.6	18.1	7.5	34.2	1.5	21.1	3.7	1	76	3.2
June	436	97.3	27.1	16.2	**5.3**	30.5	−1.0	21.7	9.2	1	76	3.6
July	469	**102.7**	26.9	**15.6**	**4.2**	35.0	**−1.5**	**22.2**	0	0	73	4.4
Aug.	**493**	98.6	29.1	18.1	7.0	35.0	0.9	**22.2**	0	0	63	**5.6**
Sep.	**533**	97.8	**31.9**	21.0	10.0	**37.8**	4.2	**21.9**	4.1	1	58	**6.2**
Oct.	**516**	88.6	**32.8**	23.1	13.2	36.5	6.0	19.6	36.1	4	53	**5.8**
Nov.	456	61.1	**29.6**	**22.7**	15.7	35.5	11.6	13.8	**186.2**	17	82	2.8
Dec.	394	42.2	27.9	22.0	16.4	32.5	11.8	11.9	**278.2**	**24**	87	1.9
Year	461	76.7	32.8	20.3	4.2	37.8	−1.5	17.3	1273.4	120	77.3	3.5

Records are for 1964–1970, except for solar radiation and insolation which are based on 1966–1970. Bold figures show extremes.

Plate 1. Two aspects of the Luiswishi Miombo at two-month interval. *Above*, 5 November 1971. The tree canopy is already well developed but the ground layer is barely started on places that had burned during the last dry season. Behind, at the right side, is a termite mount. In the center we observe a *Marquesia macroura* Gilg. tree with its typical buttresses. *Below*, 30 December 1971. The miombo is in full development, the tree canopy reaching its maximum density. The height of the herb layer is at one-half its final stage of development.

Primary Producers[1]

Few data were yet available on miombo phenology when our study was started. However, for Lupa North Forest Reserve (Tanzania), one year's observations have been compiled by Boaler (1966) for three sites in the miombo. This author recognizes five seasons (Table 2). In Lupa North Miombo, the main flowering period for trees is in the prerainy season, while suffrutex have two maxima, one in prerainy and the other in the main rainy season. From phenological considerations, the dry season and the late rainy season are periods of minimum activity.

Table 2. Number of individuals in flower: eight vegetation types observed at five seasonal periods. Lupa North Miombo Site (middle slope), Tanzania. Data from Boaler (1966).

Seasons	Canopy trees	Underwood trees	Shrubs	Suffrutex	Herbs	Bulbils	Sedges	Grasses	Total
Dry	1	3	2						6
Prerains	5	6	4	15	1	1			32
Early rains			1	6	2	2	6	2	19
Main rains			2	13			1	8	24
Late rains		1	6	2					9

In Shaba observations have been carried out at Kasapa Miombo on the University Campus, in the vicinity of Lubumbashi. A first year of observations was made in order to define for the most abundant species different phenological stages (Malaisse, 1964) such as presence or absence of leaves (buds are generally lacking in Zambezian woodland), flowers and inflorescence buds, blossoming, unripe seeds and fruits, ripe seeds, and dispersal (Lieth, 1970). At the same time a reference collection was prepared for the University Herbarium in Lubumbashi. During the second year, data were collected weekly and the different stages reached were noted whenever possible for about 50 individuals of each species.

The first data on Shaba Miombo (Fig. 2) show that tree flowering occurs mainly during the warm dry season. However, a few trees are exceptions: *Julbernardia paniculata* (Benth.) Troupin, for instance, flowers from February to April. The greatest number of flowering herbs and suffrutex is observed around New Year. However, flowers may be observed year round (Fig. 3).

We propose for Upper Shaba Miombo to distinguish five phenologically

[1] In answer to Dr. Harold C. Fritts' question about the periodical activity of cambium in the tropics, studies are in progress for the miombo. Some information is in the literature. Daubenmire (1972) gives values in percent of radial changes for 36 species of a tropical semideciduous forest in northwestern Costa Rica. Other papers dealing with periodic activity of trees are quoted in this work. For the Ivory Coast rain forest, Bernhard-Reversat *et al.* (1972) have observed the change in basal area of trees in a 800-m² plot for three years, at three-week intervals.

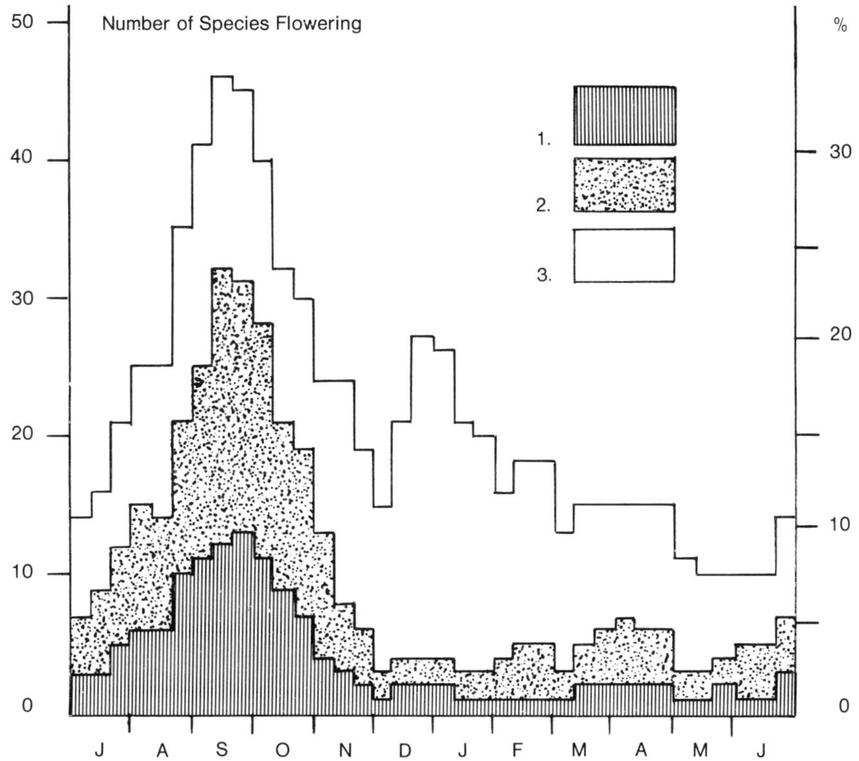

Fig. 2. Phenogram for the flowering phases of miombo in the vicinity of Lubumbashi. This figure permits comparison between maximum flowering period of trees (*1*), shrubs (*2*), and ground layer (*3*).

defined seasons: warm dry (August–September), early rains (October–November), main rains (December–February), late rains (March–April), and cold dry (May–July), corresponding roughly to the months given in parentheses. Variations from year to year may be observed in any one place, reaching as much as 30 days in amplitude.

As a first trial we have compared the dates of the beginning of rains with phenological data taken along the Kolwezi-Lubumbashi-Kasumbalesa Highway, which traverses Upper Shaba from northwest to southeast. *Haemanthus multiflorus* Martyn, which inhabits the large termite mounds and is easily recognized by its large red spherical umbel, yielded in 1971 the following results:

Site	*Latitude South*	*Haemanthus multifl. flowering period*	*Date when seasonal sum of rain after July exceeds 50 mm*
Kolwezi	10° 45′	10 Oct.–25 Oct.	30 Sept.
Lubumbashi	11° 39′	24 Oct.– 5 Nov.	15 Oct.
Tshinsenda	12° 38′	1 Nov.–13 Nov.	22 Oct.

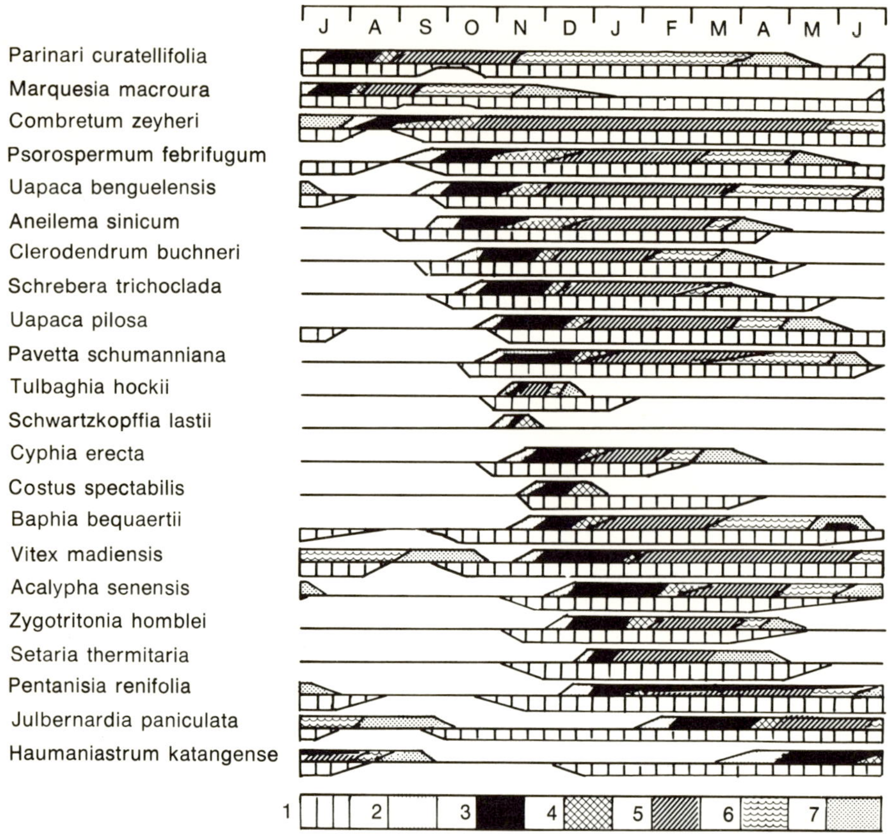

Fig. 3. Phenological spectrum for 22 selected species of a miombo at Kasapa (11° 36′ 38″ S, 27° 28′ 47″ E) in 1971–1972. The phenophase key is (1) foliation, (2) flower and inflorescence buds, (3) blossoming, (4) faded flowers, (5) unripe fruits, (6) ripe fruits, (7) scattering.

The common bamboo of our area, *Oxytenanthera abyssinica* (A. Rich.) Munro, produces a vigorous vegetative shoot every rainy season, but flowering shoots are far less frequent and appear only one year, sometimes two, after a very cold dry season.

Secondary Producers

The extreme changes in climatic conditions and the response of the vegetation induce seasonality in miombo secondary producers. This seasonality deals with developmental stages, population density, diet, specific composition, etc. Birds provide a good example of changes in specific composition. Many remain in the miombo year-round (*Rhinopomastus cyanomela* (Viellot), *Oriolus larvatus angolensis* Lichtenstein, etc); others are seasonal visitors of the dry (*Upupa africana* Bechstein) or rainy (*Poicephalus meyeri* (Cretzschman)) seasons

Some information is available on the timing of developmental stages of caterpillars. A first group of caterpillars develops first instars during the main rainy season and last instars in the late rainy season. *Elaphrodes lactea* (Gaede), *Holocerina agomensis* Karsch., etc. belong to this group. For a second group, eggs are laid around September–October and adults occur in the dry season; for instance, *Rhenea mediata* (Walker), *Xanthopan morgani* Walker, etc. Other species do not present conspicuous phenophases with sharp temporal limits, and adults may occur any time during the rainy season for instance.

Research on many groups of secondary producers inhabiting the herb layer is being carried out and the results will be available shortly.

Decomposers[2]

Decomposer activity is mainly related to litter accumulation, soil humidity, and temperature. These three factors vary differently through the year. Litter accumulation depends on litterfall, which is minimum from December to March and then increases progressively to a maximum in September (Malaisse and Malaisse-Mousset, 1970) as well as on litter breakdown, which is highest in the rainy season. Soil humidity is closely related to precipitation, whereas soil temperature depends on air temperature. We will discuss seasonal activity of pedoflora and pedofauna in miombo next.

Pedoflora. Malaisse-Mousset has studied the soil microbial flora on samples taken in various vegetation types of Upper Shaba. Methods and results about miombo, among others, have been published (Malaisse *et al.*, 1972). Figure 4 gives variations in soil microbial parameters in the miombo ecosystem for one year. These data reveal three different patterns for the early rainy season, main rainy season, and dry cold season. Algae decrease progressively; the high values of November are related to light intensity on the ground which is still high at this time before the tree canopy is fully developed and the herb layer closes again. On the other hand, aerobic digestion of cellulose is inefficient during all the rainy season but becomes important in the dry season when wood decays more quickly, as already pointed out by Hopkins (1966). The other microbial processes (proteolytic digestion, starch digestion, and ammonification) have a maximum activity in the main rainy season and appear to be highly correlated to soil moisture content.

Pedofauna. For many years Goffinet has carried on a large study on the miombo pedofauna. In recent papers (Goffinet and Freson, 1972; Goffinet, 1973) some first results are given, namely, seasonal fluctuations of Acarina (mites) and Collembola populations in miombo litter. Figure 5 shows the close dependence of Collembola on litter moisture. They disappear completely during the dry season. A local rain during the warm dry season on 8 September 1971 induced a sudden peak of Collembola and a lesser effect on the Acarina population, but a few days later

[2] In answer to Dr. F. G. Werner's question, a number of observations on termite seasonality in miombo have been carried out by Goffinet (unpublished observations). Some published observations are available (Malaisse *et al.*, 1972; Goffinet and Freson, 1972). Bouillon (1970, p. 227) discusses, for the Ethiopian region, periodism in termite activities, essentially as related to the available nutrients and to inundations during the rainy season.

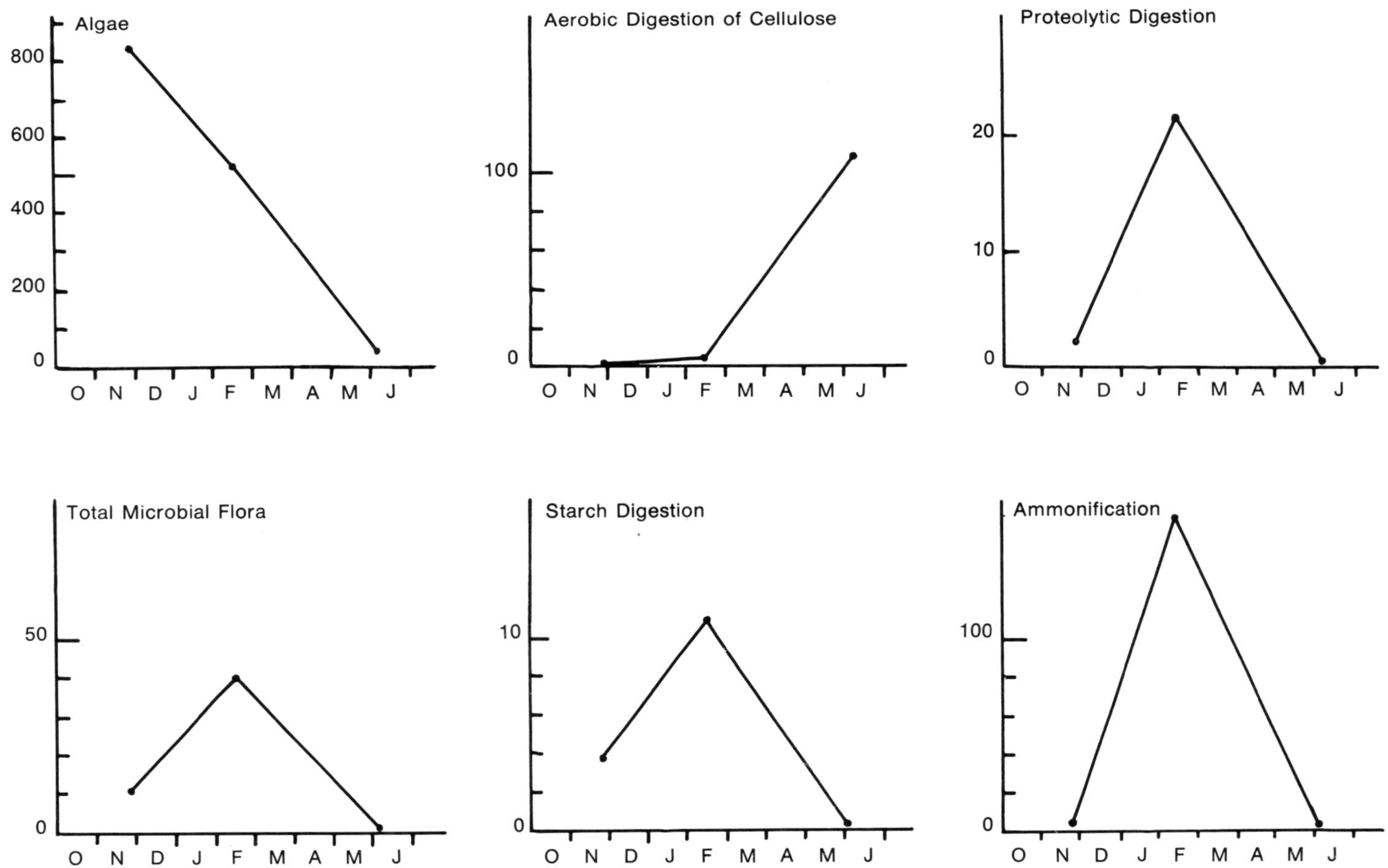
Algae
800
600
400
200
0
O N D J F M A M J
Aerobic Digestion of Cellulose
100
0
O N D J F M A M J
Proteolytic Digestion
20
10
0
O N D J F M A M J
Total Microbial Flora
50
0
O N D J F M A M J
Starch Digestion
10
0
O N D J F M A M J
Ammonification
100
0
O N D J F M A M J

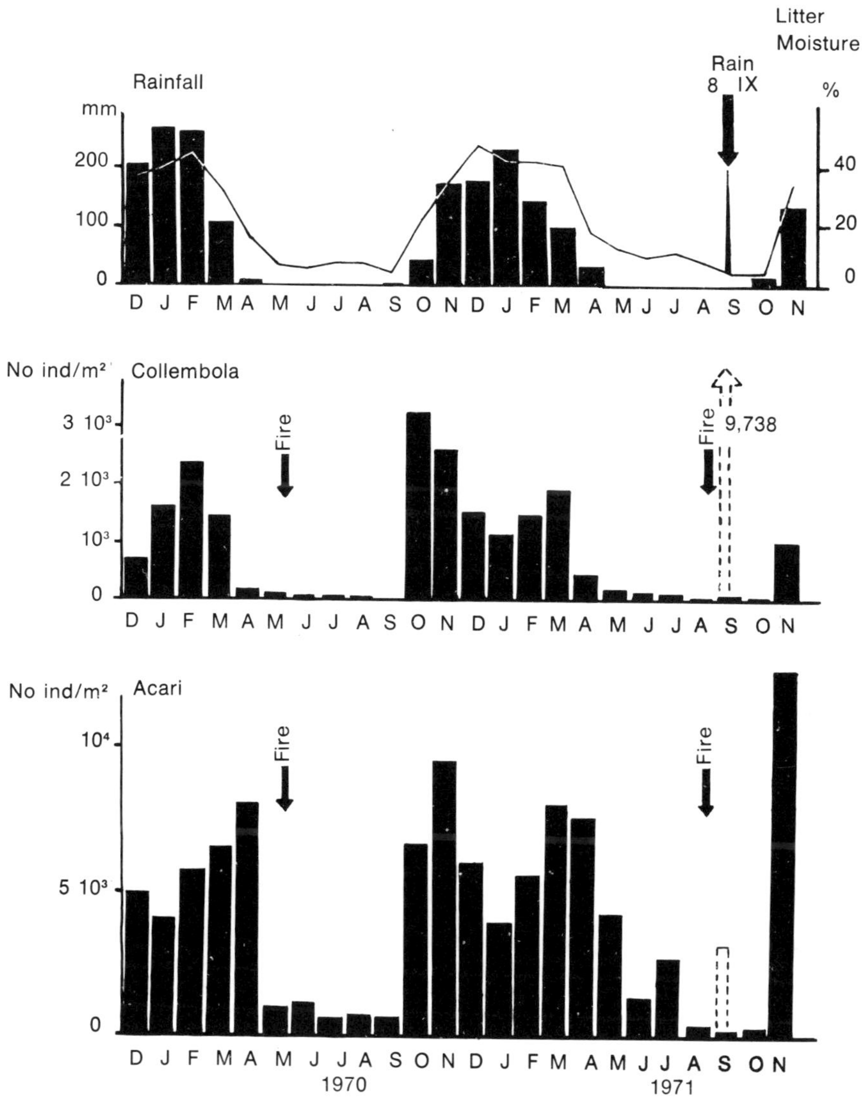

Fig. 5. Seasonal variations of Collembola and Acarina populations at litter level in miombo (rainfall-histogram; litter moisture-broken line). (After Goffinet, 1973.)

Fig. 4. Variation in soil microbial parameters in the miombo ecosystem over one year (Oct. 1968 – June 1969). All the values are given per gram of dry soil. The algae and aerobic digestion of cellulose represent absolute values, other microbial groups are values in the millions. (After Malaisse *et al.*, 1972.)

these populations had already resumed their normal density. Mites also appear to be related to litter moisture but are able to maintain a low population density through the dry season.

4 Aquatic Communities

With the exception of dambo and temporary pools, aquatic communities generally present relatively constant ecological conditions. It may be asked whether it is possible to observe seasonal variations in these communities. We review briefly for Shaba lakes, running waters, high plateau permanent pools and dambo (intermittent pools).

Lakes

Freson (1971) has studied in detail the primary production of the man-made Lubumbashi Lake (11° 39′ S., 27° 27′ 30″ E.). This lake is shallow, never exceeding 6 m in depth. Its specific conductivity varies from 135 to 310 μmhos/cm, so that it belongs to the class I of Talling and Talling (1965). The ecological characteristics present the following annual variation according to Freson (1972). Water temperature is lowest in July (17°C) and highest (22° to 25°C) from September to March. From a chemical point of view, except for SiO_2 and $SO_4^{=}$, the concentration of the components are related to the precipitation pattern, with the highest dilution occurring in March and the more concentrated waters in October. This has been shown for Ca^{++} (20 to 50 mg/liter), Mg^{++} (8 to 25 mg/liter), $Na^{+} + K^{+}$ (4.5 to 7 mg/liter), and Cl^{-} (0.8 to 2.4 mg/liter) ions. The pH varies between 7.2 (March) and 8.2 (October).

The phytoplankton inventory gives 72 species. However, the numerical variation in about 10 species controls the total rhythm. November presents a maximum of diatoms, and January and February a maximum of Chlorophyceae; and March and April are dominated by diatoms, namely, *Melosira varians* Agardh, *M. ambigua* Muller, and *Synedra ulna* (Agardh) Smith; and finally from June to September Cyanophyta assume leadership with *Anabaena flos-aquae* Brebisson, *Oscillatoria agardhii* Gomont, *O. planctonica* Woloszinska, and *O. limosa* Agardh (Fig. 6). These cyclic variations of the plankton flora are related to local climatic and ecological factors. In fact, for the Lubumbashi Lake the shift of the peak bloom of *Sphaerocystis schroeteri* Chodat, *Melosira ambigua*, and *Anabaena flos-aquae*, respectively, reflects the major patterns of seasonality.

Running Waters

In the Luanza river basin, a tributary of Luapula River with a drainage area of 531 km^2, the drainage density (length in kilometers/surface in square kilometers) varies from 1.16 for the rainy season to 0.57 for the dry season, half of the river system being temporary. The chemical characteristics (Malaisse, 1969a), as well as the various physical aspects (Malaisse 1969b), that the Luanza presents have already

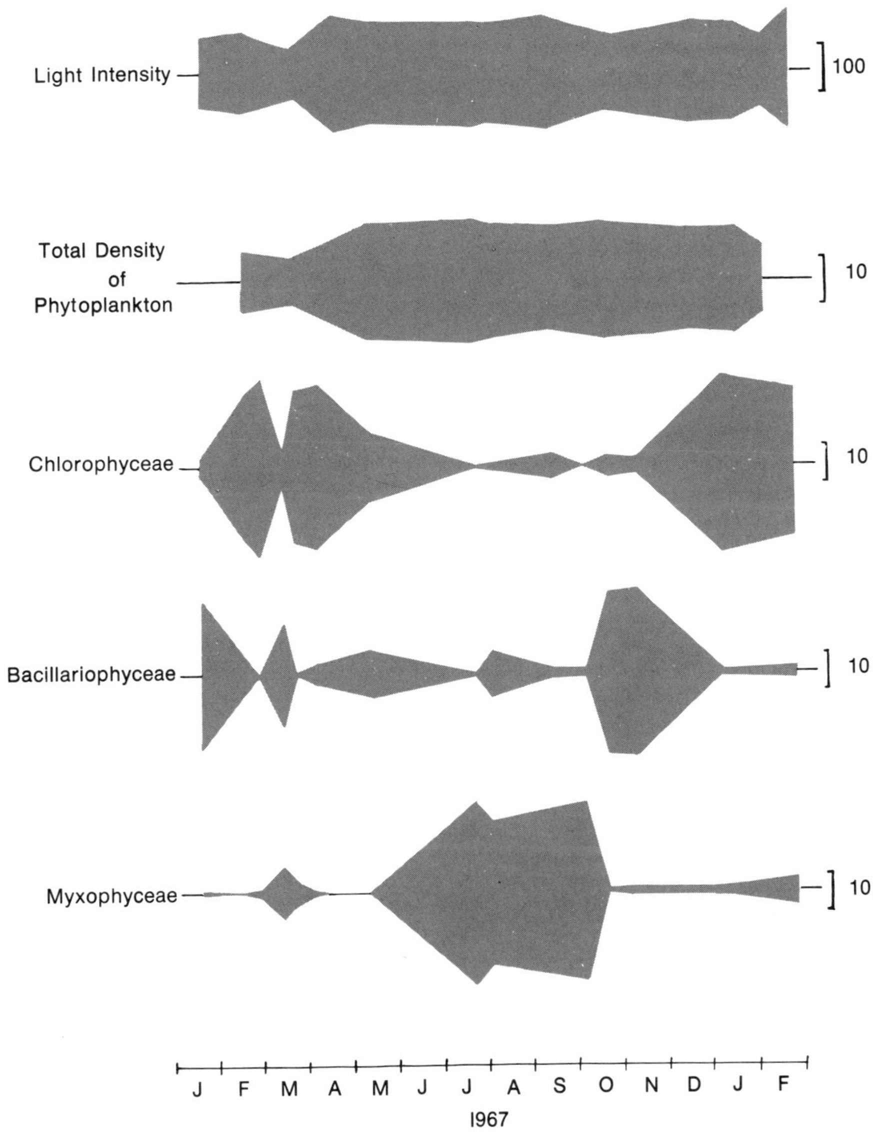

Fig. 6. Interrelation between light intensity (400–700 m) in Kerg cm/sec, total density of phytoplankton (mg chlorophyll a/m^2), and main groups of algae in percent of total population. (After Freson, 1971.)

been reported. Figure 7 gives the annual variations in the main ecological factors and in some essential biological components of the Luanza River ecosystem as observed near Kabiashia. All the aquatic phanerogams observed flower during the dry season when they are completely (*Inversodicraea congolana* Hauman) or partially emerged (*Hydrostachys triaxalis* Engl. and Gilg) or only when the water level decreases (*Eriocaulon malaissei* Moldenke). With the river depth variations, physical

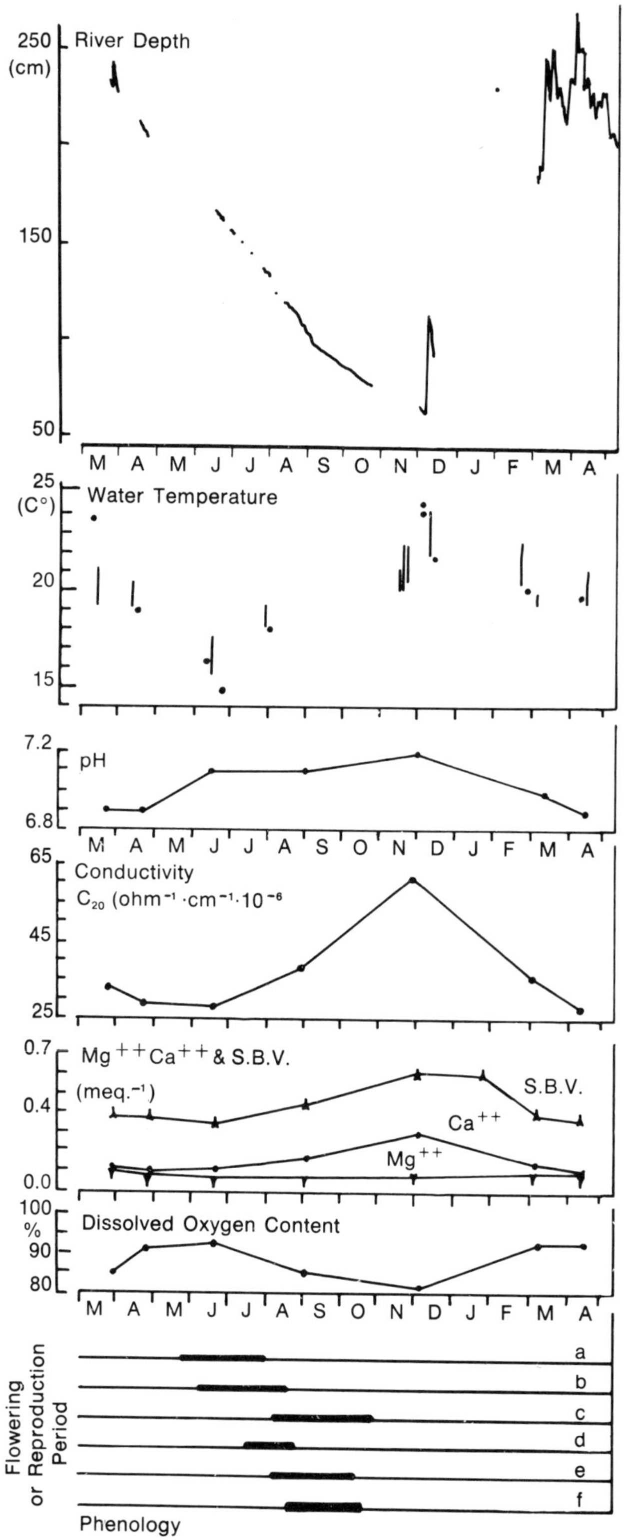

250
(cm)
River Depth
150
50
M A M J J A S O N D J F M A
25
(C°)
Water Temperature
20
15
7.2
pH
6.8
M A M J J A S O N D J F M A
65
Conductivity
C_{20} (ohm^{-1} · cm^{-1} · 10^{-6}
45
25
0.7
Mg^{++}Ca^{++} & S.B.V.
(meq.$^{-1}$)
0.4
0.0
S.B.V.
Ca^{++}
Mg^{++}
100
%
Dissolved Oxygen Content
90
80
M A M J J A S O N D J F M A
Flowering or Reproduction Period
a
b
c
d
e
f
Phenology

and chemical conditions change, and induce a similar periodicity among the animals. For instance the reproduction and laying period of many zoological groups also takes place in the dry season.

Permanent tributaries of Lufira River provide a second example. The mollusks inhabiting this area have been studied recently (Mandahl-Barth, *et al.*, 1972). Their population dynamic has been followed by Malaisse and Ripert (unpublished data) during a two-year period. Young individuals may be observed throughout the year, except during the cold dry season, with a peak density of small forms observed at the end of the warm dry season.

Dambo

Dambo are areas of gley soil sparsely scattered throughout the miombo. Beginning in December temporary pools appear in dambo after each heavy rain. At the end of December, especially around mid-January, dambos are completely under water and will remain so for two to four months. It appears to be stagnant water but in fact there is a slight discharge. At the end of May the dambo progressively dries up. Due to its high clay content, the soil presents a xeric condition during the second half of the dry season. Because of its contrasting seasonal ecological conditions, four months' total inundation and four month' dry soil, the dambo supports a special flora and fauna.

We take the Kandale dambo (10° 18′ S, 28° 09′ E) located in the Luanza River basin as an example. During the prerain and early rain seasons a first group of plants begins to spread out and flower; these plants are mostly geophytes, such as *Trochomeria macrocarpa* (Sond) Hook f., *Hypoxis angustifolia* Lam., *Temnocalyx verdickii* (De Wild. and Th. Dur.) Robyns, *Kaempferia pleiantha* K. Schum, and *Aeschynomene leptophylla* Harms.

As the rainy season advances, a second wave of plants spreads and flowers. The blossoms of December are dominated by *Triglochin bulbosa* L., *Thesium malaissei* Lawalree, and *Fuirena umbellata* Rottb. While the dambo is under water the new plants reach maturity. Some of them blossom in February–March; *Rhynchospora triflora* Vahl, *Scleria melanomphala* Kunth., *Ottelia ulvifolia* (Planch.) Walp. *Aponogeton abyssinicus* Hochst. ex A. Rich., *Eriocaulon bifistulosum* van Heurck and Muell.-Arg., *Nymphaea maculata* Schumach., *Blyxa radicans* Ridl., *Ceratophyllum demersum* L., and *Utricularia stellaris* L. f.

Others reach their maximum in May; among them are *Monademium herbaceum* Pax, *Blepharis asteracanthus* C.B.Cl., *Sphaeranthus chandleri* Ross-Craig, *Scleria foliosa* A. Rich., *Eriocaulon setaceum* L., *Monocymbium ceresiiforme* (Nees) Stapf., *Xyris capensis* Thumb., *X. straminea* Nilss, *Floscopa rivularis* (A. Rich.) Clarke, and a lot of *Utricularia* spp. (*U. subulata* L., *U. livida* E. Mey, *U. firmula* Welw. ex

---

Fig. 7. Annual variation of the main ecological factors and phenology of the Luanza River ecosystem near Kabiashia. Flowering period of *Hydrostachys triaxialis* Engl. and Gilg (*a*), *Inversodicraea congolana* Hauman (*b*), and *Eriocaulon malaissei* Moldenke (*c*); reproduction period of *Kneria wittei* Poll (*d*), *Caridina africana* Kingsley (*e*), and *Potamonautes baianus* (*f*). (After Malaisse, unpublished data.)

Oliv , *U. baouleensis* A. Chev.). Finally, from July to September, the vegetation dries up. Bush fires often occur during this last period and the dambo changes into a black-burned, forsaken land.

The dambo's fauna is mainly composed of one of the two following types: First, freshwater species that repopulate the dambo from the surrounding permanent streams, for instance *Gnathonemus angolensis* Boulenger, *Ctenopoma multispinis* (Peters), *Barbus neefi* Greenwood; and also aquatic larvae of winged adults, such as Culicidae (mosquitoes), Tendipedidae (midges), Anisoptera (dragonflies), and Zygoptera (damselflies) Second, the miombo fauna that migrate into the dambo savanna during the dry season; grasshoppers, crickets, spiders, and beetles. Some rare species make an exception to this double migration and are able to loop their complete life history in the dambo, entering diapause during the dry season. *Notobranchius* n. sp. aff. *brieni* Poll lays its eggs in the dambo mud in May. As soon as the pools fill again the fry are released. They will produce new mature fish in a few months. The lungfish *Protopterus annectens* (Owens) ssp. *brieni* Poll lives several years. It will survive the long dry seasons in the soil, at 30 cm depth in a cocoon. When the dambo is once more flooded, the lungfish resumes activity. We have experimentally established the close relation of these activities to external ecological conditions. Ostracoda (seed shrimps) and Gastropoda (freshwater mollusks) are also able to survive the dry season by burrowing into the mud. *Lymnaea natalensis* (Krauss) builds an epiphragm, *Lanistes ovum* Peters possesses an operculum, and *Sphaerium bangweolicum* (Haas) is a bivalve.

In conclusion, we see that the dambo presents several flowering waves. Locally the length of inundation differs from dambo to dambo in the miombo woodland area and its phenology is, according to the specific case, more or less condensed.

High Plateau Permanent Pools

Five high plateaus (Manika, Biano, Kibara, Kundelungu, and Marungu), with altitudes over 1500 m, are arranged in a horseshoe pattern around Lubumbashi. They are covered by a thick layer of Kalahari sands on which steppe vegetation is widely distributed (Lisowski *et al.*, 1971). Locally permanent pools are established on thickly and strongly cemented laterite crust outcrops. These pools will provide a last example of extreme seasonal vegetation change.

Depth of permanent pools varies in the course of the year. The minimum is in October and may go down to 10 cm, while the maximum is observed in March when water reaches 60 cm and sometimes 1 m or more in depth. The pH is fairly low, usually around 5.7. The phanerogamic flora presents a noticeable convergence into three main biological types, i.e. fixed hydrophytes with round, circular floating leaves, fixed hydrophytes with laciniate leaves, and rhizomatous geophytes. From a phytosociological point of view, this vegetation must be described as a new association, the *Anagalleto-Mayacetum dilunguense* Malaisse.

Most plants in these pools flower throughout the rainy season as soon as the water depth rises to 30 to 40 cm. To this group belong *Nymphaea maculata* Schumach., *Polygonum limbatum* Meissn., *Xyris capensis* Thumb., *Scirpus confervoides* Poir., *Nymphoides rautaneni* (N.E. Br.) A. Raynal, *N. tenuissima* A. Raynal,

Eleocharis variegata (Poir.) Presl., *Cyperus platycaulis* (Bak.) var. *kipasensis* (Cherm.) Kuk., *Eleocharis dulcis* (Burm. f.) Trin., and *Utricularia reflexa* Oliv. var. *reflexa*. A second group of plants blooms in the first part of the dry season, when the water level decreases, including *Mayaca baumii* Gurke, *Blyxa radicans* Ridl., *Anagalis kochii* Hess, *Eriocaulon setaceum* L., *Utricularia spiralis* Sm. in Rees var. *tortilis* (Welw. ex Oliv.) P. Taylor. Finally, some rare species blossom in the second part of the dry season, for instance *Utricularia welwitschii* Oliv., *Rotala myriophylloides* Welw. ex Hiern., and *Limnophila sessiliflora* Blume.

5 Discussion

The intertropical regions present a wide range of climatic conditions from equator to tropics. It seems useful to distinguish unequivocally between equatorial and tropical conditions and vegetation. The rain forest, usually called "tropical" rain forest, corresponds to relatively homogenous climatic conditions i.e., equatorial conditions, and is more accurately called the equatorial rain forest. We would reserve the term tropical for intertropical countries with significant seasonal change.

According to this definition, Zambezian woodland belongs to tropical conditions. Seasonal variations may be observed in all communities. Rainfall remains the principal factor, but temperature must also be taken into account. Both the cold dry season and the warm dry season supervise phenological changes in the life history of plants and animals.

Acknowledgments

The author is indebted to Professor H. Lieth for helpful discussions and Miss D. Watley for reviewing the manuscript. In particular, the writer would like to thank M. Malaisse-Mousset, G. Goffinet, and Dr. R. Freson for information and comments. The help of the staff of the Department of Botany and Climatology is gratefully acknowledged, especially the technical assistance in the field of Bulaimu and Kisimba.

I thank my wife for her assistance with the preparation of the script, Y. Lemaire-Elias for typing the manuscript, L. Lemaire for the photographic documentation, and Mulumba for the preparation of some figures.

References

Bernard, E. 1950. Aperçus fondamentaux sur la climatologie du Katanga. *Comp. Rend. Congr. Sci. Elisabethville, C.S.K., Bruxelles* 4 (2), 56–70.

Bernhard-Reversat, F., C. Huttel, and G. Lemée. 1972. Some aspects of the seasonal ecologic periodicity and plant activity in an evergreen rain forest of the Ivory Coast. In *Tropical Ecology with an Emphasis on Organic Productivity*, P. Golley and F. Golley, ed. p. 217. Athens, Ga.; Institute of Ecology.

Boaler, S. 1966. Ecology of a miombo site, Lupa North Forest Reserve, Tanzania. II. Plant communities and seasonal variation in the vegetation. *J. Ecol.* 54, 465–479.

Bouillon, A. 1970. Termites of the Ethiopian region. In *Termites.* Vol. 2, pp. 153–280. New York: Academic Press.

Daubenmire, R. 1972. Phenology and other characteristics of tropical semideciduous forest in north-western Costa Rica. *J. Ecol.* 60 (1), 147–170.

Fanshawe, D. 1971. The vegetation of Zambia. *For. Res. Bull. Min. Rural Devel. Zambia* 7, 1–67.

Freson, R. 1971. Ecologie et production primaire d'une phytoplanctocénose tropicale: Le lac de retenue de la Lubumbashi. Thèse de doctorat. pp. 1–284. Univ. Louvain.

——— 1972. Aspect de la limnochimie et de la production primaire au lac de la Lubumbashi. *Verh. Intern. Ver. Limnol.* 18, 661–665.

Goffinet, G. 1973. Contribution à l'étude de l'écosysteme forêt claire (Miombo) Note 12 Recherches sur les fluctuations saisonnières des peuplements d'Acariens et de Collemboles au niveau de la litière. *Ann. Univ. Abidjan Serie E, Écologie* 6, 257–263.

———, and R. Freson. 1972 Recherches synécologiques sur la pédofaune de l'écosystème forêt claire (miombo). *Bull. Soc. Ecol.* 3 (27), 138–150.

Hopkins, B. 1966. Vegetation of the Olokemeji Forest Reserve (Nigeria). IV. The litter and soil with special reference to their seasonal changes. *J. Ecol.* 54, 687–703.

Lieth, H. 1970. Phenology in productivity studies. In *Analysis of Temperate Forest Ecosystems*, E. Reichle, ed. pp. 29–46. New York: Springer-Verlag.

Lisowski, S., F. Malaisse, and J. J. Symoens. 1971. Une flore des hauts plateaux du Katanga. *Mitt. Bot. Staatssamml. München* 10, 51–56.

Malaisse, F. 1964. Contribution à l'étude des hêtraies d'Europe occidentale. Note 4. Quelques observations phénologiques de hêtraies en 1963. *Bull. Soc. Roy. Bot. Belgique* 98, 85–97.

——— 1969a. Ecologie et aménagement piscicole d'un cours d'eau tropical: La Luanza (Haut-Katanga, Rép. dém. Congo). *Comp. Rend. Reg. Meet. Afr. Hydrobiol. Kampala* 121–128.

——— 1969b. Les faciès d'un cours d'eau tropical: La Luanza (Haut Katanga, Rép. dém. du Congo). *Verh. Intern. Ver. Ang. Limnol.* 17, 936–940.

——— 1973. Contribution à l'étude de l'écosystème forêt claire (miombo). Note 8. Le projet Miombo. *Ann. Univ. Abidjan Serie E, Ecologie* 6, 227–250.

———, and M. Malaisse-Mousset. 1970. Contribution à l'étude de l'écosystème forêt claire (miombo): Phénologie de la défoliation. *Bull. Soc. Roy. Bot. Belgique* 103, 115–124.

Malaisse, F., J. Alexandre, R. Freson, G. Goffinet, and M. Malaisse-Mousset. 1972. The miombo ecosystem: A preliminary study. In *Tropical Ecology*, F. Golley and P. Golley, eds., pp. 363–405. Athens, Ga.: Institute of Ecology.

Malaisse-Mousset, M., F. Malaisse, and C. Watula. 1970. Contribution à l'étude de l'écosystème forêt claire (miombo). Note 2. Le cycle biologique d'*Elaphrodes lactea* (Gaede)/Noto-dontidae/et son influence sur l'écosystème "miombo." *Trav. Serv. Sylv. Pisc. Univ. Off. Congo, Lubumbashi* 8, 1–10.

Mandahl-Barth, G., F. Malaisse, and C. Ripert. 1972. Etudes malacologiques dans la région du lac de retenue de la Lufira. Distribution et écologie des mollusques aquatiques. Rôle épidémiologique des vecteurs des bilharzioses intestinale et urinaire. *Bull. Soc. Pathol. Exot.* 65, 146–165.

Sys, C. 1957. L'aménagement des sols de la région d'Elizabethville d'après leurs caractéristiques morphologiques et analytiques. *Bull. agr. Congo belge* 48 (6), 1425–1432.

Talling, J. and I. Talling. 1965. The chemical composition of African lake waters. *Intern. Rev. Ges. Hydrobiol. Hydrogr.* 50, 421–463.

White, F. 1965. The savanna woodlands of the Zambezian and Sudanian Domains. An ecological and phytogeographical comparison. *Webbia* 19 (2), 651–681.

4.7 Tropical Plant Phenology: Applications for Studies in Community Ecology[1]

Gordon W. Frankie

Department of Entomology
Texas A&M University
College Station, Texas 77843, USA

Herbert G. Baker

Department of Botany
University of California
Berkeley, California 94720, USA

and Paul A. Opler

Organization for Tropical Studies, Inc.
P.O. Box 431499, Miami, Florida 33143, USA

1 Introduction

Periodic behavior of plants in tropical (here including equatorial) environments has received little attention until very recently (Frankie *et al.*, 1974; Gibbs and Leston, 1970; McClure, 1966; Nevling, 1971; Rees, 1964). Some accounts of phenological events in the tropical literature occur only as brief notes which form small sections of larger papers (e.g., Ducke and Black, 1953; Beard, 1946). Other information on specific phenological events can be found in floristic treatments (e.g., Allen, 1956; Little and Wadsworth, 1964) and papers dealing with species of economic importance (e.g., Broekmans, 1957; Lamb, 1966, Rees, 1964). In certain studies the data on periodicity of a small percentage of species have been pooled in

[1] Supported by NSF Grants GB 7805 and GB 25592.

KEYWORDS: Phenology; community ecology; ecosystem-ecology; tropical phenology; animal-plant interactions; plant-animal interactions; pollination; frugivory; seed dispersal; plant periodicity; bee behavior; preservation; plant reproduction; African honey-bee; natural areas; Central America.

an attempt to reflect the general phenological trends of particular vegetation types (e.g., Baker and Baker, 1936; Daubenmire, 1972).

With regard to tropical forests, none of the early investigations was oriented toward the unification and analysis of periodicities for leaf, flower, and fruit of nearly all plant species of which a particular forest type is composed. However, recent studies by Boaler (1966) in deciduous miombo woodland, Tanzania; by Croat (1969) in semi-evergreen moist tropical forest, Panama; by Fournier and Salas (1966) in tropical moist forest, Costa Rica; by Hopkins (1968) in derived savanna woodland, Nigeria; by Janzen (1967) in tropical dry forest, Costa Rica; and by Nevling (1971) in elfin forest, Puerto Rico; offer phenological patterns which characterize periodicities of the flower and/or fruit for many of the tree species included in each forest. Although most of these investigations are limited as to the amount of phenological data they contain, the findings do provide us with insight on some of the major phenological patterns.

2 The Ideal Tropical Phenological Study

Ideally, a phenological investigation in a tropical ecosystem requires a relatively undisturbed area with as near complete representation of the appropriate species as possible. Adequate replication of each species allows for the detection of variation within species populations in addition to that between species populations. Observations on phenological events should be made for several years so as to account for year-to-year variation in individual species as well as community patterns that emerge. The extended period is especially necessary in order to determine the periodicities of species which do not flower and fruit on a yearly basis. This is particularly important in forest types situated in relatively nonseasonal climates (Holttum, 1953, 1968).

3 Costa Rican Study

In a paper to be published soon by Frankie *et al.* (1974), a phenological study was designed to meet most of the above requisites. The investigation, which began in Costa Rica in late 1968, compared the leafing, flowering, and fruiting patterns of tropical lowland wet and dry forest trees. In addition to the elucidation of several major and minor patterns of periodicity, the study also distinguished some patterns that were definitely related to annual climatic events from others that seem to be influenced by a complex of biotic and abiotic factors. Observations still in progress on the tree species in both forests will be continued until at least 1974.

4 Use of Tropical Plant Phenological Information

Practical Use

The practical uses of phenological information have been discussed by many workers, including Caprio (1966), Caprio *et al.* (1970), and Lieth and Radford (1971).

Uses in Community and Ecosystem Ecology

In addition to their pragmatic value, phenological data involving most members of the represented synusiae can also be used in analyses of the biological organization of communities and ecosystems. Specifically, phenological information of tropical environments can be used to study animal-plant interactions which affect pollination, seed dispersal, and seed predation. These interactions are important for plant reproduction and, reciprocally, for food for the animals involved (Baker, 1973; Janzen, 1967, 1969, 1970, 1971; van der Pijl, 1969a). Plant periodicity patterns reflect annual distributions of specific kinds of available food resources (i.e., pollen, nectar, and fruit-seed resources). In ecosystem and community analyses, data on major periodicity patterns can be broken down into several minor flower and minor fruit resource patterns. Syndromes described by Baker and Hurd (1968), Faegri and van der Pijl (1971), Percival (1965), and van der Pijl (1960–61), together with supplementary observations on visitors allow plant species to be placed in flower categories which correspond to certain pollinator types. Species can be classified according to fruit type (Dansereau and Lems, 1957; Baker, *et al.*, 1974; Smythe, 1970). After species are categorized according to flower and fruit types, they can be further divided according to season, life form, and canopy position. In some ecosystems, it is further possible to sort out component communities, based on certain characteristic flower or fruit types.

When these subpatterns of plant activity are developed, it then becomes possible to realize how the various types of resources are organized (structured) in a particular tropical environment. From these subpatterns, clues become apparent as to what animal groups may be operating at certain times of the year on particular plant life forms in particular communities. No doubt these resource patterns could be used by zoologists for investigating animal periodicities.

As adequate plant and animal data become available, combined patterns of periodicity then lend themselves to analyses of trends in plant-animal associations. Such a study was conducted by Snow (1965) in Trinidad on the frugivorous birds which feed on the fruits of trees and shrubs in the genus *Miconia* (Melastomaceae).

It is tempting to regard these association trends as the products of responsive or reciprocating evolution at the ecosystem or community level; however, these trends require further examination for statements about the nature of their evolution.

Pollination Studies

The importance of animals as pollen vectors in the tropics has been well documented (Ashton, 1969; Baker, 1973; Baker and Hurd, 1968; Dodson *et al.*, 1969; Janzen, 1967, 1971; Linsley, 1966; van der Pijl, 1954; van der Pijl and Dodson, 1966; Ramirez, 1969; Rick, 1966; Vogel, 1968–1969). Some of these studies point out the interplay of flower phenology and pollinators. Bawa (1973) has added a survey of the breeding systems of tree species in the same dry forest site where we are studying phenology. Through experimental pollinations he showed a high proportion of tree species to be self-incompatible or dioecious. Since most of these species are obligated to outcross, they depend entirely upon animals for effective transfer of pollen. In this dry forest of Costa Rica, phenological information on most of the trees and shrubs (Frankie *et al.*, 1974; Opler, Frankie, and Baker, in preparation, respectively) is followed by preliminary observations on vines and herbs (Opler, unpublished data). Most forest shrubs bloom just prior to and just after the main period of flowering for the trees (long dry season). Predominant vines and herbs flower during the long wet season. Thus overlap in flowering activity among the species of all life forms is reduced. Unsynchronized flowering among the members of two or more synusiae have been recorded by Croat (1969) and Foster (1972), working in a semideciduous forest in Panama. Boaler (1966) also recorded unsynchronized flowering among the species of several synusiae in a miombo forest in Tanzania.

In the Costa Rican dry forest the blooming periods of more than 60 dry season flowering tree species (65% pollinated by bees) are spread over 3.5 months such that flowering overlap is minimized (Frankie *et al.*, 1974). Evolutionarily, the flowering times of the tree species may have been sorted out in a manner which allows effective avoidance of competition for pollinators.

Experimental data involving floral behavior and pollinator behavior may offer insight on the bases for competition as well as for the factors which maintain it. Several plant factors known to influence bee preferences may be meaningful in this relationship. Numerous studies cited in a recent paper by Levin and Anderson (1970) note how bees prefer particular flower colors, outlines, sizes, and scents. Percival (1965), Vansell (1942), and others have shown that, when presented with a choice of nectar sources, honeybees will seek out and then concentrate upon that which provides the most concentrated solution of sugars. The recent discovery (Baker and Baker, 1973a, 1973b) that nectars from the flowers of many species contain significant quantities of amino acids may add still another factor to the apian preference for certain nectar types. Bees may also be highly specific in the pollen sources that they favor (Linsley, 1958). The relative quantity of the floral reward is also known to be important in determining a pollinator's preference (Butler, 1945; Mosquin, 1971; Free, 1968). From an energetic standpoint, Heinrich and Raven (1972) point out the importance of floral food stuffs in the relationships of flowers and pollinators. Our observations on solitary bees in Costa Rica suggest that many bee species forage at the greatest possible heights within tree canopies. A similar height preference was observed in xylocopid bees visiting flowers of *Passiflora incarnata* L. in Texas (Frankie, unpublished data).

Owing to differences in flower rewards among the complex of dry forest trees, the bees[1] may provide the greatest amount of service to the most preferred species (for reasons cited above). In effect, this differential attraction would then result in competition among plants for pollinators, which in turn may provide the force leading to a segregation of blooming periods (Levin and Anderson, 1970; Mosquin, 1971; Straw, 1972). It is conceivable that competitive interactions may also have

[1] Another possible factor that may contribute to this relationship is that bees in this and other tropical environments are believed to display relatively low host specificity (Heinrich and Raven, 1972; E. R. Heithaus, unpublished data; Michener, 1954; Opler, unpublished data).

played a role in influencing the unsynchronized flowering patterns among the four dry forest synusiae mentioned previously.

In the same Costa Rican dry forest where Frankie, Baker, and Opler are studying phenology, E. R. Heithaus and P. Heithaus of Stanford University and Opler are studying the relationship between the bats and the flowers and fruits they utilize during the year. Allen (1939) suggested that a sequence of flowering plants may provide for a necessary sustained flow of floral food stuffs to large pollinators in an area. Baker (1963, 1970), Janzen (1967), and van der Pijl (1956, 1969b) presented limited data which tends to support this suggestion. Heinrich and Raven (1972) have also considered the potential of this possible community relationship from an energetics standpoint. If it exists, the implications for future forestry practices involving a tropical monoculture system are obvious.

E. R. Heithaus is examining the patterns of utilization of floral resources by pollinators in four Costa Rican communities as they may relate to theories of diversity. His monthly assessments on flowering phenology, diversity, and abundance of flower resources and pollinators lead to measurements of two of the diversity indices, niche breadth and niche overlap.

In a more intensive study in Costa Rica, L. Wolf and F. Hainsworth of Syracuse University and F. G. Stiles of the Universidad de Costa Rica are investigating how sympatric species of hummingbirds utilize the floral resources of several species of *Heliconia* (Musaceae) in a tropical lowland wet, secondary growth community. Their studies involve a precise knowledge of flowering periodicities and floral behavior as they relate to territoriality and energetics in the hummingbirds.

Through careful examination of flowering periodicity data, it is possible to learn how specific floral behavior patterns may influence pollinator behavior and vice versa. Janzen (1971, and personal communication) draws attention to a rather distinct type of flowering behavior and its possible relationship to a suspected foraging pattern of certain large tropical bees. Certain plant species of the understory vegetation of tropical wet forests produce a few flowers daily for extended periods. The bee pollinators of these long-flowering species display what Janzen terms "traplining behavior," in which the foraging activity of the bees is such that they may visit a series of widely separated plants in order to secure adequate quantities of foodstuffs. Since many of the plants are of the same species, this movement probably serves to promote outcrossing among these species. Janzen (1971) notes that forest sphinx moths and hummingbirds may also forage in a like manner among certain widely spaced, long-flowering species. Baker (1973) presents limited information suggesting that bats of the suborder Microchiroptera also display traplining behavior among some of the species they normally pollinate in neotropical forests. Their counterparts among the Megachiroptera, which pollinate flowers in the paleotropics, do not show this behavior but concentrate their activities more on one tree at a time.

Janzen (personal communication) also recognizes a contrasting plant behavior by which large numbers of flowers are produced for relatively short time periods. The associated bee pollinators of these plant species display "opportunistic behavior," in which the generally smaller bees may spend continuous periods of time exploiting resources from single massively flowering plants. The opportunistic behavior rules against outcrossing since most of the bees probably remain relatively constant to the large numbers of flowers produced by a single tree of this type.

Phenological information helps to analyze patterns of coexistence among closely related species. For example, four species of *Guarea* (Meliaceae) occur sympatrically at the same wet forest site in Costa Rica (Frankie *et al.*, 1974). Their periodicities of flowering suggest a possible factor involved in their coexistence: Two of the species flower in the first dry season during different periods; a third species blooms in the first wet season; the fourth species flowers in three distinct periods, with each burst occurring at a different season. The separation of flowering times suggests that these four species of *Guarea* may be effectively avoiding competition for pollinators. In the same forest, several other genera contain more than one species which have different flowering times, suggesting that the displacement of flowering periods may result from competition for pollinators. Staggered blooming periods of 10 species of *Solidago* (Compositae) in a 3.5-ha plot in New York was reported by Hurlbert (1970). He interpreted this pattern as being an effective means of insuring reproductive isolation among the *Solidago* species.

Seed Dispersal

Dispersal of disseminules by animals in tropical systems is well known (Corner, 1964; van der Pijl, 1966, 1969b; Ridley, 1930). However, Snow (1971) pointed out the neglected evolutionary aspects of this habit. In the following examples, phenological information has been used to gain insight on the evolutionary impact of fruit eating and its importance at the organizational levels of the community and ecosystem.

Smythe (1970) studied fruit and seed fall in the humid forest of Barro Colorado Island, Panama Canal Zone. In addition to the climatic factors determining fruiting seasons, he also observed that small-seeded fruits (whose seeds pass through the guts of animals) ripen sequentially through the year, so that the agents of dispersal are utilized fully with minimal competition for their service. With larger-seed fruits, however, dangers of damage to the seeds by the animals is increased and, for them, synchronous fruiting within and between species ensures that some are not damaged, especially when "scatter-hoarding" animals such as agoutis are involved. Smythe points out that these animals fail to find many of the fruits they have buried and the seeds in them can germinate.

Snow (1965) studied the fruiting phenology of *Miconia* (Melastomaceae) in Trinidad. He concluded that the sequence (rather than contemporaneity) which these species show in their fruiting peaks has been arrived at through natural selection that reduces competition for the attention of the birds that disperse the seeds after eating the fruits. Yet a converse point of view may also be valid; that the birds are kept available in the community throughout the year by feeding on a sequence of *Miconia* fruits (Baker, 1972). In a later paper, Snow (1971) suggests that this relationship in Trinidad may be regarded as a symbiotic one that "may have played a significant part in the evolution of the whole ecosystem."

In more complex tropical forests, other fruiting patterns may also be involved.

For example, fruiting phenology data of the wet forest ecosystem in Costa Rica indicated that among 161 tree species at least 39 (almost all are fleshy fruited) bear fruit in any month of the year (Frankie *et al.*, 1974). However, during September and October a significant increase in fruiting activity was recorded in both the overstory and understory layers. This particular fruiting pattern and its relationship to possible activity patterns of frugivorous animals deserve further attention.

5 Conclusions

The need to conduct phenological studies in several different kinds of tropical environments is urgent. Only through the analyses of intact ecosystems or communities may we hope to understand how these systems are organized biologically and how this organization has evolved. Once these ecosystems are disturbed, the phenological patterns of the remaining elements can provide us with only a partial understanding of the original distribution of flower and fruit resources. Disturbance changes species abundances, which will reflect misleading information on the amount of resource normally available in the ecosystem. The inevitable introduction of certain exotic plant species and pollinators (e.g., the African honeybee which will soon become widespread in the New World tropics) will also complicate our understanding of these ecosystems.

The study of phenology in intact tropical ecosystems is important if we are to learn more about the dynamics of plant species evolving in particular ecosystems, as well as how they may be excluded or successful in adjacent ecosystems; it is necessary to study first how species operate (from the standpoint of reproductive biology) within the context of their respective environments. Relationships that evolve between a species and its abiotic and biotic environment may vary from one section of its geographic range to another. We should not be surprised to find ecotypic differentiation developing in some species at the extremes of range.

Most tropical plant species depend on animals for at least one phase of their reproductive cycle and offer an excellent opportunity to study such relationships. Further information on the biological organization of tropical ecosystems may then provide new insight for the study of temperature ecosystems where the relationship of plant reproductive behavior to that of animals is not as obvious, because extreme alternations of temperature and/or moisture dominate seasonal behavior of every component.

Acknowledgment

During the course of the field work reported here, support has been received from the National Science Foundation (Research Grants GB-7805 and GB-25592). The facilities in Costa Rica of the Organization for Tropical Studies have been utilized, and we offer our thanks to the administrators of that organization and our colleagues in the Ecosystem Comparison Study team of O.T.S.

References

Allen, G. M. 1939. *Bats*. Cambridge, Mass.: Harvard University Press.

Allen, P. H. 1956. *The Rain Forest of Golfo Dulce*. Gainesville: University of Florida Press.

Ashton, P. S. 1969. Speciation among tropical forest trees: Some deductions in the light of recent evidence. *Biol. J. Linn. Soc. London* 1, 155–196.

Baker, H. G. 1963. Evolutionary mechanisms in pollination biology. *Science* 139, 877–883.

——— 1970. Evolution in the tropics. *Biotropica* 2, 101–111.

——— 1973. Evolutionary relationships between flowering plants and animals in American and African tropical forests. In *Tropical Forest Ecosystems in Africa and South America: A Comparative Review*, B. J. Meggers, E. S. Ayensu, and W. D. Duckworth, ed. Washington, D.C.: Smithsonian Institution Press.

———, and I. Baker. 1973a. Amino-acids in nectar and their evolutionary significance. *Nature* 241, 543–545.

——— 1973b. Some anthecological aspects of the evolution of nectar-producing flowers, particularly amino-acid production in nectar. In *The Interrelations of Taxonomy and Ecology*, V. H. Heywood, ed. London: Academic Press.

Baker, H. G., and P. D. Hurd. 1968. Intrafloral ecology. *Ann. Rev. Entomol.* 13, 385–414.

Baker, H. G., P. A. Opler, and G. W. Frankie. 1974. Seed and seedling ecology of plants in tropical lowland wet and dry forest sites of Costa Rica (in manuscript).

Baker, J. R., and I. Baker. 1936. The seasons in a tropical rain-forest (New Hebrides). 2. *Bot. J. Linn. Soc. (Zool.) London* 39, 507–19.

Bawa, K. S. 1973. Breeding systems of tree species of a lowland tropical community. *Evolution*, (in press).

Beard, J. S. 1946. The Natural Vegetation of Trinidad. In *Oxf. For. Mem. No. 20*. Oxford: Clarendon Press.

Boaler, S. B. 1966. Ecology of a miombo site, Lupa North Forest Reserve, Tanzania. II. Plant communities and seasonal variation in the vegetation. *J. Ecol.* 54, 564–579.

Broekmans, A. F. M. 1957. Growth, flowering, and yield of the oil palm in Nigeria. *J. W. Afr. Inst. Oil Palm Res.* 2, 187–220.

Butler, C. G. 1945. The influence of various physical and biological factors of the environment on honeybee activity. An examination of the relationship between activity and nectar concentration and abundance. *J. Exp. Biol.* 21, 5–12.

Caprio, J. M. 1966. Patterns of plant development in the western United States. *Montana Agr. Exp. Sta. Bull.* 607, 1–42.

———, M. D. Magnuson, and H. N. Metcalf. 1970. Instructions for phenological observations. *Montana Agr. Exp. Sta. Circular 250.*

Corner, E. J. H. 1964. *The Life of Plants*. London: Weidenfeld & Nicolson.

Croat, T. B. 1969. Seasonal flowering behavior in central Panama. *Ann. Missouri Bot. Gardens* 56, 295–307.

Dansereau, P., and K. Lems. 1957. The grading of dispersal types in plant communities and their ecological significance. *Contr. Inst. Bot. Univ. Montreal* 71, 1–52.

Daubenmire, R. 1972. Phenology and other characteristics of tropical semi-deciduous forest in north-western Costa Rica. *J. Ecol.* 60, 147–170.

Dodson, C. H., R. L. Dressler, H. C. Hills, R. M. Adams, and N. H. Williams. Biologically active compounds in orchid fragrances. *Science* 146, 1243–1249.

Ducke, A., and G. A. Black. 1953. Phytogeographical notes on the Brazilian Amazon. *Ann. Acad. Brasileira Ciencias* 25, 1–46.

Faegri, K., and L. van der Pijl. 1971. *The Principles of Pollination Ecology*. Oxford: Pergamon Press.

Foster, R. 1972. Unpublished Ph.D. Dissertation. Durham, N.C.: Duke University.

Fournier, L. A., and S. Salas. 1966. Algunas observaciones sobre la dinámica de la floración en el bosque tropical húmedo de Villa Colón. *Rev. Biol. Trop.* 14, 75–85.

Frankie, G. W., H. G. Baker, and P. A. Opler. 1974. Comparative phenological studies of trees in tropical lowland wet and dry forest sites of Costa Rica. *T. Ecol.* (in press).

Free, J. B. 1968. Dandelion as a competitor to fruit trees for bee visits. *J. Appl. Ecol.* 5, 169–178.

Gibbs, D. G., and D. Leston. 1970. Insect phenology in a forest cocoa-farm locality in West Africa. *J. Appl. Ecol.* 7, 519–548.

Heinrich, B., and P. H. Raven. 1972. Energetics and pollination ecology. *Science* 176, 597–602.

Holttum, R. E. 1953. Evolutionary trends in an equatorial climate. *Soc. Exp. Biol Symposia* 7, 159–173.

——— 1968. The response of plants to climatic change in Singapore. *J. Ecol.* 56, 5.

Hopkins, B. 1968. Vegetation of the Olokemeji Forest Reserve, Nigeria. V. The vegetation on the savanna site with special reference to its seasonal changes. *J. Ecol.* 56, 97–115.

Hurlbert, S. H. 1970. Flower number, flowering time, and reproductive isolation among ten species of *Solidago* (Compositae). *Bull. Torrey Bot. Club* 97, 189–195.

Janzen, D. H. 1967. Synchronization of sexual reproduction of trees within the dry season in Central America. *Evolution* 21, 620–637.

——— 1969. Seed-eaters versus seed size, number, toxicity and dispersal. *Evolution* 23, 1–27.

——— 1970. Herbivores and the number of tree species in tropical forests. *Amer. Naturalist*, 104, 501–528.

——— 1971. Euglossine bees as long-distance pollinators of tropical plants. *Science* 171, 203–205.

Lamb, B. F. 1966. *Mahogany of Tropical America: Its Ecology and Management.* Ann Arbor: University of Michigan Press.

Levin, D. A., and W. W. Anderson. 1970. Competition for pollinators between simultaneously flowering species. *Amer. Naturalist*, 104, 455–467.

Lieth, H., and J. S. Radford. 1971. Phenology, resource management, and synagraphic computer mapping. *BioScience* 21, 62–70.

Linsley, E. G. 1958. The ecology of solitary bees. *Hilgardia* 27, 453–599.

——— 1966. Pollinating insects of the Galapagos Islands. In *The Galapagos. Proceedings of the Symposia of the Galapagos Intern. Sci. Project.* pp. 225–232. Berkeley: University of California Press.

Little, E. L., Jr., and F. H. Wadsworth. 1964. Common trees of Puerto Rico and the Virgin Islands. *USDA Handbook No. 249.*

Michener, C. D. 1954. Bees of Panama. *Bull. Am. Mus. Nat. Hist.* 104, 1–176.

Mosquin, T. 1971. Competition for pollinators as a stimulus for the evolution of flowering time. *Oikos* 22, 398–402.

McClure, H. E. 1966. Flowering, fruiting and animals in the canopy of a tropical rain forest. *Malay Forester* 29, 182–203.

Nevling, L. I., Jr. 1971. The ecology of an elfin forest in Puerto Rico. 16. The flowering cycle and an interpretation of its seasonality. *J. Arnold Arb.* 52, 586–613.

Percival, M. S. 1965. *Floral Biology.* Oxford: Pergamon Press.

Pijl, L. van der. 1954. *Xylocopa* and flowers in the tropics. *Proc. K. Nederl. Akad.* C57, (I), 413–423; (II) 541–551; (III) 552–562.

——— 1956. Remarks on pollination by bats in the genera *Freycinetia*, *Duabanga* and *Haplophragma*, and on chiropterophily in general. *Acta Bot. Neerl.* 5, 135–144.

——— 1960–1961. Ecological aspects of flower evolution. I–II. *Evolution* 14, 403–416; 15, 44–59.

——— 1966. Ecological aspects of fruit evolution. *Proc. K. Nederl. Akad. Wet.* C69, 597–640.

——— 1969a. Evolutionary action of tropical animals on the reproduction of plants. *Biol. J. Linn. Soc. London* 1, 85–96.

——— 1969b. *Principles of Dispersal in Higher Plants.* New York, Heidelberg, Berlin: Springer-Verlag.

Pijl, L. van der, and C. H. Dodson. 1966. *Orchid Flowers, Their Pollination and Evolution.* Miami: University of Miami Press.

Ramirez, W. B. 1969. Fig wasps: Mechanism of pollen transfer. *Science* 163, 580–581.

Rees, A. R. 1964. Some observations on the flowering behavior of *Coffea rupestris* in southern Nigeria. *J. Ecol.* 52, 1–7.

Rick, C. M. 1966. Some plant-animal relations on the Galapagos Islands. In *The Galapagos* R. I. Bowman, ed, Chap. 28, pp. 215–224. Berkeley and Los Angeles: University of California Press.

Ridley, H. N. 1930. *The Dispersal of Plants Throughout the World.* Ashford, Kent: L. Reeve & Co.

Smythe, N. 1970. Relationships between fruiting seasons and seed dispersal methods in a neotropical forest. *Amer. Naturalist* 104, 25–35.

Snow, D. W. 1965. A possible selective factor in the evolution of fruiting seasons in tropical forest. *Oikos* 15, 274–281.

——— 1971. Evolutionary aspects of fruit-eating by birds. *Ibis* 113, 194–202.

Straw, M. 1972. A markov model for pollinator constancy and competition. *Amer. Naturalist* 106, 597–620.

Vansell, G. H. 1942. Factors affecting the usefulness of honeybees in pollination. *U.S.D.A. Circular No. 650* pp. 1–31.

Vogel, S. 1968–1969. Chiroterophilie in der neotropischen Flora. *Neue Mitt.* I-III. Flora B., 157, 562–602; 158, 185–222; 158, 289–325.

PART 5

Modeling Phenology and Seasonality

Introduction

Helmut Lieth

Botany Department
University of North Carolina
Chapel Hill, North Carolina 27514, USA

A necessary component in an ecosystem model is the seasonality of the characteristic species in the three trophic levels treated in Part 3. Local species in one trophic level may never interact with those in another trophic level if their periods of activity are different, because separation in time is as absolute as geographic separation. Although modeling of seasonality is too broad a topic for a single session of a symposium, communications could be and was opened between modelers and field biologists. A panel discussion of seasonality modeling, chaired by P. E. Waggoner, showed that phenologists have been modeling perhaps longer than anyone else in ecology. Phenological modeling, however, has been pursued not for the purposes of systems ecology, but for analyzing individual species/phenophase reactions to environmental driving forces. Models presented in this volume are primarily of this type.

The types and terms of modeling are still in flux. Depending on the purpose or methods used, the same model type may be labeled in different ways. Our modeling session was replete with impressive synonyms and antonyms: deterministic versus stochastic, predictive versus simulated, spatial versus temporal, physiological versus environmental, empirical versus theoretical, numerical versus graphical, linear versus nonlinear, and so on.

In the absence of an established terminology, we may still describe the papers in this section somewhat as follows: Waggoner (Chapter 5.1) emphasizes statistical methods for building physiological simulators; Reader, Radford, and Lieth (Chapter 5.2) present spatial and temporal correlation models between plant phenophases and environmental (mainly climatic) parameters; Lettau and Lettau (Chapter 5.3) similarly analyze soil-plant-climate relations, using H. Lettau's climatonomy concept for one location; Caprio (Chapter 5.4) describes a new environmentally determined growth unit, the solar thermal unit, which may be useful in simulation, prediction, or analog modeling; James and Shugart presented the only model of animal behavior, predicting nesting of the robin (*Turdus migratorius*) across the United States. Their paper is not included here but will appear soon in *Condor*.

Although the goals of the modelers differ, their contributions complement each other. Presentation of spatial patterns appears in Chapters 5.2 and 5.4, whereas accuracy of correlation is a topic of Chapters 5.1 and 5.3. The techniques presented here should influence future work in theoretical and systems ecology.

Many other sections in this volume contain contributions to phenological modeling. This is especially true for Parts 3 and 4.

5.1 Modeling Seasonality

Paul E. Waggoner

Connecticut Agricultural Experiment Station
New Haven, Connecticut 06504, USA

with an Appendix

Analytic Solution to Model Passages Through Phenophases

Jean-Yves Parlange

Connecticut Agricultural Experiment Station
New Haven, Connecticut 06504, USA

1 Introduction

Mathematical phenological models are devices for summarizing what we know of seasonality in biology, or for mimicking the seasonality. There are formulas for discovering and for summarizing recurrence of a phenophase as a function of time only. Also there are models relating the time of the phenophase to the environment, i.e., a correlation between mean temperature and duration of a phenophase. Or the model may have parts and parameters, each with characteristic components of the real thing; we call this a simulator.

2 Mathematical Treatments of Seasonality

Discovering Seasonality of a Phenophase

The season is so obvious in most phenological studies that methods for discovering seasonality are often neglected. In tropical regions, where there is no summer-

KEYWORDS: Heatsum; fundamental period; periodic regression; normal distribution; CSMP; environmental influences; controlled experiments; soybean; hatching; fruit fly; fungus; simulator.

to-winter change, important seasonal information may nevertheless be lurking in phenometric data. Some tropical crops deviate from an annual cycle. Table 1 shows planting-to-harvest periods for different banana cultivars. Observations of seasonality suitable for model building are lacking, even in the banana, so we illustrate a mathematical test for seasonality with a test for periodicity in the activity of the crab. Crabs collected on a beach were kept in darkness at near-constant temperature while their activity was recorded (Bennett *et al.*, 1957). Bliss (1970) analyzed that activity and found the variations to more closely follow the 25-hr tidal cycle than the 24-hr diurnal cycle.

Table 1. Growing period of several cultivars of banana (platanos) in Colombia. (Adapted from Perez-Arbelaez, 1956, p. 530.)

Cultivar[a]	Months Between Planting and Cutting of Racemes
Maqueño, dominico	13
Liberal	14
Hertón (de sancocho)	13
Cachaco	13
Bocadillo	11
Guayaba	10
Tafetán morado	12
Guineo de Colí (cultivated up to 2200 m altitude)	11
Manzano Cambur (Venezuela)	13

[a] These varieties may be from different species, e.g., *Musa paradisiaca* L. or *Musa sapientum* (L.) O. Kunze.

The course of activity was summarized with the symmetrical sine curve for a variable Y:

$$Y = a_0 + a_1 \cos\left(2\pi\frac{t}{p} - \phi_1\right) \tag{1}$$

where a_0 is the mean, a_1 is the amplitude, p is the fundamental period, time is t, and ϕ is the phase angle. The sine curve was extended by adding higher terms:

$$a_2 \cos\left(2\pi\frac{2t}{p} - \phi_2\right)$$

The 25-hr fundamental period was established by analyzing the variance. The interactions of the days of observation by the harmonic terms were found smaller, while the direct effects of the harmonic terms were larger, when p was 25 rather than 24 hr. This is a straightforward detection of unusual periodicity; analysis of phenological series likewise would seem straightforward.

Detection of other periods may be required. As devised largely by J. W. Tukey and described lucidly by Panofsky and Brier (1958), obvious trends and cycles, e.g., annual cycles, are first eliminated from the phenometric observations, and the autocorrelation coefficients calculated. The spectrum of the variance attributable to each frequency is obtained by fitting a harmonic curve to the graph of autocorrelation coefficients. The variances in any region of the spectrum can then be tested for significance to help judge how much a frequency represents a material contribution to the course of events.

Summarizing Seasonality of a Phenophase

Periodic regression founded upon the sine curve with additional terms (Equation 1) seems a natural way to test the significance of humps and hollows in phenograms and in the fluctuations of phenometric quantities. Bliss (1970) employed periodic regression to confirm the significance of the daily cycle in human births and of the annual cycle in deaths by pneumonia. His tests of the significance of amplitudes and phase shifts help separate the important from the accidental in a sample. However, phenophases that can be placed exactly, or at least decisively, upon a particular day are not modeled with periodic regression; their time series consists of, e.g., zero blooms through most of the year and then a large number over a short period. Consequently, tables of dates or amounts of occurrence are usually the raw material for models summarizing our knowledge of phenophases.

We may model the variability among individual plants at a single site within a single year, or the variability of a characteristic date for the population at a site, but among several years. First, consider variation within a single year. A typical instruction to observers of a phenological network is to note "the date of first . . . bloom when one of the small flowers . . . is fully open" (Caprio, 1966). Thus a single date is generally recorded for a population rather than several dates among individual plants. This latter variability has been studied for insects; e.g., Shaw and Starr (1946) found that the logarithms of hatching times were normally distributed for one insect, while Messenger and Flitters (1958) found that this was not so in another. In time, more models will be constructed synthesizing the date on which, say, a tenth, a half, and nine-tenths of the flowers have opened or eggs have hatched.

Another problem arises in observing only the first event: It will precede the average by longer as the sample is larger. Thus the expected range increases from twice the expected standard deviation when the sample size is 4, to five times when it is 100. Shaw and Starr (1946) found that incubation measured to the first hatched insect egg was shorter as the sample was more numerous.

Variation among years, on the other hand, can be determined from existing tabulations of dates of phenophases for the population as a whole. The mean flowering date of buckeye (*Aesculus glabra*) in a 26-year record was May 15, and the standard deviation was 7.6 days (observations of Mikesell in Smith, 1915). If we know the distribution of the dates θ_i around their mean, we can simply write the model

$$\theta_i = \mu + e_i \qquad (2)$$

where μ is the mean and e_i is the random deviation of the year i. If the cumulative distribution of a normally distributed parameter is plotted on probability paper, it will be a straight line. Fig. 1 shows roughly normal distribution for the flowering

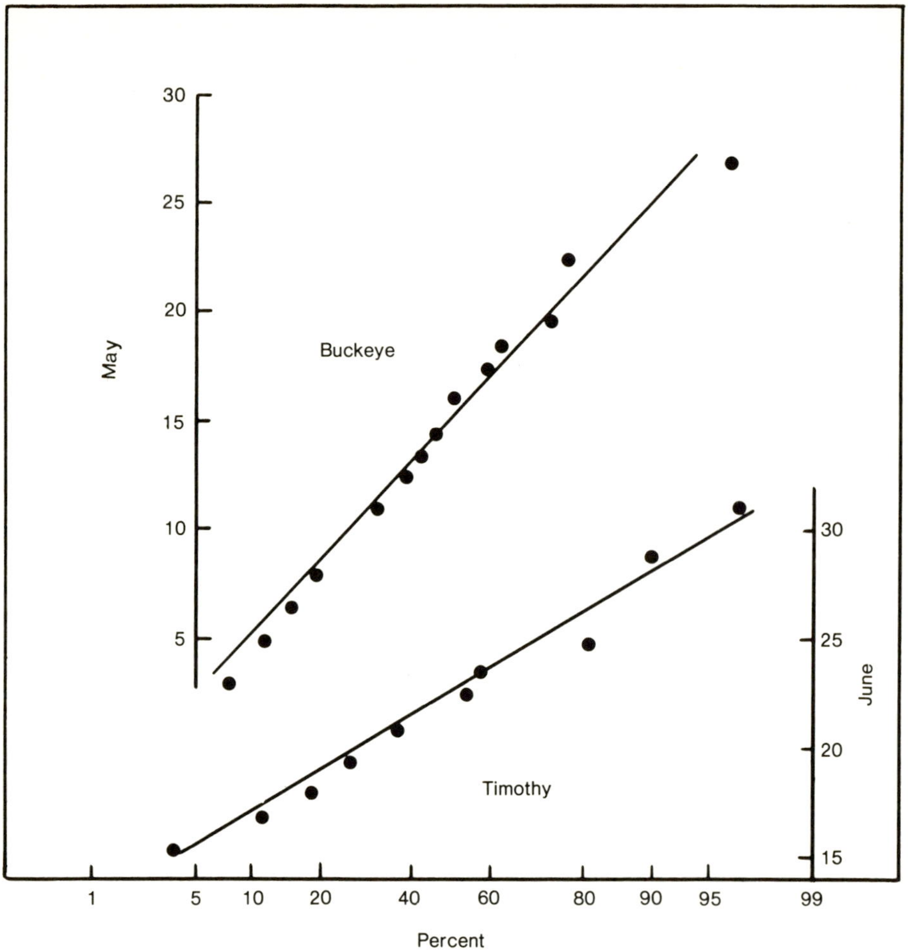

Fig. 1. The distribution of dates of flowering of buckeye and timothy at Wauseon, Ohio.

dates of buckeye and timothy. The distributions of other annual dates, i.e., of first snowfall and of last and first frost, have been thoroughly investigated and seem normally distributed (Thom, 1957; Thom and Shaw, 1958).

The time between two phenophases also deserves investigation. If the dates of the two phenophases are normally and independently distributed, the time between them will be normally distributed, with a variance equal to the sum of the variances of the boundary dates. Nuttonson (1955) has tabulated dates of spring wheat emergence and heading for 20 years at Indian Head, Saskatchewan, and for 21 years at Saratov, USSR. Unfortunately for the simple calculation of variance of the period between the dates from the variances of the dates, the dates are correlated. Late emergence tended to be followed by late heading; the square of the correlation coefficient between the dates was 0.31 at Indian Head and 0.12 at Saratov. The distributions of the periods from emergence to heading are roughly normal (Fig. 2). As expected, the standard deviations of the periods between them are not materially different from the standard deviations for the dates themselves (Table 2).

One note of caution about the use of the normal curve to summarize phenophase timing: Messenger and Flitters (1958), using data of Powsner (1935), found that hatching times of a *Drosophila* are normally distributed at 25°C but not at 19°C.

If the period between the two dates were constant, the correlation and predictability would be perfect. Our model for this is

$$\theta_{2i} = \alpha + \beta\, \theta_{1i} + e'_i \tag{3}$$

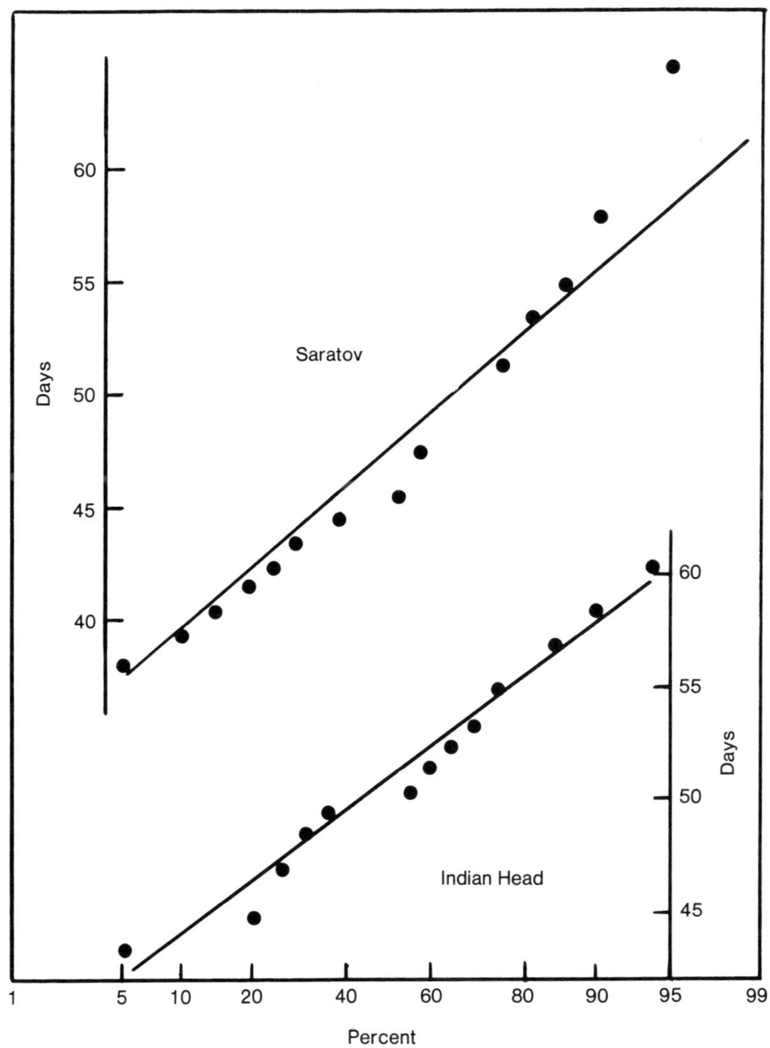

Fig. 2. The distribution of the lengths of the period from emergence to heading of spring wheat at Saratov, USSR, and Indian Head, Saskatchewan.

where the second date θ_{2i} in the ith year is a function of the first date θ_{1i}, of parameters α and β and of a random element e'_i, which will be smaller than the e_i of the distribution of θ_{2i} around its mean (Equation 2). In our sample case, the estimated variances of the e'_i for the two sites are 31 and 12% less than the variances of e_i. In fact, later emergence tends to be followed by later heading, but the period between will tend to be warmer (hence briefer) if emergence is later. The period between the dates is about a half day shorter for each day's delay in emergence at both sites. (The relations between length of periods and emergence dates explain 39 and 26% of the variability in the periods.)

Table 2. Standard deviations of dates of phenophases.

Species	Phenophase	Sample size	Standard Deviation
Aesculus glabra	Flowering	26[a]	7.6
Aquilegia canadensis	Flowering	28[a]	7.5
Phleum pratense	Flowering	28[a]	4.4
Syringa vulgaris	Flowering	25[a]	7.2
Zea mays	Flowering	27[a]	6.6
Triticum vulgare	Emerged	20[b]	6.2
Triticum vulgare	Emerged	21[c]	6.6
Triticum vulgare	Headed	20[b]	5.1
Triticum vulgare	Headed	21[c]	6.9
Triticum vulgare	Emerged to headed	20[b]	5.4
Triticum vulgare	Emerged to headed	21[c]	7.6

[a] Wauseon, Ohio (Smith, 1915).
[b] Spring wheat, Indian Head, Saskatchewan (Nuttonson, 1955).
[c] Spring wheat, Saratov, USSR (Nuttonson, 1955).

Seasonality and Environment: The Use of CSMP

General Considerations. The premier model of the relation of a phenophase to environment is the heatsum. Reaumur (18th century) conceived that flowers would bloom when the integral of spring temperature over time had reached a necessary value. We have learned since that this simple rule must be modified by taking into account the effects of light or moisture, the ineffectiveness of temperatures below a certain threshold, and that temperatures above an optimum are less effective than their numerical value may imply. Still, it is reasonable to say that for a given place and for an organism, a plant, that takes the temperatures of its surroundings, a phenophase will appear when the integral L has reached a critical value, C degree-days.

$$L = \int_{t_0}^{t} (T - T_h)\, dt \tag{4}$$

where T and T_h are the current temperature and the threshold temperature. The initial time t_o is the visible beginning of a previous phenophase, e.g., sowing or any time during a long period of temperatures below T_h. In practice, the integral is approximated by adding the daily mean temperatures, hence called heatsum with units in degree-days.

The mathematical model relating the phenophase P_h to environment is simply $P_h = 0$ when $L < C$ and $P_h = 1$ when $L \geq C$. The physical model corresponding to this mathematical model must be something as follows: e.g., florigenic substances or "initials" may be conceived as entering a "box" at a rate $(T - T_h)$. When the level L in the box has reached C, flowers appear.

Figure 3 shows a more elaborate physical model of the rate of progress toward a phenophase as controlled by the environment. [This system was developed for

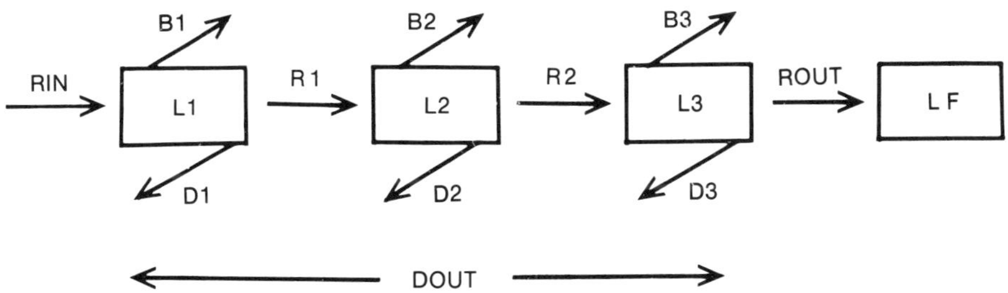

Fig. 3. A series of three states $L1$, $L2$, and $L3$ on the way to a final state LF. The contents may be destroyed at rates $B1$, $B2$, and B3 or diverted to DOUT at rates $D1$, $D2$, and $D3$. They may also pass through the states at rates $R1$, $R2$, and ROUT.

simulation of sporophore and spore formation in a pathogenic fungus (Waggoner and deWit, 1972).] Boxes 1, 2, and 3 represent invisible states during the barren phenophase; box F contains the flowers. The level L in each box is the number of flower initials or flowers themselves per hectare in each state. At time t_o and rate RIN, floral initials enter the first box or integral. As time passes, the rates $R1$, $R2$, and ROUT of flowers per hectare per day carry the initials through the invisible states and on to their destination of flowers. Environmental influences are reflected in the rates.

Generality is achieved by providing two additional exits from the boxes. Exit rates $B1$, $B2$, and $B3$ banish or extinguish a portion of the initials; exit rates $D1$, $D2$, and $D3$ divert or detour all the initials to DOUT. DOUT may be placed elsewhere.

Use of the Continuous System Modeling Program (CSMP). The calculation of the progress toward LF is simply specified in the programming language CSMP (IBM' 1969). Among other features, the CSMP has subroutines called MACRO. The calculations of rates are presented as MACRO (Fig. 4). The three invisible states in floral initials per hectare are written as

$$L1 = \text{INTGRL}\ (.1,\ \text{RIN} - R1 - B1 - D1)$$

where INTGRL indicates integration, and the first argument .1 is the negligible level at t_o.

The rates in floral initials per hectare per day between the boxes are calculated from the mean residence time, HOLT days, in each box. The variable W is 0 or 1 and determines whether progress continues or stops. For example $R1$ is

$$R1 = L1/\text{HOLT}*W$$

The relative extinction rate B per day causes the loss of floral initials per hectare per day and $B1$ is

$$B1 = L1*B$$

```
* HOLD DELAYS A PROCESS THROUGH 2, 3, 4 OR 6 BOXES, L.
MACRO  ROUT,DOUT=HOLD(RIN,HOLT,BOX,W,B)
        L1=INTGRL(.1,RIN-R1-B1-D1)
        L2=INTGRL(.1,R1-R2-B2-D2)
        L3=INTGRL(.1,R2-R3-B3-D3)
        L4=INTGRL(.1,R3-R4-B4-D4)
        L5=INTGRL(.1,R4-R5-B5-D5)
        L6=INTGRL(.1,R5-R6-B6-D6)
* MEAN RESIDENCE IN EACH BOX IS HOLT. ADVANCE ONLY WHEN W=1.
        R1=L1/HOLT*W
        R2=L2/HOLT*W
        R3=L3/HOLT*W
        R4=L4/HOLT*W
        R5=L5/HOLT*W
        R6=L6/HOLT*W
        RX=INSW(BOX-3.,R2,R3)
        ROUT=FCNSW(BOX-4.,RX,R4,R6)
* DURING RESIDENCE, B/TIME ARE DESTROYED.
        B1=L1*B
        B2=L2*B
        B3=L3*B
        B4=L4*B
        B5=L5*B
        B6=L6*B
* WHEN W=0, CONTENTS ARE COLLECTED IN DOUT.
        D=1.-W
        D1=EMPTY(L1,D)
        D2=EMPTY(L2,D)
        D3=EMPTY(L3,D)
        D4=EMPTY(L4,D)
        D5=EMPTY(L5,D)
        D6=EMPTY(L6,D)
        DX=INSW(BOX-3.,DOUT2,DOUT3)
        DOUT=FCNSW(BOX-4.,DX,DOUT4,DOUT6)
        DOUT2=D1+D2
        DOUT3=D1+D2+D3
        DOUT4=D1+D2+D3+D4
        DOUT6=D1+D2+D3+D4+D5+D6
ENDMAC

* EMPTY THE CONTENTS FROM AN INTEGRAL.
MACRO RATE=EMPTY(L,COND)
        RATE=INSW(L-1.E-50,0.,L/DELT*COND)
ENDMAC
```

Fig. 4. The CSMP MACRO's for delaying or holding a process, as the development of flowers, through several preparatory stages.

Finally when W is 0, D is 1, and the boxes are emptied and diverted to DOUT floral initials per hectare per day. $D1$ is

$$D1 = \text{EMPTY}\ (L1,\ D)$$

The command EMPTY brings MACRO into play. MACRO produces a rate that empties the box in the calculation interval DELT if the contents are greater than 10^{-50}. If the contents are less than 10^{-50}, the rate is set at zero. The calculations for $L1$, $R1$, $B1$, and $D1$ are amplified to allow 2, 3, 4, or 6 boxes or invisible states. The system for moving the initials of a phenophase through the antecedent states is brought into play by

$$\text{ROUT, DOUT} = \text{HOLD (RIN, HOLT, BOX}, W, B)$$

The new parameter is BOX—the number of boxes, states, or integrals. By changing the number of boxes, the dispersion in the passage of individuals into LF is varied (Fig. 5). Jan Goudriaan (unpublished) has shown that the standard deviation

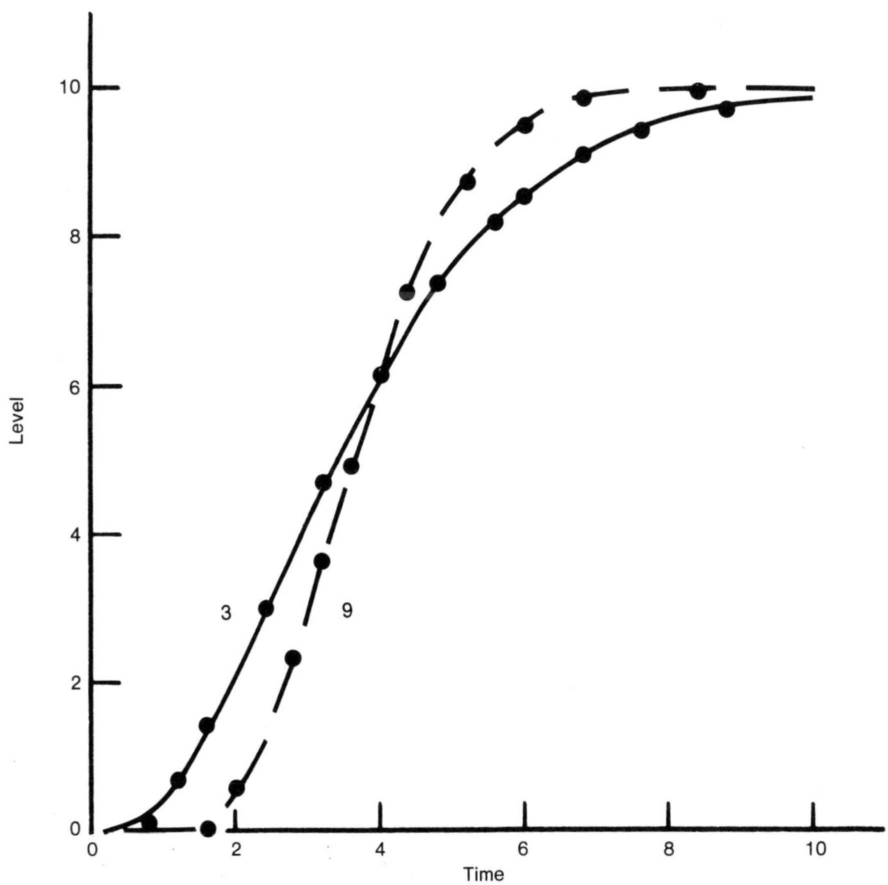

Fig. 5. Effects of 3 and 9 boxes upon the distribution of emergence from a series of boxes.

of the appearance time equals the mean residence time in the entire system divided by the square root of the number of boxes.

We introduced this model of boxes with the simple integral of temperature (Equation 4), based on a conception of initials passing through antecedent states at a rate proportional to the warmth $(T-T_h)$ above the threshold T_h. At a constant temperature T, the phenophase would be reached in $C/(T-T_h)$ days. In the CSMP it would be reached in HOLT*BOX days. Thus the heatsum calculation is equivalent to a model of initials passing through BOX invisible states at the relative rate P per day of

$$P = 1/\text{HOLT} = \text{BOX} * (T-T_h)/C \quad (5)$$

What advantages have been gained by these complications upon Reaumur's simple heatsum? For those who worry about employing a single temperature to represent a whole day, the computer program can specify a computation interval to match the speed of temperature change and thus approach more nearly the integration implied by the integral for L. Given the uncertainty in biology, however, this advantage is for the overly fastidious. The greater advantage is that Fig. 3 and the accompanying computer program have evoked a physical model. Once we think of rates through antecedent states, we begin to search for these states or for accumulations of florigenic substances, for example. Then we can go beyond models that merely summarize field observations and toward simulators with parts identified with components of the real plant, and tested against the real plant's development in the environment.

3 Correlations of Seasonality with Environment

The word "correlation" evokes the multiple correlation of the date of appearance of some phenophase such as flowering with a range of such environmental factors as temperature, radiation, etc. The classic examples, however, have not taken this form. One good reason is the difficulty of deciding which among many ways of expressing, say, temperature, one should choose. We know, for example, that flowering is the culmination or integral of many events of the past and not just yesterday's temperature. Thus our common-sense discourages a thoroughly empirical attack to obtain a relation between the appearance of a phenophase and a plethora of environmental variables.

A Single Environmental Factor

Phenological observations of winter wheat in Saskatchewan and Russia introduced earlier can be used to show the correlation between the period from emergence to heading and a heatsum. Nuttonson (1955) calculated the critical C degree-days above a threshold T_h of 40°F by averaging samples of 20 or 21 years. As his criterion of success, he employed the coefficient of variation, 100 times the standard deviation divided by the means of the critical degree-day sum C. The coefficients of variation of C_i at the two sites were between 7 and 8%, which is smaller than the

11 to 16% coefficients of variation of the days themselves. Put another way, the standard deviations in the critical heatsums were about 70 degree-days (°F). At a conservative daily rate of 20 degree-days these standard deviations of C_i are equivalent to 3 to 4 days, which is shorter than the 5- to 7-day standard deviation of the days themselves (Table 2).

A mathematical model (Equation 6) relating the period between the two dates in the *i*th year to the critical heatsum C, is, however, more consistent with Equations 2 and 3 than is the examination of coefficients of variation.

$$(\theta_2 - \theta_1)_i = \alpha + \beta\ C_i + e'_i \tag{6}$$

Analyzed in this fashion, the variability in $(\theta_2 - \theta_1)_i$ is decreased 22 and 8% by C_i and the model (Equation 6). In a different sort of plant and place, the variability in dogwood flowering dates is decreased 40 to 60% by relating them to heatsums calculated from January 1 to various cutoff dates in the spring (Radford, 1971). As Reaumur observed centuries ago, phenophase models founded upon heatsums have smaller random elements than models using the mean date.

Several Environmental Factors

Thornthwaite's development units, invented to schedule pea planting, incorporate the influences of several environmental factors (Thornthwaite and Mather, 1954). These development units are calculated in the same way as is transpiration. Using transpiration to calculate developmental rate is quite ingenious, as we can appreciate if we review the models preceding Thornthwaite's, and also his equation for evaporation.

Lissner (in Smith, 1920) attempted to reconcile the differences in heatsums required to reach the same developmental stage in different climates. He employed the sum of the effective temperature for a certain phase of development divided by the sum of the effective temperature for the entire year, reducing the variability in critical heatsums found between warm and cool climates. Lissner assumed that since plants at southern latitudes are subject to much heat for the entire year, they simply lengthen all their phases of development.

With their discovery of photoperiodism, Garner and Allard (1920) injected daylength into phenology. Nuttonson (1948) found that the time between phenophases is more closely related to the average daylength times the degree-day sum than it is to either the daylength or the degree-day sum taken separately. Similarly, Hopkins' (1938) Bioclimatic law states that spring phenophases proceed over the eastern United States at a rate according to latitude, longitude, and altitude. This completed the foundation for a model of phenophase length that would include degree-days, average degree-days of a locality, and daylength.

Equation 7 gives Thornthwaite's (1944) expression for potential evaporation and also expresses the developmental unit. This complicated expression was obtained from observations of monthly temperature and evaporation from well-watered drainage basins and from lysimeters.

$$1.6\ \ell\,(10\ T_a/I)^m \text{ cm/month} \tag{7}$$

where $I = \sum_{1}^{12}(\overline{T}_a/5)^{1.5}$, ℓ is a daylength and month factor, T_a is air temperature for the month, $\overline{T}_a$ is the climatologically normal temperature for each of the 12 months, and m is a function of the term I. Note the appearance here of factors already found important in determining phenophase timing, i.e., Nuttonson's daylength ℓ and Lissner's mean temperature $\overline{T}_a$.

This correlation of a state with environmental factors is not the analysis of dependent and independent variables that a statistician would attempt. Rather it is a reasonable combination of important phenological factors into an equation. Thornthwaite took great pleasure, I imagine, in the same equation's being used to schedule both planting and irrigation.

A correlation of a state with several environmental variables in a form the statistician expects is provided by Radford (1971), who found that adding precipitation to heatsum increased the precision of predictions of dogwood flowering. Benacchio and Blair (1972) investigated the time to appearance of the first leaf in four perennial plants and found better correlation of leafing with soil, rather than air, temperature. Waggoner *et al.* (1960) found that warming the soil (but not the air) around strawberry plants by a translucent mulch speeds flowering.

4 Simulators of Phenophases from the Results of Controlled Experiment on Environmental Effects

Biology is organized in the levels of molecules, cells, organs, organisms, and, finally, organisms together in the field. The level we have examined so far is the population in the field. Phenologists have predicted severe difficulty for those attempting to relate natural phenology to controlled laboratory experiments on single organisms (Smith, 1920). We have further restricted our discussion to relation rather than process. Relating flowering dates to degree-days or transpiration implies process of course, such as van't Hoff's law or photoperiodism, but the core of existing models is that some combination of weather factors through a phenological period is related to the length of that period. Generally, phenologists have not sought to employ knowledge of a physiological process to construct a simulator of phenological events in the field.

But some beginnings can be cited: (1) an early attempt to relate corn growth in liquid culture in a laboratory to corn growth in the field, (2) an extrapolation from soybeans in the growth chamber to soybeans in the field, (3) a study of egg hatching in the Oriental fruit fly, (4) an attempt to simulate the sporulation of a microscopic, pathogenic fungus. These are discussed below.

The Time to Flowering of Maize and Soybean

Using measurements of the elongation of *Zea mays* seedlings at steady temperatures, B. E. Livingston (Smith, 1920) built a model of plant development in the field. We may charitably think of his model as incorporating the concept of process if we say that development is the outcome of the process of shoot elongation. A similar, early use of observations in the laboratory to anticipate events outdoors is found in Shelford's (1927) hatching of codling moth (*Carpocapsa pomonella*) eggs at steady temperatures.

The heart of Livingston's model is a graph of physiological indices versus temperature obtained from measurements of shoot growth. The rate of shoot elongation increased slowly from 4° to 16°C, increased rapidly to a maximum rate at 32°C, and then slowed precipitously as temperatures increased further. His model, or temperature efficiency index, was the sum of these indices of elongation.

Livingston realized the deficiencies of his method, and his caution bears quotation:

> While it is quite apparent that . . . physiological indices here described [are] far superior . . . to other[s] . . . it is equally clear that . . . they are based upon tests of only a single . . . species . . . derived from the growth of seedlings [in a single environment] . . . with . . . 12 hours' [of steady] temperatures.

Nevertheless, Livingston had devised a logical simulator.

Brown (1960) evolved an index of physiological development using Van Schaik and Probst's (1958) observations of *Soja max* in controlled environments with varied daylength and temperature. Brown noted that nearly the same number of night hours was required to reach flowering for all photoperiods at any one temperature. Therefore he defined units Y of developmental rate as the reciprocal of the night hours from planting to flowering with the decimal point shifted to make whole numbers, i.e., $Y = 10{,}000/\text{night hours to flowering}$. The relation between these soybean flowering rates and temperature was different from the rate of growth of maize shoot, as Livingston had cautioned, but the rates for two varieties of soybeans were not significantly different from each other.

Brown and Chapman (1960) tested the model in the field. The developmental units derived from the flowering of two soybean varieties in the laboratory predicted flowering time of three other varieties in the field at six places in Ontario and Iowa in several years. Needless to say, Brown's units were applicable to the field only when the field's moisture resembled the ample moisture of the chambers in which the soybeans were grown. The units derived from observations of preflowering time in the laboratory did not explain variation in times between two other phenophases, verifying another caution of Livingston.

Let us summarize the progress toward a physiological simulator. After examining some correlations between phenophases and field environment, we turned to simulators suggesting processes and certain relevant variables derived from controlled experiments. Livingston made a logical simulator of development from observations of temperature and maize elongation, described its weaknesses, and left it largely untested. A generation later, Brown built a simulator of soybean development from observations of daylength, temperature, and the length of the preflowering period in a growth chamber. He went on to test the simulator in fields of several regions and varieties and found it worked during the preflowering phase but not during other phases or during drought. We now suggest a soybean simulator in a form Brown might well have used had he made his simulator in 1972.

In a CSMP program (Fig. 6) the number of blooms is calculated as the integral of ROUT blooms per hectare per hour coming from a system of antecedent states, as in Fig. 3. The preflowering period is conceived as a series of 12 invisible phenophases through which the flower must pass. The 12 is an arbitrary choice for the number of boxes that would give a reasonable period between the opening of the first and last blossoms. We start the succession through the phases by placing the

```
* BLOOMS INCREASE AS BLOOM INITIALS LEAVE PRECURSOR STATE AT RATE ROUT.
        BLOOMS=INTGRL(O.,ROUT)

* ENTER 100 BLOOM INITIALS IN FIRST OF PRECURSOR STATES AT START TIME.
        RIN=INSW(TIME-.5,100./DELT,0.)
* MOVE INITIALS THRU TWICE 6 STATES OR BOXES WITH MEAN HOLD TIME HOLT IN EACH.
        RIN2,DOUT2=HOLD(RIN,HOLT,BOX,W,B)
        ROUT,DOUT=HOLD(RIN2,HOLT,BOX,W,B)
PARAM BOX=6., W=1., B=0.

* HOLD TIME IS FUNCTION OF LITE.
        HOLT=FCNSW(LITE,DARKT,DARKT,LITET)
* IN LITE, NO DEVELOPMENT.
PARAM LITET=1.E50
* IN DARK, MEAN RESIDENCE IN EACH BOX IS 1.E4/ BROWN UNITS/ NUMBER OF BOXES.
        Y=4.39*TEMP-.0256*TEMP**2-155.18
        DARKT=1.E4/Y/BOX/2.
PARAM TEMP=80.

* LITE AND DARK DETERMINED FROM DAYL DAY LENGHT.
        LITE=DAYL-AMOD(TIME,24.)
PARAM DAYL=(12.,16.)

LABEL OPEN BLOOMS(HOLT)
PRTPLT BLOOMS(HOLT)
```

Fig. 6. The simulator of development of soybean flowers.

beginnings of 100 flowers in the first box, in the first calculation interval DELT, by means of a rate RIN of 100/DELT (Fig. 6). The HOLD process, introduced earlier, is invoked twice to obtain 12 boxes:

RIN2, DOUT2 = HOLD(RIN, HOLT, BOX, W, B)
ROUT, DOUT = HOLD(RIN2, HOLT, BOX, W, B), primordia per hour per hectare

We shall have no use for DOUT here, and it is a dummy. Since we want the primordia to move through 12 boxes, and HOLD is invoked twice, BOX is 6. Making W equal 1 prevents the primordia from being detoured to DOUT, and making B equal zero prevents their destruction.

PARAM BOX = 6., W = 1., B = 0.

Brown conceived that flowers develop differently in the dark and light, and the following statement chooses DARKT as the mean residence in each box when LITE is negative or zero, and it chooses LITET when LITE is positive.

HOLT = FCNSW (LITE, DARKT, DARKT, LITET), hours

Brown conceived that no development occurred in the light, i.e., LITET is a very long time.

$$\text{PARAM LITET} = 1.\text{E}50, \text{ hours}$$

Brown conceived that development in the dark proceeded at a rate Y that was inversely proportional to the hours of darkness required for flowering at a given temperature TEMP in °F:

$$Y = 4.39*\text{TEMP} - 0.0256*\text{TEMP}**2 - 155.18, 10^4/\text{hr}$$

We convert Y into a mean residence time in each of the 12 or 2*BOX boxes:

$$\text{DARKT} = 1.\text{E}4/(Y*2.\ *\text{BOX})$$

The temperature is established:

$$\text{PARAM TEMP} = 80., \text{°F}$$

Twelve DAYL hours of light are imposed by

$$\text{LITE} = \text{DAYL} - \text{AMOD}(\text{TIME}, 24.)$$

$$\text{PARAM DAYL} = 12.$$

When the CSMP simulator (Fig. 6) is submitted to the computer, it of course simulates the development of soybeans growing in controlled environments. Since it performs as Brown's model, it also mimics the observations in the field.

Certain attributes of the CSMP simulator should be noted. First, the simulated flowers do not appear simultaneously but over a period as in Fig. 5, and their times of appearance have a standard deviation that is one over the square root of 12, or 29% of the mean time to flowering. To shorten this spread of the flowering times, the number of boxes must be increased.

Second, varying temperatures can be employed in the simulator to test its validity against behavior in chambers with varying temperatures, or against behavior in the naturally changing outdoor temperatures. A valuable simulator would employ developmental rates from observations in controlled, steady environments and temperatures taken from the natural environment. When the output of this simulator is compared to actual development in natural environments, we can learn whether we should share Livingston's fear of models built on laboratory observations.

Finally, attention is naturally drawn to the arbitrary states that I have conceived in the simulator between planting and flowering. Here is a place in phenology for the scientist who will ask what these precursor states are and whether the rates between them are similarly affected by environment. For example, one would find that the emergence of the soybean seedlings from the soil was affected by soil temperature and would make HOLT in the first box a function of soil temperature, not of light and air temperature.

The Duration and Success of Incubation of Insect Eggs

A simulator of the variable proportion of eggs hatched and of the different lengths of incubation caused by different temperatures can also be accomplished with CSMP. Eggs of the Oriental fruit fly *Dacus dorsalis* were incubated at steady temperatures of 54° to 98°F by Messenger and Flitters (1958). The median hatching time varied from 26 to 240 hr, and the hatch varied from 0 to 84%. The range of hatching time at a given temperature increased from 4 to 54 hr, generally in step with the length of the median hatching time.

The HOLD with 12 boxes is used again, the parameter B causes the eventual hatch to be less than 100%, and HOLT largely controls the incubation time. Since the median hatching time $t_{1/2}$ is about BOX*HOLT, HOLT is estimated as $t_{1/2}$ divided by 12. We call the relative rate 1/HOLT by the name P.

To estimate B, examine Fig. 3: the fraction of $L1$ that reaches $L2$ will be

$$\frac{R1}{R1+B1}=\frac{P}{P+B}$$

The hatching percentage, or the proportion of $L1$ that finally passes successfully all the way to LF, is

$$\left(\frac{P}{P+B}\right)^{\text{BOX}}$$

One can now estimate a P (or HOLT) and a B for each of Messenger and Flitters' observations of incubation time and hatching percentage at constant temperature, and compile tables of P and B as functions of temperature.

The rates in and out of the 12 boxes are stated as in the simulator of soybean flowers (Fig. 6) except that the units are now eggs per batch per hour. The HOLT and B are interpolated from the tables by the arbitrary function-generating facility of CSMP with an argument of temperature. The simulator mimics Messenger and Flitters' observations of hatching time and percentage at steady temperature.

Messenger and Flitters (1959) went on to observe both hatching times and percentages at varying temperatures and thus provided excellent data for testing our simulator. Incubation temperatures were varied smoothly and diurnally in imitation of outdoor temperatures. Messenger and Flitters (1959) estimated $t_{1/2}$ quite well by summing development units derived from their earlier experiments at steady temperatures. In their first four experiments at varying temperatures (Table 3), the full range of variation lies within the range for which we can estimate P and B.

Before attempting the simulation, however, the failure to simulate the observed small standard deviations must be corrected. The range of observed hatching times at steady 70°F implies a standard deviation s of slightly more than 2 hr, while the simulator with 12 boxes produces an s of 16 hr. According to the formula

$$\text{BOX} = (t_{1/2}/s)^2 \tag{8}$$

a $t_{1/2}$ of 52 and an s of 2.2 indicates fully 558 boxes!

An analytic solution for many boxes by J.-Y. Parlange appears as an appendix to this paper. To calculate $t_{1/2}$ for the limit of very many boxes and variable temperatures, he states

$$\int_0^{t_{1/2}} P\,dt = \text{BOX} \tag{9}$$

The proportion of eggs hatching can be calculated from

$$\exp\left(-\int_0^{t_{1/2}} B dt\right)$$

That is at $t_{1/2}$ the integral of P will reach the number of boxes, and the percentage hatch will equal the exponential of an integral of past destruction rates B. The integrals of P and B were obtained numerically for the courses of temperature employed by Messenger and Flitters. The standard deviation is a function of P at time $t_{1/2}$:

$$s = \text{BOX}^{1/2}/P_{1/2} \tag{10}$$

In the isothermal case

$$P = \text{BOX}/t_{1/2}$$

and Equation 10 for s reduces to Equation 8 used to calculate BOX from observations $t_{1/2}$ and s.

The outcome of the simulations is shown in Table 3. The $t_{1/2}$ were simulated about as closely as those Messenger and Flitters achieved by adding development

Table 3. Observed and calculated median hatching time $t_{1/2}$ and percentage hatch with incubation at five temperature regimes. The eggs of the Oriental fruit fly *Dacus dorsalis* were incubated at temperatures that varied smoothly within ranges of 10°, 20° or 30° F.

Temperature, °F Mean	Range	$t_{1/2}$ Obs.[a]	Calc.[b]	Percentage Hatch Obs.[a]	Calc.[b]
70	20	52.5	53.3	81	77
80	10	32.0	33.6	72	80
80	20	31.5	33.5	75	79
80	30	33.0	33.5	76	74
90	20	30.0	35.3	72	22

[a] Messenger and Flitters (1959).
[b] Results of CSMP simulator.

units. When the mean temperatures were 70° and 80°F and the full diurnal range was within the range of observations for which we could estimate *P* and *B*, the simulations of percentage hatch were adequate. In these four cases it should be pointed out that the temperatures of 65° to 95°F encountered when the means were 70° or 80°F did not encompass great changes in the hatching percentages derived from experiments at steady temperatures.

In the last simulation in Table 3, the temperature ranged 20°F around a mean of 90°F and daily exceeded the 98°F that killed all eggs when incubation was at a steady temperature. In the simulation I specified that the hatching percentage was 0.4% at 97.5°F and hotter; the 0.4 was, in fact, observed at 97.5° only and hatching was nil at 98°F in the incubations at steady temperatures. In this case, the hatch was fully 72% rather than the simulated 22%. This confirms the conclusion already reached by Messenger and Flitters: brief exposure to harmful temperatures is much less harmful than expected from the incubations at steady temperatures.

The model of Fig. 3 has now been employed to simulate the percentage of individuals that succeed, as well as the time required to reach the phenophase. Furthermore, an analytic solution has permitted the use of many boxes and thus the simulation of a relatively brief standard deviation of the times of hatching, i.e., arrival at *LF*. When this simulator and estimates of the rates of progress and destruction derived from experiments at steady temperatures were applied to experiments at varying temperatures, both hatching time and percentage were

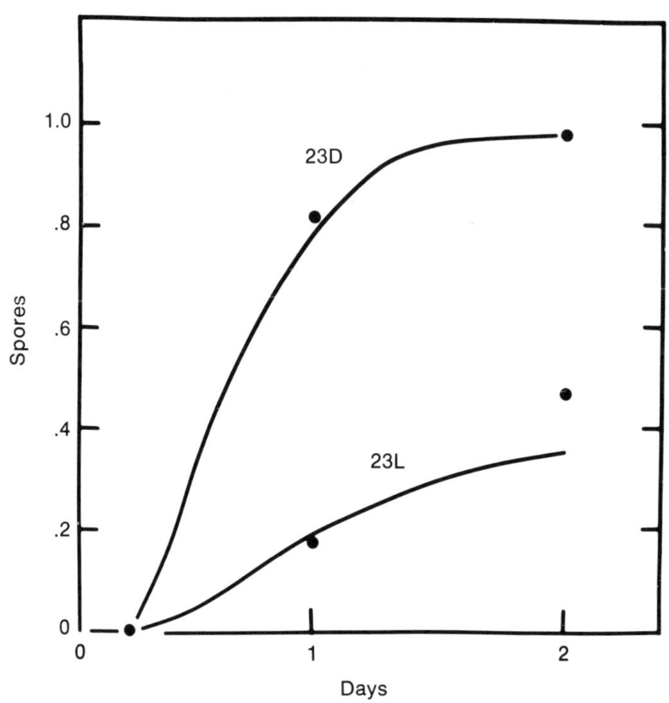

Fig. 7. Sporulation of *Helminthosporium maydis* sporophores at 23°C in the light (*L*) and dark (*D*) (Waggoner and de Wit, 1972).

mimicked reasonably well. Simulation did fail, however, when the varying temperatures became very hot, because the eggs could evidently survive brief exposure to temperatures that were lethal on longer exposure.

The Sporulation of a Pathogenic Fungus

A final example taken from plant pathology illustrates the use of the DOUT or diversion rate in HOLD. The fungus *Helminthosporium maydis* caused the great southern corn leaf blight epidemic in 1970. In 1971 a Fortran simulator EPIMAY (Waggoner *et al.*, 1972) was employed with some success to anticipate the course of the disease (Felch and Barger, 1971). Subsequently, the simuator was composed more lucidly in CSMP (Waggoner and deWit, 1972). Although the seasons are only hours long, the simulator is replete with phenophases.

One of the stages in the fungal life cycle, sporulation on sporophores or "stalks" that have been dried, is depicted in Fig. 7 as two different courses, one at 23°C in the light and the other at the same temperature but in the dark. About 6 hr after they become wet, the stalks begin to sporulate, more rapidly in the dark and more slowly in the light. It is assumed that if they are dried at any time, the condition of the stalks that have not yet sporulated returns to the initial condition and requires another 6 hr of wet before they again begin sporulating. Because sprinkling water on the stalks decreases sporulation, it is assumed that the beating of rain destroys some.

The census CDE of dried stalks that are effective in sporulation is the number of stalks that have been wet 6 hrs and are ready to fruit or sporulate. CDE is, of course, an integral.

CDE = INTGRL (0., RDH–BDE–DDE–RDS), effective stalks per hectare.

The recruits RDH come from a HOLD:

RDH, DDH = HOLD(RGD, HGD, BOX, WET, BETR), stalks per hectare per day

Since the mean delay BOX*HGD is 6 hr

PARAM BOX = 4., HGD = 0.0625, days

The RGD stalks per hectare per day that enter the hold is the sum of the DDH stalks dried while in the hold, plus DDE dried from the effective CDE, plus others that need not concern us here.

The parameter WET that either propels the stalks toward fruition or returns them to the initial state is a weather observation. WET is an argument of HOLD and also determines DDE:

DDE = EMPTY (CDE, 1.–WET)

BETR, which destroys a fraction of the stalks in HOLD, is derived from rainfall observations. BETR also controls the destruction BDE of CDE:

BDE = CDE*BETR.

In the end, successful stalks sporulate on wet leaves at the rate RDS, according to temperature, light, and wetness.

RDS = PDS*CDE*WET, spores per hectare per day

PDS is interpolated from a table or series of temperatures TEMP and relative rates derived from the observations of Fig. 7. The table or series called DSDT is chosen in the dark and the table called DSLT is chosen in the light. Then the arbitrary function generating facility of CSMP is employed for the interpolation. This choice and interpolation are accomplished by the statement

PDS = INSW(LITE, AFGEN(DSDT, TEMP),
AFGEN(DSLT, TEMP)), spores per stalks per day

The curves of Fig. 7, which mimic the observations, were computed by the simulator. Drying, of course, interrupts the sporulation and diverts the stalks to DOUT, here called DDH. The DDH and DDE are returned to the initial box.

This example has illustrated a period before the phenophase in which destructions and diversions occur, and a fruition that proceeds according to three environmental factors: wetness, temperature, and light.

5 Simulator of the Phenophases in an Entire Life Cycle

We have reviewed models or simulators for some of the constituent stages of life cycles. Now we try to integrate these constituent stages into full life cycles using the plant pathogenic fungus, *Helminthosporium maydis*, introduced in the preceding section.

Fungi have some practical advantages for phenology. Their life cycles revolve rapidly, large populations can be seen at once through a microscope, they can be subjected to controlled environments in simple incubators, and they display marked phenophases. Thus a worker can conveniently compose a simulator of the entire cycle of phenophases of an important organism.

In *H. maydis*, the lesions, sporophores, spores, and germ tubes are the most visible states of the life cycle. The effect of temperature, moisture, and light upon each stage may be measured to provide the rates at which to move the simulator. Figure 7 illustrates the sort of data that must be obtained concerning each phenophase or state (Waggoner *et al.*, 1972).

In the CSMP simulator of *H. maydis* (Waggoner and deWit, 1972) are a set of states or integrals called "censuses" for a hectare of land.

Lesions Visible. This census grew as infections completed incubation, and the rate depended solely upon the number of infections because experiments revealed no effect of environment upon the duration or success of incubation. The number of lesions, with the area of the lesions, shows the injury to the host and sets the stage for the production of offspring.

Area of Lesions. Since experiments showed that temperature affects enlargement of lesions, a proportional rate was selected according to the current temperature. Since each lesion seems to grow toward a maximum rather than indefinitely, the proportional rate was multiplied by the difference between the attained lesion area per hectare and a ceiling. The ceiling was the product of the number of lesions times the observed maximum size for single lesions.

Opportunities Effective for Stalk Formation. We invented this invisible stage on the way to sporophore (stalk) formation for two reasons. First, since the number of stalks seems limited to a maximum of about $300/mm^2$, we conceived a consumable thing, the opportunities, to enable simulation of the approach to a limit or maximum. Second, stalk formation does not begin as soon as lesions are wet, rather a lag of about 6 hrs is evident between wetting and the appearance of the first stalk. Therefore, when a lesion is wetted, invisible opportunities are started through a HOLD. They enter the HOLD at a rate equal to the product of the enlargement of lesion area times the maximum number of stalks per lesion area. If the lesion remains wet long enough, the opportunities emerge from the HOLD as effective opportunities and begin forming stalks. If the lesion dries, the opportunities are caused to return to the beginning of the HOLD. Once the opportunities have become effective, i.e., completed their wet incubation, stalks appear at a rate determined by current temperature and light so long as wetness continues. Since detrimental environments cause the number of stalks to approach lower limits per lesion area than optimal environments do, a rate of extinction as well as fulfilment of opportunities is subtracted from the census of effective opportunities; the rate of extinction as well as fulfilment is determined by light, temperature, and wetness.

Green Stalks. If stalks remain moist or "green" once they are formed, sporulation begins after a brief lag. Thus the barren stalks freshly formed upon opportunities are made effective after a brief hold and then caused to sporulate at a rate determined by current light and temperatures so long as wetness continues. Here the depletion of the supply of barren green stalks causes the sporulation to level off at 100%.

Dried Stalks. When green stalks are dried and then rewet, they sporulate more slowly and often less completely than they would had they remained moist or green. The lag and other features of the sporulation have already been treated in the preceding section and depicted in Fig. 7.

Spores on Stalks. This census receives recruits as both green and dried stalks sporulate and loses them as they are blown or washed away. Wind is caused to disperse the spores according to the stress, i.e., the square of the wind speed. To obtain a plausible aerial concentration of spores, the number dispersed is simply divided by the speed of the dispersing and diluting. Dispersal by rain is still too poorly treated to deserve examination in this summary.

Spores on Foliage. Another plausible rate, this one involving foliage area as well as wind speed, is used to simulate the trapping of the airborne spores on susceptible leaves. The trapped spores are permitted to survive persistent drought or be killed upon drying after wetting. When they are wet, however, they germinate at rates determined by temperature. Since experiments showed that germination proceeds to less than 100% in unfavorable environments, just as stalk formation proceeded to lower limits in deleterious environments, a rate of extinction as well as germination is chosen according to the temperature and subtracted from the census of spores on foliage.

Germ Tubes. After germination creates these microscopic tubes, they invade or infect healthy foliage. Again, rates of accomplishment and extinction are determined by the temperature and wetness. Finally, a HOLD between infection and the appearance of visible lesions simulates incubation.

When the parts described above are put together, they make a comprehensive simulator of a life cycle, called EPIMAYB. It simulates with some success the period between phenophases of *H. maydis* and also produces the phenogram of lesions, stalks, spores, and germ tubes. Figure 8 shows the simulated course of sporulation and of airborne spores during two days; the wet night is the season of sporulation and the dry day is the season of blowing of spores.

EPIMAYB is a relatively simple yet realistic simulator of an entire life cycle constructed from laboratory observations. Eight, generally visible, states or phenophases are identified as integrals. Where lags or incubations occur, the HOLD convention is invoked. The rates are made proportional to the unexpended antecedent state. The proportional rates of accomplishment are chosen according to the current moisture, temperature, or light, and according to the results of controlled experi-

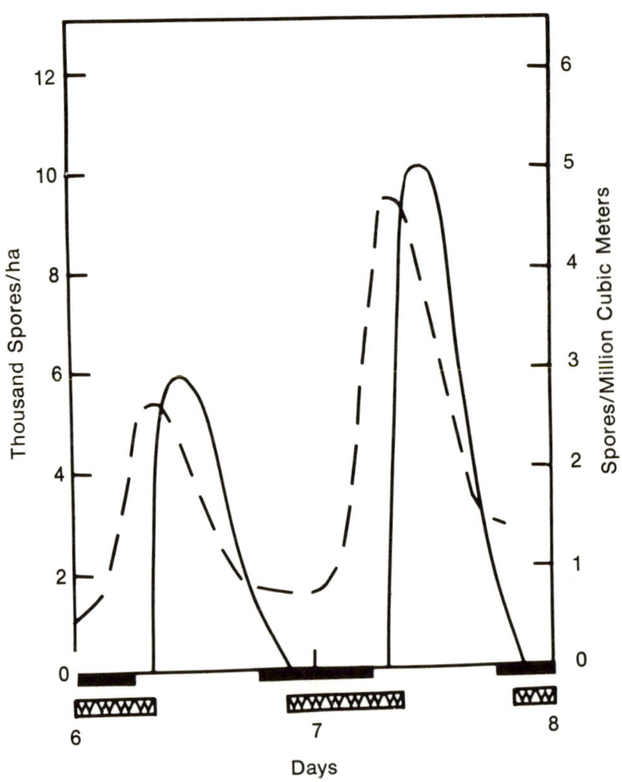

Fig. 8. Simulated course of sporulation – – – and of concentration of airborne spores ——— of *H. maydis* (Waggoner and de Wit, 1972).

ments. CSMP can change the calculation interval to match the changeability of the weather or of the fungus.

A predecessor, EPIMAY, was used in 1971 in a national network, providing weekly estimates of the progress of southern corn leaf blight (Felch and Barger, 1971). Such practical uses of seasonality models are discussed in Chapter 6.4.

6 Appendix: Analytic Solution to Model Passages Through Phenophases

We consider a series of F boxes; in Equation 5 F was called BOX. Objects move toward the final box at a relative rate of P and are lost at a relative rate of B. If the sum of the rates B plus P is A, then

$$dL_n/dt + AL_n = P\,L_{n-1} \tag{A1}$$

where $n = 1, \ldots, F$, P, and B are functions of time and assumed to be given. L_n is the concentration in the nth box. At the beginning

$$L_1 = 1 \tag{A2}$$

$$L_n = 0,\ F \geq n \geq 2 \tag{A3}$$

Finally, we are interested in the successful departures L from the Fth box

$$L = \int_0^t P(\alpha)\, L^F(\alpha) d\alpha \tag{A4}$$

We are also interested in the eventual success, which is called the hatching percentage in the text, and becomes L/L_∞ here. The half time $t_{1/2}$ is the time required for L/L_∞ to reach 0.5. The solution of Equation A1 with boundary conditions A2 and A3 is

$$L_1 = \exp\left[-\int_0^t A(\alpha)\, d\alpha\right] \tag{A5}$$

$$L_n = \left[\int_0^t P(\alpha_{n-1}) d\alpha_{n-1} \int_0^{\alpha_{n-1}} P(\alpha_{n-2}) d\alpha_{n-2} \ldots \int_0^{\alpha_2} P(\alpha_1) d\alpha_1\right]$$

$$\exp\left[-\int_0^t A(\alpha) d\alpha\right];\ F \geq n \geq 2 \tag{A6}$$

Consequently

$$L = \int_0^t P(\alpha_F)d\alpha_F \left[\exp - \int_0^{\alpha_F} A(\alpha)d\alpha\right]\left[\int_0^{\alpha_F} P(\alpha_{F-1})d\alpha_{F-1} \ldots \int_0^{\alpha_2} P(\alpha_1)d\alpha_1\right] \quad \text{(A7)}$$

Repeated integrations by part yield

$$(F-1)!\,L = \int_0^t P(\beta)d\beta \left[\exp - \int_0^{\beta} A(\alpha)d\alpha\right]\left[\int_0^{\beta} P(\alpha)d\alpha\right]^{(F-1)}, \quad \text{(A8)}$$

If P and A do not depend on time, Equation A8 can be simplified and becomes

$$(F-1)!\,L = (P/A)^F \int_0^{At} \beta^{F-1} \exp(-\beta)d\beta \quad \text{(A9)}$$

or

$$L/L_\infty = \int_0^{At} \beta^{F-1} \exp(-\beta)d\beta/(F-1)! \quad \text{(A10)}$$

The integral in Equation A9 is the standard incomplete gamma function and

$$L_\infty = (P/A)^F \quad \text{(A11)}$$

Equation A9 could also be written in terms of the probability integral of a χ^2–distribution. Of particular interest is the case $F \gg 1$, when L/L_∞ approaches the normal distribution

$$L/L_\infty \simeq (2\pi)^{-1/2} \int_0^{(At-F)F^{-1/2}} \exp(-\alpha^2/2)d\alpha \quad \text{(A12)}$$

and

$$t_{1/2} \simeq F/A, \quad \text{(A13)}$$

the standard deviation s being $s \simeq t_{1/2}/F^{1/2}$ (A14)

If P and A are known functions of time but F is sufficiently large, then Equation

A8 can still be simplified. We first rewrite Equation A8 as

$$(F-1)!\,L = \int_0^t d\beta \left[P(\beta) \Big/ \int_0^\beta P(\alpha)d\alpha \right]$$

$$\left\{ \exp\left[-\int_0^\beta A(\alpha)d\alpha + F\ln\int_0^\beta P(\alpha)d\alpha \right] \right\} \tag{A15}$$

The exponent has a maximum for a value of β which will turn out to be equal to $t_{\frac{1}{2}}$ and is given by

$$A_{1/2}\int_0^{t_{1/2}} P(\alpha)d\alpha = F\,P_{1/2} \tag{A16}$$

where $A_{1/2}$ and $P_{1/2}$ are the values of A and P at $t_{1/2}$. Note that if P and A are constant in Equation A16, then the latter reduces to Equation A13. Expansion of the exponent in Equation A15 around $\beta = t_{1/2}$, yields approximately

$$(F-1)!\,L \simeq F^{F-1}(P_{1/2}/A_{1/2})^{F-1}P_{1/2}\left[\exp - \int_0^{t_{1/2}} A(\alpha)d\alpha\right]$$

$$[\exp\{-(\beta - t_{1/2})^2/2s^2\}d\beta] \tag{A17}$$

where

$$s^{-2} = (dA/dt)_{t_{1/2}} - (A_{1/2}/P_{1/2})\,(dP/dt)_{t_{1/2}} + A_{1/2}^2/F \tag{A18}$$

Equation A17 indicates that the distribution of L is normal. The approximation of Equation A17 holds as long as P and A can be expressed by a Taylor expansion around $t_{1/2}$, so that P and A must not vary too rapidly between the times $(t_{1/2}-s)$ and $(t_{1/2}+s)$. Equation A17 and Stirling's formula yield

$$L_\infty = (P_{1/2}/A_{1/2})^{F-1}\,P_{1/2}sF^{-1/2}\exp\left[\int_0^{t_{1/2}} \{A_{1/2}(P/P_{1/2}) - A\}d\alpha\right] \tag{A19}$$

Equations A16 to A19 apply whatever the relative magnitude of A and P is. In practice, A is slightly larger than P. It is then possible to simplify these equations further when $A-P \ll P$. Equation A16 becomes

$$\int_0^{t_{1/2}} P(\alpha)d\alpha \simeq \mathbf{F} \tag{A20}$$

Equation A18 becomes

$$s^{-2} \simeq P_{1/2}^2/F \tag{A21}$$

Equation A19 becomes

$$L_\infty = \exp\left[-\int_0^{t_{1/2}} (A-P)dt\right] \tag{A22}$$

Equations A20, A21, and A22 are used in this chapter. These equations are very simple and are adequate to describe most cases of interest as long as P and A are known functions of time.

References

Benacchio, S. S., and B. O. Blair. 1972. A new approach to phenological research—Relationships between environmental factors and days to appearance of the first leaf in four perennial species. *Agron. J.* 64, 297–302.

Bennett, M. F., J. Shriner, and R. A. Brown. 1957. Persistent tidal cycles of spontaneous motor activity in the fiddler crab, *Uca pugnax*. *Biol. Bull.* 112, 267–276.

Bliss, C. I. 1970. *Statistics in Biology*. II, 639 pp. New York: McGraw-Hill.

Brown, D. M. 1960. Soybean ecology. I. Development-temperature relationships from controlled environment studies. *Agron. J.* 52, 493–496.

———, and L. J. Chapman. 1960. Soybean ecology. II. Development-temperature-moisture relationships from field studies. *Agron. J.* 52, 496–499.

Caprio, J. M. 1966. Pattern of plant development in the western United States. *Montana Agr. Exp. Sta. Bull. 607*. 42 pp.

Felch, R. E., and G. L. Barger. 1971. EPIMAY and southern corn leaf blight. *Weekly Weather Crop Bull.* 58 (43), 13–17.

Garner, W. W., and H. A. Allard. 1920. Effect of the relative length of day and night and other factors of the environment on growth and reproduction in plants. *J. Agr. Res.* 18, 553–606.

Hopkins, A. D. 1938. Bioclimates—A science of life and climatic relations. *U.S. Dept. Agr. Misc. Publ. 280*. 188 pp.

IBM. 1969. *System/360 Continuous System Modeling Program Users Guide*. H20-0367-3. 61 pp.

Messenger, P. S., and N. E. Flitters. 1958. Effect of constant temperature environments on the egg stage of three species of Hawaiian fruit flies. *Ann. Ent. Soc. Am.* 51, 109–119.

——— 1959. Effect of variable temperature environments on egg development of three species of fruit flies. *Ann. Ent. Soc. Am.* 52, 191–204.

Nuttonson, M. Y. 1948. Some preliminary observations of phenological data as a tool in the study of photoperiodic and thermal requirements of various plant material. In *Vernalization and Photoperiodism—A Symposium*, A. E. Murneek, and R. O. Whyte, ed. pp. 129–143. Waltham, Mass: Chronica Botanica.

——— 1955. Wheat-climate relationships and the use of phenology in ascertaining the thermal and photo-thermal requirements of wheat. 388 pp Washington D.C. *Am. Inst. Crop Ecol.*

Panofsky, H. A., and G. W. Brier. 1958. *Some Applications of Statistics to Meteorology*. 224 pp. University Park: Pennsylvania State University.

Perez-Arbelaez, E. 1956. *Plantas Utiles de Colombia*. 831 pp. Bogota: Camacho Roldan.

Powsner, L. 1935. Effects of temperature on the duration of the developmental stages of *Drosophila melanogaster*. *Physiol. Zool.* 8, 474–520.

Radford, J. S. 1971. Phenological delimitation of the growing season in North Carolina. Computer mapping and correlation of spring, fall and length of growing season with environmental parameters. 158 pp. Unpublished MSc. Thesis, Chapel Hill: University of North Carolina.

Shaw, J. G., and D. F. Starr. 1946. Development of the immature stages of *Anastrepia serpentina* in relation to temperature. *J. Agr. Res.* 72, 265–276.

Shelford, V. E. 1927. Experimental investigation of the relations of the codling moth to weather and climate. *Ill. Nat. Hist. Surv. Bull. 16*, 307–440.

Smith, J. W. 1915. Phenological dates and meteorological data recorded by Thomas Mikesell at Wauseon, Ohio. *Monthly Weather Rev. Suppl.* 2, 23–93.

——— 1920. *Agricultural Meteorology*. 304 pp. New York: Macmillan.

Thom, H. C. S. 1954. The rational relationship between heating degree days and temperature. *Monthly Weather Rev.* 82, 1–6.

———1957. Climatological analysis of snowfall thresholds. *Arch. Meteorol. Geophysi. Bioklim.* 8, 195–201.

———, and R. H. Shaw. 1958. Climatological analysis of freeze data for Iowa. *Monthly Weather Rev.* 86, 251–255.

Thornthwaite, C. W. 1944. Report of the committee on transpiration and evaporation, 1943–44. *Trans. Am. Geophys. Union* 25, 683–693.

———, and J. R. Mather. 1954. Climate in relation to crops. *Meteorol. Monograph* 2(8), 1–10.

Van Schaik, P. H., and A. H. Probst. 1958. Effects of some environmental factors on flower production and reproductive efficiency in soybeans. *Agron. J.* 50, 192–197.

Waggoner, P. E., and C. T. de Wit. 1972. A more lucid simulator of plant disease. In manuscript.

Waggoner, P. E., J. G. Horsfall, and R. J. Lukens. 1972. EPIMAY. A simulator of southern corn leaf blight. *Conn. Agr. Exp. Sta. Bull. 729*. 84 pp.

Waggoner, P. E., P. M. Miller, and H. C. de Roo. 1960. Plastic mulching: Principles and benefits. *Conn. Agr. Exp. Sta. Bull. 634*. 44 pp.

5.2 Modeling Important Phytophenological Events in Eastern North America[1,2]

Richard Reader[3]

Ecology Curriculum
University of North Carolina
Chapel Hill, North Carolina 27514, USA

John S. Radford

IBP Desert Biome
Botany Department
Utah State University, Logan, Utah 84321, USA

and Helmut Lieth

Botany Department
University of North Carolina
Chapel Hill, North Carolina 27514, USA

1 Introduction

Investigators need to know the calendar dates of the initiation and cessation of vegetative activity to schedule their fieldwork and, later, to simulate their field observations by means of synchronized computer models. The specific objective of

[1] Contribution No. 89 from the Eastern Deciduous Forest Biome, US/IBP.

[2] Research supported by the Deciduous Forest Biome Project, IBP, funded by the National Science Foundation under Interagency Agreement AG-100, 40-193-69 with the U.S. Atomic Energy Commission-Oak Ridge National Laboratory. Computer time was supplied by the University of North Carolina.

[3] Present address: Department of Botany, University of Guelph, Guelph, Ontario, Canada.

KEYWORDS: Phenology; seasonality; bioclimate; bud break; flowering patterns; leaf coloration; leaf emergence; growing season; photosynthetic period; deciduous forest; computer mapping; correlation modeling; observer network; North America.

our work in North Carolina is, first, to determine the length of the natural photosynthetic period in the eastern deciduous forest biome of North America, and then to correlate the observed spatial and temporal variation in its length with environmental variation. This article reports some results of the descriptive and analytical phases of our project.

2 Methods

Selection of Indicator Species and Phenophases

We approximate the length of the photosynthetic period for hardwoods as the difference between the date of fall leaf coloration and the date of leaf emergence in the preceding spring. It would be very time consuming to record leaf emergence and coloration dates for all deciduous species, but selected phenophases could be easily recognized at a number of locations for several indicator species. Only native species were considered. While all individuals may not be genetically homogeneous, they are physiologically attuned to surrounding environmental conditions and therefore offer a distinct advantage over the use of planted or introduced species. In addition, native plants are located at a greater number of sites than would be those species introduced or planted for the same purpose.

A 1969 investigation conducted on the former IBP site at Research Triangle in the Piedmont region of North Carolina indicated that the dates of first flowering of the native redbud (*Cercis canadensis* L.) and dogwood (*Cornus florida* L.) occurred during the period of spring leaf emergence (Lieth and Radford, 1971). The opening of the first flower in these well-known species is both obvious and distinct. The date of red maple (*Acer rubrum* L.) first flowering was observed as well, since the range of dogwood and redbud does not extend to the extreme boundaries of the deciduous forest biome.

Red maple and tulip poplar (*Liriodendron tulipifera* L.) were selected as fall indicator species. The date of 50% leaf area coloration of a branch preselected for observation was recorded to indicate the end of the photosynthetic period.

Observer Network

Recording spatial and temporal variation inherent in the initiation of seasonal events requires the help of individuals at various locations in the deciduous forest biome. An observer network for North Carolina was established in 1969 through direct communication with garden clubs and biology classes. Other groups, including professional biologists and weather station operators, were later contacted by mail. Appeals through television and newspapers increased volunteer participation. Cooperating observer networks were subsequently established by other workers in New Jersey, Tennessee, and Virginia (Kramer, 1972) to gather the same information being collected in North Carolina.

The number, type, and distribution of individuals who have participated in the North Carolina Phenology Project since its inception are presented in Table 1,

Table 1. Tabulation of the number and type of individuals participating in the North Carolina Phenology Project (Through Spring 1972).

	Professional Biologist	Biology Class Member	Experiment Station Personnel	Weather Station Operator	Garden Club Member	Interested Individual	N. C. Counties Represented, and % of total	Total No. of Participants
North Carolina								
Spring 1970	25	45	14	0	32	197	75 (75%)	313
Fall 1970	19	53	9	0	7	151	57 (57%)	239
Spring 1971	26	47	8	0	4	162	77 (77%)	247
Fall 1971	15	52	9	34	3	103	55 (55%)	216
Spring 1972	13	39	6	24	3	119	62 (62%)	204
Eastern deciduous forest biome other than North Carolina								
Spring 1970	21	0	23	0	0	7	16 (38%)[a]	51
Fall 1970	15	1	11	0	0	1	11 (26%)	28
Spring 1971	25	1	9	0	0	1	12 (29%)	39
Fall 1971	17	1	8	0	0	0	11 (26%)	26
Spring 1972	22	4	7	0	11	4	17 (40%)	48

[a] These numbers represent American states and Canadian provinces represented, and percentage of total.

together with the statistical information necessary to judge the degree of reliability of our maps and models.

Computer Mapping of Phenological Patterns

A map is the most convenient display for summarizing the flowering and leaf coloration information that is recorded by the observer network. These phenophase dates are transformed into yeardays, with January 1 as yearday 1, grouped by county, and averaged. The average county yeardates are plotted on the map with the county centers as data points. The difference between the earliest and latest recorded dates is divided into seven-day intervals, and isophenes are drawn. The resulting patterns delimit areas in which the event occurred during the same seven-day period. Selection of a seven-day interval was convenient for the total length of the phenophase period and the number, distribution, and accuracy of the reports received.

To map the data we used the computer program SYMAP (see Chapter 2.4; Dudnik, 1972) which employs the high-speed line printer to spatially depict irregularly spaced qualitative and quantitative data. A symbol representing the appropriate mapping interval is assigned to all points for which no observations are reported. The program assumes that the values at these points are a weighted average of the values known for surrounding data points. The effect of surrounding data points on intervening points is inversely proportional to the square of the distance between the known data and unknown data points. To assign a value to the point in question, the program has default conditions for search-radius size and for number of data points to be used for interpolation. In order to compare the computer maps, certain information must appear in the map legend: number of data points or the search-radius size used for interpolation, number of points for which actual data were available, total range of the values (actual data) mapped, and number and length of the mapping intervals.

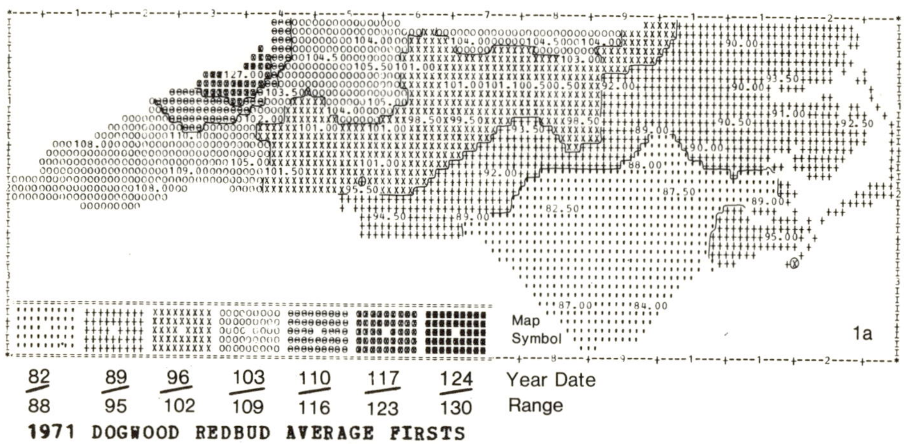

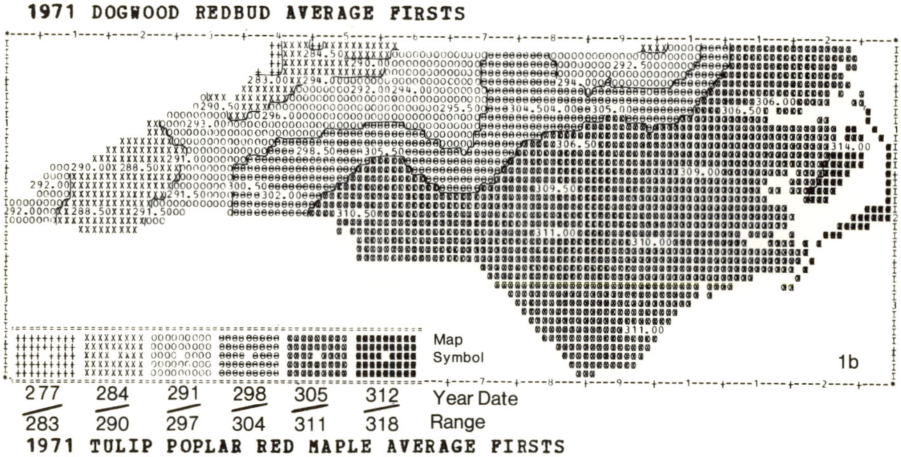

Fig. 1a. 1971 flowering yeardate of (dogwood + redbud)/2 in North Carolina. Total number of data points appearing on the map = 50 county averages. Range of the data values = 82.5 to 127.5. Number of contour intervals appearing on the map = 7. Contour interval length = 7 days. Number of points used for the interpolation = 7. Search radius = 1.6 in. on a 5 × 13 in. map.

Fig. 1b. 1971 yeardate of 50% leaf area coloration for (tulip poplar + red maple)/2 in North Carolina. Total number of data points appearing on the map = 37 county averages. Range of the data values = 283 to 314. Number of contour intervals appearing on the map = 6. Contour interval length = 7 days. Number of points used for the interpolation = 7. Search radius = 1.9 in. on a 5 × 13 in. map.

3 Results

Descriptive Phenology

Mapping the Arrival of Spring and Fall. We consider the average of the first flowering dates of the dogwood and redbud to indicate the beginning of spring leaf emergence, and the average of dates of 50% leaf area coloration for tulip poplar and red maple to coincide with the end of the vegetation period. The timing and the geographic distribution of spring and fall arrival have been plotted (Fig. 1) from reports sent to us by observers in various parts of the eastern deciduous forest biome. The mean dates of dogwood-redbud flowering in North Carolina as a whole were similar in 1970 and 1971 ($t = 0.23$, $p > 0.8$ of finding the observed difference by chance), while the mean dates of fall leaf coloration differed significantly ($t = 5.0$, $p < 0.001$). Both the biome and the North Carolina maps show a south-north and east-west trend in the progression of spring, with fall's movement being the direct opposite.

Mapping the Length of the Vegetation Period. An estimate of the length of the photosynthetic period, or potential growing season, in North Carolina was obtained by plotting the difference between the mean date of tulip poplar-red maple leaf coloration and the mean date of dogwood-redbud first flowering for all counties reporting the dates of both spring and fall events (Fig. 2). The mean length of the vegetation period in the state as a whole was significantly longer in 1971 (198 days) than in 1970 (191 days) ($t = 4.3$, $p < 0.001$). Similar growing season maps and comparisons for the biome as a whole cannot be made yet because of a lack of data.

The North Carolina maps presented here represent one of the first attempts to determine the length of the potential natural vegetation period in North America based on observations of native plants. Previous estimates of the length of the growing season have been based on air temperature data, with the freeze-free period considered to be roughly equivalent to the potential growing season, especially for agricultural crops. A preliminary comparison was made between the length of the freeze-free (0°C) period in 1970 and 1971, and the 1970 and 1971 vegetation periods calculated by the method described in the legend for Fig. 2. These differences were mapped (Fig. 3). In both years the freeze-free period underestimated the length of the natural vegetation period in the western, mountainous region of the state and overestimated the vegetation period length in the Piedmont and Coastal Plain regions. This interesting comparison will again be made when phenophase data become available from a greater number of locations in the state.

Analytical Phenology

The analysis of spatial and temporal variation in the occurrence of spring and fall phenological events is the subject matter of analytical phenology. The basic premise of analytical phenology is that the spatial and temporal variation observed in the occurrence of an event is directly related to environmental differences and rhythms. Biotic factors, including endogenous timing mechanisms, are considered

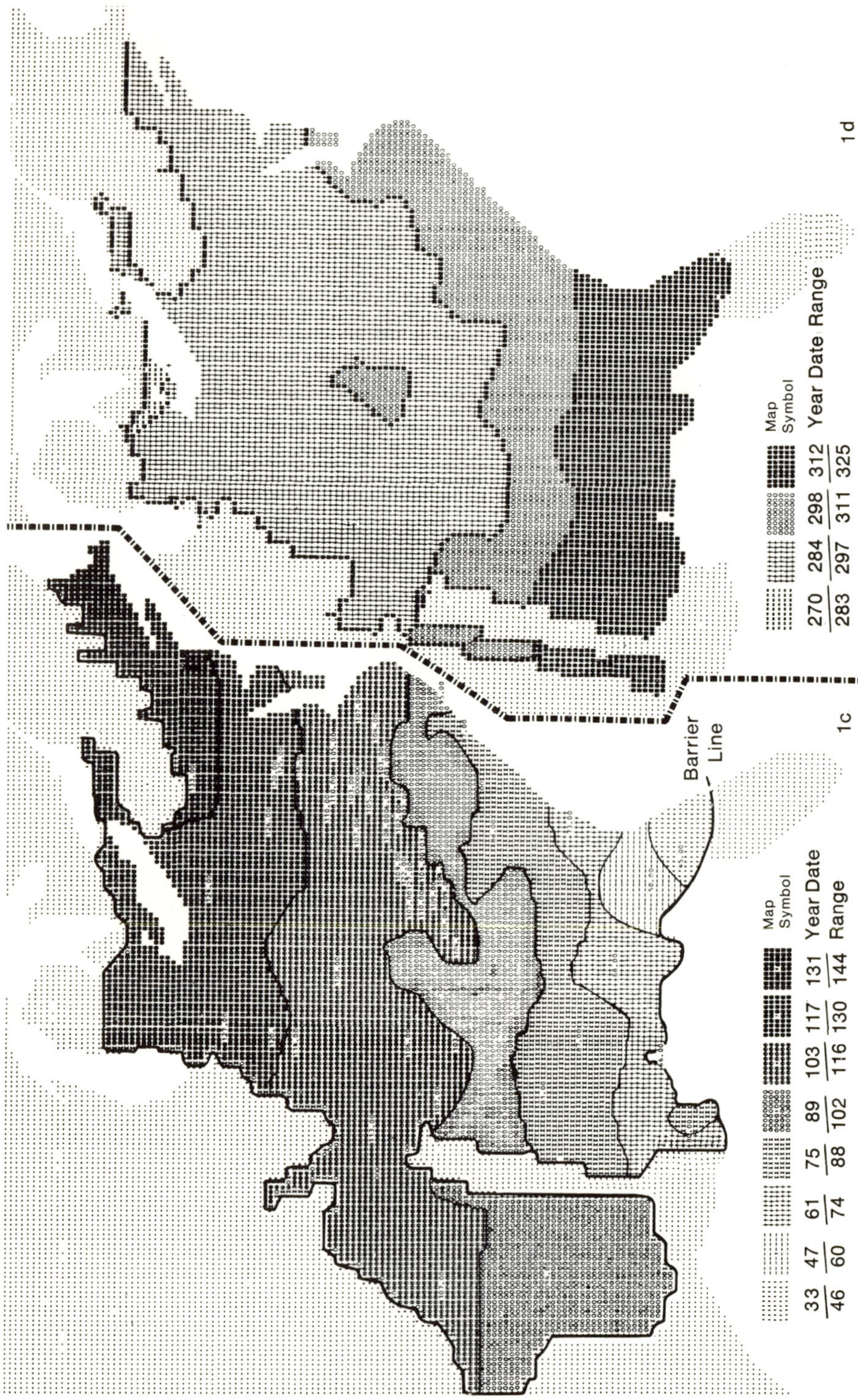
1c
1d
Barrier Line
33 46
47 60
61 74
75 88
89 102
103 116
117 130
131 144
Map Symbol
Year Date Range
270 283
284 297
298 311
312 325
Map Symbol
Year Date Range

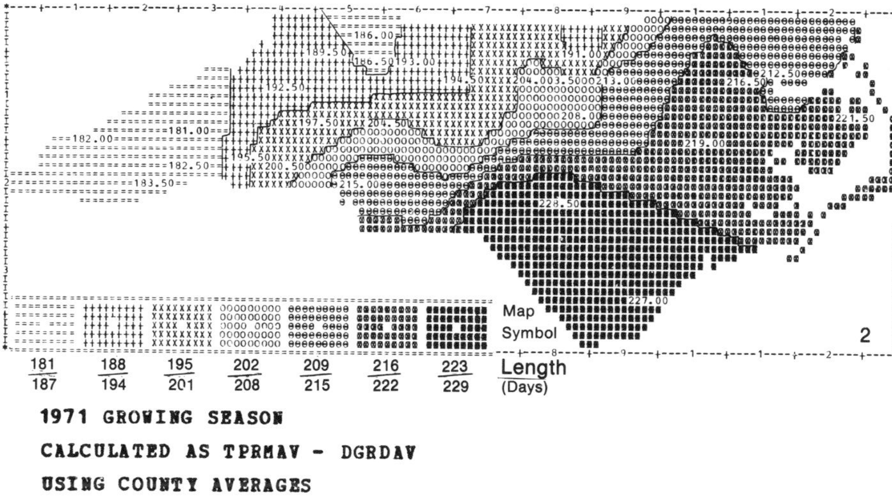

Fig. 2. Length of the natural vegetation period in 1971, estimated as the difference between the leaf coloration date for (tulip poplar + red maple)/2 and the flowering date for (dogwood + redbud)/2. Total number of data points appearing on the map = 26 county averages. Range of the data values = 181 to 228.5 days. Number of contour intervals appearing on the map = 7. Contour interval length = 7 days. Number of points used for the interpolation = 7. Search radius = 2.4 in. on a 5 × 13 in. map.

to play an indirect role in phenological timing. The objectives of the analysis are to determine which factors are significantly related to the timing of the events, and to formulate models which reflect this understanding of the nature of the controlling mechanisms. This analytical approach will be illustrated using the dogwood, redbud, and red maple flowering dates reported in 1970, 1971, and 1972.

Flowering Date Versus Geographic Coordinates. The date of spring flowering in the three species named above was dependent to a large extent on a plant's location, with southern and eastern plants blooming before plants in more northerly and westerly locations. The strength of the relationship between flowering date and the latitude, longitude, or altitude at which a plant is growing can be expressed quantitatively using Pearson's product-moment correlation coefficients, which measure the linear relationship between two randomly fluctuating variables.

Fig. 1c. 1970–1971 averaged flowering yeardate of (dogwood + redbud)/2 within the native range of dogwood. This range is delimited by the barrier. Total number of data points appearing on the map = 73. Range of the data values = 45 to 133. Number of contour intervals appearing on the map = 8. Contour interval length = 14 days. Number of points used for the interpolation = 3. Search radius = 1.7 in. on a 13 × 18 in. map.

Fig. 1d. 1970–1971 averaged yeardate of 50% leaf area coloration for (tulip poplar + red maple)/2 within the native range of tulip poplar. Total number of data points appearing on the map = 50. Range of the data values = 284 to 325. Number of contour intervals appearing on the map = 4. Contour interval length = 14 days. Number of points used for the interpolation = 3. Search radius = 1.2 in. on a 12.5 × 11 in. map.

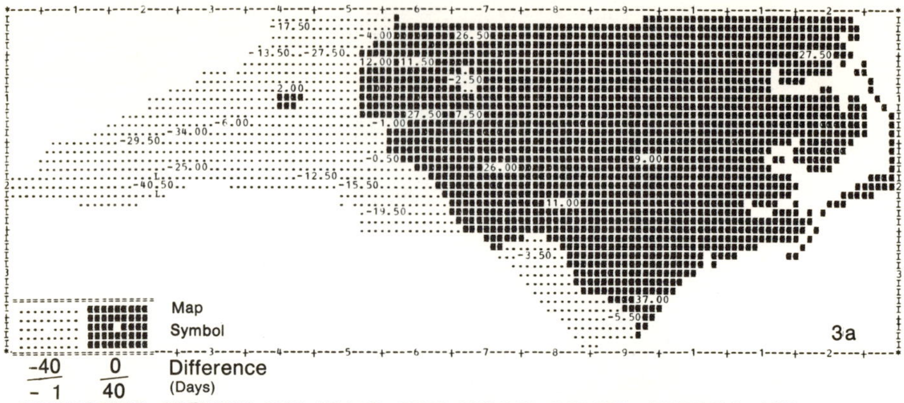

DIFFERENCE BETWEEN THE FROST FREE PERIOD LENGTH (TLENG)) AND
THE VEGETATION PERIOD TPRMAV - DGRDAV, I.E. PLENG
T = TLENG - PLENG 1970

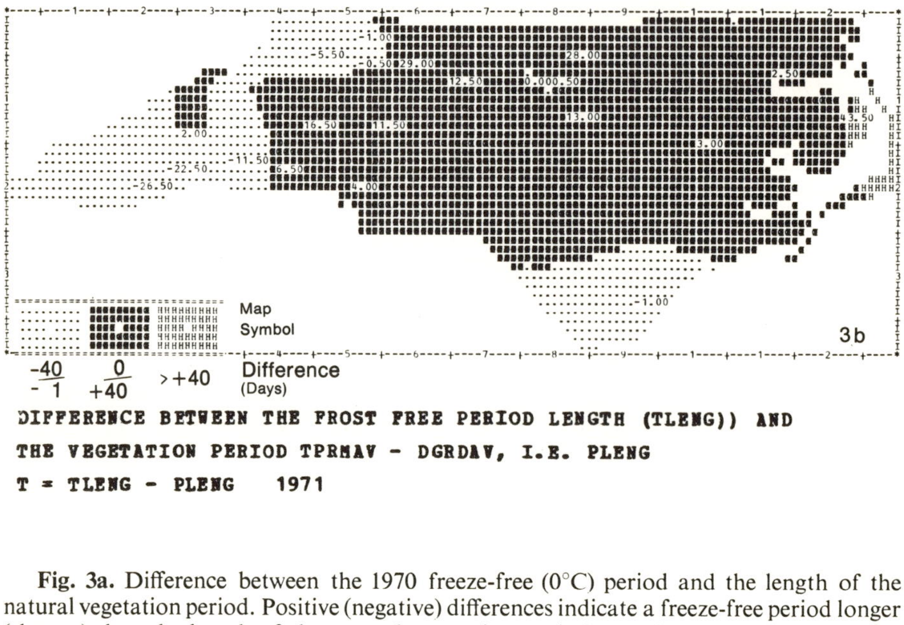

DIFFERENCE BETWEEN THE FROST FREE PERIOD LENGTH (TLENG)) AND
THE VEGETATION PERIOD TPRMAV - DGRDAV, I.E. PLENG
T = TLENG - PLENG 1971

Fig. 3a. Difference between the 1970 freeze-free (0°C) period and the length of the natural vegetation period. Positive (negative) differences indicate a freeze-free period longer (shorter) than the length of the natural vegetation period. Total number of data points appearing on the map = 28 county averages. Range of the data values = −40.0 to + 37. Number of contour intervals appearing on the map = 2. Contour interval length = 40 days. Number of points used for the interpolation = 7. Search radius = 2.2 in. on a 5 × 13 in. map.

Fig. 3b. Difference between the 1971 freeze-free (0°C) period and the length of the natural vegetation period. Positive (negative) differences indicate a freeze-free period longer (shorter) than the length of the natural vegetation period. Total number of data points appearing on the map = 21 county averages. Range of the data values = −26.5 to 43.5. Number of contour intervals appearing on the map = 3. Contour interval length = 40 days. Number of points used for the interpolation = 7. Search radius = 2.7 in. on a 5 × 13 in. map.

Table 2. Matrices of zero-order Pearson correlation coefficients based on 1970, 1971, and 1972 flowering reports for dogwood, redbud, and red maple from North Carolina and the Eastern deciduous forest biome. A perfect correlation is indicated by ±1.0.

North Carolina: Matrix based on 90 reports. All correlations between flowering dates and latitude, longitude, and altitude are significant at the 0.001 level with 88 d.f.

	Latitude	Longitude	Altitude	Dogwood	Redbud	Red Maple	Dogwood-Redbud Avg.
Latitude	1.000						
Longitude	0.0452	1.000					
Altitude	0.219	0.794	1.000				
Dogwood	0.399	0.588	0.881	1.000			
Redbud	0.430	0.680	0.885	0.891	1.000		
Red Maple	0.401	0.549	0.868	0.892	0.897	1.000	
Dogwood-Redbud Avg.	0.427	0.663	0.907	0.958	0.983	0.919	1.000

Eastern deciduous forest biome, excluding North Carolina: Matrix based on 31 reports. All correlations are significant at the 0.001 level with 29 d.f., except those between flowering date and altitude, which are significant at the 0.05 level.

	Latitude	Longitude	Altitude	Dogwood	Redbud	Red Maple	Dogwood-Redbud Avg.
Latitude	1.000						
Longitude	−0.637	1.000					
Altitude	0.208	−0.093	1.000				
Dogwood	0.955	−0.582	0.274	1.000			
Redbud	0.928	−0.540	0.359	0.972	1.000		
Red Maple	0.861	−0.574	0.283	0.827	0.864	1.000	
Dogwood-Redbud Avg.	0.947	−0.564	0.320	0.992	0.993	0.850	1.000

The correlation matrices (Table 2) indicate a significant relationship between the recorded flowering dates and the three location attributes: latitude, longitude, and altitude. The strength of these relationships varies, with altitude being the most significant locational parameter in North Carolina (ranging from 0 to 1200 m between stations), and latitude being the most significant over the biome as a whole (ranging from 29° to 41°N).

The strength of these correlation coefficients served as the basis for the construction of several spatial models. Spatial modeling involves the characterization of the time of phenophase occurrence, various environmental parameters, or combinations of these, as functions of the independent variable, geographic location. Our spatial models took the form of linear regression equations, with the value of the dependent variable, averaged dogwood-redbud flowering date (Y), considered to be equal to a linear combination of the independent variables, latitude (X_1), longitude (X_2), and altitude (X_3), plus a random error term (e). All three locational parameters were significant contributors to the regression equation when the entire range of the species was considered (Table 3), while only altitude and latitude were found to be significant for the phenophases in North Carolina. The percentage of flowering date variation accounted for by locational parameters was 75.4% in North Carolina and 89.1% in eastern North America.

In a similar study conducted in 1963, Kaszkurewicz and Fogg (1967) showed

Table 3. Spatial models constructed with average dogwood-redbud flowering date (Y) as a function of latitude (X_1), longitude (X_2), altitude (X_3), and red maple, flowering date (X_4), based on flowering dates reported in 1970, 1971, and 1972.

The general form of the linear model is: $Y = \beta_0 + \beta_1 X_1 + \beta_2 X_2 + \beta_3 X_3 + \beta_4 X_4 + e$

Region of Model's Applicability	Total *d.f.*	Constant β_0	Latitude, in degrees β_1	Longitude, in degrees β_2	Altitude, in feet and meters β_3	Red maple flowering date, in yeardays β_4	Coefficient of Determination r^2
			Coefficients of Independent Variables ± a Standard Error				
North Carolina	427	−127.81	6.19[a] ±0.47	n.s.	0.00806[a] ±0.00026 (ft)	Not entered	0.754
		−125.5	6.12[a] ±0.46	n.s.	0.0266[a] ±0.0008 (m)		
North Carolina	89	−81.44	4.16[a] ±1.01	n.s.	0.00565[a] ±0.00073 (ft)	0.386[a] ±0.069	0.910
		−79.28	4.13[a] ±0.99	n.s.	0.0194[a] ±0.0022 (m)	0.367[a] ±0.068	
Eastern North America	133	−149.03	6.28[a] ±0.26	0.231[b] ±0.17	0.00606[b] ±0.0011 (ft)	n.s.	0.891
		−148.97	6.28[a] ±0.26	0.231[b] ±0.17	0.0198[b] ±0.003 (m)	n.s.	

[a] Significant at the 0.01 level.
[b] Significant at the 0.05 level; n.s. = not significant.

that 84.8% of the variation in the dates of bud break for sycamore (*Platanus occidentalis*) and 78.3% of those for cottonwood (*Populus deltoides*) could be explained by the quadratic term in a polynomial linear regression equating bud break dates and latitude. Multiple correlation coefficients of the same magnitude were also found by Caprio (1967), who correlated the variation in lilac (*Syringa vulgaris* L.) first bloom dates in the western United States with the three locational parameters. However, none of the 16 combinations of regression coefficients reported by Caprio matched those found in this study.

It appears that the influence of these three parameters varies among species. When the relationship between location and phenophase timing is known for additional species, it may be possible to establish a set of mean values which reflect the relative influences of these locational factors. Hopkins (1938) conducted studies of the relation of latitude and altitude to the difference in timing of the emergence of the Hessian fly, and its attack on wheat sown at given times. He proposed that the timing of an event differs between two locations by an average of four days for every degree's difference in latitude, by 0.8 day for every degree's difference in longitude, and by one day for every 100-ft difference in altitude. Significant deviations were apparent when these suggested mean values were entered into a spatial model and the predicted dates were compared with actual dogwood-redbud averaged flowering dates in North Carolina (Fig. 4). According to Hopkins, the deviation between the recorded quantity and his proposed country-wide constants represents a measure of the intensity of the local and regional influences operating on the event.

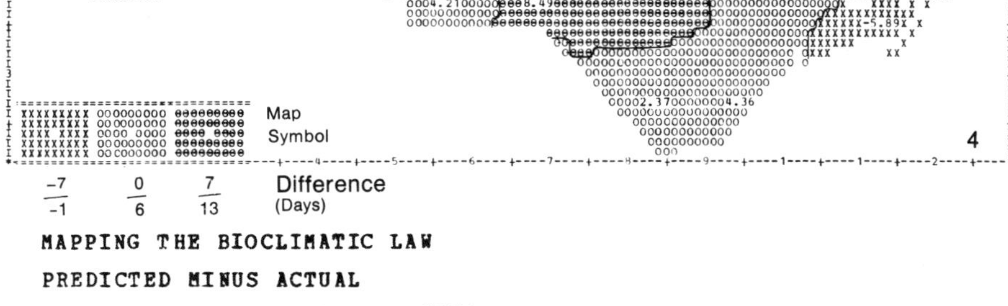

Fig. 4. Comparison of the (dogwood + redbud)/2 flowering date in 1971 and the date dredicted from Hopkins' law. Positive (negative) differences indicate that the predicted date is later (earlier) than the observed date. Differences were significant at the 0.01 level ($t = 6.6$ with 48 degrees of freedom). Total number of data points appearing on the map = 50. Range of the data values = −5.89 to 12.68. Number of contour intervals appearing on the map = 3. Contour interval length = 7 days. Number of points used for the interpolation = 7. Search radius = 1.5 in. on a 5 × 13 in. map.

It may be possible to reduce the remaining unexplained variation in flowering dates by including the effects of local and regional influences, such as the slope of the ground and its aspect, in spatial models. Physical and chemical edaphic characteristics, as well as various attributes of the trees observed (age, genetic constitution, and vitality), could also be considered. The study of these matters requires experimentation. We have therefore planted native saplings of the same age at a number of locations in the North Carolina Botanical Garden (Chapel Hill) to evaluate the effects of microhabitat differences on the dates of leaf emergence and coloration.

Flowering in Relation to Other Phenological Events. If the main objective of the spatial modeling is prediction, then factors other than locational may be better related to the event of interest and should also be entered into the equation. For example, in North Carolina the occurrence of the first flowering of the red maple was found to be significantly correlated ($p = 0.01$) with the averaged flowering dates of dogwood and redbud, i.e.

$$r\ \text{dogwood-redbud, red maple/lat., long., alt.} = 0.51$$

The linear regression between the averaged dogwood-redbud flowering date and red maple flowering date, including the effects of latitude, longitude, and altitude, accounted for 91% of the variation in the dogwood-redbud flowering date (Table 3). Red maple flowering precedes dogwood and redbud flowering by approximately one month in North Carolina, and red maple is more widely distributed (especially in high altitudes) than either of the other species. In future surveys we will test the accuracy of estimating the dogwood-redbud averaged flowering date, i.e., the period of leaf emergence, from previously reported red maple flowering dates.

Flowering in Relation to Meteorological Measurements. Another approach which needs more investigation is the construction of models with flowering date as a function of biologically relevant climatic parameters. Models incorporating the timing of seasonal changes in temperature, moisture, and insolation regimes at specific locations are likely to be better predictors of the time of occurrence of phenological events than are models based on geographic constants.

The construction of models involving biologically meaningful parameters is a difficult task. The date of occurrence of a phenological event can be correlated with any one of a number of aspects of the temperature, moisture, and insolation regimes, e.g., the July average temperature (Perry and Wang, 1960) or averaged daily minimum temperature for the three months preceding the event (Table 4); yet this correlation does not necessarily imply that a cause-and-effect relationship exists between the variables considered. A theoretical basis for the modeling of phenological timing in terms of significant biological factors must first be established. Our basic assumption is that a certain energy accumulation or activation level must be reached for each of one or more factors before the phenological event will occur. This might require, for instance, the achievement of a certain number of hours above the threshold temperature for enzymic activation, and a number of darkness periods of a certain minimum length, while a positive moisture balance exists, etc. Experimentation under controlled and natural conditions often is necessary to determine both the threshold and total activation amounts needed for the initiation

Table 4. Some correlation models tested in the North Carolina Phenology Project (from Radford, 1971).

Dependent Variable	Independent Variable	Number of Data Points	Coefficient of Determination, r^2
Arrival of spring[a]	County average minimum temperature (Jan.–Mar.)	70	0.46
Arrival of spring	County average minimum temperature (Jan.–Mar.) Total rainfall to date of spring's arrival Date predicted using Hopkins' law	69	0.61
Dogwood flowering date	Date predicted by heat summation from January 1 using a temperature threshold of 3.3°C	23	0.56
Dogwood flowering date	Date predicted by heat summation Total rainfall to date of dogwood flowering	23	0.72
Dogwood flowering date	Date predicted by heat summation Total rainfall to date of dogwood flowering Date predicted by Hopkins' law	23	0.83

[a] Arrival of spring = (dogwood flowering date + redbud flowering date)/2.

of the event of interest. Details of the type of experimentation involved are described in Chapter 2.5. The degree of experimentation required depends on the type of indicator species being employed. Minimal experimentation is necessary when genetically homogeneous plants of the same age are used. More elaborate experimentation is required to determine the effects of age and genetic constituency of native plants on the threshold temperatures and activation energies. Spatial models with threshold and total activation energies as dependent variables, and latitude, longitude, and altitude as independent variables, need to be considered when native plants are being observed.

Agricultural meteorology frequently uses values derived from meteorological measurements, such as temperature or heatsum, precipitation, insolation, etc., or combinations of these variables, as functions of the independent variable time. Such a temporal model is used to determine the number of days required to reach the activation level or amount, and the integral form of a time-series function describing the temperature, moisture, or insolation regimes. A number of curve-fitting techniques described by Stringer (1972) are being investigated for the construction of the time-series functions.

The ability to predict the timing of future phenological events successfully will not only be of significance for natural resource management, but also will reflect a new and greater understanding of the biological basis of phenological timing.

References

Caprio, J. M. 1967. Phenological patterns and their use as climatic indicators. In *Ground Level Climatology*, R. H. Shaw, ed. Washington, D.C.: AAAS Publications, pp. 17–43.

Dudnik, E. E. 1972. *SYMAP User's Reference Manual for Synagraphic Computer Mapping*. Rep. No. 71–1. 114 pp. offset. Chicago College of Architecture and Art, University of Chicago.

Flint, H. 1974. Phenology and genecology in woody plants. p. 83. In : *Phenology and seasonality modeling*. H. Lieth, ed. New York: Springer-Verlag.

Hopkins, A. D. 1938. Bioclimatics—A science of life and climate relations. *USDA Misc. Publ. 0280*. 188 pp.

Kaszkurewicz, A., and P. J. Fogg. 1967. Growing seasons of cottonwood and sycamore as related to geographic and environmental factors. *Ecology* 48, 789–793.

Kramer, R. J. 1972. Mapping the seasons in Virginia—You can help. *Virginia J. Sci.* 23(1), 21–26.

Lieth, H., and J. S. Radford. 1971. Phenology, resource management and synagraphic computer mapping. *BioScience* 21, 62–70.

Perry, T. O., and C. W. Wang. 1960. Genetic variation in the winter chilling requirement for date of dormancy break for *Acer rubrum*. *Ecology* 41, 790–794.

Radford, J. S. 1971. Phenological delimitation of the growing season in North Carolina: Computer mapping and correlation of spring, fall, and length of growing season data with environmental parameters. 158 pp. MSc. Thesis. Chapel Hill: Botany Department, University of North Carolina.

Reader, R., and UNC Applied Programming Group. 1972. SYMAP (version 5.16A) instruction manual. *Document No. LSR-139-0*. Research Triangle Park. N. C. Triangle Universities Computation Center. 65pp.

Stringer, E. T. 1972. *Techniques of Climatology*. San Francisco: W. H. Freeman.

5.3 Modeling of the Annual Cycle of Soil Moisture[1]

Katharina Lettau

Center for Climatic Research
University of Wisconsin, Madison, Wisconsin 53706, USA

with an Appendix

Evapotranspiration Climatonomy of Drainless Areas

Heinz Lettau

Department of Meteorology
University of Wisconsin, Madison, Wisconsin, 53706, USA

1 Introduction

Primary productivity depends strongly on seasonal variations of soil moisture and evaporation. It is also known that the amount of soil moisture and its rate of exchange, both indispensable factors of any plant-climate model, are difficult to assess. For example, Lemon *et al.* (1971) mention specifically that soil moisture variations and evaporation processes are not very successfully simulated by the so-called "SPAM" model with which he describes the "Sun's Work in a Cornfield." This steady state model is an example of the microstructure approach with emphasis on the complexities of a specific plant canopy. On the other end of the spectrum are global-scale models as exemplified by Lieth and Box (1972), who simulate world-wide relationships between primary productivity and annual averages of evapotranspiration in the "Thornthwaite Memorial Model."

[1] This research was supported by Grant (GA-10651X) Section on Atmospheric Science, National Science Foundation.

KEYWORDS: Steppe environment; annual cycle; global radiation; precipitation; runoff; evapotranspiration; soil moisture storage; water preservation; residence time of soil moisture; surface albedo; evaporivity; steppe plant communities; plant-climate model; climatonomy.

In the following I would like to describe an application of a model developed by H. Lettau (1969, see Appendix) which is based on the annual variations of energy and mass supply as input and generates the constituents of the hydrological cycle as output. The basic water balance equation is solved by rigorous numerical integration following parameterization of the process of evapotranspiration.

2 Community Types and Evapotranspiration Climatonomy in a Steppe Environment

We know that the moisture factor is most crucial in semiarid climates. Figure 1 shows a set of data from the semiarid region of eastern Washington as reported by Daubenmire (1972) in his discussion of annual cycles of soil moisture and temperature as related to grass development in a steppe environment. The actual soil moisture distributions in eight climax steppe communities demonstrate significant differences in the soil moisture profiles of these plant communities, although the abiotic input, i.e., precipitation and global radiation, was presumably the same. Daubenmire concludes that the plant communities themselves influence the soil moisture status—in other words, some communities conserve water better than others—but he cautions that these conclusions should be validated by observations in different years and in various stands.

We can try to compare Daubenmire's observations with an evapotranspiration climatonomy model to see what levels of soil moisture can be predicted if macroclimatic factors are the same but with the plant communities having different hydrological characteristics.

Briefly, the model makes use of two parameters. The first is called evaporivity, and denotes that fraction of precipitation which is returned to the air via evaporation so quickly that it cannot contribute to soil moisture storage changes. That portion of the precipitation which is actually stored underground may be spent either quickly or slowly, and this variable decay rate is measured by the second parameter called residence time. (For further details and mathematical formulations, see the Appendix.)

Figure 2 shows the annual variation of precipitation and global radiation for Spokane, Washington, from August 1962 to August 1963, which was, according to Daubenmire, not an unusual year. Because there is no runoff in the area of Daubenmire's observations, the annual total evaporation must be equal to the annual total precipitation, which was 408 mm. In spite of significant differences in evaporivity and residence time, the calculated monthly evapotranspiration values show little contrast, less than 10 mm per month. But the soil moisture regimes are significantly different (by a factor of 2) as shown in Fig. 3. In comparison with Daubenmire's measurements, remember that the model calculations produce soil moisture totals which correspond to vertically integrated soil moisture per unit volume. Nevertheless, a certain threshold value of m will correspond to the wilting coefficient and another will indicate the approach to field capacity (compare Equations 1 and 4 in the Appendix). For Daubenmire's cases these threshold values are about 55 and 110 mm of water column, respectively. Evaporivity and residence time affect the seasonal variations of the soil moisture regime very distinctly. The soil with high evaporivity and low residence time can be compared with a frivolous spender who dispenses immediately of most of what he receives and disposes of the rest within a short time. The differences in the hydrological

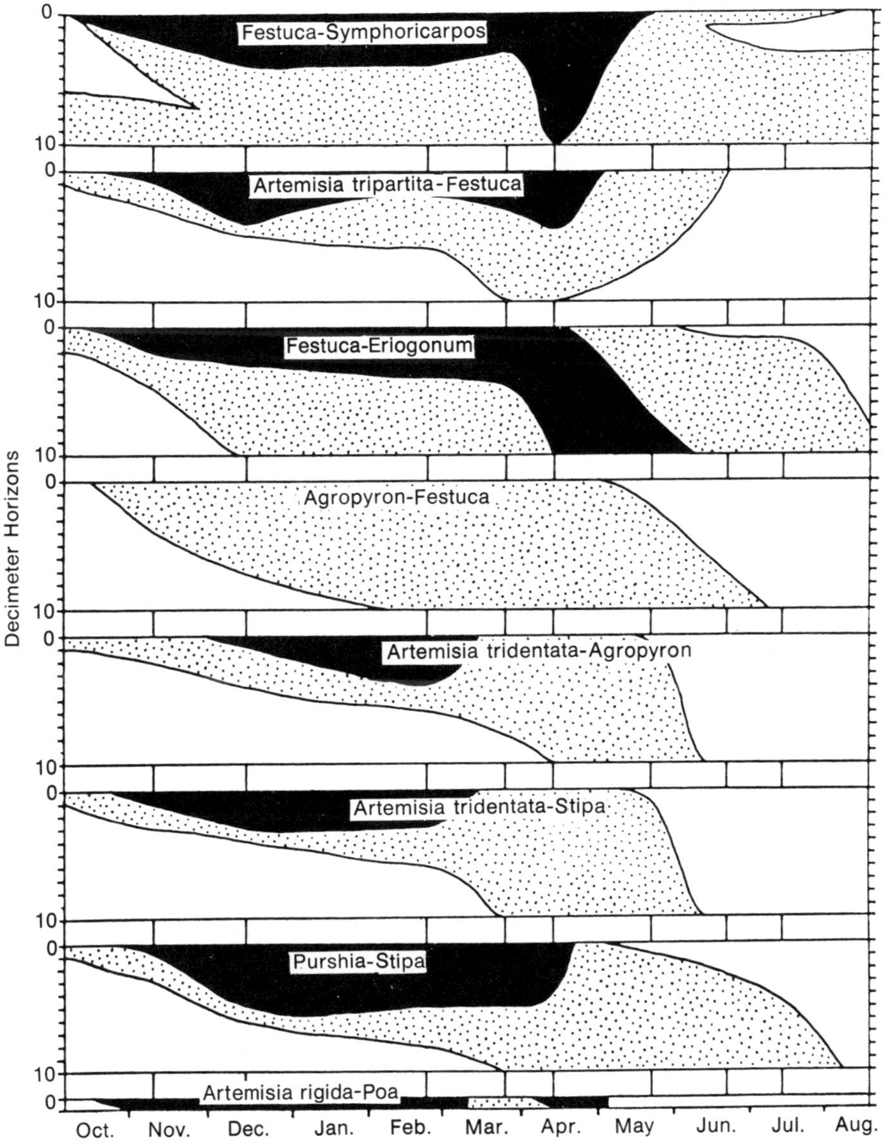

Fig. 1. Soil moisture status in eight climax steppe communities in eastern Washington from October 6, 1962 to August 17, 1963. Limits of decimeter horizons shown on ordinates. Black areas indicate water content at or in excess of the field capacity; stippled areas indicate water content between field capacity and wilting coefficient; unshaded areas lacked growth water. Vertical lines between horizontal panels show actual dates of sampling. (After Daubenmire, 1972.)

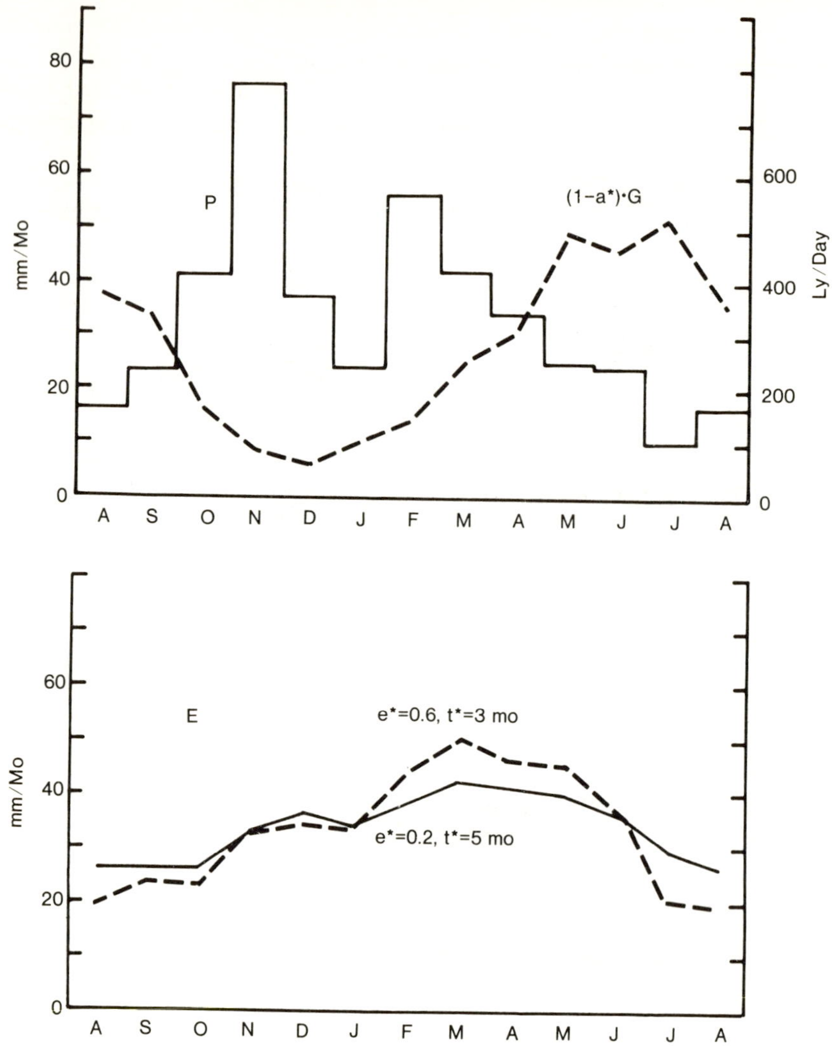

Fig. 2. *Upper*: Observed monthly means of precipitation P and global radiation $(1\text{-}a^*) \cdot G$ for Spokane, Washington, in 1962–1963. *Lower*: Calculated evapotranspiration E (mm/mo) for two sets of parameter values of evaporivity e^* and residence time t^*.

cycles documented by the three calculated curves are significant in terms of threshold values for wilting point and water saturation, and we conclude that vegetation differences could significantly control the hydrological cycle.

3 Computer Simulation of Seasonal Soil Water Changes

Daubenmire emphasized that his measurements should be repeated in different stands and under different climatic conditions. The advantage of a quantitative

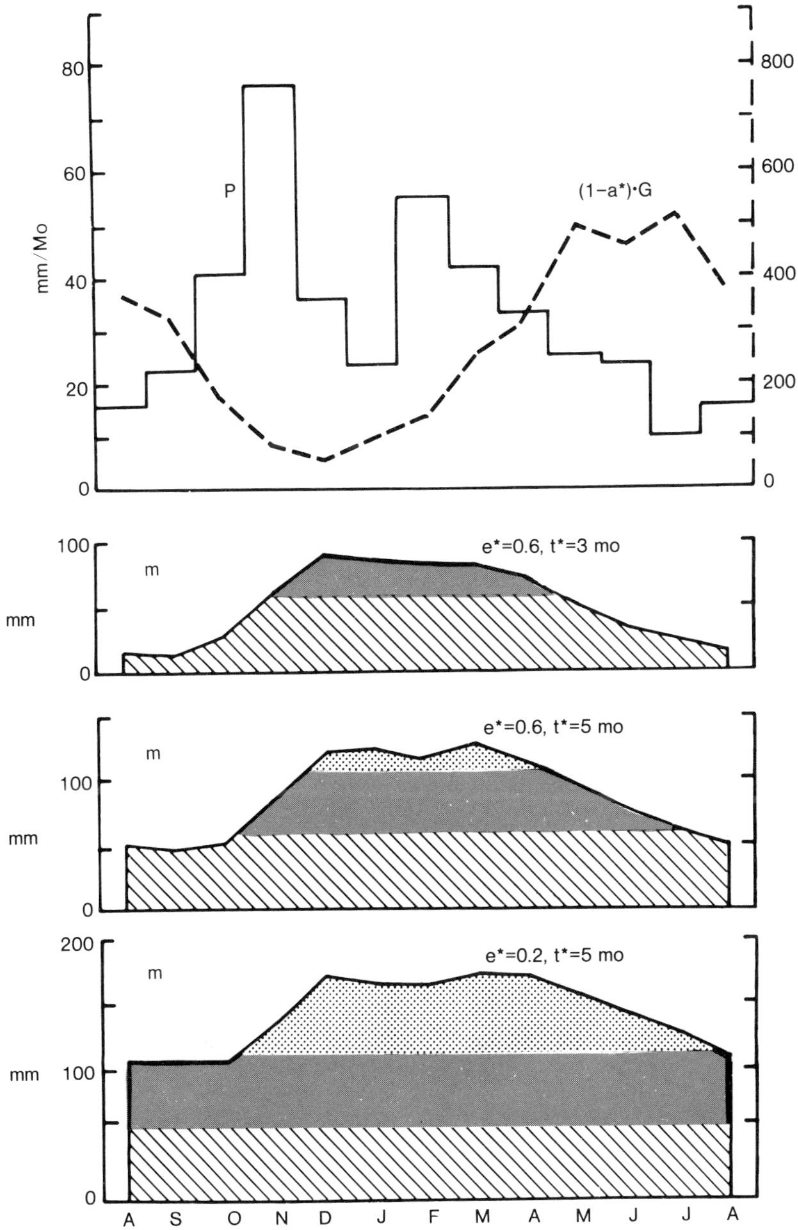

Fig. 3 *Upper*: Observed monthly means of precipitation P and global radiation $(1\text{-}a^*)\cdot G$ for Spokane, Washington, as in Fig. 2. *Lower*: Calculated monthly means of exchangeable soil moisture m for three sets of parameter values of evaporivity e^* and residence time t^*. The horizontal lines at 55 and 110 mm soil moisture indicate approximate threshold values for wilting and field capacity, respectively, as suggested by Daubenmire in Fig. 1.

model is that such effects can be simulated by the computer. Figure 4 illustrates such an application. For a sequence of several years the annual insolation curve is assumed to be equal, whereas the monthly precipitation values were doubled for one year and halved for another year with "normal" years before, in between, and after. The soil moisture curves were calculated with the same set of parameters as before; evidently water is carried forward from "fat" to "lean" years if the soil parameters favor water preservation. Model calculations of this type could serve to simulate conditions in different plant communities and could perhaps be related to variations in productivity. Further research is necessary to clarify problems which may be due to adaptation and feedback processes.

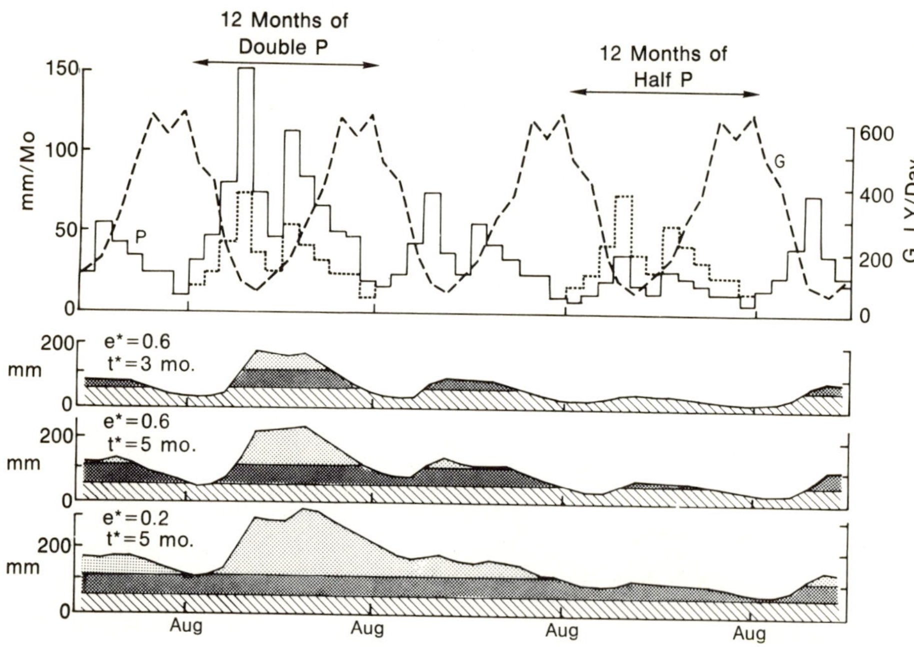

Fig. 4. Results of a model experiment showing the effects on soil moisture variations for a sequence of four years for three sets of parameter values (as in Fig. 3). The seasonal variation of insolation is assumed to be the same each year as it was in 1962–1963, whereas the precipitation of August 1962 to August 1963 is doubled during one 12-month period and halved during another 12-month period.

Nothing has been said about soil and air temperature. Climatonomical models of the complete energy budget at the soil-air interface are available which can predict soil and air temperature by parameterizing the processes of radiation, conduction, and convection. Considering that latent heat is such an important contributor to the total energy budget, and that apparently modeling of soil moisture cycles presents problems in presently available plant-climate modeling, it seemed justified to concentrate explicitly on the hydrological cycle.

4 Appendix: Evapotranspiration Climatonomy of Drainless Areas

Evapotranspiration climatonomy (Lettau, 1969) has recently been extended by the consideration of seasonal variations of the parameters (Lettau and Baradas, 1973) which are used to express the processes of evaporation and runoff. Presently, the model is being extended to predict water level changes of inland lakes and reservoirs. However, the principle of the concept and the mathematical characteristics of the model algorithm are readily demonstrated by an application to the drainless plains of a semiarid region. A steppe environment as discussed in the preceding text offers an instructive example for utilization of climatonomy.

To present the theoretical background for K. Lettau's discussion of abiotic forcing in the Spokane region, and to establish a basis for future applications, a brief summary of the algorithm appears to be in order. We utilize neither regression equations nor other statistical formulas for predicting evapotranspiration. We start out with the basic water balance equation of a soil column and solve it by rigorous numerical integration, following a suitable physical parameterization of the process of evapotranspiration.

For a drainless area the water balance equation has only three terms, which represent input, process, and balancing response functions:

$$P-E = dm/dt \tag{1}$$

where t = time = independent variable (with differentials dt normally interchanged by finite increments Δt, for example, 1 Mo = one month); m = exchangeable soil moisture (mm H_2O) = dependent variable; P = precipitation (including, if applicable, irrigation) = forcing function of mass input (mm/Mo); and E = evapotranspiration = process function (mm/Mo). The variable $m(t)$ is the response function (a function of time) because the mass input P must be prescribed as a time series. The derivative dm/dt (or, alternately, the difference quotient $\Delta m/\Delta t$) represents the storage term. Moisture can be stored either on the surface (snow pack, rain puddles, etc.) or in pores and other cavities within the soil, or by vegetation (within roots, stem, fruits, or leaves, etc.). Soil moisture measurements from layer to layer refer to a volume or weight basis, and such samplings must be vertically integrated (between the surface and the depth of vanishing changes over all seasons) to yield the value of m (expressed as millimeters of H_2O). Namely, exchangeable soil moisture m is the total water column involved in any climatic cycle, including multiannual climatic trends, at a given locality.

The main characteristic of climatonomy is a parameterization of the process function to the effect that the moisture balance Equation 1 can be rigorously integrated. To achieve this, we separate the process E into two additive terms:

$$E = E'+E'' \tag{2}$$

assuming that during any time increment Δt, the first term E' is directly proportional to input of mass (P) as well as to absorption of solar energy, while the remainder term $E'' = E - E'$ varies in proportion to the variable soil moisture m. The two parameters necessary to express these two proportionalities are $e^* =$ evaporivity (dimensionless) and $t^* =$ soil moisture residence time (expressed in units of Δt). Equations of definition are

$$E' = e^* P(1-a^*)\, G/\overline{(1-a^*)G} \tag{3}$$

$$E'' = m/t^* \tag{4}$$

where $G =$ global radiation $=$ forcing function of solar energy interception; and $a^* =$ surface albedo so that $(1-a^*)G =$ absorbed solar energy. Both quantities a^* and G must be representative of the same locality for which P is prescribed. In Equation 3, and in the following formulas the overbar denotes an annual average, so that in Equation 3 the units for G are canceling out. Normally, langleys per day will be used, where one langley (1 ly) is defined as one gram calorie per square centimeter.

In view of Equations 3 and 4 it is convenient to define the reduced input function P':

$$P' = P - E' = P \cdot [1 - e^* \cdot (1-a^*)G/\overline{(1-a^*)G}] \tag{5}$$

and, furthermore, a dimensionless time τ, using t^* as a scaling unit:

$$d\tau = \frac{dt}{t^*}; \quad \text{or,} \quad \tau = \int_0^t \frac{dt}{t^*} \tag{6}$$

With the aid of Equations 5 and 6 we obtain from equation 1 the transformed balance equation

$$dm/d\tau = -m + t^* P' \tag{7}$$

Rigorous integration of Equation 7 yields the response function

$$m = \mathrm{e}^{-\tau}(m_0 + \int_0^t P'\, \mathrm{e}^{+\tau} dt) \tag{8}$$

where m_0 is the initial value of m at $t = 0$. Once the time series of m has been calculated on the basis of the given inputs P and G, together with prescribed parameters e^* and t^*, we use Equations 4 and 2 to determine the time series of E'' and finally E.

Equations 5 through 8 serve to clarify the physical meaning of the two parameters e^* and t^*. Equation 5 shows that evaporivity e^* determines for any time increment Δt the fraction of P that is returned "quasi-immediately" to the atmosphere due to energy absorption during the same period Δt. Consequently, only the reduced input $P' = P - E'$ is responsible for actual storage changes from one Δt to the next Δt. In this respect the parameter e^* resembles a^*, since albedo also determines the fraction of an intercepted flux (in this case of G) that is returned immediately to the atmosphere. Like albedo, evaporivity can only be determined empirically, from flux measurements. Tentative analyses thus far indicate that e^* for natural watersheds may vary between 0.2 and 0.8. Factors which produce relatively high evaporivity are, for example, rains of the shower type which leave puddles to be acted upon by intermittent sunshine; or, a soil structure that prevents rain water from seeping quickly down to the lower subsoil layers, etc.

The physical meaning of the parameter t^* is most clearly illustrated if we consider Equation 8 for the special case of a dry spell, during which $P = P' = 0$ for a sequence of Δt periods, after an initial m_0 value had been established by earlier rains. In this case the m function decays as $m_0 \cdot e^{-\tau}$ which would be a truly exponential decay if, additionally, t^* would be a constant. In any case the t^* parameter is related to the "half-time scale" similar to that of other natural decay processes. Results obtained thus far (Lettau and Baradas, 1973) indicate that t^* has averages of two to four months and may show a seasonal range between 1.5 and about 10 months. Environmental factors which produce relatively large t^* values are, for example, dense tall vegetation, a high porosity of the soil, or a frozen state of ground water.

In a characteristic departure from conventional models [including the methods by Thornthwaite, Penman, Budyko, or others, as reviewed in standard textbooks, for example, Sellers (1965)], the water balance equation is here solved by integration rather than by algebraic accounting procedures, and climatonomy directly produces the annual average and the seasonal variation of exchangeable soil moisture as well as the "natural" value of m_0 for the month at which the study of the annual cycle is begun. A flexible computer program has been developed by P. Guetter at the University of Wisconsin which can be used to perform the integrations and process calculations for sequences of several years, once the variable time series for forcing functions and parameter values are prescribed.

The program is also flexible with respect to the possible consideration of runoff and run-in processes. These could be important even in a steppe region, which is drainless on the large scale, but may show horizontal movement of moisture from convex to concave terrain structures on a relatively small scale. It is important that the product of $P \cdot G$ (inputs of mass and energy) determines the magnitude of P' in Equation 5 to the effect that the average and the annual variations of m depend significantly on the covariance (for annual cycles: the phase difference) of the inputs P and G. Computer experiments can be performed which demonstrate that the soil moisture regime for a climate with winter rains (Mediterranean type) differs significantly from that with summer rains (monsoon type) even though averages of P and G may be nearly identical. Once the parameters of a region have been established, the model lends itself directly to the prediction of soil moisture variations during periods of either excessive rains or drought. The advantage of the model is that water utilization is limited by the natural supply of mass and energy, and that somewhat artificial concepts of "potential evapotranspiration" or "evaporation under conditions where water is not a limiting factor" are avoided. The major problem areas remaining are the determination of parameter values from primary information and the consideration of possible feedback between soil moisture levels and either parameters or process.

References

Daubenmire, R. 1972. Annual cycles of soil moisture and temperature as related to grass development in the steppe of eastern Washington. *Ecology* 53, 419–421.

Lemon, E., D. W. Stewart, and R. W. Shawcroft. 1971. The sun's work in a cornfield. *Science* 174, 371–378.

Lettau, H. 1969. Evapotranspiration climatonomy. I. A new approach to numerical prediction of monthly evapotranspiration, runoff, and soil moisture storage. *Monthly Weather Rev.* 97, 691–699.

———, and M. W. Baradas. 1973. Evapotranspiration climatonomy. II. Refinement of parameterization, examplified by application to the Mabacan River watershed. *Monthly Weather Rev.* 101, 636–649.

Lieth, H., and E. Box. 1972. Evapotranspiration and primary productivity. *Papers on Selected Topics in Climatology, Thornthwaite memorial volume 2*, J. R. Mather ed. p. 37–46. Elmer, N. J.

Sellers, W. 1965. *Physical Climatology*. 272 pp. Chicago: University of Chicago Press.

5.4 The Solar Thermal Unit Concept in Problems Related to Plant Development and Potential Evapotranspiration

Joseph. M. Caprio

Plant and Soil Science Department
Montana State University
Bozeman, Montana 59715, USA

1 Introduction

Since 1735, when Reaumur first attempted to relate temperature to plant development, the degree-day or heat-unit concept has been applied to problems of plant-climate relations. This paper is a contribution to the discussion of this problem. It describes our solar thermal unit concept, which attempts to explain why different quantities of heat units are sometimes required in different locations and seasons for a given stage of plant development to occur.

The solar thermal unit theory (Caprio, 1971) is based on the analysis of 10 years of phenological data on the purple common lilac recorded at hundreds of locations throughout the western United States. The equation which describes the accumulation of solar thermal units to the time when lilacs begin to bloom is

$$\mathrm{STU} = \sum_{Te \geqq 0}^{BB} \left[\frac{\mathrm{cal}\ (0.3\text{–}4\ \mathrm{micron}\ \lambda)}{\mathrm{cm}^2\ \mathrm{day}}\right]\left[\left(\frac{T_{\max}+T_{\min}}{2}\right) - 31.0\right] = 380{,}000$$

where $\left(\frac{T_{\max}+T_{\min}}{2}\right)$ = mean daily temperature computed from daily maximum temperature (°F) and daily minimum temperature (°F) as recorded in standard Weather Service shelters at the height of approximately 5 ft (1.8 m). λ = wavelengths of solar radiation measured on a horizontal surface. Σ = summation of daily solar thermal units from the time when mean daily temperature becomes effective ($Te > 0$) to the time when lilacs begin to bloom (BB). $Te = \left(\frac{T_{\max}+T_{\min}}{2}\right) - 31.0$ = effective temperature.

KEYWORDS: Phenology; evapotranspiration; radiation; temperature; lilac; bloom; flower; ecotype; solar thermal unit; North America.

Independent variables in this equation are mean daily temperature and the daily total amount of solar radiation measured in langleys. The rate of plant development is the dependent variable. Related to the dependent variable is an index of plant development, expressed as solar thermal units per day. At a given mean daily temperature above the threshold of 31°F (−0.6°C), the number of solar thermal units increases in direct proportion to the amount of solar radiation. Likewise, at a given solar radiation, solar thermal units accumulate at the same rate as the mean daily temperature above the threshold of 31°F. Solar radiation refers to the total incoming energy (0.3- to 4-micron wavelength), both direct and indirect, received on a horizontal surface.

2 Development of the Solar Thermal Unit Theory

Determination of Base Temperature

The solar thermal unit scale was formulated after research had established that the purple common lilac begins to develop in the spring when the mean daily temperature reaches about 31°F (day temperatures >0°C). In order to determine this threshold temperature of the purple common lilac both phenological and temperature records were required for many locations. Suitable phenological data on purple common lilac bloom and temperature were available for the 10-year

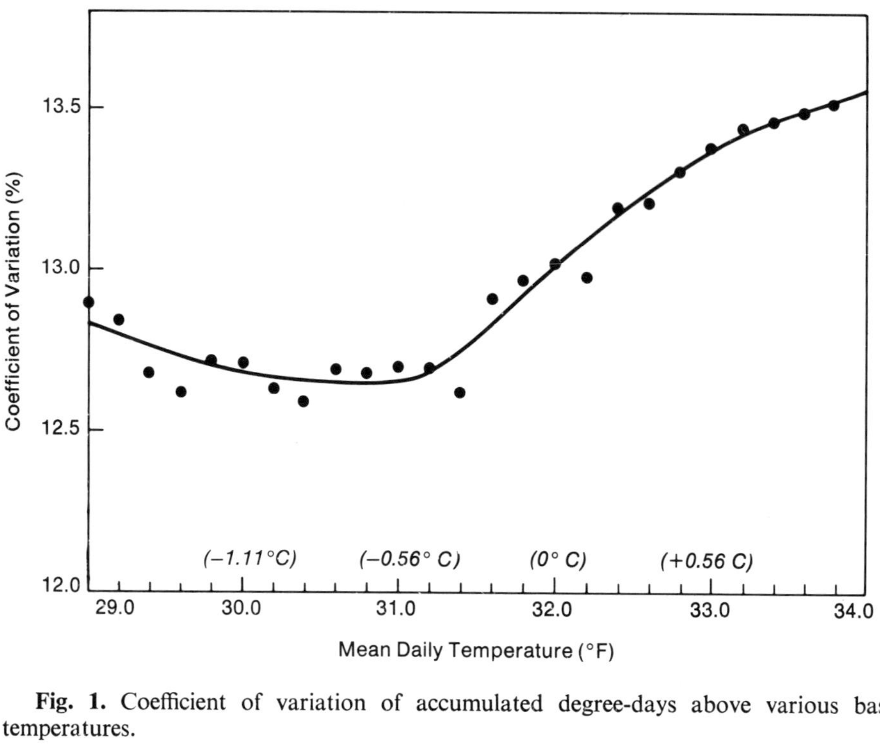

Fig. 1. Coefficient of variation of accumulated degree-days above various base temperatures.

period 1957–1966 at 251 locations in 8 of the 12 western states: Colorado, Idaho, Montana, Nevada, Oregon, Utah, Washington and Wyoming. Phenological stations in three states of the western region (Arizona, California, and New Mexico) were not included in the analysis because many locations in these southern states do not always have enough winter cold to break the plant rest.

The median date of first lilac bloom at each of the sites served as the dependent variable in the analysis. Mean daily temperatures for each station were computed from third-degree polynomials fitted to long-term (1951–1960) mean monthly maximum and minimum temperatures for the seven-month period January through July. Coefficients of variation of accumulated degree-days above base temperatures varying from 29° to 34°F (−1.7° to 1.1°C) by intervals of two-tenths (~1/10°C) of a degree were determined, as described by Arnold (1959).

The results are presented in Fig. 1. The curved line, fitted by eye, reaches a low plateau when the mean daily temperature drops to about 31°F (−0.6°C), marking this as the most convenient threshold temperature. The coefficient of variation at this threshold temperature was found applicable for smaller geographic zones within the eight-state area. Accumulation of daily mean temperatures above 31°F to median date when lilacs begin to bloom averaged 807 degree-days within the eight-state study area. Similar analyses of daily maximum and daily minimum temperatures each yielded coefficients of variation much larger than those found for mean daily temperature.

Evidence in Support of Temperature-Solar Radiation Interaction

A comparison of observed dates of lilac bloom with bloom dates estimated by the date of accumulation of 807 degree-days above the 31°F threshold temperature disclosed that lilac blooms later than estimated in some areas and earlier in other areas. It was apparent that areas where more degree-days are required to bring about bloom are characterized by less sunshine in the spring. A consideration of known lilac bloom dates in Norway reported by Batta (1970) and Lauscher and Printz (1959) suggested that the daylength-temperature interaction, as portrayed by the photothermal unit concept of Nuttonson (1955), does not account for the difference in lilac bloom dates over wide geographic areas. However, it appeared that the occurrence of late bloom in Norway relative to the date predicted by the heat-unit model might also be linked to the low intensity of solar radiation in spring.

We therefore computed the product of the number of degree-days of mean daily temperature above 31°F and the daily amount of solar radiation incident on a horizontal surface (in langleys) for those places in the northern parts of the West where data are available on both solar radiation and temperature, and where daily minimum temperatures in midwinter fall below 31°F. The total number of these daily units which accumulated to the time when lilacs begin to bloom consistently totaled about 380,000 langley–°F–days. The time interval from begin bloom to end bloom required about 280,000 ly–°F–day, regardless of the duration of lilac bloom stage, which varies from less than 18 days in the extreme eastern part of the western region to more than 30 days along the Pacific coast. These units of ly–°F–days were dubbed solar thermal units (STU).

3 Potential Evapotranspiration and Solar Thermal Units

Solar thermal units were compared with various empirical equations that relate potential evapotranspiration to mean daily temperature and daily solar radiation (Fig. 2). A number of these empirical equations are given in Jensen (1966). Among these are empirical equations developed by Jensen and Haise, Makkink, Pruitt, Stephens, and Stephens and Stewart. Data collected by van Bavel on evapotranspiration of alfalfa in Arizona are also given in Jensen's paper. The multiplication of

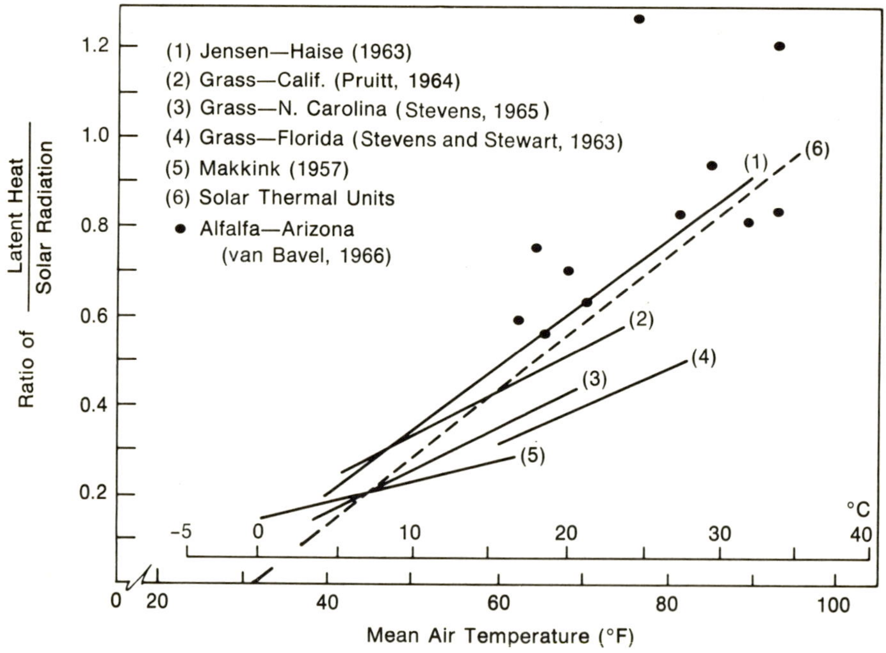

Fig. 2. Comparison of solar thermal unit method for estimating potential evapotranspiration with some empirical equations and single-day values for alfalfa in Arizona.

solar thermal units by 10^{-5} provides a convenient estimate of potential evapotranspiration in inches that is near the mean of estimates derived by field experiments. Associated with the time when lilac begins to bloom in the spring are STU $\times 10^{-5} =$ 3.80 in. of accumulated potential evapotranspiration. Thus solar thermal units are indicative of plant development and also provide an estimate of potential evapotranspiration.

Figure 3 shows in a nomogram the relation between daily values of solar radiation, mean daily temperature, and solar thermal units. When daily temperature is equal to or less than 31°F, or if solar radiation were to be zero, no solar thermal units would accumulate. The use of the nomogram is demonstrated in Fig. 3. When solar radiation accumulated over a given day totals 500 langleys and the mean daily temperature is 61°F, the total number of solar thermal units is $500 \times (61-31)$ or 15,000, and the potential evapotranspiration is 0.15 in. (solar thermal units $\times 10^{-5}$).

While the formula appears to be applicable over extensive areas, it is likely that refinements can be developed with further study. Environmental factors, such as extremes of temperature, atmospheric humidity, and wind speed, are likely to impose limits on the applicability of the model. For example, we know that inadequate cold during the winter months can result in delayed foliation in extreme southern areas. Also, the concept can be applied to other phenophases and species other than lilac bloom.

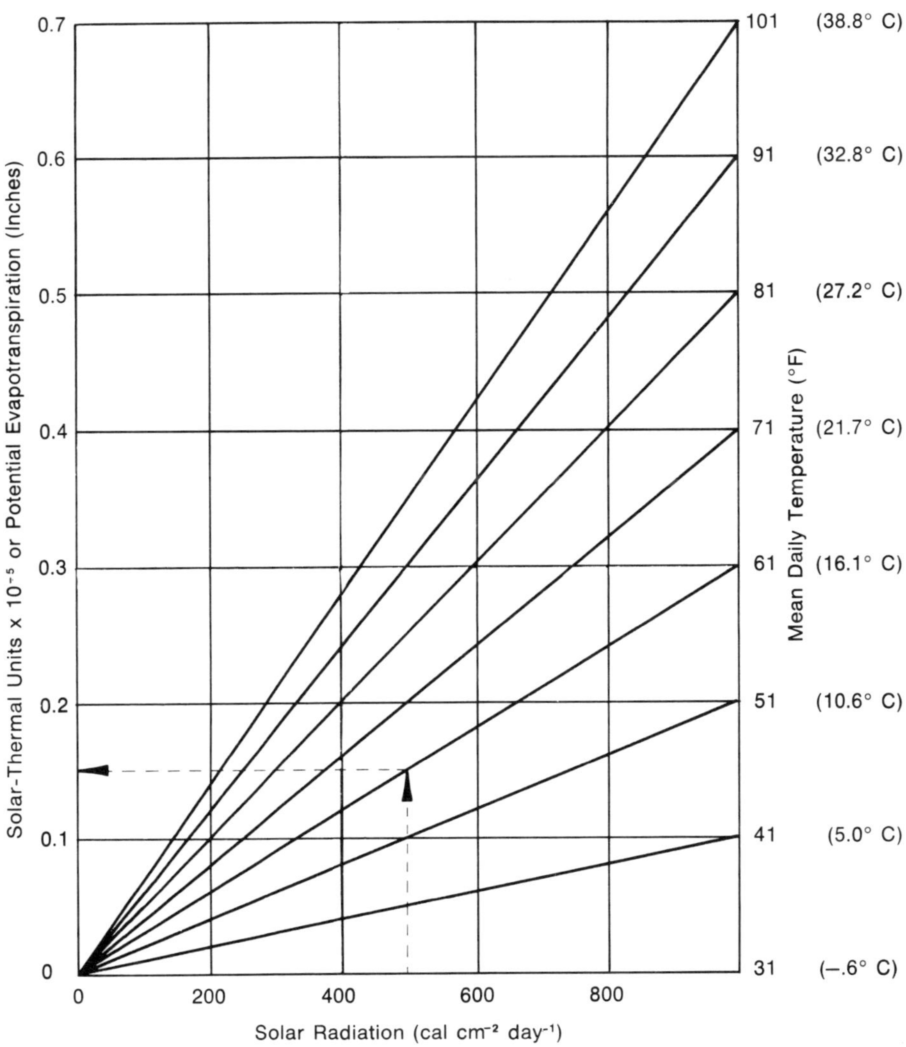

Fig. 3. Relation between daily values of solar radiation, mean temperature, and solar thermal units.

Table 1. Dates of phenological events as estimated by STU accumulations for various locations

Location	Latitude	Longitude	Mean Dates When Given STU Have Accumulated					
			Lilac First Leaf	Lilac Full Leaf	Lilac Begin Bloom[a]	Honeysuckle Begin Bloom	Lilac Peak Bloom	Lilac End Bloom
			150,000	*200,000*	*380,000*	*430,000*	*520,000*	*660,000*
Matanuska, Alaska	61°34′N	149°16′W	5-16	5-23	6-11	6-15	6-24	7-6
Reykjavik, Iceland	64°08′N	21°54′W	5-17	5-26	6-23	6-30	7-11	7-28
Sapporo, Japan	43°03′N	141°20′E	5-3	5-10	5-30	6-4	6-12	6-23
Tromsø, Norway	69°39′N	18°57′E	6-5	6-14	7-10	7-16	7-28	8-21
Verkhoyansk, USSR	67°33′N	133°23′E	6-8	6-12	6-28	7-2	7-9	7-21
Wakkania, Japan	45°25′N	141°41′E	5-10	5-18	6-10	6-15	6-24	7-8

[a] Lilacs usually do not produce a bloom when STU accumulation of 380,000 is not reached by about June 15.

4 Some Applications of the Solar Thermal Unit Concept

In order to test the general applicability of the STU concept, one can attempt to predict phenophases in other geographic locations. Table 1 gives normal dates on which various accumulations of solar thermal units are reached at six locations. All are situated in the northern hemisphere at latitudes greater than 60°N, except for the two stations in Japan. The levels of accumulated solar thermal units in the table correspond approximately to the time when lilacs reach the five phases being observed by phenological observers in the West, while the accumulation to 430,000 STU corresponds approximately to the time of begin bloom of the two honeysuckles being used as phenological indicator plants in the West, *Lonicera tatarica* Arnold Red and *Lonicera korolkowii* var. *zabelii*.

In the western United States wherever the accumulation of solar thermal units does not reach 380,000 by about June 15, lilacs usually do not grow or they do not regularly show annual blooming. If this rule could be generalized to include other parts of the world, lilacs would not be well adapted at Reykjavik, Tromsø, or Verkhoyansk. Bloom on honeysuckle, however, is known to have a further northern extent than lilacs and might occur at most stations in Table 1.

5 Data to Test the Solar Thermal Unit Concept

Table 2 gives information from selected locations in the western United States which are characterized by different amounts of springtime solar radiation. Information for Ås, Norway, is also included. In the case of Moro, Oregon, solar radiation was estimated from surrounding stations. This station is included because it is one of the few places in the West which has both mean temperature below 31°F in midwinter and low springtime solar radiation. Bloom dates for all stations are based on averages of a number of stations in the area, not on observations of one plant.

Table 2 shows that the number of degree-days required to bring lilacs into bloom varies greatly between locations, whereas the number of solar thermal units remains relatively constant near 380,000 at all locations. When lilacs are beginning to bloom, about 25 degree-days accumulate each day, while the normal daily accumulation of solar thermal units at time of bloom averages about 12,000. Thus, for the locations given in Table 2, the solar thermal unit method provides estimates within about a day or two of the bloom date, while an error of more than 20 days occurs using the degree-day method.

In the Seattle-Tacoma area where the mean daily temperature in midwinter does not fall below 31°F, the time to begin accumulating solar thermal units must be determined from knowledge of the rest period. Accumulation of solar thermal units backward in time from the date of begin bloom to November 30 comes to 380,000 STU. Data are presented in Table 2 assuming a November 30 completion of rest and the onset of plant development. Many ornamental shrubs in the Northwest break their rest in the fall period. Even if this assumed date of rest break at

Table 2. Degree-days and solar thermal units accumulated to begin lilac bloom date at several locations characterized by different solar radiation regimes.

Station	Solar Radiation	Degree-Days to Begin Bloom (31°F Base)	STU × 10^{-5} to Begin Bloom	Date of Begin Bloom
Glasgow, Montana	High	780	3.90	May 17
Ås, Norway	Low	987	3.76	June 1–2[a]
Moro, Oregon	Low	1032	3.71	May 6
Seattle-Tacoma	Very low	1724[b]	3.80[b]	April 27

[a] Based on average of two studies (Batta, 1970; Lauscher and Printz, 1959).
[b] Assumes rest is broken on November 30.

Seattle is in error by several weeks, the number of solar thermal units accumulated to the time of bloom stage would not differ greatly from the STU value indicated in Table 2, due to cool winter weather and the small amount of solar radiation received at this time of year.

The solar thermal unit theory states that the number of degree-days required to bring lilacs to bloom stage is inversely related to the amount of solar radiation. Thus large differences may exist in the number of degree-days required between geographic locations, while the number of solar thermal units remains near 380,000 at all of these locations.

Solar Thermal Units Required During Bloom Stage

Table 3 gives information for several locations in the West on the number of degree-days above 31°F and on the number of solar thermal units required to advance the lilac through its bloom stage, i.e., from begin bloom to end bloom dates. Bloom dates for these locations are based on averages of a number of stations in the general area. Glasgow and Great Falls, Montana, both located in a region of

Table 3. Degree-days and solar thermal units required to advance lilacs from begin bloom to end bloom at several locations characterized by very short and very long bloom stages.

Location	Date of Begin Bloom	Date of End Bloom	Length of Bloom Stage (days)	Degree-Days Above 31°F	Solar Thermal Units × 10^{-5}
Glasgow, Montana	May 17	June 4	19	497	2.84
Great Falls, Montana	May 17	June 5	20	494	2.73
Eureka, California	April 4	May 7	34	680	2.80
Tatoosh Island, Washington	May 14	June 14	32	689	2.86

high springtime solar radiation, require only 19 and 20 days, respectively, to complete the bloom stage. Eureka, California, and Tatoosh Island, Washington, both located in a region of very low solar radiation, require 34 and 32 days, respectively, to complete the bloom stage. Springtime temperatures at the time of bloom are cooler along the coast than in the interior, which also contributes to a longer bloom stage at locations near the coast.

While the number of solar thermal units accumulated during the bloom stage under both high and low radiation regimes remains relatively constant at about 280,000, the number of degree-days accumulated during the bloom stage varies from less than 500 under sunny conditions to nearly 700 in cloudy areas.

Relating the Development of Different Species Through the Solar Thermal Unit Concept

It is possible to relate the developmental status of different species through the solar thermal unit concept. For example, when lilacs begin to bloom the development of alfalfa toward one-tenth bloom (harvest stage) is about 50% complete since begin bloom of lilacs requires 380,000 STU and one-tenth bloom of alfalfa requires roughly 800,000 STU. By the end of lilac bloom, which requires about 660,000 STU, alfalfa has completed about 80% of its development toward harvest stage.

Possible Evidence of Ecotypic Difference in Solar Thermal Unit Requirements

Clausen *et al.* (1940) demonstrated through clonal transplant experiments in California that when ecotypes of a given species indigenous to different elevations are grown at the same site, the time of bloom in the spring is generally earlier on those plants which originated at higher elevations. Thus one might expect that ecotypes of a given species from higher elevations bloom with the accumulation of fewer solar thermal units compared to those plants of the same species which originated at lower elevations. Bloom on plants transplanted from higher elevations will tend to be more exposed to springtime freezing weather than plants transplanted from lower elevations. The hazard in transplanting plants from lower to higher elevations tends to arise from the inability of the plant to complete its entire life cycle in the shorter growing season.

Records of average begin bloom dates have been published for a large number of species growing at Manhattan, Kansas (Hulbert, 1963), and Bluffton, Indiana (Lindsey and Newman, 1956), located about 700 mi. apart. Information is available for six species at both locations, all or most of which are probably native to the respective areas. Table 4 lists five of these species along with the number of solar thermal units associated with begin bloom phase of each species. Solar radiation for Bluffton had to be estimated from surrounding stations in order to compute solar thermal units. The number of solar thermal units associated with begin bloom phase is about 28% greater at Manhattan than at Bluffton for all except the one species, *Cercis canadensis*, which exhibited the reverse relation and is not

Table 4. Average dates of begin bloom for five species at Bluffton, Indiana, and Manhattan, Kansas, and accumulated solar thermal units to bloom date and to time of first fall freeze.

Scientific Name	Common Name	Manhattan		Bluffton		STU Ratio
		Date	STU × 10^{-5}	Date	STU × 10^{-5}	M/B
Aesculus glabra	Ohio buckeye	April 22	3.82	April 30	3.14	1.22
Aquilegia canadensis	American columbine	May 4	5.55	May 8	4.15	1.34
Isopyrum biternatum	Atlantic isopyrum	April 6	2.07	April 14	1.60	1.29
Ribes odoratum	Clove currant	April 18	3.32	April 24	2.49	1.33
Staphylea trifolia	American bladdernut	April 27	4.50	May 5	3.75	1.20
					Average	1.28
Average first freeze:		October 12	42.02	October 8	33.62	1.25

included in the table. Local microclimatic conditions, transfer of seed or plant material from great distances, or other factors at one or at both locations might account for the anomalous behavior of *Cercis canadensis*. The fact that five of the species show about the same relation raises the question of whether ecotypic variation may indeed exist between plants of each of the five species growing at these two sites.

The accumulated number of solar thermal units up to the time of the average date of the first freeze in the fall is 42.02×10^5 at Manhattan and 33.36×10^5 at Bluffton. The ratio of the total accumulated STU for the season at Manhattan and Bluffton is 1.25, which is almost the same as the ratio of STU required at these locations to bring the same species into bloom. Thus, as also suggested by the study of Clausen *et al.* (1940), the number of solar thermal units required to bring ecotypes of the same species into bloom appears to be in some way directly related to the total solar thermal units available for the entire growing season.

While this small sample supports the above hypothesis, more studies are needed to ascertain cause-and-effect relations. Information in Table 4 suggests that the development of natural populations of a given species growing in diverse climatic habitats will tend to be associated with the accumulation of different numbers of solar thermal units.

References

Arnold, C. Y. 1959. The determination and significance of the base temperature in a linear heat unit system. *Proc. Am. Soc. Hort. Sci.* 74, 430–445.

Batta, A. J. 1970. Blomstringstid og Blomstermengde hos Lignoser. Institute for Dendrologi og Planteskoledrift. Unpublished data (personal communication).

Caprio, J. M. 1966. Pattern of plant development in the western United States. *Montana State Univ. Agr. Exp. Sta. Bull. 607* 42 pp.

——— 1967. Phenological patterns and their use as climatic indicators. In *Ground Level Climatology*. AAAS Publ. No. 86, pp. 17–43. Washington, D.C.: American Association for the Advancement of Science.

——— 1971. The solar-thermal unit theory in relation to plant development and potential evapotranspiration. *Montana State Univ. Agr. Exp. Sta. Circular 251* 10 pp.

Clausen, J., D. D. Keck, and W. M. Hiesey. 1940. Experimental studies on the nature of species. I. Effect of varied environments on western North American plants. *Carnegie Inst. Washington Publ. No. 520* 452 pp.

Hulbert, L. C. 1963. Gates' phenological records of 132 plants at Manhattan, Kansas, 1926–1955. *Trans. Kansas Acad. Sci.* 66, 82–106.

Jensen, M. E. 1966. Empirical methods of estimating or predicting evapotranspiration using radiation. In *Conf. Proc. Evaportanspiration and Its Role in Water Resources Management*. Dec. 5 and 6, pp. 49–56. Chicago: American Society of Agricultural Engineers.

———, and H. R. Haise. 1963. Estimating evapotranspiration from solar radiation. *J. Irrig. Drain.* 89 (IR4), 15–41.

Lauscher, A. F., and H. Printz. 1959. *Die Phänologie Norwegens*. Teil II, 176 pp. Phänologische Mittelwerte Fur 260 Orte. Oslo: H. Aschehong & Co.

Lindsey, A. A., and J. E. Newman. 1956. Use of official weather data in springtime temperature analysis of an Indiana phenological record. *Ecology* 37, 4, 812–823.

Makkink, G. F. 1957. Testing the Penman formula by means of lysimeters. *J. Inst. Water Eng. London* 11, 277–288.

Nuttonson, M. Y. 1955. Wheat-climate relationships and the use of phenology in ascertaining the thermal and photo-thermal requirements of wheat. 388pp. Washington D.C. *Am. Inst. Crop Ecol.*

Pruitt, W. O. 1964. Cyclic relations between evapotranspiration and radiation. *Trans. AEAE* 7, 3, 271–275, 280.

Reaumur, R. A. F. 1735. Thermetric observations made at Paris during the year 1735, compared to those made below the equator on the Isle of Maurituis, at Algiers and on a few of our American islands. Paris Memoirs. Acad. Sci. *Memoirs Acad. Sci., Paris.* 545.

Stephens, J. C. 1965. Discussion: Estimating evaporation from insolation. *J. Hydrol. Div., Amer. Soc. Civil Eng. Proc.* 91 (HY5), 171–182.

———, and E. H. Stewart. 1963. A comparison of procedures for computing evaporation and evapotranspiration. *Intern. Assoc. Sci. Hydrol.–Intern. Union Geod. Geophys. Publ. No. 62.* 123–133.

PART 6

Applications of Phenology

Introduction

Richard J. Hopp

Department of Plant and Soil Science
University of Vermont
Burlington, Vermont 05401, USA

and Helmut Lieth

Department of Botany
University of North Carolina
Chapel Hill, North Carolina 27514, USA

The preceding sections have described the methods and scientific goals of phenology. The significance of a science is not adequately demonstrated, however, without a thorough discussion of its application. Why are we so concerned with phenology and seasonality? Why do we make phenological observations from a local scale right to an international scale? What are some of the applications of phenology now and in the foreseeable future?

A phenological concept was employed long before the term *phenology* existed. Ancient Chinese and Romans used phenological calendars to guide agricultural activities. These calendars no doubt were known or revived during the Middle Ages, and nature's own weather bureau was active before organized meteorological observations became known. Early New England farmers followed an Indian custom by planting corn according to the foliage development of trees. Planting time was when the elm leaves were the size of a squirrel's ear or the oak leaves the size of a mouse's ear. An old saying among Vermont fishermen is: "When the shadberry blooms the trout will rise to a fly." Forestry has had its own phenological applications: a phenological network existed in Germany from 1885 to 1894.

Today phenology offers a necessary counterview to the mechanical outlook in biological resource management. Increasingly, we find poor understanding of biological systems in the very people devoted to such management. Is it possible to revive in the agricultural and silvicultural engineers a feeling of concern? Ecosystems are our true GNP, the "growth national product," to be exploited, necessarily, but also to be loved and served. It is no accident that a renewed interest in phenology and its applications has appeared during the last 20 years.

Phenology can assist greatly in the careful management of ecosystems, e.g., in avoiding most use of pesticides. In Montana farmers making the first cutting of alfalfa within 10 days of lilac bloom eliminate the overwintering brood of alfalfa weevil before the eggs hatch. Emergence of the European corn borer appears to coincide with the blooming date of *Viburnum dentatum*. On Long Island, county agricultural agents use forsythia bloom as indicator to predict arrival of cabbage maggot adults, while in Massachusetts the average date of first observance of cabbage maggot eggs was reported to coincide with the full pink stage of McIntosh apple blossom buds.

Agricultural and silvicultural research have uncovered important applications for phenology. Many of these are presented by Wielgolaski (Chapter 6.1), Schnelle and Volkert (Chapter 6.2), and Nienstaedt (Chapter 6.3). Waggoner (Chapter 6.4) demonstrates some uses of phenological models in resource management. Public health problems are often seasonal in their appearance. Schlichting (Chapter 6.5) and Kramer (Chapter 6.6) show how correlations between environmental parameters and seasonal germ development and spore distribution can lead to understanding of epidemic and seasonal illnesses.

The use of indicator organisms to alert us to simultaneous or subsequent events is possibly the most important aspect of phenology. Biological seasonality may be monitored in the future by satellites like *ERTS-1*. Such technological developments stimulate further research, e.g., for ground truth data to calibrate the satellite sensors (see Chapter 2.3). During the symposium Byron Blair gave a summary of the current *ERTS* ground truth program.

Finally, Stearns (Chapter 6.7) gives a moving argument for the incorporation of phenology into educational and community programs, including suggestions on how this might be done.

6.1 Phenology in Agriculture

Frans-Emil Wielgolaski

Botanical Laboratory
University of Oslo
Blindern-Oslo 3, Norway

1 Introduction and Review

The agricultural phenologist studies the impact of climate and soil conditions on the timing of biological events in plants of commercial and ornamental importance. The studies may include forests and pastures, as well as food crops. Phenophases observed include economic elements like harvesting and marketing of commercial plants; and in this respect his interests are broader than those of purely biological phenologists.

Phenology has been used in agriculture for several purposes; traditional agricultural practices usually reveal some type of understanding of phenological relationships. An old practice in many countries was to begin grazing domestic animals, especially cattle, at the time of apple blossoming. Ihne (1913) found in Germany that early potatoes traditionally were grown commercially only in districts that showed the earliest beginning of the growing seasons (and warmest temperatures) on his phenological maps.

Schübeler (1886–1888) collected many notes in Norway on the flowering time and winter survival of mainly horticultural plants. In continental Europe the collection of phenological data, especially for agriculture, has been widespread, perhaps inspired by Ihne (1884), Rosenkranz (1951), and Schnelle (1955).

Phenology is used most often in agriculture to predict the date of a particular event from an earlier phenophase of the same or of another species in the same year. Hentschel (1953) found that flowering of winter rye was predictable from phenological events occurring at the beginning of May. Often the time of *Galanthus* flowering is used to predict later events (Hertlein, 1954; Pearce and Preston, 1954; Hoffmann, 1969). Nuttonson (1953, 1955) and others have studied the heatsums (adjusted for other environmental factors) to predict different stages in the development of wheat, especially the time for harvesting. Ottosson (1958) and Balvoll and

KEYWORDS: Agricultural phenology; phenological predictions; agricultural events; agricultural planning; phenometry; basic temperature; starting dates; day and night temperature influences; precipitation influence; soil factors; climatic factors.

Bremer (1965), as well as others, have used studies of environmental factors to predict the best harvesting date for the pea, as Holmes and Robertson (1959) have done for this and several other agricultural plants. An intensive phenological study was done in many European countries for some varieties of barley during 1957–1963 (Lang, 1966). One of the main purposes of the study was to correlate phenophases and quality of barley with environmental factors in order to find where the best barley was grown for use in beer breweries.

The average dates for critical developmental stages of a plant of interest, taken over several years, indicate whether that plant should be grown commercially in a certain area. The stations with earliest ripening dates are not the only ones of interest. It might be important to grow the plants at stations with varying dates of ripening to facilitate harvesting at intervals suitable to commercial canning practices (Holmes and Robertson, 1959). In Norway, where the strawberry is grown both for the home market and for export, it is as important to find regions where berries will ripen in August-September and still have good winter survival, as it is to find areas where the harvesting can start in June.

The developmental rate may also influence the amount and type of fertilizers used in the fields (Schnelle, 1955). Newer studies found it useful to relate the seasonal variation of chemical element levels in the soil in different climates to particular phenophases of crop species. These examples demonstrate that agricultural planning may be based on phenological evidence, including the selection of plant varieties best suited to an area and/or a commercial purpose.

Even though temperature is often the most important single factor for development of agricultural plants, other climatic factors, and soil conditions, may also be of importance (Nuttonson, 1955; Ottosson, 1958; Strand, 1965). In the classical work *Vegetable Statistics* Stephen Hale (Gregory, 1952) presented quantitative data on the growth rate of plant stems and leaves. Reamur was the first (in 1735) to study mathematically the relationship between temperature and the rate of biological processes (Belehradek, 1935). He expected heatsums above 0°C to be a constant factor for the time between specific stages of each species. About 1860 Burckhardt and de Candolle suggested that basic temperatures (lowest mean temperatures for development of plants) other than 0°C should be used in the heatsums. Abbe (1905) has referred to older studies. Physiological studies (Sachs, 1865) showed that 24-hr temperatures are not the best ones to use in the comparison with plant growth and development, and that temperatures in various phenological periods could have a different influence on plants (Went, 1957; Fritts, 1960; Bassett *et al.*, 1961; Wielgolaski, 1966).

Sachs (1865) was the first to find the curvilinear relationship between plant development and growth and temperatures, which means that, strictly speaking, the heatsum concept is not quite realistic. Several attempts have been made to compare growth and development of plants with curvilinear expressions of temperature (Robertson, 1907; Livingston and Livingston, 1913; Mitscherlich, 1919; V. Blackman, 1919; West *et al.*, 1920; Sanderson, 1954; G. Blackman *et al.*, 1955; Salter, 1960). One of the more realistic curvilinear relationships between plant development and temperature was suggested by Dahl and Mork (1959) using what what they call "respiration equivalents" for Norway spruce at various temperatures.

Brown (1960) compared the real growth and development of soybean plants with various mathematical curvilinear temperature expressions and found the best correlations with the logarithm of temperature increase above 3.8°C (40°F). For short time intervals Arzt and Ludwig (1950) found that the straight line could be used in the analysis.

Not only the classic phenological stages, such as leaf and flower bud, flowering, leaf fall, and so on, should be included in phenological plant studies, but also length of shoots and number of shoots at different times of the vegetation period, as well as number of leaves and leaf areas at various stages. This is usually done in the IBP phenology studies and has also been done, more or less, by several workers in agricultural phenology, e.g., Lang (1966) for barley; Kurihara and Okubo (1960), Engel and Raeuber (1961), and Raeuber and Engel (1966) for potato; Higgins (1952) and Went (1957) for peas; Salter (1960) for cauliflower; and Utricht (1956) for sugarbeets. The detailed phenological studies of vegetative and reproductive stages performed by Higgins (1952), Newman and Beard (1962), and partly by Wielgolaski (in manuscript) has been called phenometry after Schnelle (1955, p. 49). An excellent application of the phenometric approach was executed by Thornthwaite (1952) when he developed his phenological slide rule for crop development on Seabrook Farms in New Jersey.

2 Use of Various Environmental Parameters in Phenology

To get the best phenological knowledge it is essential to study the development of plants in as much detail as possible and, simultaneously, to study climatic and soil parameters significant to the plant of interest. It is often difficult, however, to find the most important factors for various species. Even varieties within the same species may show differences in the relationships to critical environmental factors. Correlation tests between environmental factors and the phenology of plants may be found using stepwise multiple regressions. Principal component technique and discriminant analysis have many advantages when used in broad-scale phenological studies.

Basic Temperature and Starting Dates

One important factor for phenological studies in relation to temperature is the basic temperature, the lowest mean temperature for visible plant development. In studies of perennial herbaceous plants and woody plants it is also important to know the correct date in spring to use as a starting point for correlations between development of plants and environmental factors. In North America basic temperatures are often calculated from a method developed by Arnold (1959) and Bassett *et al.* (1961) in which the basic temperature is assumed to be the intersection with the x axis in rectilinear regressions between developmental rates of plants and mean temperatures. Schrödter (1952) calculated the basic temperature on the basis of a hyperbolic function between number of days to the phenological event of interest, and the mean temperatures. This method has been adopted by many workers in Central Europe. The two methods will give the same basic temperature when the correlation coefficient is 1. With weaker correlations, however, the American method is found to give lower basic temperatures than the European method. The author has tried to combine the two methods, which are strongly related (Wielgolaski, in manuscript), in an attempt to make the basic temperature calculations independent of the correlation.

The basic temperatures are not the same throughout all stages of development for one species. Possibly the basic temperature is highest in the period of strongest metabolism in the plant. Most herbaceous plants have high growth and development activity around the time of bud-with-color. Species that develop fruits also have a high metabolic rate after flowering, and thus a high basic temperature in that period as well. Balvoll and Bremer (1965) found that for peas the basic temperatures increased with the development of the plants. This was also found in studies by the author for peas as well as other annual herbaceous plants, e.g., from a basic temperature for peas at 3° to 4°C in the first period after germination to 10° to 11°C at flowering, and similarly for potato from 5° to 6°C in the earliest periods to 10° to 11°C at flowering (basic temperatures based on 24-hr mean temperatures). Unger and Schneider (1956) found 8°C to be a mean basic temperature between two stages of pea ripening. The basic temperature of perennial herbaceous plants was also found to increase with development (Abrami, 1972; Wielgolaski, in manuscript). However, in species which do not develop fruits, it was lower after flowering than in the period between bud-with-color and flowering, regardless of the flowering time in the season. This confirms the dependency of basic temperature on the physiological conditions of the plant.

In woody plants low basic temperatures were found in the author's study for development of organs more or less initiated in the fall the year before, e.g., flowering in fruit trees. Basic temperature for development of plum flowers in the Victoria variety was found to be about −3°C and for flowering of lilac (*Syringa vulgaris*) −1°C (starting date April 1 in southern Norway). Leaf budding of woody plants was defined as the first day the leaf stalk could be seen, and thus is a stage preceded by new vegetative growth. This caused the basic temperature to be somewhat higher than in the period to flowering; e.g. for the plum variety Victoria, about 0°C (starting date April 1).

The basic temperatures calculated in the author's study were usually lower than those found in Central Europe. The basic temperature in the period April 1 to flowering of apple was about −0.5°C, while Hertlein (1954) found it to be 3.2°C and Voigts (1949) and Hoffmann (1969) are using 6°C. It is striking that for most of the woody plants studied in Norway, around April 1 was found to be the best starting time for the analysis. For the perennial herbaceous plants studied in Norway, however, the best starting date was generally found to be February 15, possibly because of the higher temperatures below the snow cover than in the air during winter, which allow the vegetation period to start rather early.

Normally basic temperatures and usually the phenophases will be reached later with higher latitudes (Hopkins, 1938). In Germany (50°N) the highest correlation between developmental time to flowering of pears and apples and air temperatures was found using March 11 as the starting date, 40 days before flowering of pears and 46 days before flowering of apples (Hertlein, 1954). In Norway, at 60°N, the best starting date for the same stages was found to be April 3, giving a mean of 57 days before flowering of pears and 62 days before flowering of apples.

Day and Night Temperatures

Temperature at various periods of 24-hr intervals may have different meanings to the phenology in various stages of plants. For most of the agricultural species studied in Norway (Wielgolaski, in manuscript), there was a noticeable difference in the correlations between developmental rate in different periods and the temperatures day and night, respectively. Temperature conditions during the night (Fig. 1) were much more important than temperatures during the day in the periods from April 1 to leaf budding and flowering of woody plants (fruit trees). The same was found for herbaceous plants from February 15 of perennial plants, and from sowing of annual plants up to germination (sometimes even in the first period afterwards). This was found in total correlations with temperature and in partial correlations where the effects of other climatic factors and some soil factors were eliminated. The temperature during the night was also most important at flower fading, fruit setting, and fruit ripening, as was found for ripening of red currant in partial correlations with temperature. In partial correlations between plant development and environmental factors in these periods, maximum temperatures were some-

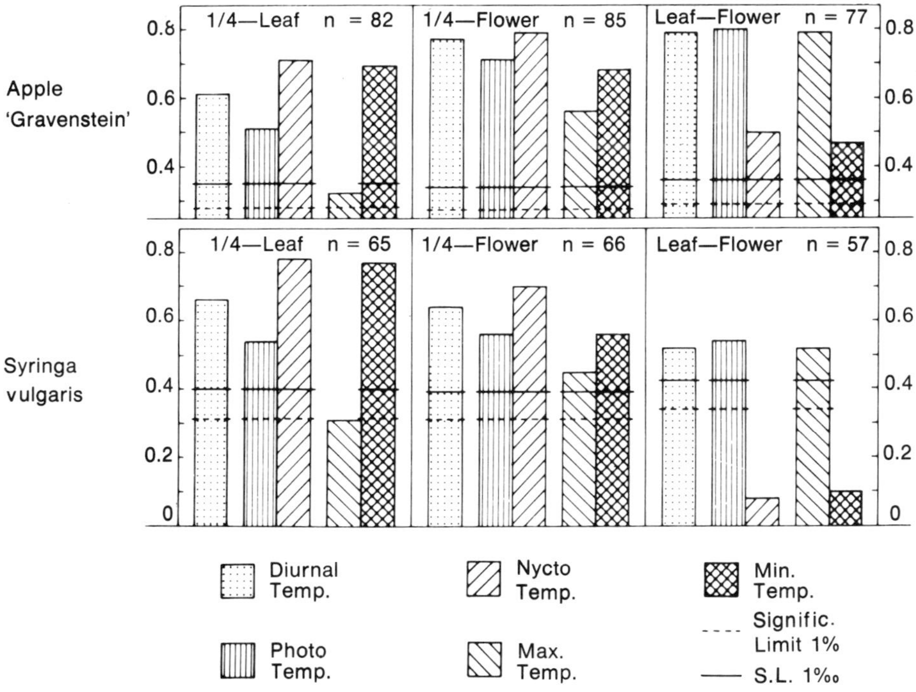

Fig. 1. Correlation coefficients for correlation of developmental rate in various phases of apple and *Syringa vulgaris* with different temperature parameters, starting date taken as April 1.

times even found to have a negative effect when minimum temperature, precipitation, and some soil factors were kept constant.

On the other hand, day temperature had the highest positive influence on development of woody plants in the period from leaf budding to flowering, e.g., in plum and apple. In herbaceous plants, both annual and perennial, the highest correlation was nearly always found between developmental rate and day temperature (both in total and partial analyses) during the periods from germination, or shortly afterward, to flowering or just before that stage. The effect of minimum temperature in these periods was slightly negative when other climatic and soil parameters were kept constant. Most of the results from field studies were confirmed by experiment under controlled conditions in climatic chambers in standard soil conditions.

Generally it seems that in periods with strong vegetative development, and thus a high production of new leaves, increasing day temperatures are most favorable for high photosynthesis, as long as light intensity is not limiting production. However, in periods where development mostly occurs from energy taken from storage organs (in spring) or when the energy bound by assimilation is mostly used for initiating reproductive organs or is translocated to storage organs, the night temperature is most important (Fig. 1). The photosensitive pigment enzyme, phytochrome, partly determines the initiation of reproductive organs (Siegelman and Hendricks, 1964). The balance between the plant hormones might also be important for flower formation, and so might formation and transformation of hypothetical flowering hormones, which take place especially in the dark and are somewhat temperature dependent (Salisbury, 1963). This may explain why sometimes night and sometimes day temperatures have been found to be most important for phenological studies (literature review in Wielgolaski, 1966).

Such differences in day and night temperature dependence in various developmental periods has to be taken into account in phenological studies. It was found that maximum temperature could be used as a substitute for phototemperature and, similarly, minimum temperature could be used for nyctotemperature. It was also found in the author's studies that curvilinear expressions of temperature seldom gave significantly better correlations with development than rectilinear expressions, which confirmed Arzt and Ludwig's statement (1950). Therefore, for rough estimates the heatsum concept could be used. Heatsums could be estimated above basic temperatures, estimated separately for maximum and minimum temperatures as well as for 24-hr temperature means. Maximum temperature should then be used for heatsums during the typical vegetative phases; in other periods, minimum temperature should be used.

Precipitation

In dry areas precipitation may be more important for phenology than temperature, but even in humid temperate areas, it has some effect. From Norwegian studies (Wielgolaski, in manuscript) it seems that precipitation, like temperature, may have different effects in various phenophases. For herbaceous plants high precipitation usually had a small accelerating influence up to time of germination. In the fastest vegetative growth period, high precipitation usually reduced the developmental rate. For most species studied, however, both herbaceous and woody, increasing precipitation was more or less favorable to rapid plant development again in the latest periods, i.e., from flowering onward. This was true for potato, and may support the findings by others that good water supply and moderate temperature are favorable to development of potato tubers (Went, 1957; Holmes and Robertson, 1959; Engel and Raeuber, 1961; Ivins and Milthorpe, 1963).

Soil Factors

Soil factors may greatly influence the development of most plants (Conrad, 1959; Strand, 1965), even though the effect is usually smaller than the climatic effect (Caprio, 1966). Variation in soil seems to be important in that soil types influence soil moisture (Nuttonson, 1955, p. 382), possibly due to the changes in soil temperatures with soil moisture (Schnelle, 1955, p. 246). Stepwise multiple regressions were performed in the Norwegian agricultural studies (Wielgolaski, in manuscript) among developmental rate of woody plants and climate, and some of the most significant soil factors (Fig. 2). From five groups with a total of about 90 variables, 10 to 15 independent variables were selected for this final analysis, taken as the most significant factors in stepwise multiple regressions among plant development and the various groups of environmental factors.

These soils factors accounted for up to 40% of the variation that could be explained by those factors found to have significant influence on the development. However, most often in development of fruit trees and berries, two to four soil factors had a significant influence on the analyses and accounted for 10 to 20% of the variation (Fig. 2*A*, *B*). Usually the physical soil parameters were most important: e.g., often a positive influence was found on developmental rate with an increase in bulk density at a soil depth of 35 to 40 cm (Fig. 2*A*).

In annual herbaceous plants also two to four soil parameters were found to exert the most significant influence by the same sort of stepwise multiple regressions, accounting again for 10 to 20% of the variation. None of the soil parameters in these analyses explained more than about 10% of the variation in development. A 10% influence was found e.g., for the negative effect of an increase in the amount of total nitrogen in the soil on the developmental rate of peas in the period between germination and height 10 cm, for the positive effect of an increase in the amount of water at pF = 4.2 at 5 to 10 cm soil depth on the development of peas in the period between bud-with-color to faded flowers, and for the positive effect of an increase in specific gravity of the soil at 5 to 10 cm on development of potato in the period between germination and 10 cm height. In the annuals, as mentioned earlier for woody plants, the significance of soil parameters pertaining to water conditions was highest for the developmental rates. These examples show the importance of including soil parameters in detailed phenological studies in agriculture. For practical purposes, however, these studies are too complicated and thus most comparisons are done only between development and temperature.

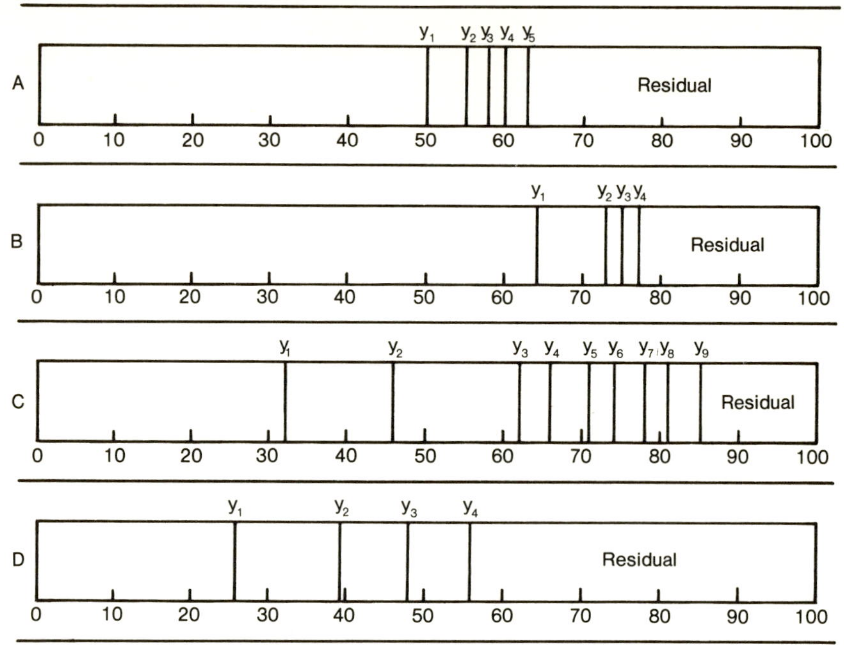

Fig. 2. Stepwise multiple regressions between developmental rate and environmental factors and t-tests for the significance of each factor.

A. Apple Gravenstein leaf budding. $R = 0.79, n = 72$.

$$y_5 = 0.27 + 0.22\,x_3 - 0.002\,x_{42} + 1.56\,x_{80} + 0.045\,x_{81} - 0.007\,x_{20}$$
$$t_3 = 7.33.\ t_{42} = -3.66.\ t_{80} = 3.60.\ t_{81} = 2.64.\ t_{20} = -2.14.$$

B. Apple flowering. $R = 0.88, n = 73$.

$$y_4 = 1.56 + 0.09\,x_3 - 0.10\,x_{10} - 0.0003\,x_{42} - 0.004\,x_{83}$$
$$t_3 = 8.96.\ t_{10} = -6.25.\ t_{42} = -2.74.\ t_{83} = -2.20.$$

C. Raspberry Preussen flowering. $R = 0.92, n = 40$.

$$y_9 = 1.90 + 0.020\,x_2 - 0.007\,x_{46} - 0.064\,x_{10} - 0.032\,x_6 - 0.0002\,x_{42} + 0.057\,x_{26} + 0.001\,x_{38} - 0.27\,x_{75} - 0.01\,x_{52}$$
$$t_2 = 2.11.\ t_{46} = -6.03.\ t_{10} = -6.21.\ t_6 = -4.60.\ t_{42} = -4.05.\ t_{26} = 2.96.\ t_{38} = 3.72.\ t_{75} = -3.44.\ t_{52} = -2.65.$$

D. Raspberry ripening. $R = 0.75, n = 47$.

$$y_4 = 0.52 + 0.09\,x_{13} + 0.016\,x_3 - 0.0001\,x_{42} + 0.001\,x_{20}$$
$$t_{13} = 3.71.\ t_3 = 3.28.\ t_{42} = -3.75.\ t_{20} = 2.58.$$

x_2 = photo temperature, average daytime temperature, (°C). x_3 = nycto temperature, average night-time temperature, (°C). x_6 = temperature amplitude, between the maximum and minimum temperatures. x_{10} = precipitation in millimeters, sum in periods studied. x_{13} = hours of sunshine in the middle of the day from 9:30 AM to 2:30 PM. x_{20} = incoming radiation at the site during afternoon, after 2:30 PM (cal/cm²/day). x_{26} = pH in soil 5 to 10 cm deep. x_{38} = soil P 5 to 10 cm deep, mg/100 g dry soil (readily soluble by Egnér's lactate method). x_{42} = soil K 5 to 10 cm deep, mg/100 g dry soil (soluble in 1N HNO_3 method of Reitemeier). x_{46} = soil Ca 5 to 10 cm deep, mg/100 g dry soil (readily soluble by Egnér's lactate method). x_{52} = ratio soil K/Mg 5 to 10 cm deep. x_{75} = specific gravity in soil 5 to 10 cm deep, (g/cm³). x_{80} = bulk density in soil 35 to 40 cm deep, (g/cm³). x_{81} = percentage of soil water at pF = 4.2, 35 to 40 cm deep. x_{83} = available water in soil 35 to 40 cm deep.

Table 1. Simple linear correlations between earlier and later phenophases in a three year study in Norway.

Early Phases:	*Salix* Flower 113 days from Jan. 1			*Betula* Leaf 125 days from Jan. 1			Apple Leaf 129 days from Jan. 1		
Later Phases:	No. days between phases	*n*	*r*	No. days between phases	*n*	*r*	No. days between phases	*n*	*r*
Red currant flower	25.0	40	xxx 0.75	13.0	77	xxx 0.90	7.5	72	xxx 0.86
Plum flower	34.2	17	xx 0.64	22.3	45	xxx 0.67	18.9	53	xxx 0.72
Pear flower	35.4	19	x 0.49	22.8	51	xxx 0.81	19.4	62	xxx 0.85
Apple flower	38.2	37	xxx 0.73	27.2	70	xxx 0.83	22.5	84	xxx 0.88
Syringa vulgaris flower	44.9	36	xxx 0.63	33.6	59	xxx 0.83	28.8	64	xxx 0.80
Phleum pratense spike	51.0	19	xx 0.60	41.6	47	xxx 0.78	36.0	55	xxx 0.78
Raspberry flower	56.2	23	x 0.50	45.1	42	xxx 0.73	39.7	46	xxx 0.76
Phleum pratense flower	78.5	15	0.36	66.3	35	xxx 0.64	61.5	37	xxx 0.58
Red currant ripe	87.0	33	xx 0.47	77.3	70	xxx 0.54	74.5	66	xxx 0.65
Raspberry ripe	95.0	26	xxx 0.59	83.7	51	xxx 0.58	78.4	55	xxx 0.67

n = sample size; *r* = correlation coefficient; *x* = significance at 0.05 level; *xx* = significance at 0.01 level; *xxx* = significance at 0.001 level.

Phenological Predictions.

Earlier phenological stages can give predictions of the time of important later stages (Seyfert, 1966). Our Norwegian studies showed several correlations between time for early phenological events and time for later interesting stages of woody plants, especially fruit trees and berries. The time for flowering of willow (*Salix caprea*) was important for many later stages, mostly to flowering of red currant (which as a mean occurred 25 days later), but also to flowering of apple (mean 38 days later), flowering of lilac (*Syringa vulgaris*) (mean 45 days later), and even ripening of raspberries (mean 95 days later) (Table 1). The time for ripening of raspberries was most strongly correlated with time for leaf budding of bird cherry (*Prunus padus*) $r = 0.76$, and so also was the time for ripening of red currant, $r = 0.70$ (mean 83 days later). Most of the later stages were strongly correlated with the time of leaf budding of birch (*Betula pubescens*) and of apple. Ripening of red currant was very much dependent on time of leaf budding of *Syringa vulgaris*, $r = 0.71$, $n = 57$ (mean 73 days later) and so was flowering of raspberry, $r = 0.77$, $n = 39$ (mean 38 days later), and also time for visible spike of timothy (*Phleum pratense*) besides flowering of *Syringa*. Flowering of raspberry and time for visible spike of *Phleum* also depended on times for leaf budding of ash (*Fraxinus excelsior*) and flowering of *Prunus padus*. Common for early development of all four stages are high temperature and moderate precipitation. In the Netherlands and in Germany it is found that time for leaf budding of horse chestnut may be a very good indicator for the later flowering of fruit trees (Post and Kramer, 1951; Hertlein, 1954).

A study done by the author (in manuscript) correlated the time for stages of *Aster novi-belgii* one year and time for flowering of fruit trees and *Syringa vulgaris* the next year. The correlations between bud-with-color, as well as flowering of *Aster novi-belgii* one year and flowering of red currant and apple and *Syringa vulgaris* the next year, were strong with coefficients of about $r = 0.6$, at a sample size between 30 and 40. Bud with color and flowering of *Aster* occurred in late September. Possibly this was an important period for flower initiation of red currant, apple, and *Syringa*. The correlations were much lower between the time for the stages of *Aster* mentioned and the time of leaf budding of fruit trees and *Syringa*, and also between ealier stages of *Aster* and flowering the next year.

On the whole there is much work to be done in the study of phenology in agriculture. To better coordinate the work internationally, guidelines should be prepared and tried for future studies, to include both methods and plants with exactly defined stages which could easily be studied in many countries. Preferably, plants which have earlier been used for phenological mapping in many countries should be used, e.g., *Syringa vulgaris*, apple (without variety specification), peas (*Pisum sativum*), and barley.

References

Abbe, C. 1905. A first report on the relations between climates and crops. U.S. Government Printing Office, Washington, D.C. Bulletin 36, W.B. No. 342, 386 pp.

Abrami, G. 1972. Optimum mean temperature for plant growth calculated by a new method of summation. *Ecology* 53 (5), 893–900.

Arnold, C. Y. 1959. The determination and significance of the base in a linear heat unit system. *Proc. Am. Soc. Hort. Sci.* 74, 430–445.

Arzt, T., and W. Ludwig. 1950. Alte Probleme der Phänologie in neuer Beleuchtung. *Met. Rdsch.* 2, 450–459.

Balvoll, G., and A. H. Bremer. 1965. Varmesum og planteavl i samband med vekst og utvikling av ymse grønsakvokstrar. (The heat-unit system and plant production in connection with growth and development of different vegetables). *Meldinger Norges Landbrukshøgskole* 44, 1–18.

Bassett, I. J., R. M. Holmes, and K. H. Mackay. 1961. Phenology of several plant species. *Can. J. Plant Sci.* 41, 643–652.

Belehradek, J. 1935. *Temperature and Living Matter*. 277 pp. Berlin: Borntraeger.

Blackman, V. H. 1919. The compound interest law and plant growth. *Ann. Bot.* 33, 353–360.

Blackman, G. E., J. N. Black, and A. W. Kemp. 1955. Physiological and ecological studies in the analysis of plant environment. X. An analysis of the effects of seasonal variation in daylight and temperature on the growth of *Helianthus annuus* in the vegetative phase. *Ann. Bot.* 19, 527–548.

Brown, D. M. 1960. Soybean ecology. I. Development-temperature relationship from controlled environment studies. *Agr. J.* 52, 493–496.

Caprio, J. M. 1966. A statistical procedure for determining the association between weather and non-measurement biological data. *Agr. Meteorol.* 3, 55–72.

Conrad, C. 1959. Über Schwankungen der Vegetationsperiode und über Variationen ausgewählter phänologischer Phasen von Kulturpflanzen. *Ang. Met.* 3, 193–205.

Dahl, E., and E. Mork. 1959. Om sambandet mellom temperatur, ånding og vekst hos gran. (On connection between temperature, respiration and growth in Norway spruce). *Medd. Skogfors. vesen* 16, 83–93.

Engel, K.-H., and A. Raeuber. 1961. Beitrage zur Phänometrie der Kartoffel. *European Potato J.* 4, 152–164.

Fritts, H. C. 1960. Multiple regression analysis of radial growth in individual trees. *Forest Sci.* 6, 334–349.

Gregory, F. G. 1952. The control of growth and development by external factors. *Rept. XIIIth. Intern. Hort. Congr.* 96–105.

Hentschel, G. 1953. Phänologisch-singuläre Prognostik der Winterroggenblüte. *Ang. Met.* 1, 285–320.

Hertlein, G. 1954. Phänologisch-meteorologische Untersuchungen zur Blühvorhersage bei der Obstgehölzen. 97 pp. Diss. Hohenheim.

Higgins, J. J. 1952. Instructions for making phenological observations of garden peas. *John Hopkins Univ. Lab. Climat. Seabrook, N.J.* 8 pp.

Hoffmann, K. 1969. Agrarmeteorologische Untersuchungen zur Standortforschung im Obstbau auf der Grundlage phänologische Beobachtungen. *Abh. Met. Dienst. DDR* 12, 1–124.

Holmes, R. M., and G. W. Robertson. 1959. Heat units and crop growth. *Can. Dept. Agric. Publ. 1042*, 3–31.

Hopkins, A. D. 1938. Bioclimatics: A science of life and climatic relations. *USDA Misc. Publ. No. 280* 188 pp.

Ihne, E. 1884. Geschichte der Pflanzenphänologischen Beobachtungen in Europa in Beiträge zur Phänologie. Giessen.

Ihne, E. 1913. Der feldmässige Gemüsebau in Grossherzogtum Hessen im Jahre 1912 nach seiner Anbaufläche und Verteilung auf die klimatische-phänologischen Zonen. *Phän. Mitt.* 31, 37.

Ivins, J. D., and F. L. Milthorpe. Ed. 1963. *The Growth of Potato.* 328 pp. London: Butterworths.

Kurihara, H., and T. Okubo. 1960. Studies on the growth process of the potato plant. *Proc. Crop Sci. Japan* 29, 117–120.

Lang, J. 1966. Der Einfluss von Klimafaktoren auf Wachstums- und Entwicklungsverlauf sowie Ertrags- und Qualitätsleistung der Braugerste, dargestellt anhand spezifischer internationaler Anbauversuche. 205 pp. (Barley Committee, European Brewery Convention). Diss. Weihenstephan.

Livingston, B. E., and G. J. Livingston. 1913. Temperature coefficients in plant geography and climatology. *Bot. Gaz.* 56, 349–375.

Mitscherlich, E. A. 1919. Das Gesetz des Pflanzenwachstums. *Landwirtsch. Jahrb.* 53, 167–182.

Newman, J. E., and J. B. Beard. 1962. Phenological observations: The dependent variable in bioclimatic and agrometeorological studies. *Agr. J.* 54, 399–403.

Nuttonson, M. Y. 1953. Phenology and thermal environment as a mean for a physiological classification of wheat varieties and for predicting maturity dates of wheat. *Am. Inst. Crop Ecol. Washington.* 108 pp.

——— 1955. Wheat climate relationships and the use of phenology in ascertaining the thermal and photothermal requirements of wheat. *Am. Inst. Crop Ecol. Washington* 388 pp.

Ottosson, L. 1958. Growth and maturity of peas for canning and freezing. *Växtodling* 9, 112 pp.

Pearce, S. C. and A. P. Preston. 1954. Forecasting the time of blossoming in apple trees from temperature records. *Ann. Rept. East Mall. Res. Sta. 1953.*

Post, J. J., and C. Kramer. 1951. Blooivoorspelling op grond van de bladontplooiing van de wilde kastanje. *De Fruitteelt* 41, 278.

Raeuber, A., and K.-H. Engel. 1966. Untersuchungen über den Verlauf der Massen zunahme bei Kartoffeln (*Sol. tuberosum* L.) in Abhängigkeit von Umwelt- und Erbeinflüssen. *Abh. Met. Dienst. DDR.* 10, 1–117.

Robertson, T. B. 1907. On the normal rate of growth of an indivudual and its biochemical significance. *Arch. Entwicklungsmechanik.* 25, 581.

Rosenkranz, F. 1951. *Grundzüge der Phänologie.* 69 pp. Wien: Fromme.

Sachs, J. 1865. *Handbuch der Experimental-Physiologie der Pflanzen.* 514 pp. Leipzig: Engelmann.

Salisbury, F. B. 1963. The *Flowering Process.* 234 pp. Oxford: Pergamon Press.

Salter, P. J. 1960. The growth and development of early summer cauliflower in relation to environmental factors. *J. Hort. Sci.* 35, 21–33.

Sanderson, F. H. 1954. *Methods of Crop Forecasting.* 259 pp. Cambridge, Mass.: Harvard University Press.

Schnelle, F. 1955. *Pflanzenphänologie.* 299 pp. Leipzig: Akademische Verlagsgesellschaft Geest & Portig.

Schrödter, H. 1952. Untersuchungen über die 'Temperatursummenregel' an Hand der phänologischen Beobachtungen in Wernigerode 1854–1884. *Ang. Met.* 1, 225–234.

Schübeler, F. C. 1886–1888. *Viridarium Norwegicum. I–II.* 610 and 587 pp. Christiania.

Seyfert, F. 1966. Ein Beitrag zur Prognose phänologischer Daten. *Abh. Met. Dienst. DDR* 10, No. 80.

Siegelman, H. W., and S. B. Hendricks. 1964. Phytochrome and its control of plant growth and development. *Advan. Enzymol.* 26, 1–33.

Strand, E. 1965. Forelesning i plantekultur ved Norges landbrukshögskole 1–111. 225 pp.

Thornthwaite, C. W. 1952. Climate in relation to planting and irrigation of vegetable crops. In *Proceedings, VIII General Assembly-XVII Congress, International Geographical Union.* pp. 290–295. Washington, D.C.

Unger, K. and A. Schneider. 1956. Über Zusammenhänge zwischen Reifeentwicklung von Gemüse-Erbsen und bestimmten mikro-meteorologischen Faktoren. *Züchter* 26, 12–22.

Utricht, A. 1956. The influence antecedent climates upon the subsequent growth and development of the sugar-beet plant. *J. Am. Soc. Sugar Beet Technol.* 9, 97–109.

Voigts, H. 1949. Wärmestundensummen und phänologische Erscheinungen *Arch. Met. Ser. B.* 1, 410–421.

Went, F. W. 1957. *The Experimental Control of Plant Growth*, 343 pp. Waltham, Mass.: Chronica Botanica Co.

West, C., G. E. Briggs, and F. Kidd. 1920. Quantitative analysis of plant growth. *New Phytol.* 19, 200–207.

Wielgolaski, F. E. 1966. The influence of air temperature on plant growth and development during the period of maximal stem elongation. *Oikos* 17, 121–141.

6.2 International Phenological Gardens in Europe The Basic Network for International Phenological Observations

Fritz Schnelle

Deutscher Wetterdienst, Zentralamt (retired)
605 Offenbach/Main

and Erik Volkert

Forstliche Fakultät Univ. Göttingen
351 Hann. Münden, W. Germany

1 Introduction

A major goal of phenology is the understanding of the effects of climate on plant development. The usual phenological network established within a region may elucidate local patterns to some extent, but the information is not precise enough to evaluate research hypotheses. A network of observers will not report on the same representatives of a species each year, so location is only vaguely indicated when the data are compared from year to year. Variations in plant development may arise from hereditary factors, as well as from factors of location. Also, every species has distinct relationships to the environment, and some are more useful than others for research purposes.

In order to minimize nonclimatic influences on plant development, a network of phenological gardens may be set up. This is not a network of observers, but a true network of plants. Each garden contains the same selection of species and the same selection of clonal strains among each species. The observers at the gardens have detailed, illustrated instructions that describe exactly the phenophases to be reported. Phenological gardens, therefore, assure a maximum of exact observations with controlled internal plant conditions. An adjacent meteorological station is essential for later correlations among the results from a phenological garden, and for possible physiological modeling.

The attractiveness of an international network of phenological gardens is evident. Because national networks are more limited in outlook, their programs

KEYWORDS: Phenological gardens; Europe; locations; plants, trees, shrubs; observations.

Table 1. International Phenological Gardens in Europe

NORWAY

1 Harstad—Kvaefjord

2 Trondhjem—Kvithamar

3 Bergen—Fana

4 Oslo-Ås-Vollebekk

5 Grimstad—Dømmesmoen

SWEDEN

6 Stockholm—Bogesund

7 Svalöf—Ekebo

FINLAND

8 Turku—Piikkiö-Yltöinen

9 Insel—Seili

10 Punkaharju

11 Oulu—Pelsonsuo

12 Apukka—Rovaniemi

13 Utsjoki—Kevo

DENMARK

14 Tylstrup

15 København—Hørsholm

16 Hofmannsgave

17 København—Højbakkegård

IRELAND

18 Dublin—National Botanical Gardens

19 Wexford—Johnstown Castle

20 Wexford—J. F. Kennedy Park

21 Kerry—Valentia Observatory

ENGLAND

22 London-Farnham—Headly P.

SCOTLAND

23 Edinburgh—Bush Estate

NETHERLANDS

24 Wageningen

BELGIUM

25 Gent—Melle

26 Bastogne—Michamps

FRANCE

27 Caen—Lieury

GERMANY, FEDERAL REPUBLIC OF (BRD)

28 Hamburg—Schmalenbeck

29 Hann. Münden—Steinberg

30 Offenbach

31 Trier

32 Stuttgart—Hohenheim

33 Stuttgart—Weilimdorf

34 Freiburg-Kaiserstuhl—Liliental

35 München—Grafrath

36 Freising—Weihenstephan

37 Hamburg—Wulfsdorf

38 Freiburg—Schauinsland

39 Donaueschingen

40 Freiburg—Feldberg

41 Freyung—Lusen

42 Freyung—Waldhäuser

43 Freyung—Klingenbrunn

44 Freyung—Schönbrunn

GERMAN DEMOCRATIC REPUBLIC (DDR)

45 Quedlinburg

46 Tharandt—Hartha

POLAND

47 Mikołajki

48 Poznán—Kórnik

49 Brzezna—Chodorowa

SWITZERLAND

50 Zürich—Birmensdorf

AUSTRIA

51 Innsbruck—Rinn

52 Wien-Stadt—Hohe Warte

53 Wien—Obersiebenbrunn

CESKOSLOVENSKA REPUBLICA (ČSR)

54 Slepĉany—Mlynany

55 Zvolen—Kysihybel-Vulh

HUNGARY

56 Sárvár

57 Gödöllö

58 Püspökladány

Table 1—*continued*

YUGOSLAVIA	BULGARIA
59 Ljubljana	67 Sofia
60 Zagreb—Krizevci	68 Velingrad—Yundola
61 Beograd—Zeleno Brdo	69 Vidin—Božuritza
62 Sombor	
63 Sarajewo—Ivan Sedlo	GREECE
64 Dubrovnik—Trsteno	70 Saloniki-Serrai—Lailia
65 Bar	
66 Skopje	PORTUGAL
	71 Porto

The names of the gardens are underlined. The nearest larger city is added where necessary to better locate the position of a garden.

Mailing addresses may be obtained from the authors or from the editor of this book. The numbers in the table refer to the numbers in Fig. 1.

are subject to change. With an international network, problems of wider interest are examined and the need for long-range data collection can be met. The gardens may be placed more strategically with international cooperation.

2 The network of Existing Gardens

Presently there are more than 70 International Phenological Gardens in 20 European countries (Fig. 1, Table 1). Clones for most of the species originated in Central Europe, but several came from extreme provenances in the border districts of Europe. This collection allows the study of large climatic differences, both in the environment and reflected in the genetic acclimatization, upon the development of plant populations. The standard selection of plants in the gardens is shown in Table 2.

Table 2. Standard selection of plants in the International Phenological Gardens.

Conifers (3 species)	Deciduous trees (9 species)
Larch (*Larix decidua*)	Birch (*Betula pubescens*)
Norway spruce (*Picea abies*)	Beech (*Fagus sylvatica*)
Scotch pine (*Pinus sylvestris*)	Gray poplar (*Populus canescens*)
	Aspen (*Populus tremula*)
Shrubs (7 species)	Bird cherry (*Prunus avium*)
Alpine currant (*Ribes alpinum*)	Oak (*Quercus robur*)
Various willows: *Salix acutifolia*	Black locust (*Robinia pseudoacacia*)
Salix aurita	Mountain ash (*Sorbus aucuparia*)
Salix glauca	Linden (*Tilia cordata*)
Salix smithiana	
Salix viminalis	
Black elder (*Sambucus nigra*)	

3 Expected Results

These gardens make possible not only exact investigations of phenological-climatic relationships, but also of agricultural and silvicultural problems. Data taken from the gardens may serve to identify ecophysiographic regions for resource management purposes, based on seasonal rhythms in different parts of a continent.

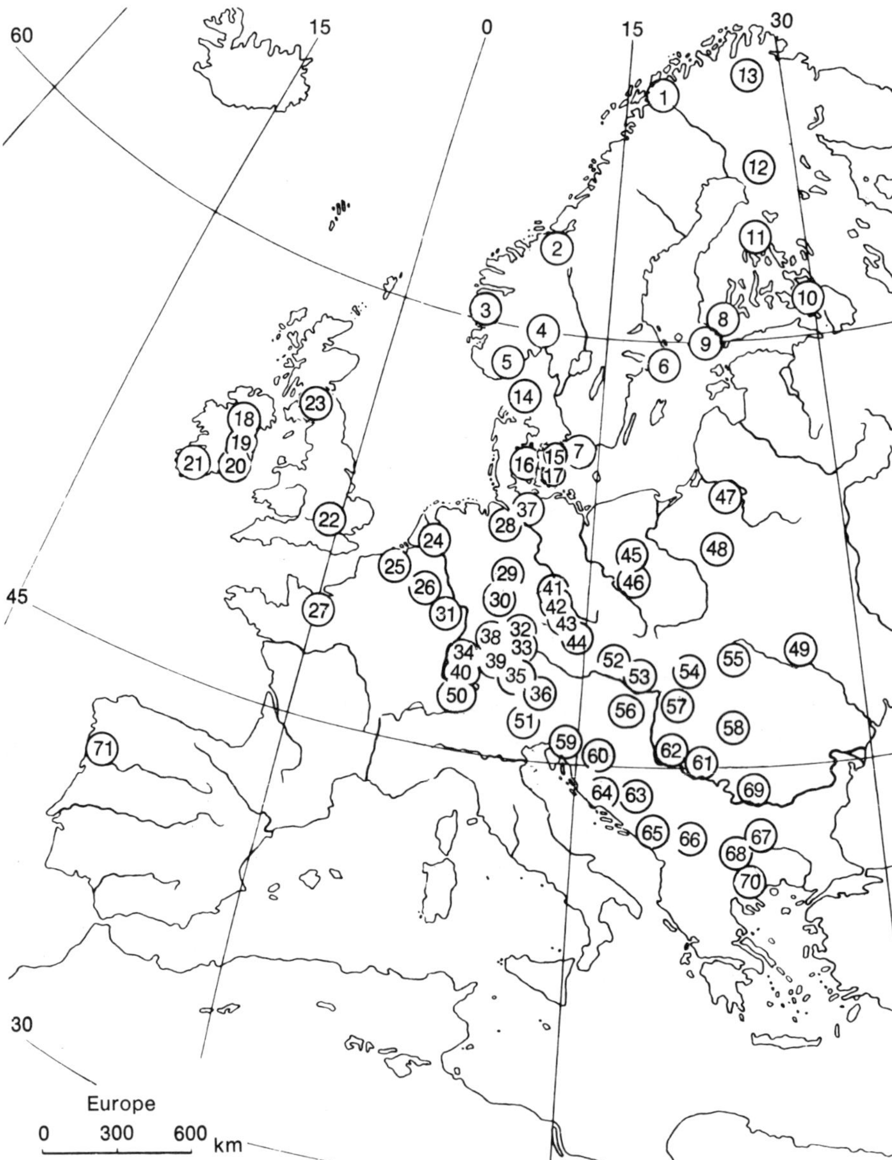

Fig. 1. Locations of the International Phenological Gardens. The numbers refer to the names in Table 1.

A forest botanist may be interested in the acclimatization of the needles of conifers under various environmental conditions (see Chapter 6.3). Timber specialists expect information on the effects of the environment on timber quality. Geneticists, especially forest geneticists, study a wide variety of congenital physiological properties from plants of the same hereditary disposition planted at different locations, and from plants of different provenances put together at the same locations.

In such studies, the longer the series of observations is, the more valuable and conclusive are the results. Ten to twelve years of observations exist for the oldest gardens, and five to eight years for later additions to the network. The results of observations from all gardens are published as yearly reports in *Arboreta Phaenologica, Information of the Working Group of International Phenological Gardens*, and are sent to all interested persons.

References

Schnelle, F., and E. Volkert. 1964. *Internationale Phänologische Gärten*. Vol. 1, No. 1, pp. 22–29. Amsterdam: Agricultural Meteorology.

———. Report on the results of phenological observations in the International Phenological Gardens in Europe.

Arboreta Phaenologica. Yearly. Information of the Working Group of International Phenological Gardens. Hann. Münden and Offenbach.

6.3 Genetic Variation in Some Phenological Characteristics of Forest Trees

Hans Nienstaedt

North Central Forest Experiment Station
Rhinelander, Wisconsin 54501, USA

1 Introduction

During the night between June 9 and 10, 1972, the temperature at Rhinelander dropped to −3.5°C. Severe frost damage of new shoot growth of both conifers and hardwoods resulted (Plate 1). Since that dismal June night we have been evaluating the damage to different spruce species, populations within species, and progencies of individual parent trees within populations. At all levels there are significant differences in the amounts of damage. Some differences are caused by actual resistance to frost; the actively growing tissues can withstand freezing temperatures without showing damage (Kiellander, 1953, 1962), but much of the variation is due to frost avoidance. Seasonal development differs among types, and since the phenophases differ in susceptibility to frost, the differences in damage can be seen. Thus frost damage is closely related to the time of the initiation of growth.

The genetic variation in phenological characteristics of individual trees, clones, or families of related individuals has been studied in only a few species. Significant variation of up to three weeks has been reported in the time of budbreak in some species; where it has been possible to determine heritabilities, estimates have been high, ranging from $h^2 = 0.30$ to about $h^2 = 0.90$.

Information on height growth cessation has rarely been considered from the standpoint of genetic variation among individual trees. One study of *Picea glauca* indicated three weeks between the date the first and the last clone stopped height growth; repeatability was high.

The amount of growth in a particular season is determined by the rate and duration of growth. The cessation of shoot and cambium activity is a determining factor, and the more fully the plant can utilize the growing season without suffering from spring and fall frost, the greater the amount of annual growth, final harvest,

---

KEYWORDS: Population variation; clines; ecotypes; races; variation within populations; heritability; growth initiation; growth cessation; temperature effects; photoperiodic effects; growth prediction.

and return on the investment in planting stock, planting, and maintenance. Much of the interest in forest tree phenology stems from concern for these practical questions. They are considerations of special importance in long-lived crops in which the inability of the plant to use the growing season fully accumulates from year to year until harvest—a period of perhaps 30 to 100 years.

Forest biologists have studied many other aspects of tree phenology, such as characteristics relating to seed production and harvest. This has included studies of time of pollen shedding and female receptivity, as well as seed shedding itself. However, for the reasons suggested above, the bulk of the work has been related to the periodicity of vegetative growth. In this paper the emphasis will be on initiation and cessation of growth. These characteristics will be discussed in terms of genetic variation among populations within species and variation among individual trees within populations. The magnitude of the differences, and particularly the patterns of variation, will be discussed. With few exceptions the discussion will relate to North American species.

2 Variation Patterns in Phenological Characteristics Among Populations of Trees

Growth Initiation

Studies of transplanted tree populations have given some indication of common patterns of variation. Sugar maple (*Acer saccharum* Marsh.) exemplifies one of them. In a study at Wooster, Ohio, of 18 populations originating as far north as northern Wisconsin and as far south as Georgia, the northern collections started growth first. Gradually they were followed by more southerly seed sources, and plants from the most southern reaches of the range in North Carolina, Oklahoma, and Georgia started growth last. The variation was continuous or clinal in nature (Kriebel, 1957) a common pattern of variation. Yellow birch from northern sources generally flushes early (Clausen and Garrett, 1969); for example, seedlings of Quebec seed flushed out approximately four weeks before seedlings raised from Tennessee seed.[1] A similar pattern was shown in *Pinus strobus* L. (Mergen, 1963).

Other species show a reversed trend: Trees from southern sources flush first and are gradually followed by those from more northerly seed sources. Bey and co-workers have demonstrated this pattern in black walnut (*Juglans nigra* L.). In one study, seedlings of seed sources from Texas flushed as much as 10 days earlier than seedlings of Minnesota origin (Bey *et al.*, 1971a). At one location in southern Illinois another study of one-year-old seedlings from 20 sources indicated that flushing is delayed one day for every 63 mi north of the planting site seed has been collected (Bey *et al.*, 1971b). And in tests at five locations in Illinois, Ohio, and Michigan, flushing began, on the average, one day earlier for every 85 mi south of the planting site seed was collected. In all plantings, the regressions for date of leaf flush over latitude were highly significant (Bey, 1972).

[1] Data on file, U.S. Forest Service, Institute of Forest Genetics, Rhinelander, Wisconsin.

Plate 1. Severe frost injury on Norway spruce resulting from late spring frost. The four trees in the immediate foreground and center left are an early flushing provenance from Salzburg, Austria. They show heavy damage to new growth, which is shrivelled, drooping and red (the defoliation of the tops is unrelated damage caused by winter desiccation). The late flushing Ural Mts., U.S.S.R. provenance in the center background and right, escaped damage.

Similar early flushing of the seedlings from more southern origins has also been demonstrated in studies of yellow poplar (*Liriodendron tulipifera* L.) (Sluder, 1960; Farmer *et al.*, 1967), sweetgum (*Liquidambar styraciflua* L.) (Winstead, 1968), and American sycamore (*Platanus occidentalis* L.) (Schmitt and Webb, 1971), and in some but not all studies of cottonwood (*Populus deltoides* Bartr.). Eldridge *et al.* (1972) concluded that southern seed foliated earlier, but another study of cottonwood seedlings from six locations from Louisiana to Wisconsin followed the pattern expected for seed sources as far north as southern Illinois. There was essentially no difference between material from northern and southern Illinois, but Minnesota-Wisconsin sources, in one year of observation, were earlier than material from Mississippi and the following year were as early as material from Missouri.[2]

Considering that variation in time of flushing is adaptive in nature, the north-south and south-north variation patterns clearly are oversimplifications. This is particularly evident in a study of sitka spruce *Picea sitchensis* (Bong.) Carr. by Burley (1966a, 1966b). In 47 provenances representing a latitudinal spread from California (41° 34′ N) to Alaska (60° 45′ N), there was no direct relation between latitude, per se, and flushing during the spring one year after seeding. However, "when flushing is considered in relation to the nature of the spring temperature distribution at the point of seed origin, a systematic pattern of flushing is observed among the seed sources." The difference between the earliest and the latest average day of flushing was 15 days. Burley divided the seedlings into nine groups on the basis of seed source origin; five of these groups fall into a north-south pattern, the southern sources being late flushing. The exceptions are the two most southerly sources from California where spring frosts are rare. Burley suggests that in this group "there has been little selection for late flushing." The seventh group, four sources from Alaska, shows a spread of 10 days. These sources may represent two early flushing coastal types evolved in regions protected from low temperatures in late spring, and two late flushing provenances adapted to regions exposed to flows of cold air from inland ice fields. The two remaining groups similarly represent provenances adapted to distinct local climates.

In *Liriodendron tulipifera* L. budbreak was correlated with length of growing season ($r = 0.90$) and with date of last killing frost ($r = 0.81$), both functions of temperature. The correlation with latitude, which is closely related to photoperiod, was only $r = 0.57$ (Sluder, 1960).

Morgenstern (1969a) analyzed populations of black spruce, *Picea mariana* (Mill.) B.S.P., and showed that temperature and temperature timing-related characteristics of the seed origin accounted for the greatest amount of variation in the timing of bud flushing. The pattern of growth initiation was considered clinal (Morgenstern, 1969b). Burley (1966a) has suggested that selection against early flushing does not take place in continental climates where the spring rise in temperature occurs rapidly and late, and subsequent spring temperatures rarely fall below freezing. However, Morgenstern (1969a, 1969b) worked exclusively with provenances that have evolved in a typical continental climate characterized by

[2] Data on file, U.S. Forest Service, Southern Hardwoods Laboratory, Stoneville, Mississippi (courtesy Dr. C. B. Briscoe).

frequent high-pressure systems of Arctic air and severe night frosts occurring as much as a month after the general warming trend has brought about growth initiation.

Douglas fir is a much studied species, yet the relation between the various phenological stages and seed source is still not clear. Sweet (1965) pointed out an interesting relation between burst of terminal versus side buds. In all the 32 provenances he studied in the nursery at New Zealand, the terminal bud bursts after the lateral buds. The lag between the two ranged from 1 to 23 days. No relations could be demonstrated between the climate or geography of the seed source and bud burst of either terminal or lateral buds. The time lag between the two bud types, however, did relate to length of frost-free season, date of last spring frost, mean temperature of the coldest month, and the altitude of the seed source. The more variable the date of last spring frost in a provenance, the greater was the time lag between lateral and terminal bud burst. Sweet suggests that the time lag has evolved as a survival mechanism in response to the selective action of late spring frosts.

Temperature is one of the environmental factors that controls the beginning of vegetative growth, but it is probably not the only one. Douglas fir shows considerable variation in flushing—three weeks in a study of sources from New Mexico and as far north as Montana and Vancouver Island (Irgens-Moller, 1968)—and although seedlings from high-altitude seed sources generally are later, there is no clearly identifiable pattern (Ching and Bever, 1960). Irgens-Moller (1967) suggested that earliness of one source was an adaptation to summer drought, and lateness in other material an evolved avoidance of a *Botrytis* spp. fungus that attacks new foliage in early spring when air humidity is high. While the latter idea is not as far-fetched as it may sound, it is the only suggestion I have found for such a relation. In loblolly pine, for instance, there is no relation between fusiform rust infection and the phenology of budbreak.[3]

Long photoperiods can hasten bud bursting in material that has received insufficient chilling, as has been demonstrated in white spruce (Nienstaedt, 1966) and other species. Other than studies of photoperiod as a factor compensating for chilling, there is no critical research demonstrating the effect of photoperiod on bud opening. Still an effect cannot be ruled out. Pauley and Perry (1954) pointed out that light, per se, is required for budbreak in some species, but concluded that in *Populus* neither light nor its periodicity appears to be directly concerned in the break of dormancy. Morgenstern (1969a), in his principal components analysis, considered the temperature regime, photoperiod, and soil moisture, and definitely concluded that temperature was the major factor determining the initiation of growth in black spruce. Irgens-Moller (1957) concluded that Douglas fir seedlings from high-altitude seed sources are sensitive to photoperiod, while those from low-altitude seed sources are not. As a result, the long photoperiod seed source seedlings would flush later in the season when the danger of night frost is low. An effect of photoperiod has been suggested by other authors such as Kaszkurewicz and Fogg (1967), working with cottonwood and American sycamore. Critical studies unequivocally establishing the possible role of photoperiod and its inter-

[3] Data on file, U.S. Forest Service, Institute of Forest Genetics, Gulfport, Mississippi, made available by Dr. Osborn O. Wells.

action with temperature are needed, and possible studies designed to determine the relations will be discussed later.

Growth Cessation

Cessation of growth also varies among populations within species. The magnitude of the differences generally is greater than that for initiation. Thus, while growth initiation may vary as much as a month, differences in growth cessation of almost four months have been recorded for populations of species such as *Populus trichocarpa* Torr. and Gray (Pauley and Perry, 1954). The patterns of variation, however, are more uniform. Trees from southern or low-altitude seed sources, which have evolved in the milder climates, are the last to set buds in the fall. Transplanting studies have shown this to be the case for all species on the following partial list:

Acer saccharum Marsh. (Kriebel, 1957)
Juglans nigra L. (Bey *et al.*, 1971b)
Larix laricina (Du Roi) K. Koch (Rehfeldt, 1970)
Liquidambar styraciflua L. (Williams and McMillan, 1971; Winstead, 1968)
Liriodendron tulipifera L. (Sluder, 1960)
Picea sitchensis (Bong.) Carr. (Burley, 1966a, 1966b)
Picea mariana (Mill.) B.S.P. (Morgenstern, 1969b)
Pinus banksiana Lamb. (Holst and Yeatman, 1961)
Pinus contorta Dougl. (Critchfield, 1957)
Pinus elliottii Engelm. var. *elliottii* (Bengtson *et al.*, 1967)
Pinus strobus L. (Mergen, 1963)
Populus deltoides Bartr. (Pauley and Perry, 1954)
Populus tacamahacca Mill. (Pauley and Perry, 1954)
Populus trichocarpa Torr. and Gray (Pauley and Perry, 1954)
Pseudotsuga menziesii (Mirb.) Franco (Irgens-Moller, 1958, 1967, 1968; Sweet, 1965)
Quercus rubra L. (Kriebel, 1965)
Tsuga canadensis (L.) Carr. (Nienstaedt and Olson, 1961)

Photoperiod has been regarded as the major factor in the control of the cessation of growth. Morgenstern's (1969a) analysis showed that daylength and growth cessation were continually coupled. He emphasized the agreement with results of other investigators, and concluded that in *Picea mariana* cessation of growth also depends largely on photoperiod. Southern provenances of most species, when they are moved north, respond to the increased daylength by growing longer into the fall in spite of low temperatures. The result may be injury from frost.

In all species there is pronounced scattering around the regression line for date of growth cessation plotted over latitude. Burley (1966a), for example, determined an r value of -0.862 in *Picea sitchensis*. This may indicate that factors other than photoperiod also are involved. Morgenstern (1969a) considered temperature the most important secondary factor in *Picea mariana*. Burley (1966a) also interpreted some of the variation in his sitka spruce material as genetic adaptation to unique local temperature conditions resulting from the flow of cold air from Alaskan glaciers. The overall pattern of variation in *Picea sitchensis* is clinal, but ecotypes may have been developed in response to local discontinuous environmental conditions; and smaller clines at one latitude may have developed in response to factors confined to a particular locale and varying continuously, e.g., altitudinal gradients (Burley, 1966a). Few studies are as detailed as Burley's, and most investigators of even the more detailed studies have considered only very general variation patterns. However, as Burley's discussion suggests, the variation patterns are probably very complex.

3 Variation Patterns in Phenological Characteristics Among Individual Trees Within Populations of Trees

Important Variation in an Individual Tree

Before discussing the genetic variation in growth initiation and cessation among individual trees within a population, it may be pertinent to mention a few important phenological characteristics of the individual tree itself.

In general, buds open later and later as the young tree matures. In Douglas-fir, for example, growth began progressively later during four years of observations of trees two and eight years old (Irgens-Moller, 1967). In that species young plants may be as much as three weeks earlier than older ones (Irgens-Moller, 1957). Eventually, stable conditions are reached as mentioned for *Picea abies* (L.) Karst, by Büsgen and Münch (1929); a gradual retardation develops with aging up to 16 to 25 years, at which time flushing may be delayed at least four weeks.

All the buds within the crown of an individual tree do not open at the same time. The lag time in Douglas-firs terminal bud shows a distinct variation pattern as already mentioned (Sweet, 1965). In *Tsuga heterophylla* (Raf.) Sarg. leaders start growing before laterals in most individuals; in *Thuja plicata* D. Don growth commences simultaneously throughout the crown of an individual, while *Pinus monticola* Dougl. follows the pattern of Douglas-fir (Walters and Soos, 1963). Late flushing of the terminal bud is also typical for Norway spruce (*Picea abies*) (Büsgen and Münch, 1929), white spruce, and many other conifers. In Douglas-fir the lag time may be as much as 23 days; it functions as a protective mechanism against frost (Büsgen and Münch, 1929; Sweet, 1965).

Initiation of Growth

In a number of species the variation in initiation of growth among individual trees within a population is large and under strong genetic control. In two consecutive years, clones of seven Douglas-fir trees flushed in the same sequence. The genetic component accounted for 94 and 96% of the variation in the time of bud breaking in the two years; nearly all the variation was genetic. In three clones from one population, the difference between the early and the late clone was two weeks (Silen, 1962). The magnitude of the difference can change from year to year; over a 14-year period, dates of growth initiation for two trees at Portland were separated by as much as 12 days and as little as two days (Morris *et al.*, 1957).

In *Picea mariana* the mean date of budbreak of 121 families from nine different provenances differed by 23 days. Sixty-one percent of the total variance was accounted for by provenance variation and 8% by family variation; both variances were significant at the 0.1% level (Morgenstern, 1969b).

Two to three week differences in growth initiation have been observed among *Tsuga heterophylla* individuals (Walters and Soos, 1963).

Heritability (h^2) estimates are available for a few species. In black walnut one study determined $h^2 = 0.67$ for 17 open-pollinated families representing six provenances (Bey *et al.*, 1971a); another study found $h^2 = 0.67$ and 0.96 in two different plantings (Bey, 1972). These values are similar to those determined for white spruce, *Picea glauca* (Moench) Voss: h^2 ranged from 0.278 to 0.706, depending on experimental conditions, age, and method of computations (Nienstaedt, in press; Nienstaedt and King, 1970).

Growth Cessation

The total growth period from flushing to cessation, first of height growth and subsequently of cambial activity, has been studied on an individual tree basis in many North American species, but little information on genetic variation is available. In *Populus*, Pauley and Perry (1954) showed large differences among individual clones originating at the same latitude, but it is not clear from their data to what extent the differences reflect differences between populations or difference among individual clones within populations. They did show consistent behavior of the individual clones from year to year, and that growth cessation was delayed when the photoperiod was artificially extended.

Current research with 16 late flushing and nine early flushing grafted clones of *Picea glauca* indicates large genetic differences. Approximately three weeks separate the first and the last clone to stop growth (growth cessation taken as the date at which 95% of the height growth is completed). The range changes little from year to year: from 19 to 23 days over four years.[4] Repeatability from year to year is high with the correlation coefficient exceeding $r = 0.84$ (d.f. 23) in all cases. The date of flushing of the 25 clones showed differences of similar magnitude (from 17 to as much as 21 days) and repeatability. Early flushing clones also terminated growth early so the period of growth is similar for the two groups, i.e., 31.7 to 44.1 days for the late flushing clones and slightly longer, 35.3 to 47.2 days, for the early flushing clones.[5]

4 Phenological Observations Used as Predictors of Growth Responses in Forest Trees

Recommendations for the distribution of seed of selected fast-growing populations have been made on the basis of transplanting studies, and seed zones have been

[4] Data on file, U.S. Forest Service, Institute of Forest Genetics; Rhinelander, Wisconsin.
[5] *Ibid.*

delineated in order to prevent losses from planting of seedlings from poorly adapted seed sources. In nearly all cases, the recommendations have been based on observations in established tests, and have been used only in areas circumscribed by the tests. Data have not been used to develop predictors, per se.

An attempt at prediction has been made on the basis of the *Picea glauca* clones described above, and controlled pollinated progenies observed under growth room and greenhouse conditions (Nienstaedt and King, 1970). Using heritability estimates for the progenies and degree-day requirements for the flushing of parent clones, the reduction in frost damage that may be achieved through breeding was determined. Degree-day (d.d.) requirements ranged from 663 d.d. (above +40°F) for the latest clone to 384 d.d. for the earliest, with a mean value of 519 d.d. for all clones. Selecting the two best clones with a mean requirement of 645 d.d., the selection differential (i) was 125 d.d..With heritability estimated at $h^2 = 0.705$, the genetic gain (ih^2) is estimated at 89 d.d.

Table 1. Number of times during the years of record that a temperature of − 2.2°C[a] or lower occurred later than the date the degree-day class was reached.

Degree-Days[b]	Laona, Wis. (15 years of records)	Long Lake, Wis. (20 years of records)
480–499	7	7
500–519	7	6
520–539	5	5
540–559	5	5
560–579	4	5
580–599	4	5
600–619	4	4
620–639	4	4
640–659	4	4
660–679	3	2
680–699	2	2

[a] The actual temperature at which injury to newly flushed white spruce shoots occurs is not known. It is probably not above − 2.2°C (28°F).
[b] Degree days in °F above base 40°F (4.4°C).

A table of the probability of spring frosts occurring after a certain number of degree-days had accumulated was constructed for selected weather stations (Table 1). On the basis of the table, for a clone that flushes after 515 d.d. have accumulated, frost injury could be expected 7 out of 15 years at Laona, and 6 out of 20 years at Long Lake, Wisconsin. This is comparable to the average for all cones (519 d.d.). Breeding with the two selected clones may, on the basis of the estimated genetic gain, bring this up to 608 d.d. Risk of frost (using the data in Table 1) would be reduced to 4 out of 15 years at Laona and 4 out of 20 years at Long Lake. Additional data are now being gathered on clones as well as progenies to verify these preliminary predictions.

5 An Evaluation of Past Phenological Observations in Provenance Studies and Future Research Needs

Most studies of populations of North American species have tried to establish gross patterns of variation by relating yield characteristics, such as total height or diameter growth, dry matter production or wood specific gravity, and in fewer cases phenological characteristics, to readily available climatic parameters, soil characteristics, or geographic components. The independent variables of the environment have been relatively crude, i.e., average temperatures for certain months or seasons, total or seasonal precipitation, pH or the nature of soil parent material, latitude, altitude, aspect, and in some cases geographic isolation such as individual watersheds in mountainous country. They have demonstrated great genotypic variation in all the measured characteristics in almost all species; variation commonly is adaptation to local environmental conditions. The selective forces that have molded species into large-scale patterns of clines or races can be overruled by local forces that can select adapted genotypes with responses that deviate from the overall variation patterns. It is essential, therefore, that phenological studies take genetic variation of native tree species into consideration.

The studies have indicated that, although temperature plays a major role in the control of phenological response during the spring, other factors may also be important. One of these, photoperiod, varies systematically with season and latitude. For phenological observations on native trees to be reliable indicators and predictors of seasonal events, the effects of the interacting environmental factors must be understood. This will require the use of better parameters for the environment. Mean temperature for months and seasons will not be enough; some expression of accumulated biologically effective heat will be needed. The *Picea glauca* studies discussed above have indicated how the use of degree-days may lead to more practical interpretations of genetic information. Other studies that cannot be discussed because of space limitations suggest other applications of heat summations in forest biology (Sarvas, 1967, 1970a, 1970b; Boyer, 1970; Dietrichson, 1964).

The review of the literature on phenological observations in provenance studies has suggested some of the limitations to the use of heat summation:

1. Other environmental factors besides temperature may also be involved; of these, photoperiod, because it varies systematically with season and latitude, is important. Research is needed to establish:
   a. Whether the seasonal increase in photoperiod can modify heat requirements during the spring.
   b. Whether the latitudinal increase in photoperiod can modify heat requirements.

   Answering these questions should also clarify the interpretation of Linsser's principle in "marginal regions" (for details see Sarvas, 1970a).
2. Heat requirements change with age in trees. Research is needed to determine the nature of this change.
3. Wang (1960) has raised other questions regarding the heat-unit approach,

such as the determination of the best base temperature and problems relating to biological reaction rates, particularly at high and low temperatures. These problems need clarification as they relate to forest trees.

There is a need for critical phenological observations in the major existing provenance tests. This will require establishment of continuously recording weather stations on the test sites and, preferably, of several test sites along climatic (photoperiodic) gradients. These studies should be supplemented with *in situ* studies of the provenances and, where possible, studies under controlled environmental conditions. In addition to budbreak and cessation of height growth, observations should include the period of maximum height growth and the termination of cambial activity. Results should be used to develop precise predictors of the possible range extensions for the high-yielding provenances.

References

Bengtson, G. W., W. H. D. McGregor, and A. E. Squillace. 1967. Phenology of terminal growth in slash pine: Some differences related to geographic seed source. *Forest Sci.* 13, 402–412.

Bey, Calvin F. 1972. Leaf flush in black walnut at several midwest locations. *Proc. 19th Northeastern Forest Tree Improvement Conf.* 47–51.

———, N. L. Hawker, and P. L. Roth. 1971a. Variations in growth and form in young plantation black walnut trees. *Proc. South. Forest Tree Improvement Conf.* 11, 120–127.

———, J. R. Toliver, and P. L. Roth. 1971b. Early growth of black walnut trees from twenty seed sources. *U.S. Forest Serv. Res. Note NC–105*, 4 pp.

Boyer, W. D. 1970. Shoot growth patterns of young loblolly pine. *Forest Sci.* 16, 472–482.

Büsgen, M., and E. Münch. 1929. *The Structure and Life of Forest Trees.* 436 pp. New York: John Wiley.

Burley, J. 1966a. Genetic variation in seedling development of Sitka spruce, *Picea sitchensis* (Bong.) Carr. *Forestry* 39, 68–94.

——— 1966b. Provenance variation in growth of seedling apices of Sitka spruce. *Forest Sci.* 12, 170–175.

Ching, K. K., and D. Bever. 1960. Provenance study of Douglas-fir in the Pacific Northwest Region. I. Nursery performance. *Silvae Genetica* 9, 11–17.

Clausen, K. E., and P. W. Garrett. 1969. Progress in birch genetics and tree improvement. In *Birch Symp. Proc.* pp. 86–94. U.S. Forest Service; Northeastern Forest Expnt. Sta.

Critchfield, W. B. 1957. Geographic variation in *Pinus contorta. Maria Moors Cabot Foundation Publ. 3*, 118 pp.

Dietrichson, J. 1964. Proveniensproblemet belyst ved studier av vekstrytme og klima. *Medd. Norske Skogforsøksvesen* 19, 499–656.

Eldridge, K. C., A. R. Rout, and J. W. Turnbull. 1972. Provenance variation in the growth pattern of *Populus deltoides. Australian Forest Res.* 5, 45–50.

Farmer, R. E., Jr., T. E. Russell, and R. M. Krinard. 1967. Sixth-year results from a yellow-poplar provenance test. *Proc. Ninth Southern Conf. on Forest Tree Improvement.* 65–68.

Holst, M. J., and C. W. Yeatman. 1961. A provenance study in *Pinus banksiana* Lamb. *Recent Advan. Bot.* 2, 1612–1616.

Irgens-Moller, H. 1957. Ecotypic response to temperature and photoperiod in Douglas-fir. *Forest Sci.* 3. 79–83.

——— 1958. Genotypic variation in the time of cessation of height growth in Douglas-fir. *Forest Sci.* 4, 325–330.

——— 1967. Patterns of height growth initiation and cessation in Douglas-fir. *Silvae Genetica* 16, 56–58.

——— 1968. Geographical variation in growth patterns in Douglas-fir. *Silvae Genetica 17*, 106–110.

Kaszkurewicz, A., and P. J. Fogg. 1967. Growing seasons of cottonwood and sycamore as related to geographic and environmental factors. *Ecology 48*, 785–793.

Kiellander, C. L. 1953. Om kontinentgranen. Härdighet, rasförädling och fröförsörjning. *Svensk Papperstidning No. 23 and 24.* (Reprint, 34 pp.).

——— 1962. *Picea, Abies, Pseudotsuga.* In *Handbuch der Pflanzenzüchtung.* 2 Ed., VI, pp. 854–873. Berlin and Hamburg: Verlag Paul Parey.

Kriebel, H. B. 1957. Patterns of genetic variation in sugar maple. *Ohio Agr. Exp. Sta. Res. Bull. 791*, 56 pp.

——— 1965. Parental and provenance effects on growth of red oak seedlings. *Proc. 4th Central States Forest Tree Improvement Conf.* 19–25.

Mergen, F. 1963. Ecotypic variation in *Pinus strobus* L., *Ecology* 44, 716–727.

Morgenstern, E. K. 1969a. Genetic variation in seedlings of *Picea mariana* (Mill.) B.S.P. I. Correlation with ecological factors. *Silvae Genetica* 18, 151–161.

——— 1969b. Genetic variation in seedlings of *Picea mariana* (Mill.) B.S.P. II. Variation patterns. *Silvae Genetica* 18, 161–167.

Morris, W. G., R. R. Silen, and H. Irgens-Moller, 1957. Consistency of bud bursting in Douglas-fir. *J. Forestry* 55, 208–210.

Nienstaedt, H. 1966. Dormancy and dormancy release in white spruce. *Forest Sci.* 12, 374–384.

——— Degree day requirements for budbreak in white spruce—Variation and inheritance. *Proc. 8th Central States Forest Tree Improvement Conf.* In press.

——— and J. S. Olson. 1961. Effects of photoperiod and source on seedling growth of eastern hemlock. *Forest Sci.* 7, 81–96.

——— and J. P. King. 1970. Breeding for delayed budbreak in *Picea glauca* (Moench) Voss. Potential frost avoidance and growth gains. *Second World Consultations of Forest Tree Breeding, FAO Proc.* I, 61–80.

Pauley, S. S., and T. O. Perry. 1954. Ecotypic variation of the photoperiodic response in *Populus. J. Arnold Arboretum* 35, 167–188.

Rehfeldt, G. E. 1970. Genecology of *Larix laricina* (du Roi) K. Koch in Wisconsin. I. Patterns of natural variation. *Silvae Genetica* 19, 9–16.

Sarvas, R. 1967. Climatological control of flowering in trees. *Proc. 14th IUFRO Congr.* III, Sect. 22, 15–30.

——— 1970a. Genetical adaptation of forest trees to the heat factor of the climate. *Second World Consultations of Forest Tree Breeding, FAO, Proc.* I, 187–202.

——— 1970b. The annual developmental cycle of forest trees. IUFRO Sect. 22, Working Group Meeting. *Proc. Sexual Reproduction of Forest Trees.* 16 pp.

Schmitt, D. M., and C. D. Webb. 1971. Georgia sycamore seed sources in Mississippi plantings: Site adaptability a key factor. *Proc. Eleventh Conf. on Southern Forest Tree Improvement*, 113–119.

Silen, R. R. 1962. A study of genetic control of bud bursting in Douglas-fir. *J. Forestry* 60, 472–475.

Sluder, E. R. 1960. Early results from a geographic seed source study of yellow poplar. U.S. Forest Service; *Southeastern Forest Exp. Sta. Res. Note No. 150*, 2 pp.

Sweet, G. B. 1965. Provenance differences in Pacific Coast Douglas-fir. I. Seed and seedling characteristics. *Silvae Genetica* 14, 46–56.

Walters, J., and J. Soos. 1963. Shoot growth patterns of some British Columbia conifers. *Forest Sci.* 9, 73–85.

Wang, J. Y. 1960. A critique of the heat unit approach to plant response studies. *Ecology.* 41, 785–790.

Williams, G. J. III, and C. McMillan. 1971. Phenology of six United States provenances of *Liquidambar styraciflua* under controlled conditions. *Am. J. Bot.* 58, 24–31.

Winstead, J. E. 1968. Ecotypic differentiation in *Liquidambar styraciflua* L. IX + 138 pp. Ph.D. Thesis. Austin: University of Texas.

6.4 Using Models of Seasonality

Paul E. Waggoner

Connecticut Agricultural Experiment Station
New Haven, Connecticut 06504, USA

1 Introduction

The fascination of the unfolding blossom or the emerging chrysalis cannot be doubted; it is the compensation of the countless volunteers who man phenological networks. In addition to fascination, however, one can ask if phenology and models of seasonality have utility. This paper, therefore, first provides some prerequisites for utility and then cites some examples.

A voluminous account of phenological applications compiled by Holmes and Robertson (1959) points out that in many crops, for example, strawberries, the flavor and aroma, which depend upon time of harvest rather than on sheer quantity, determine value. They also say that the simple remainder index or degree-day system of heatsums has been enthusiastically received, perhaps because growth increases nearly linearly for many crops over the range of temperatures encountered in temperate regions. The enthusiastic reception may also be increased by the ease of obtaining the climatic degree-day sum for a locality from the mean temperature (Thom, 1954).

Holmes and Robertson then reported surveys and correspondence concerning application of heat unit models between 1949 and 1957. They related critical heatsums found by practical people for peas, corn, and beans. In 1949, about a third of the pea and sweet corn acreage in the United States was scheduled so that a series of fields would provide an orderly supply of exactly mature and flavorable produce for processing. Anticipating the maturity of fruit seems a mixed business. In some cases the period from bloom to maturity was fairly constant and constituted the best index, while in other cases mean temperature during the period seemed important. Holmes and Robertson summarized these investigations but cited no specific applications.

Four prerequisites for the practical use of models of seasonality come to mind. Obviously the subject of the model must have considerable importance to mankind. The importance can be manifested economically as in the case of food or legally

KEYWORDS: Model applications; heatsum; crops; pest control; planting; weather data; probabilities; climatology.

as in pesticide schedules. A phenological model concerning, e.g., pea planting or one concerning an insecticide schedule that avoids the contamination of apples certainly is useful.

The second prerequisite to utility is variability in the phenophase that is modeled. If the phenophase always occurs on the same date or if it is almost of constant length, as pregnancy in many mammals, this information is undoubtedly important, but its knowledge has been at hand longer than science itself; phenology scarcely needs intervene. In the plants and insects that have no warm blood and hence are wards of the environment, on the other hand, phenophases are often variable, and the complications of a phenological model relating phase to environment may be worthwhile.

Needless to say, a phenological rule must be accurate to be useful. A rule that employs several independent variables and is barely statistically significant may have academic interest, but practical utility may require both simplicity and a resounding improvement in the ability to anticipate a phenophase.

The possibility of action is the fourth aid, if not a prerequisite for utility. Thus, knowing when the main brood of the destructive Hessian fly will emerge is useful because the sowing of winter wheat can be delayed. The likelihood of useful action is, of course, increased if a forecast is made. Thus a rule that forecasts the date of full bloom of apricots about three months in advance (Buys and Kotze, 1963) permits more action than a heatsum that reaches a critical value on the same day that the orchardist looks out the window and sees the blossoms.

This list of prerequisites or aids to utility suggests that most applications of phenology will be found in crop production. Thus, not only have my examples in the modeling section (see Chapter 5.1) concerned plants and their pests, but the applications that follow will also concern crops and their pests.

2 Applications in Pest Control

In 1914 in Illinois an unusually hot and dry summer accelerated the development of a destructive insect infesting apples. The sprays for controlling this insect, the codling moth, were effective only if applied early in the larval period. The accelerated development put the standard spray program out of step with the larval periods, and heavy losses were suffered by the "most intelligent and careful" orchardists. Subsequently a serious study of the changing development rate of the insect under different environments was undertaken. Development of eggs, pupae, and larvae was observed at controlled temperatures and humidities, and the speed of development was calculated. A chart of developmental units as a function of temperature and humidity was presented, together with rules for predicting the phenophases critical in spraying (Shelford, 1927).

The importance of the problem and the thoroughness of the investigation, the report, and the explanation of application seem exemplary, even a half century later. Nevertheless, the method seems not to have been used generally (Uvarov, 1931). Perhaps the complications of the calculations were too forbidding. Whatever the reasons for Shelford's work not being widely applied, it is a standard for other phenologists. A phenologist who dreams of applications but who is not deluded must plan something different from Shelford's exemplary work.

Something different has apparently been done in the biological control of insects (Messenger, 1970). If a pest insect is to be controlled by a parasitic insect that destroys the pest, the parasite obviously must prosper in the environment of the pest. "But when both host and natural enemy undergo diapause, there occurs the added complication of life cycle synchrony, which is very often subject to climatic control." Messenger goes on to illustrate with the example of a parasite that is ineffective because it emerges as an adult before the host or pest does. Thus phenology must be important in selecting a parasite to control a pest. The complexities of phenology may be worth mastering in this case, for this analysis must be undertaken only once upon the selection of the parasite rather than daily as in Shelford's method for calculating the cycle of the codling moth.

The argument can be advanced that the current desire to decrease pesticide application may make schemes like Shelford's more attractive. Ideally, less pesticide would be needed if its application could be timed precisely by phenology rather than made more frequent to guarantee the death of the pest whenever it becomes susceptible. This argument has, I believe, yet to be illustrated by a practical and successful example.

3 Applications in Plant Management

Scheduling planting seems the most common use of phenology, but Buys and Kotze (1963) have suggested some novel additional uses. For example, "timing . . . sprays to counteract delayed defoliation [of fruit trees] is vital, especially since they have to be applied some 6–7 weeks before full bloom. . . . It, therefore, automatically follows that . . . access to a technique for full bloom forecast . . . eliminates the uncertainty of the time at which sprays must be applied."

They also mention, "the value of full bloom forecasts . . . where frost damage occurs. In years during which a late bloom is expected, arrangements concerning . . . fuel . . . and smudge pots can be modified." They add that the plowing under of cover crops that aggravate frost can be timed better if full bloom is forecast.

The timing of planting, however, seems the chief use of phenology. Every gourmet gardener who has employed a mixture of planting dates and varieties to get a steady supply of perfect sweet corn understands the goal and the rudiments of the game. I have mentioned Thornthwaite's (1944, 1952) prediction of pea maturity. (see page 311).

4 Weather Information for Modeling

To employ a model of seasonality of the sort I have described generally requires a specification of the weather. Some models are influenced by day-length or latitude and longitude, e.g., Nuttonson's (1955) or Hopkin's (1938) models of flowering;

and the time from flowering to fruit maturity may be independent of environment, but most of the models require information about temperature and even other weather factors. The question before us now is, "What weather information shall we feed to the model?"

Let us begin with a verified model that employs a single factor, degree-days or heat units. If the mean date of the modeled phenophase is desired for the region where the model was tested, the mean date itself is undoubtedly the most direct estimate. If, on the other hand, the mean date is wanted for another region it can be estimated by feeding climatic values of degree-days themselves to the model; such a climatology exists for the Northeastern United States (Dethier and Vittum, 1963). Alternatively the degree-days can be estimated from mean temperatures (Thom, 1954). Presumably the user of phenological information will be interested in the probabilities of a phenophase before certain days and want the distribution as well as the mean of phenological dates. The user therefore will likely employ the distribution as well as means of degree-days in the phenological model.

Some models that employ the single variable of temperature will not employ it in the simple form of degree-days. For example, Livingston's (Smith, 1920) phenological index of corn development or the rates of development and destruction of fruit fly eggs derived from Messenger and Flitters' (1958) observations must be calculated from temperatures themselves rather than degree-days. Presumably mean indices or rates and thus length of phenoperiods could be calculated from climatic values. Because of the fluctuation of temperatures beyond linear portions of the relation between temperatures and rate, however, it seems wiser to feed the models actual series of temperatures rather than means. The correlations between temperatures at nearby times also argue for using real rather than synthetic weather data. Finally, the brevity of a period like the incubation of the fly eggs relative to the period of, say, a monthly mean also argues for the use of real temperature series. In other words, the simple and wise course seems to be feeding the model real temperature observations at intervals much shorter than, say, an incubation period, and then obtaining the distribution of periods from the outcome of the model calculations rather than feeding it climatology.

When we turn to models that depend upon more than one factor, the wisdom of using actual rather than mean values seems even greater. Thus Brown's (1960) model of soybean development concerns temperatures in the dark only. A precise application of his model would therefore require a climatology of temperatures after dark, an unlikely thing to find. One could also conceive the difficulties in synthetic or climatic values for both radiation and temperature or precipitation and temperature. Only a satisfactory model of the atmosphere itself, not the usual climatic tables, would seem to produce values with the correlations among factors and the correlations between succeeding values of the same factor that are actually encountered outdoors.

My conclusion is, therefore, that simulators that relate biological knowledge about recognizable parts of development to the environment and mimic natural behavior seem the most satisfying models; when they are realistic, they can transform actual or variable rather than mean weather parameters into phenological statistics.

References

Brown, D. M. 1960. Soybean ecology. I. Development-temperature relationships from controlled environment studies. *Agron. J.* 52, 493–496.

Buys, M. E. L., and A. V. Kotze. 1963. Forecasting of full bloom. *Deciduous Fruit Grower* 13, 356–360.

Dethier, B. E., and M. T. Vittum. 1963. Growing degree days. *N.Y. State Agr. Exp. Sta. Bull. 801* 84 pp.

Holmes, R. M., and G. W. Robertson. 1959. Heat units and crop growth. *Can. Dept. Agr. Publ. 1042* 31 pp.

Hopkins, A. D. 1938. Bioclimates—A science of life and climatic relations. *U.S. Dept. Agr. Misc. Publ. 280* 188 pp.

Messenger, P. S. 1970. Bioclimatic inputs to biological control and pest management programs. In *Concepts of Pest Management*, R. L. Rabb, and F. E. Guthrie, eds. pp. 84–102 Raleigh: North Carolina State University.

———, and N. E. Flitters. 1958. Effect of constant temperature environments on the egg stage of three species of Hawaiian fruit flies. *Ann. Ent. Soc. Am.* 51, 109–119.

Nuttonson, M. Y. 1955. Wheat-climate relationships and the use of phenology in ascertaining the thermal and photo-thermal requirements of wheat. *Am. Inst. Crop Ecol.* 388 pp.

Shelford, V. E. 1927. Experimental investigation of the relations of the codling moth to weather and climate. *Illinois Nat. Hist. Surv. Bull. 16* 307–440.

Smith, J. W. 1920. *Agricultural Meteorology* 304 pp. New York: Macmillan.

Thom, H. C. S. 1954. The rational relationship between heating degree days and temperature. *Monthly Weather Rev.* 82, 1–6.

Thornthwaite, C. W. 1944. Report of the committee on transpiration and evaporation, 1943–44. *Trans. Am. Geophys. Union* 25, 683–693.

——— 1952. Climate in relation to planting and irrigation of vegetable crops. In *Proc. VIII General Assembly-XVII Congress Intern. Geogr. Union.* pp. 290–295. Washington, D.C.

Uvarov, B. P. 1931. Insects and climate. *Ent. Soc. London Trans.* 79, 247 pp.

Waggoner, P. E., and J.-Y. Parlange. 1974. Modeling Seasonality, with appendix "Analytic solution to model passages through phenophases". In *Phenology and Seasonality Modeling*, H. Lieth ed. p. 301, New York: Springer Verlag.

6.5 Periodicity and Seasonality of Airborne Algae and Protozoa

Harold E. Schlichting, Jr.[1]

Department of Botany
North Carolina State University
Raleigh, North Carolina 27607, USA

1 Introduction

Only a few studies in the history of science have been concerned with airborne algae and protozoa. Those pioneers studying protozoa from the atmosphere were Spallanzani (1777), Miquel (1883), and Puschkarew (1913). Those initiating studies of airborne algae were Ehrenberg (1844), Salisbury (1866), van Overeem (1937), and Pettersson (1940). More recently, Brown *et al.* (1964, 1971) conducted long-term studies on airborne algae in Texas and Hawaii, while long-term studies on both airborne algae and protozoa were conducted by Schlichting in Michigan, Texas, and North Carolina (1961, 1964, 1969) and by Smith in North Carolina (1973). Short-term studies of algae in the atmosphere have been conducted by Gregory *et al.* in Great Britain (1955), Hamilton in Great Britain (1959), Stevenson and Collier in Texas (1962), Chang in Taiwan (1967), Luty and Hoshaw in Arizona (1967), and Maynard in Florida (1968a). Mahoney (1968) studied both airborne algae and protozoa in Texas. Additional studies of airborne algae are in progress in Brazil, Florida, and Ohio.

Investigators have sampled the atmosphere by drawing air through cotton or glass wool, bubblers and impingers, membrane filters, the Rotorod and various sequential samplers, as well as exposing glass plates, slides, and culture media. Sampling rates have varied from 2.8 liters/min (0.1 ft^3) to 15 liters/min (420 ft^3). As of 1969 over 189 taxa of airborne algae and protozoa had been identified. A review of the studies of algae and protozoa in the atmosphere, including methods used and organisms collected, is given by Schlichting (1969).

Specific origins of the airborne algae and protozoa collected have not been demonstrated. Many algae and protozoa sampled could have originated from the

[1] Present address: H. E. Schlichting, Jr., President BioControl Company, P.O. Box 43, Port Sanilac, Michigan 48469, USA.

KEYWORDS: Airborne algae; allergies; dispersal; micrometeorology; periodicity; protozoa; public health; sampling; seafoam; seasonality.

rooftop or tower from which they were sampled or from the trees, buildings, soil, or water surfaces of the adjacent area. While a number of algae have been found to be resistant to desiccation for long periods of time (Parker *et al.*, 1969), algae are not always found in aerial samples. Sampling in Michigan near Lake Huron over a one-year period did not yield any more planktonic freshwater algae than did sampling in north-central Texas over one year. Stevenson and Collier (1962) reported a few small marine diatoms and flagellates collected on media-coated slides placed a few feet above the shoreline on the Texas coast, while Maynard (1968a) collected two viable marine microalgae in wind nets on a tower about 25m (75 ft) above the ground and about 65 Km (40 mi) from the nearest marine habitat. Sampling over the open Pacific and Atlantic oceans by ships and aircraft produced negative results (Schlichting, 1969; Brown, 1971).

2 Periodicity and Seasonality of Airborne Algae

Greater diversity of species and larger numbers of airborne algae and protozoa seemed to be found in semiarid areas where winds commonly raise dust into the air, as indicated in Texas sampling when compared to Michigan or North Carolina (Schlichting, 1969). Smith (1973) also found very close correlations between particulate matter in the air and numbers of algae and protozoa sampled in North Carolina (Fig. 1).

Tropical areas with abundant soil and subaerial algae may also allow for great diversity and numbers. The tropical study of Brown (1971) on mid-Pacific islands (Hawaii) may pinpoint specific origins of the airborne organisms.

Algae and protozoa do not show definite diurnal variation as do some fungi (Kramer *et al.*, 1964) and bacteria (Pady and Kramer, 1967). So few studies have

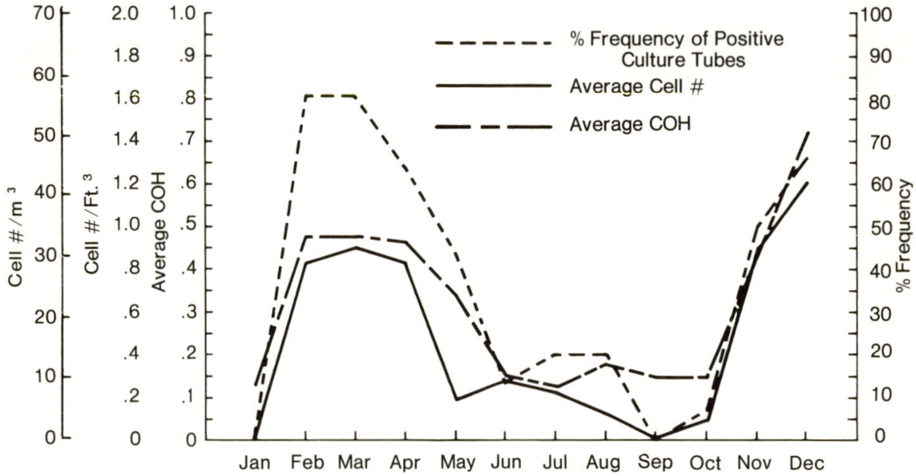

Fig. 1. Percent frequency of positive culture tubes and average cell number per unit air volume, compared to the average COH.

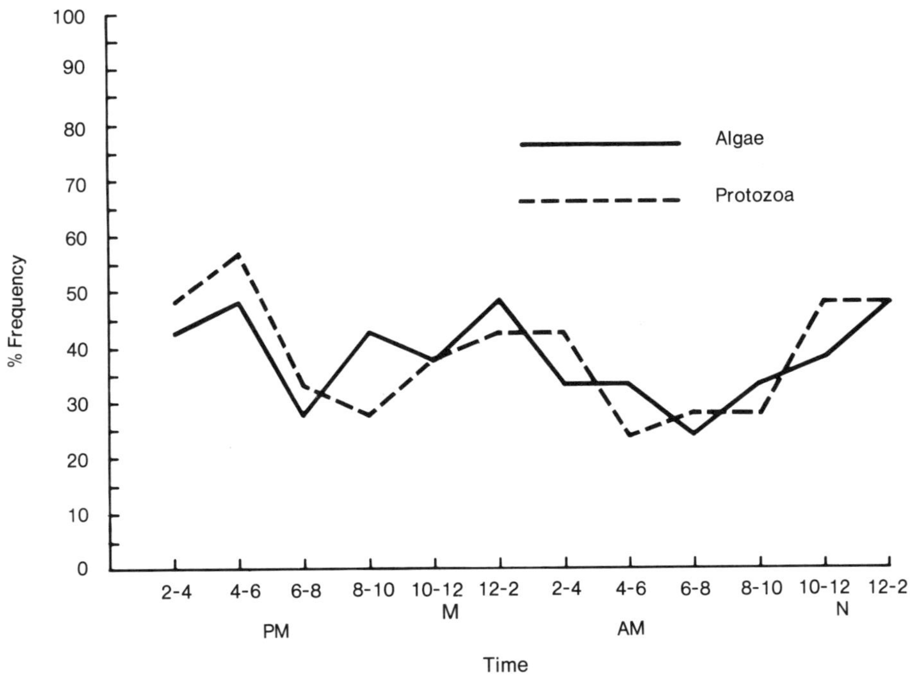

Fig. 2. Percent frequency of positive culture tubes during the diurnal sequence of 2-hr. exposures.

been made, that generalizations relating phenological events to seasonality of airborne algae and protozoa are difficult. Little diurnal variation occurred in Michigan (Schlichting, 1961), while in Texas (Schlichting, 1964) nearly twice as many algae and protozoa were obtained from the atmosphere between 12:00 and 24:00 hr as were collected from 24:00 to 12:00. Smith (1973) found more algae and protozoa in the North Carolina atmosphere between 08:00 and 12:00, 14:00 and 18:00, and 20:00 and 04:00 and least during the hours of 04:00 and 08:00 and 18:00 and 20:00 (Fig. 2).

In Texas a greater variety of algae and protozoa were sampled in February than in March through August while in Michigan a greater variety was collected in June through August (Schlichting, 1964). Smith (1973) found the most algae and protozoa in the air during February through April and fewer during the summer months in central North Carolina. Bernstein and Safferman (1972a, 1972b) reporting on airborne algae as a cause of human allergies, noted increased reactions in patients during November.

3 Standardization of Sampling Methods and Simultaneous Assessment of Environmental Parameters

Diurnal and seasonal variations of algae and protozoa in the atmosphere suggest a need for better standardized sampling to understand the importance of the aerial biota to man's welfare. Weather bureau data tracing the movement of air masses are insufficiently accurate to give the micrometeorological conditions within a given air mass. The minimum on-site meteorological data recorded should be rainfall, air temperature, relative humidity, radiation, wind velocity, and wind elevation angle.

Although greater wind velocity might suggest more species and greater numbers of airborne algae and protozoa, Schlichting (1964) found most species in prevailing winds of 22 to 29 km/hr (gusts to 55 km/hr) but fewer in prevailing winds of 51 to 56 km/hr (with gusts up to 85 km/hr). The wind elevation angle is also important in relating wind speed to the nature of the aerial biota (Schlichting, 1964). Smith (1973) also found very close correlations between the average wind speed and average cell numbers and percent frequency of positive cultures obtained (Fig. 3).

Air quality measurements above specific geographic locations require large volumes of air sampled sequentially over a minimum of 24 hr. Short periods of 2 hr, within the 24 hr are also necessary because of rapidly changing environmental conditions and the possibility of toxic materials becoming concentrated on membrane filters and tape samplers or in bubblers.

The closer to the ground the sampler is located, the more algae and protozoa are likely to be found. A sampler on a hill may collect many different algae, but in lesser numbers, than one located in a valley. A station in an urban industrial area would be less apt to collect viable organisms than one in the country. A global network of strategically located sampling stations would permit long-term studies to aid in gaining a better knowledge of the phenology of microorganisms (Schlichting *et al.*, 1971, 1972).

Airborne algae may cause respiratory disorders (Salisbury, 1866) and respiratory and dermatological allergies (McElhenney *et al.*, 1962; McGovern *et al.*,

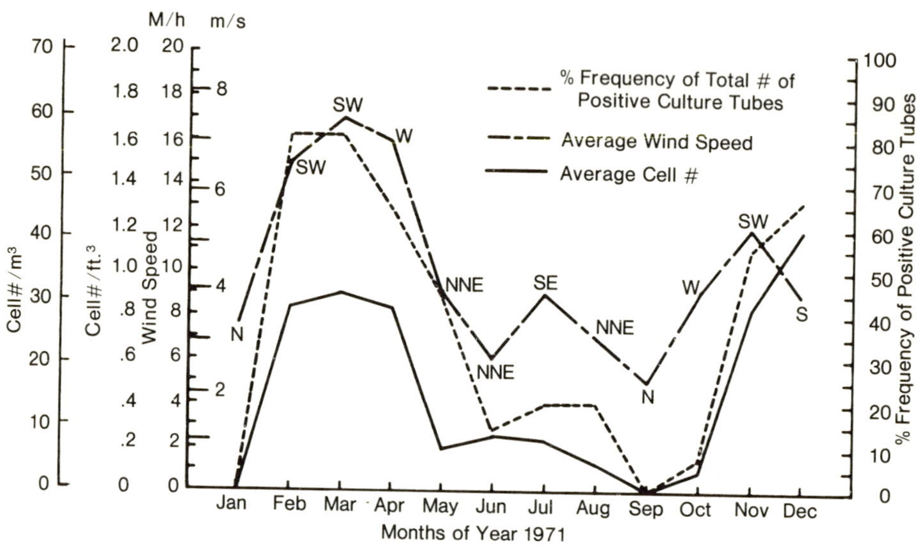

Fig. 3. Average wind speed and wind direction compared to percent frequency of positive culture tubes and cell numbers per unit air volume.

1965, 1966). Bernstein and Safferman (1966) reported on the sensitivity of skin and bronchial mucosa to green algae, and later (1970) used *Chlorella* and *Chlorococcum* as test algae because of their world-wide prevalence in house dust and in other atmospheric samples. It is not surprising that algae can cause respiratory difficulties since they are high in protein and are a suitable size to be inhaled and retained in the lungs. Man, at rest, inhales about 7 liters (0.25 ft^3) of air per minute. Up to eight algal cells per 14 liters of clear air (Schlichting, 1964) would allow the inhalation of a minimum of 1500 cells per day.

Woodcock (1948) suggested that aerosols are formed by bursting bubbles from breaking waves and that microorganisms were the particles, 0.3 to 30 μ in diameter, in atmospheric water droplets which could be dispersed by air currents. Mahoney (1968) showed that airborne algae, protozoa, and bacteria originated from trickling filters and sewage aeration tanks in Texas and Michigan. Man's responses to airborne algae may be enhanced by other irritants, such as dust, sulfur dioxide, ozone, and nitrogen oxides as well as bacteria, viruses, and fungi associated with them. Maynard (1968b) stressed the importance of foams as ideal habitats for microorganisms, and Schlichting (1971) showed the great variety and large number of algae and protozoa in sea foam. Schlichting (1974) demonstrated that bursting bubbles did thrust these organisms up to 13 cm into still air.

4 Airborne Algae and Public Health

It has been known since the advent of atomic energy that algae can take up radioactive elements and concentrate them on and within their cells within a few minutes. The concentration of pesticides by airborne or soil algae is poorly understood since most research has been concerned with aquatic ecosystems. The study of algae in house dust by Bernstein and Safferman (1972a) opens an entirely new dimension to medical allergenicity studies. Related to this is the importance of unstudied algal deposition from the air into drinking water supplies and their accumulation and growth in industrial cooling towers, air-filtering systems, and even in swimming pools. Not only may man be allergic to the algal cells and their adhering chemicals, bacteria, or viruses, but algae may also be capable of giving off organic volatile compounds into the air from filtering systems that may adversely affect the health of man.

Cultures of airborne algae and protozoa may be used to monitor the air for an assay of various concentrations of air pollutants such as oxides of nitrogen, sulfur dioxide, ozone, and metals (Schlichting *et al.*, 1972). Some airborne organisms may act as indicators of air pollution, especially in regard to their growth rate and pleomorphic changes in cytology and morphology as related to even very low concentrations of chemicals in the atmosphere. Maynard (1968b) and Schlichting (1969), suggested that specific algae collected in air samples may allow the detection of spot origins of hurricanes and tornadoes prior to their detection by radar. Studies relating air chemistry and radioactivity to numbers and species of algae and protozoa in the atmosphere have yet to be attempted, but might add to the understanding of the aerial biota surrounding our planet.

References

Bernstein, I. L., and R. A. Safferman. 1966. Sensitivity of skin and bronchial mucosa to green algae. *J. Allergy* 38, 166–183.

———, 1970. Viable algae in house dust. *Nature* 227, 851–852.

———, 1972a. Sampling living particulates in the atmosphere. In *US/IBP Aerobiology Handbook*. II, pp. 441–445. Ann Arbor: University of Michigan.

———, 1972b. A model system for studying the aerobiology and biomedical impact of green algae. In *US/IBP Aerobiology Handbook*, III, pp. 1–4. Ann Arbor: University of Michigan.

Brown, R. M., Jr. 1971. Studies of Hawaiian freshwater and soil algae. I. The atmospheric dispersal of algae and fern spores across the Island of Oahu, Hawaii. In *Contributions in Phycology*. pp. 175–188. Lawrence, Kan.: Allen Press.

———, D. A. Larson, and H. C. Bold. 1964. Airborne algae: Their abundance and heterogeneity. *Science* 143, 583–585.

Chang, T. 1967. A preliminary survey on air-borne algae in the Taipei atmosphere. *Taiwania* 13, 1–9.

Ehrenberg, C. G. 1844. Bericht über die zur Bekanntmachung geeigneten Verhandlungen der Königl. *Preuß. Akad. Wiss. Berlin* 9, 194–197.

Gregory, P. H., E. D. Hamilton, and T. Sreeramula. 1955. Occurrence of the alga *Gloeocapsa* in the air. *Nature* 176, 1270.

Hamilton, E. D. 1959. Studies on the air spora. *Acta Allergol.* 13, 143–175.

Kramer, C. K., S. M. Pady, J. J. Wiley. 1964. Kansas aeromycology. XIV. Diurnal studies 1961–1962. *Trans. Kansas Acad. Sci.* 67, 442–459.

Luty, E. T., and R. W. Hoshaw. 1967. Airborne algae of the Tucson and Santa Catalina Mountain areas. *J. Arizona Acad. Sci.* 4, 179–182.

Mahoney, J. L., III. 1968. A qualitative survey of the airborne algae, protozoa and bacteria at the Denton sewage treatment plant. M.Sc. Thesis, Denton: North Texas State University.

Maynard, N. G. 1968a. Significance of airborne algae. *Z. Allg. Mikrobiol.* 8, 225–226.

——— 1968b. Aquatic foams as an ecological habitat. *Z. Allg. Mikrobiol.* 8, 119–126.

McElhenney, T. R., H. C. Bold, R. M. Brown, Jr., and J. P. McGovern. 1962. Algae: A cause of inhalant allergy in children. *Ann. Allergy* 20, 739–743.

McGovern, J. P., T. R. McElhenney, and R. M. Brown, Jr. 1965. Airborne algae and their allergenicity. I. Air sampling and delineation of the problem. *Ann. Allergy* 23, 47–52.

McGovern, J. P., T. J. Haywood, and T. R. McElhenney. 1966. Airborne algae and their allergenicity. II. Clinical and laboratory multiple correlation studies with four genera. *Ann. Allergy* 24, 145–151.

Miquel, M. P. 1883. *Les organismes vivants de l'atmosphere*. Paris: Gautheir-Villars.

Overeem, M. A., van. 1937. On green organisms occurring in the lower troposphere. *Rec. Trav. Botan. Neerl.* 34, 389–439.

Pady, S. M., and C. L. Kramer. 1967. Diurnal periodicity in airborne bacteria. *Mycologica* 59, 714–716.

Parker, B. C., N. Schanen, and R. Renner. 1969. Viable soil algae from the herbarium of the Missouri botanical garden. *Ann. Missouri Bot. Garden* 56, 113–119.

Pettersson, B. 1940. Experimentelle Untersuchungen über die euanemochore Verbreitung der Sporenpflanzen. *Acta Bot. Fennica* 25, 1–102.

Puschkarew, B. M. 1913. Über die Verbreitung der Süsswasserprotozoen durch die Luft. *Arch. Protistenk.* 28, 323–362.

Salisbury, J. H. 1866. On the cause of intermittent and remittent fevers, with investigations which tend to prove that these affections are caused by certain species of Palmellae. *Am. J. Med.* 51, 51–75.

Schlichting, H. E., Jr. 1961. Viable species of algae and protozoa in the atmosphere. *Lloydia* 24, 81–88.

——— 1964. Meteorological conditions affecting the dispersal of airborne algae and protozoa. *Lloydia* 27, 64–78.

——— 1969. The importance of airborne algae and protozoa. *Air Pollution Control J.* 19, 946–951.

——— 1971. A preliminary study of algae and protozoa in seafoam. *Bot. Marina* 14, 24–29.

———, G. S. Raynor, and W. R. Solomon. 1971. Recommendations for aerobiology sampling in a coherent monitoring system: Algae and Protozoa in the atmosphere. In *US/IBP Aerobiology Handbook*. I, pp. 60–61. Ann Arbor: University of Michigan.

Schlichting, H. E., Jr., R. M. Brown, Jr., and P. E. Smith. 1972. Airborne algae of Hawaii: A model for coordinated aerobiological research. In *US/IBP Aerobiology Handbook*. III, pp. 63–68. Ann Arbor: University of Michigan.

Schlichting, H. E., Jr. 1974. Ejection of microalgae into the air via bursting bubbles. *J. Allergy and clinical Immunology*. In press.

Smith, P. E. 1973. The effects of some air pollutants and meteorological conditions on airborne algae and protozoa. *Air Pollution Control J.* 23, 876–880.

Spallanzani, L. 1777. Opuscules de Physique animale et vegetale. *Trad. l'italien Jean Sennebier.*

Stevenson, R. E., and A. Collier. 1962. Preliminary observations of the occurrence of airborne marine phytoplankton. *Lloydia* 25, 89–93.

Woodcock, A. H. 1948. Note concerning human respiratory irritation associated with high concentrations of plankton and mass mortality of marine organisms. *J. Mar. Res.* 7, 56.

6.6 Seasonality of Airborne Fungi

Charles L. Kramer

Division of Biology
Kansas State University
Manhattan, Kansas 66502, USA

1 Introduction

The microbiology of the atmosphere has been studied in many parts of the world using a variety of methods and approaches. Among the first to sample continuously over long periods of time was Miquel (1885) in the late 1800s at the Observatorie Municipal Montsouris, Paris. However, his and other studies undertaken throughout the first half of the 20th century were qualitative but not quantitative.

More recent studies using quantitative sampling devices such as the Hirst spore trap (Hirst, 1952) and the Kramer-Collins spore sampler (Kramer and Pady, 1966) have shown how the concentration and composition of the population of air spora vary greatly with the season of the year, time of day, environmental conditions, location, and human activity.

2 Methods

Two primary approaches have been used in studying the seasonal variations in the population of fungi in the atmosphere. The method used most often has been to trap the spores and other particles on adhesive-coated surfaces such as glass slides or on the surface of culture media in Petri dishes. Both methods have advantages and disadvantages. Identification of spores trapped on glass slides is limited to a comparatively small number of fungi because of the general lack of distinctive morphological characteristics. Petri dishes have the advantage of allowing specific identification of the colonies obtained, but they are selective for only living organisms and for only those that will develop colonies on the particular medium being used. Thus specific procedures used in a study of the air spora must be determined by the specific questions being asked.

KEYWORDS: Seasonality; fungi; *Cladosporium*; *Homodendrum*; *Alternaria*; *Penicillium*; *Aspergillus*; air spora; ascospores; basidiospores; Circadian variation; environmental controls; concentration and composition; spore traps.

Diverse media were tested before selection of Farrow's (1954) glucose-peptone preparation containing rose bengal and streptomycin. The rose bengal restricts the size of the fungal colonies while allowing fruiting structures and spores to develop naturally so that they may be identified. This material was very effective

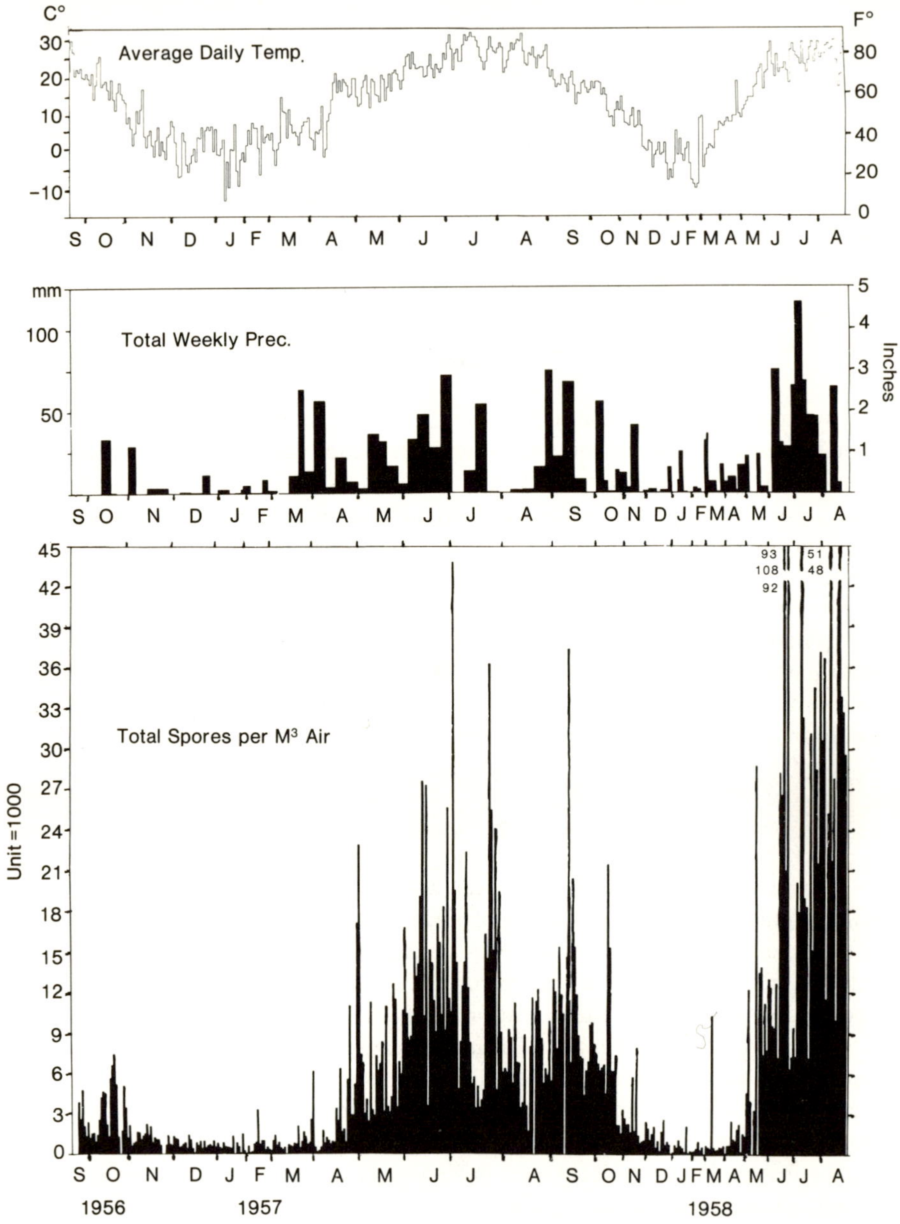

Fig. 1. Number of all spores collected per cubic meter of air at 9:00 AM at Manhattan, Kansas, from September 16, 1956 through August 30, 1958. Spores were collected on silicone slides exposed in a Pady-Rittis slit sampler.

in that occasionally over 100 colonies were obtained on a single Petri plate, and these could be identified and counted with ease.

Studies of the phenological variations of the composition and concentration of the population of air spora in Kansas were begun by Pady, Rogerson, and Kramer in 1956 (Kramer *et al.*, 1959c). These investigations involved collecting fungi trapped on both slides and Petri dishes from the roof of a building on the campus of Kansas State University.

3 Results

Figure 1 illustrates the total number of spores obtained each day from samples taken over a two-year period from September 1956 through August 1958 (Kramer *et al.*, 1959c). This includes most days during the growing season except Saturdays and Sundays, but only once or twice each week during the winter months. The average frost-free period at Manhattan, Kansas, extends from April 23 to October 12. Because of the time required in the processing of the samples taken during one hour, exposures were made only once each day at 9:00 AM.

As expected, numbers of fungi in the atmosphere were much more abundant during the growing season than during winter. However, the magnitude of these differences varies with moisture conditions of the surrounding area. The fall of 1956 was comparatively dry with less than 1.5 in. of precipitation occurring during September and October; August was also extremely dry. Comparison of the number of spores obtained during September and October of 1956 with those of the same months of 1957 shows the effects of the high rainfall during the latter part of August, September and October of 1957. Effects of rainfall can also be seen during July and August of 1957 and 1958. In 1957 a dry period decreased the number of spores obtained. Extremely wet June, July, and the first part of August 1958 provided excellent growing conditions for fungi in the surrounding areas, resulting in extremely high numbers of spores being released into the atmosphere throughout the entire period.

Total number of colonies obtained from the exposure of culture media in Petri plates showed the same general trends (Kramer *et al.*, 1959c), but numbers of colonies were much lower than spores (Fig. 2). Not all spores present in the atmosphere are viable and there may be some selectivity to the media used. During the two-year study from 1956 to 1958, a total of 89 genera were obtained on the exposed plates (Kramer *et al.*, 1959c). *Cladosporium* (= *Homodendrum*) was the most frequently collected fungus, comprising approximately 45% of the total colonies obtained. *Alternaria* (12.6%) was second in occurrence, with *Penicillium* (6.1%) and *Aspergillus* (5.4%) next in order of the most frequently collected fungi. The majority of the genera, 63, belonged to the Fungi Imperfecti while 13 were Ascomycetes, 7 Phycomycetes, and 6 Basidiomycetes.

Cladosporium has been reported repeatedly to be one of the most common of airborne spores (Fig. 3). Gregory and Hirst (1957) found it to account for 46% of the total spore catch of the summer airspora at Rothamsted in 1952. Harvey (1967) compared the *Cladosporium* spore count of urban and rural sites in Wales and found it to comprise 42% of the total spore count at the rural site and 25.5% of the total at the center of the city of Cardiff. Similar results have been reported from Israel (Barkai-Golan, 1958), Hawaii (Myers, 1956), Nigeria (Dransfield, 1966), and New Zealand (Dye and Vernon, 1952). At times it has been reported to reach as high as 96% of the total spore count in an area (Hyde and Williams, 1953).

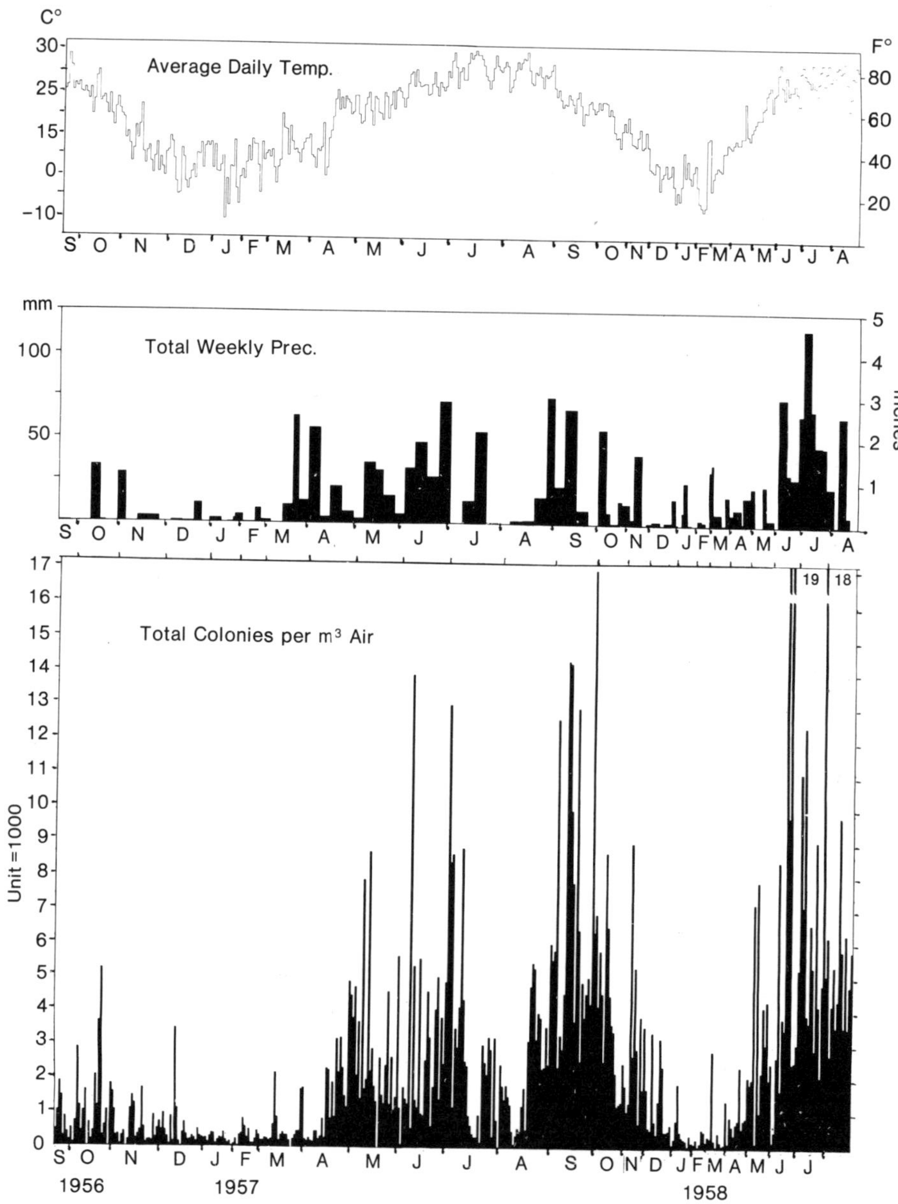

Fig. 2. Average number of all colonies of fungi collected per cubic meter at the same time and under the same conditions as for the spores in Fig. 1.

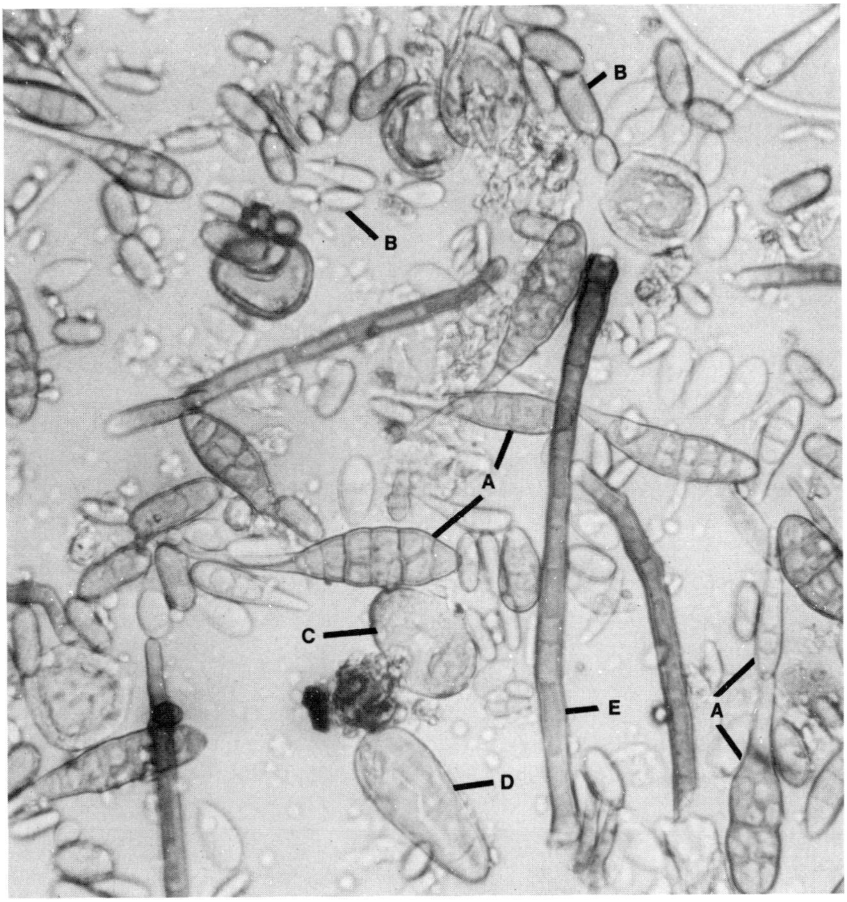

Fig. 3. Airborne fungal spores and hyphal fragments collected on an exposed slide. *a*: *Alternaria; b*: *Cladosporium; c*: leaf rust urediospore; *d*: stem rust urediospore; *e*: hyphal fragment.

Cladosporium has a distinct seasonal periodicity. In Kansas, spores present throughout the winter begin to increase in April or May and remain high throughout the growing season (Kramer *et al.*, 1959a). In England Richards (1954, 1956), Harvey (1967), and Gregory and Hirst (1957) found *Cladosporium* to occur in high numbers (up to 25,000/M^3 between June and October. Kramer *et al.*, (1959a) identified four species with *C. cladosporioides* comprising 53% of the total colonies of *Cladosporium*; *C. herbarum*, 29%; *C. macrocarpum*, 17%; and *C. sphaerospermum* 1%. These four species also were found to comprise most of the total *Cladosporium* catch in England but there *C. herbarum* was the most frequently collected species (Harvey, 1967; Gregory and Hirst, 1957).

Alternaria (Fig. 3), which is known to be responsible for causing allergenic respiratory problems in man as well as many plant diseases, is important and exhibits a marked phenological periodicity similar to that of *Cladosporium*. In Kansas numbers began to increase in May and remained high throughout the growing season (Kramer *et al.*, 1959b). However, as in the case of *Cladosporium* and most other spore types, numbers fluctuate with environmental conditions. In England, Hyde and Williams (1946) and Sreeramulu (1959) found *Alternaria* to be absent during the winter months and occurring from June through September. Gregory and Hirst (1957) reported it to be most abundant during August. Hyde and Williams (1946) further stated that the absence of *Alternaria* spores in the air of the countryside around Cardiff correlated closely with the wheat harvest. Durham (1938) stated that the incidence of these spores seemed to vary with the availability of the straw of cereals. Perhaps this would account for the occurrence of *Alternaria* throughout the year at our collecting sites at Manhattan, Kansas, which is located in the center of the northern Flint Hills prairies. In a study of the air spora of two contrasting ecological sites near Manhattan (Long and Kramer, 1972), we have found that numbers of *Cladosporium* and *Alternaria* at a prairie site are higher on dry windy days following periods of moisture than they are during periods with rainfall and high humidity. Spore release in these organisms is facilitated by drying and strong winds.

In Kansas all of the fungi, with the possible exception of *Aspergillus* and *Penicillum*, have distinct phenological periodicities (Kramer *et al.*, 1960) with maxima occurring during the growing season. In 1957 and 1958 fewer spores were collected of these genera during late winter and spring than during the fall and early winter. Richards (1954) reported *Aspergillus* to have a rather indistinct phenological periodicity, while *Penicillum* was "nonseasonal" in its occurrence in the air in Cardiff. He also stated that although most spore types were more abundant in rural areas than within urban areas, *Penicillum* seemed to be an exception and was more abundant in the air of large cities.

Apparently most fungi which occur saprophytically on a variety of organic substrata are present in the atmosphere to a greater or lesser extent, depending on moisture conditions, throughout the growing season. However, those that are parasitic on specific plant hosts occur only during the periods when the hosts are being actively parasitized. *Erysiphe* conidia were found chiefly in spring in our Kansas studies (Kramer and Pady, 1960). In England, Gregory and Hirst (1957) found them to be most common during the latter part of June and early July. Highest concentrations of *Ustilago* also occurred during June and July in both England and Kansas (Pady and Kramer, 1960). However, substantial numbers were also present during the fall in Kansas.

Urediospores of *Puccinia recondita* begin to occur in the air at Manhattan, Kansas, about May 1 and are present through June (Pady and Johnston, 1959). Maxima are reached during the first two weeks of June correlated with maturation and harvesting of the wheat in that area. To the south, in Oklahoma and northern Texas, these dates would be earlier, while to the north they would become progressively later. The appearance of *Puccinia graminis* in the area of Manhattan, Kansas, is generally about two weeks later than that of *P. recondita* because of its preference for warmer weather.

Gregory and Hirst (1957) reported the occurrence of *Phytophthora infestans*-type spores at the Rothamsted Experiment Station to occur during August. However, the trap was placed near a potato field with an abundant source of this organism, and its occurrence in the atmosphere is atypical of the district as a whole. In a study of the air spora of an abandoned orchard near Manhattan, Kansas (Pady *et al.*, 1967), we were able to obtain a number of spore types in comparatively large numbers that were rarely collected at the roof top site on the University campus, including *Hypoxylon*, *Sordaria*, basidiospores of *Geaster* and *Coprinus*, and aecidiospores of *Gymnosporium juniperi-virginiana*. Thus location of the sampling site must always be considered in relation to the surrounding environment when undertaking a study of the air spora.

In addition to the location of the collecting site, sampling methods, and prevailing environmental conditions, the time of day when samples are obtained must also be taken into consideration when undertaking a study of airborne fungi. Both the concentration and composition of the air spora may vary with the time of day (Hirst, 1953; Pady *et al.*, 1962).

When conditions are favorable for growth and sporulation of fungi in the area surrounding a sampling site, the occurrence of spores through a 24-hr period generally would be similar to that represented in Fig. 4. This histogram was compiled from data obtained at three different collecting sites within and near the city

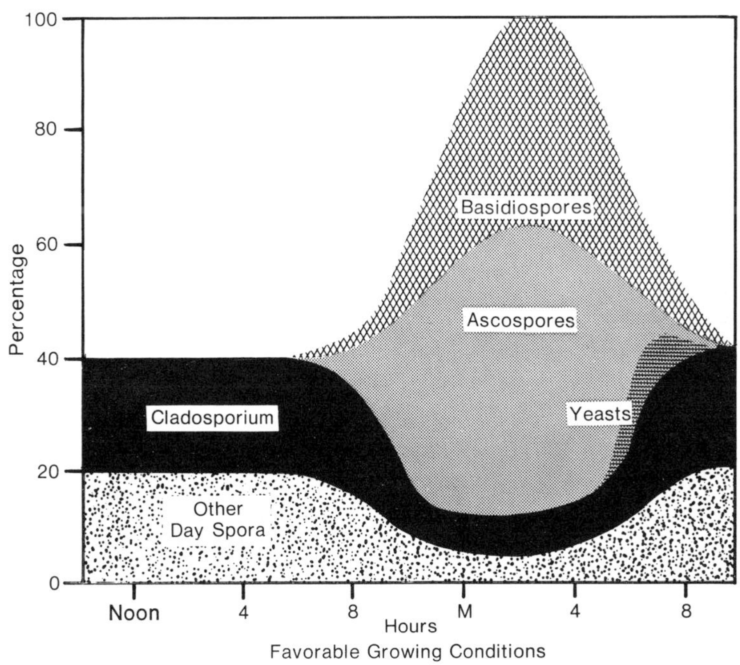

Fig. 4. Compiled from data obtained at three different collecting sites within and near thg city of Manhattan, Kansas. Concentrations at a given hour are expressed as percentages of the peak geometric mean concentration. The peak, represented by 100%, generally occurs around 2:00 AM on days when environmental conditions are favorable for growth and sporulation of fungi in the areas surrounding the collecting sites.

of Manhattan, Kansas (Long and Kramer, 1972). Concentrations at a given hour are expressed as percentages of the peak geometric mean concentration. The magnitude of peak concentrations, which generally occur around 2:00 AM will vary depending on a number of factors, such as the time of year, location, and environmental conditions during the previous few weeks. In Kansas maximum concentrations have been found to exceed $100{,}000/M^3$ (approximately $3000/ft^3$) of air. The proportions of these various components may also vary somewhat from day to day and from one location to another, but, in general, this is the distribution that one might expect at Manhattan, Kansas, on a day when optimum conditions for growth and development of fungi prevail.

During periods when environmental conditions have been dry, providing poor conditions for growth and development of fungi in the area surrounding the sampling site, there is little change in either the concentration or composition of the air spora throughout a 24-hr period (Fig. 5). Concentrations range from 5 to 25% of those during periods of favorable growing conditions. The composition of the population of air spora during dry periods is composed primarily of day spora through a 24-hr period with a complete absence of basidiospore and ascospore maxima. Since growth and development of fungi are curtailed during periods of dry environmental conditions, a high percentage of the spores present in the air at

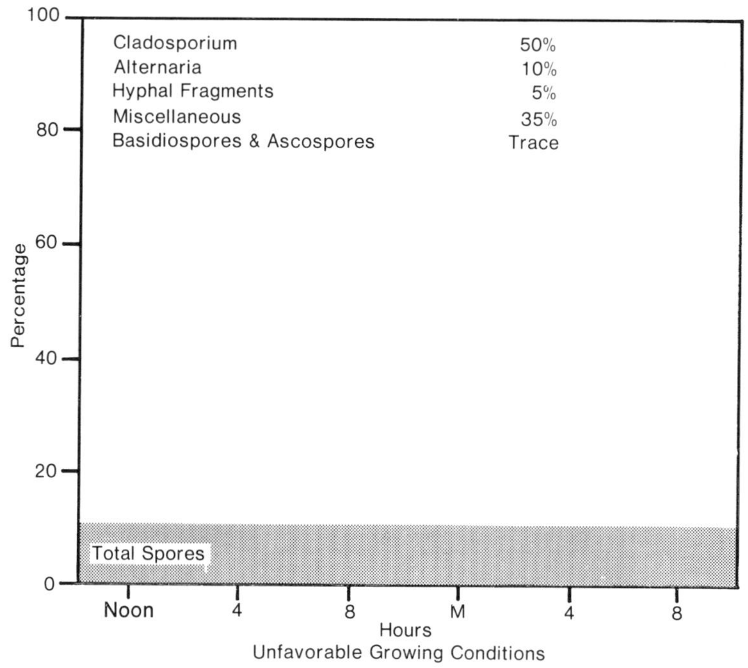

Fig. 5. The percentage of total spores in the air during periods that are unfavorable for the growth and sporulation of fungi is approximately 10% of the peak concentration as shown in Fig. 4. The percentages of the various spore types present during these periods are shown in the graph.

such collecting sites is undoubtedly a result of the spores being transported there from remote sources. As spores are transported long distances, they become dispersed within the atmosphere and the effects of their time of release are not evident, resulting in the absence of maxima occurring at specific times of the day.

4 Summary and Conclusions

The concentration of fungal spores in the atmosphere exhibits a distinct seasonal periodicity. In Kansas, numbers begin to increase in April, reach maxima during June, July, and August and then decrease during October. Although present in low numbers, spores do occur throughout the winter months. At other locations in the temperate regions, this pattern may vary somewhat depending on seasonal temperatures and moisture conditions. The magnitude of the concentration may vary from day to day depending on environmental conditions of the surrounding areas with rainfall being the most important factor.

The concentration and composition of the air spora may also vary with the location of the sampling site. The surrounding habitat greatly influences the amount and kinds of fungi that develop and release spores into the atmosphere. The occurrence of the spores of plant pathogens coincides with the occurrence of their specific host plants throughout the growing season.

The composition and concentration of the air spora may also vary with the time of day. During periods with favorable conditions for growth and development of fungi in the surrounding areas, the composition of the spores present in the atmosphere at night is distinct from that of the day. During periods of unfavorable growing conditions, such a difference may not occur.

References

Barkai-Golan, R. 1958. A study of airborne fungi in Israel. *Bull. Res. Conv. Israel* 6D, 247–258.

Dransfield, M. 1966. The fungal air-spora at Samaru, Northern Nigeria. *Trans. Brit. Mycol. Soc.* 49, 121–132.

Durham, O. C. 1938. Incidence of airborne fungus spores. II. *Hormodendrum*, *Alternaria* and rust spores. *J. Allergy* 10, 40–47.

Dye, M. H., and T. R. Vernon. 1952. Airborne mould spores, a New Zealand survey. *New Zealand J. Sci. Technol.* 34, 118–127.

Farrow, W. M. 1954. Tropical soil fungi. *Mycologia* 46, 632–646.

Gregory, P. H., and J. M. Hirst. 1957. The summer air-spora at Rothamsted in 1952. *J. Gen. Microbiol.* 17, 135–152.

Harvey, R. 1967 Air-spora studies at Cardiff. I. *Cladosporium*. *Trans. Brit. Mycol. Soc.* 50, 479–495.

Hirst, J. M. 1952. An automatic volumetric spore trap. *Ann. Appl. Biol.* 39, 257–265.

——— 1953. Changes in atmospheric spore content: Diurnal periodicity and the effects of weather. *Brit. Mycol. Soc. Trans.* 36, 375–393.

Hyde, H. A., and D. A. Williams. 1946. A daily census of *Alternaria* spores caught from the atmosphere at Cardiff in 1942 and 1943. *Trans. Brit. Mycol. Soc.* 29, 78–85.

Hyde, H. A. 1953. The incidence of *Cladosporium herbarum* in the outdoor air at Cardiff 1949–1950. *Trans. Brit. Mycol. Soc.* 36, 260–266.

Kramer, C. L., and S. M. Pady. 1960. Kansas aeromycology. IX. Asomycetes. *Trans. Kansas Acad. Sci.* 63, 53–60.

——— 1966. A new 24-hour spore sampler. *Phytopathology* 56, 517–520.

———, and C. T. Rogerson. 1959a. Kansas aeromycology. III. *Cladosporium. Trans. Kansas. Acad. Sci.* 62, 200–207.

——— 1959b. Kansas aeromycology. IV. *Alternaria. Trans. Kansas. Acad. Sci.* 62, 252–256.

———, and L. G. Ouye. 1959. Kansas aeromycology. II. Materials, methods and general results. *Trans. Kansas Acad. Sci.* 62, 184–199.

Kramer, C. L., S. M. Pady, and C. T. Rogerson. 1960. Kansas aeromycology. V. *Penicillum* and *Aspergillus. Mycologia* 52, 545–551.

Long, D. L., and C. L. Kramer. 1972. Air spora of two contrasting ecological sites in Kansas. *J. Allergy* 49, 255–266.

Miquel, M. P. 1885. Annuaire de l'observatoire de Montsouris pour l'an. Paris.

Myers, M. A. 1956. Airborne molds in Honolulu. *J. Allergy* 27, 531–535.

Pady, S. M., and C. O. Johnston. 1959. Cereal rust aerobiology and epidemiology in Kansas in 1958. *Plant Disease Rept.* 43, 607–612.

Pady, S. M., and C. L. Kramer. 1960. Kansas aeromycology VII. Smuts. *Phytopathology* 50, 332–334.

———, and R. Clary. 1967. Diurnal periodicity in airborne fungi in an orchard. *J. Allergy* 39, 302–310.

Pady, S. M., C. L. Kramer and B. J. Wiley. 1962. Kansas aeromycology. XII. Materials, methods and general results of diurnal studies 1959–1960. *Mycologia* 54, 168–180.

Richards, Merfyn. 1954. Seasonal periodicity in atmospheric mould spore concentrations. *Brit. Mycol. Soc. Trans.* 37, 357–366.

——— 1956. A census of mould spores in the air over Britain in 1952. *Brit. Mycol. Soc. Trans.* 39, 431–441.

Sreeramulu, T. 1959. The diurnal and seasonal periodicity of spores of certain plant pathogens in the air. *Trans. Brit. Mycol. Soc.* 42, 177–184.

6.7 Phenology and Environmental Education

Forest Stearns

Department of Botany
University of Wisconsin, Milwaukee, Wisconsin 53201, USA

This volume has presented many scientific and applied aspects of phenology; almost the whole of biology from algal blooms through hibernating mice and larval mosquitoes to germinating zoospores. Authors have indicated wide public interest in biological events, from the first robin to the last green leaf. This interest has motivated the many lay observers who serve as the backbone of the major phenological networks. Probably reports from lay observers are generally more reliable and of better quality than if the reports were received from scientifically trained observers engaged in distracting studies.

The information contained in many articles indicates the potentially large scientific input to and output from phenology. Phenological knowledge has many applications in human welfare, agriculture, forestry, and public health.

My purpose is to suggest that we, as educators, may have neglected one of the most valuable tools in environmental education and thus one of the most important applications of phenology. The trend toward more and larger cities will continue. Urbanization, technology, and population growth together exert increasing stresses upon man. Today environmental education is the most important single area of learning; it is concerned directly with species survival. However, environmental education is not a prerequisite for employment or profit nor for obtaining today's food or today's recreation. Instead, it speaks of tomorrow. "Tomorrow! tis a sharper who stakes his penury against thy plenty." (N. Cotton, in Edwards, 1933). Environmental education must consider the longer time frame, and for this reason it has been slow to receive acceptance.

Then, too, much of what passes as environmental education has been directed toward the immediate problems brought on by suburbanization and urban growth —loss of cherished natural areas, pollution of waters and air, and disposal of solid waste. There has been much talk about urgent problems and technological solutions; environmental education must relate this to an understanding of basic biological principles, and the creation of a mass ecological conscience. As Clay

Contribution No. 121 of the Eastern Deciduous Forest Biome US/IBP.

KEYWORDS: Environmental education; phenology; high school teacher training; North America; youth groups; college courses.

Schoenfeld points out (1972) "we are environmental illiterates and . . . we are slow really to appreciate the all-encompassing nature of the man-land community concept."

What has environmental education to do with phenology? I believe that the two are inextricably related. As the science of biological events relating to climate, weather, and other environmental factors, phenology is useful at all levels of education to make relevant the complexity of organism and environmental interactions. Observations demonstrate diverse responses of organisms and can be made pertinent to man's activities. Phenology may serve as a link between the urban dweller and the natural world just as it now serves the rural resident. In the classroom, phenology can stimulate and maintain interest in natural things and develop the skills of observation by touching and seeing. Environmental data may be useful in developing capability in deduction.

A good farmer is a natural phenologist—he knows when to drop the pitchfork for the fishing rod. But there have been no articles on phenology in the *American Biology Teacher* in the past several years. In 1969 the *Science Teacher* carried only a brief paper by David Gates (Gates, 1969). Reports which have pointed out opportunities for phenological studies in adult education have been few and brief. Here then is a major challenge for the teacher and the phenologist: How best can we stimulate interest and expand the value of phenology to bring contact with natural reality back into daily life?

Unconsciously we value plants and animals other than man as parts of our healthy environment—essential to our psychological and mental health. The natural landscape can serve in a major way to relieve stress, and its aesthetic values are widely shared.

In addition to the psychological, social, and aesthetic needs of man, we now understand in general terms the close interrelationship of all organisms on this earth and the precarious future of many of our natural systems. Phenological concepts are essential to an understanding of these relationships and should be exploited.

In elementary schools phenology already plays a strong role. The changing seasons are used as themes in teaching with or without realizing their basic appeal. They appear in art classes, the music sung, the literature read, and the poems learned. We learned the simple songs "Spring Is Here" or "Oats, Peas, Beans and Barley Grow" or "Falling Leaves." Could we buttress this beginning with specific observations of plants and animals on the school grounds or in the neighborhood? Children enjoy keeping charts and recording events. They like to watch the lists of events grow, particularly if they can relate this activity to changes in the weather, effects upon themselves, and the unspoken feeling of the continuity of life.

Observations of plants and animals may promote respect for nature. It would be a traumatic event for the entire third grade class if some vandal destroyed the tree whose buds they were watching.

Phenological observations in the upper elementary grades and junior high school might well help to reduce vandalism, a serious problem today in urban streets and parks. To provide material for observation, schools could profitably create small gardens of wild and cultivated plants arranged to change in color and aspect with the season. Grounds and garden maintenance could be reduced by planting long-lived perennials. Weeds could be controlled during the vacation period by use of mulch. Several gardens in different exposures would provide contrasts in development as affected by microclimate. Events in the school garden and in the neighborhood can be related readily to a wide variety of elementary school activities. Animal and bird events should also be incorporated into the program.

At the junior high and high school levels the student is attempting to resolve his relationships with others and to understand himself, and the job of the teacher is often difficult. He or she must use great ingenuity to stimulate continued observation, as by relating instruction and chronology of life to the specific interests of the adolescent. In rural areas and small towns there is no question of the relationship between frost and the end of the tomato crop or between summer heat and corn growth. Phenological events are not so obvious during the winter, but other natural events, such as freezing and thawing of lakes and of soil, will serve. In most areas of the northern hemisphere it is possible to follow winter changes in natural events; color changes in the coat of the whitetailed deer and in bark color of the dogwood or willow are examples.

In high school, phenology can be used to relate biology to the other sciences—to chemistry through the chemical changes in plants during spring and fall; to physics in the effects of freezing—and to an understanding of relationships between climate and life. Phenological observations and special projects would serve as a focus outside the classroom and could even be used to test predictions based on earlier experience, meteorology, and a bit of mathematics. Would this help to relieve the tedium and spring restlessness engendered by the long school year?

Youth groups could make especially good use of phenology in their programs. The Boy and Girl Scouts and the 4-H clubs seek contact with and understanding of nature in order to assist young people to find themselves, and to develop physical and observational skills, feelings of independence and self-confidence, and strength of character. The Scout merit badge programs and the 4-H nature study guides should include a strong phenological component.

When the student moves from the high school to the university he functions largely in an artificial world, one which puts a strong emphasis upon intellectual values, technology, and abstract ideas. Yet chronology of human biology remains important. Biological studies in the universities are often dully detailed and microscopically or chemically oriented. The physiological response resulting from interaction between environment and organism can be related to knowledge which the student already has or is gaining in such diverse courses as meteorology, physics, and chemistry. Likewise, with adequate background in observation and understanding of phenological events, the student will be better equipped to relate the impact of the natural world on traditions of literature, art, and music or even to economics and engineering.

The timing of developmental events may play a major role in isolating populations, thus permitting speciation. The study of evolution, ecology, and behavior would benefit if they were taught together in a field setting.

In his pioneering course, "Reading the Landscape," Zimmerman makes effective use of 28 years of phenological data as the basis for exercises that excite interest and force students to think about results. Every student is required to make field observations. Phenological study, especially in high school and college, requires substantial observational data on which to base questions and interpretation. A school can rapidly accumulate such information. Student projects, such as the observation of specific trees in different areas, are useful. Records of local Audubon Society members may be valuable, as may contacts with the local weather observer. Once a beginning is made, the observational projects may be and should be carried along on a continuing basis from year to year. "Such observations will require care and feeding by the teacher to develop the interest and the self discipline needed to keep accurate records and to interpret them." (J. Zimmerman, personal communication). Zimmerman points out that the teachers and students need ready access to a diverse biota, both to develop interest and to make observations easy. A school natural area or school forest is essential, although observations can be made throughout the community.

Teacher training and refresher courses could utilize phenological observations and understanding to a far greater extent than they do today. One must first sensitize the teacher before one can sensitize the student. To quote Hugh Iltis, "We will never get to the root of our environmental crisis by picking up tin cans. . . . The biology teacher, more than any other professional, has the answer to many of today's problems." (Iltis, 1972). Phenological studies are difficult because of the physical growth of our urban universities, but this is no excuse. Plants and animals in city parks, gardens, and weedy lots provide ample opportunities for observation.

Environmental education cannot stop with formal schooling, even with the assumption that the schools have done their job. For his own sanity the adult should maintain contact with natural things, and phenology provides him an easy and pertinent contact whether he be a city or country dweller. Phenology can be a lifetime hobby available to everyone. As he grows older the adult often becomes increasingly aware of and interested in watching the clocks of life. Our newspapers frequently report phenological events (normal and especially abnormal), an indication that both readers and reporters are interested. The most ardent bird watchers and gardeners are adults who have found in these activities physical recreation and relief from stress.

Environmental education is also closely linked with the activities we euphemistically call "outdoor recreation." Phenology is of vital importance to participants in outdoor recreation no matter what their ages. In planning a hunting trip, one needs to know something of the timing of mosquitoes and blackflies. The swimmer is interested not only in the warmth of the water but also in the presence of swimmer's itch and of algal blooms. The fisherman and the hunter unconsciously (at least) recognize the relationship between the weather and the activity of their prey. The bird watcher knows what birds will come to his feeder in November or February and he follows with interest the goings and comings throughout the season. As Leopold wrote, "In October my pines tell me by their rubbed-off bark when the bucks are beginning to 'feel their oats'." Again, "There are two times to hunt in Adams: ordinary times, and when the tamaracks are smoky gold." (Leopold, 1949).

There are dollar signs which suggest that natural history and phenology remain as expressions of human subconscious need. The resort economy of the northern

Lake States receives a shot in the arm each year by capitalizing on the autumnal event of leaf coloration. Avenues of fall color have added a month or more to the tourist season. Within every major city a large contingent of the population follows the passing seasons through flower shows at conservatories and parks. A substantial component of the economy of the Great Smokies now depends upon dogwood flowering each April. Some celebrations of phenological events, such as the cherry blossom and rhododendron festivals, have persisted for many years.

Phenology clearly has great value in environmental education and can be used at all levels of sophistication from grade school to college. It promotes observation and self-discipline and forces the observer to consider the interrelatedness of events. In what other area can we combine education and pleasure so happily?

Phenology "worms," "flys," and "bursts" its way into all aspects of human life, whether we stalk the wild asparagus, hunt the wild goose, cover our tomato plants from the late frost, pay a higher price for Florida grapefruit, thrill to the sandhill crane in the marsh, or watch the starling on the chimney pot. These events and many more help to bring the natural world closer and relate it to human activity. Phenology provides a contact point between urban man and natural life. Let us extend our knowledge and the science of phenology, but let us not be so concerned with "science" that we ignore the meaning of natural events and their place in the environmental education of mankind.

References

Edwards. T., ed. 1933. *Useful Quotations*. p. 922. New York: Grossett and Dunlap.

Gates, D. 1969. Phenology program of the IBP. *Science Teacher* 36, 28–30.

Iltis, H. 1972. The biology teacher and man's mad and final war on nature. *Am. Biol. Teacher*.

Leopold, A. 1949. *The Sand County Almanac*. 226 pp. New York: Oxford Press.

Popow, N. W. 1950. *Phenological Observations in Schools*. Moscow (in Russian).

Schoenfeld, C. 1972. Are we environmental illiterates? *J. Environ. Education* 3, 65.

Subject Index

Species Index

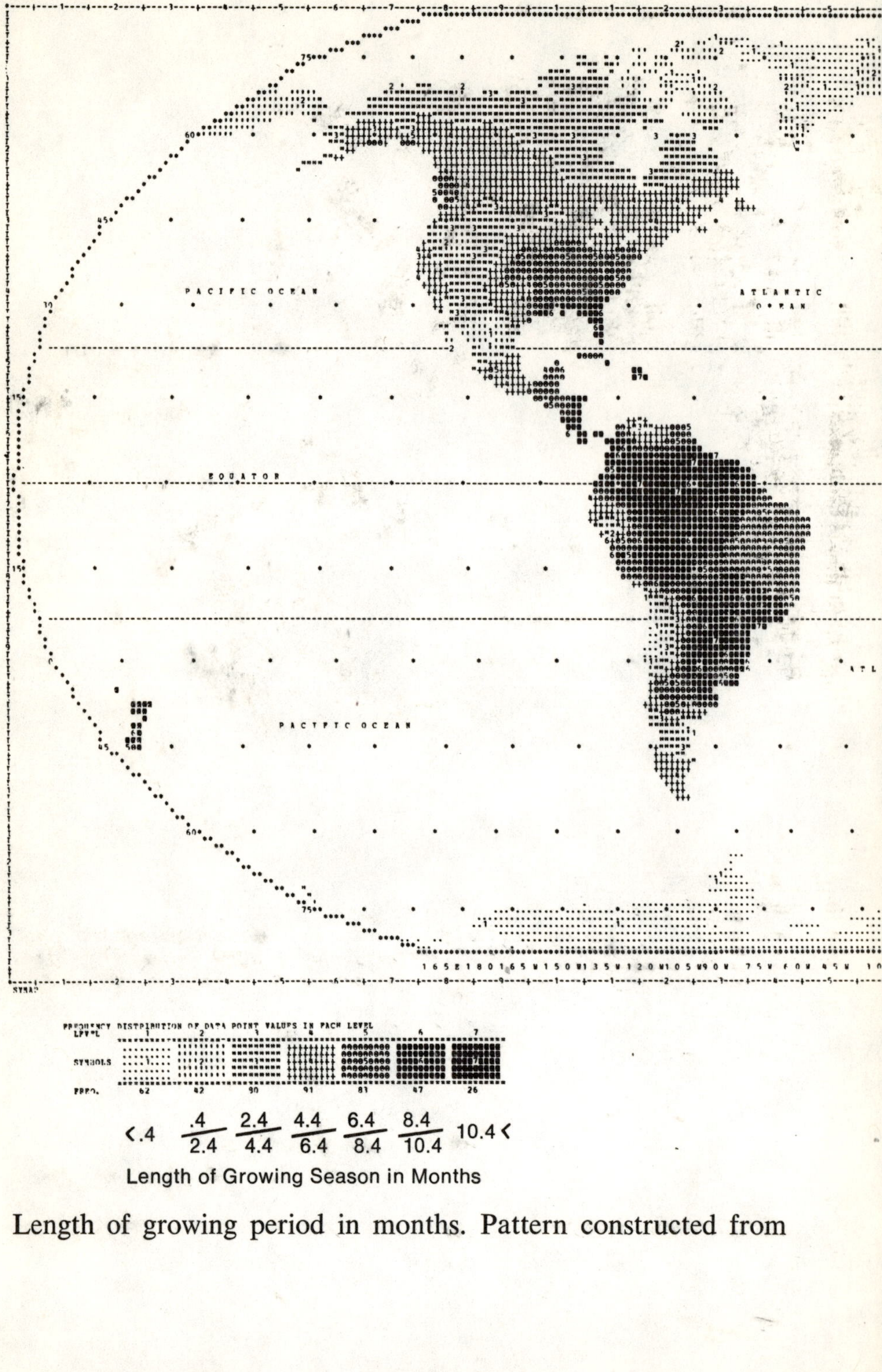

Length of growing period in months. Pattern constructed from